Minerals, Metals and Sustainability

Minerals, Metals and Sustainability

Meeting Future Material Needs

W.J. Rankin, CSIRO

National Library of Australia Cataloguing-in-Publication entry

Rankin, W. J.

Minerals, metals and sustainability: meeting future material needs/by W.J. Rankin.

9780643097261 (pbk.)
9780643097278 (epdf)
9780643104228 (epub)

Includes bibliographical references and index.

Exploitation.
Conservation of natural resources.
Mines and mineral resources – Management.
Mineral industries – Management.
Sustainability.

333.85

Published exclusively in Australia and New Zealand by

CSIRO PUBLISHING
150 Oxford Street (PO Box 1139)
Collingwood VIC 3066
Australia

Telephone: +61 3 9662 7666
Local call: 1300 788 000 (Australia only)
Fax: +61 3 9662 7555
Email: publishing.sales@csiro.au
Web site: www.publish.csiro.au

Published exclusively throughout the world (excluding Australia and New Zealand) by CRC Press/Balkema, with ISBN 978-0-415-68459-0

CRC Press/Balkema
P.O. Box 447
2300 AK Leiden
The Netherlands
Tel: +31 71 524 3080
Website: www.balkema.nl

Front cover: Iron ore reclaimer, Port Hedland, Western Australia (iStockphoto)

Set in 10/13 Adobe Minion and ITC Stone Sans
Edited by Adrienne de Kretser, Righting Writing
Cover design by Samantha Duque
Text design by James Kelly
Index by Russell Brooks
Typeset by Desktop Concepts Pty Ltd, Melbourne
Printed in China by 1010 Printing International Ltd

CSIRO PUBLISHING publishes and distributes scientific, technical and health science books, magazines and journals from Australia to a worldwide audience and conducts these activities autonomously from the research activities of the Commonwealth Scientific and Industrial Research Organisation (CSIRO). The views expressed in this publication are those of the author(s) and do not necessarily represent those of, and should not be attributed to, the publisher or CSIRO.

Original print edition:
The paper this book is printed on is in accordance with the rules of the Forest Stewardship Council®. The FSC® promotes environmentally responsible, socially beneficial and economically viable management of the world's forests.

Contents

Preface xv
Acknowledgements xvii

1 Introduction 1

2 Materials and the materials cycle 5

2.1 Natural resources 5
2.2 Materials, goods and services 6
2.3 The material groups 9
2.3.1 Biomass 9
2.3.2 Plastics 10
2.3.3 Metals and alloys 10
2.3.4 Silicates and other inorganic compounds 10
2.4 The materials cycle 12
2.5 The recyclability of materials 14
2.6 Quantifying the materials cycle 15
2.6.1 Materials and energy balances 16
2.6.2 Material flow analysis 16
2.7 References 23
2.8 Useful sources of information 24

3 An introduction to Earth 25

3.1 The crust 25
3.2 The hydrosphere and biosphere 26
3.2.1 Life on Earth 27
3.2.2 The Earth's biomes 28
3.2.3 Ecosystem services 30
3.3 Some implications of the basic laws of science 31
3.3.1 Thermal energy flows to the biosphere and hydrosphere 32
3.3.2 The greenhouse effect 32
3.3.3 The Sun as driver of both change and order 33
3.4 The biogeochemical cycles 34
3.4.1 The carbon and oxygen cycles 35
3.4.2 The water cycle 36
3.4.3 The nitrogen cycle 37
3.4.4 The phosphorus cycle 38

3.4.5 The sulfur cycle 38
3.5 References 40
3.6 Useful sources of information 40

4 An introduction to sustainability 41

4.1 The environmental context 42
4.1.1 The state of the environment 42
4.1.2 The ecological footprint 43
4.1.3 The tragedy of the commons 46
4.2 A brief history of the idea of sustainability 47
4.2.1 The rising public awareness 47
4.2.2 International developments 47
4.2.3 Corporate developments 48
4.3 The concepts of sustainable development and sustainability 49
4.3.1 Alternative definitions of sustainability 49
4.3.2 Interpretations of sustainability 51
4.3.3 Responses to the challenge of sustainability 52
4.4 Sustainability frameworks 53
4.4.1 Triple bottom line 54
4.4.2 Eco-efficiency 54
4.4.3 The Natural Step 54
4.4.4 Natural Capitalism 55
4.4.5 Biomimicry 55
4.4.6 The five capitals model 55
4.4.7 Green chemistry and green engineering 56
4.4.8 Putting the frameworks into context 56
4.5 A model of sustainability 58
4.6 References 60
4.7 Useful sources of information 61

5 Mineral resources 63

5.1 Formation of the Earth 63
5.2 The geological time scale 65
5.3 Formation of the crust 66
5.3.1 Continental crust 67
5.3.2 Oceanic crust 68
5.3.3 The distribution of elements 68
5.4 Minerals and rocks 71
5.4.1 Mineral classes 72
5.4.2 Rock classes 75

5.4.3 The rock cycle 81

5.5 Mineral deposits 82

5.5.1 Formation of mineral deposits 83

5.5.2 Common forms of mineral deposits 84

5.5.3 The distribution of base and precious metal deposits 85

5.6 Resources and reserves 86

5.7 Extracting value from the crust 89

5.7.1 Physical separation 90

5.7.2 Chemical separation 92

5.7.3 The effect of breakage on the surface area of materials 93

5.7.4 By-products and co-products 94

5.7.5 The efficiency of extraction 94

5.8 References 94

5.9 Useful sources of information 95

6 The minerals industry 97

6.1 Mineral commodities 97

6.1.1 Traded commodities 97

6.1.2 Mineral commodity statistics 100

6.1.3 Reserves and resources of mineral commodities 101

6.2 How mineral commodities are traded 105

6.2.1 Mineral and metal markets 105

6.2.2 The complexities of trading mineral commodities 107

6.3 The economic value of mineral commodities 109

6.3.1 Hotelling's rule 109

6.3.2 Limitations of Hotelling's rule 110

6.4 The mining project cycle 112

6.4.1 Exploration 113

6.4.2 Evaluation and development 113

6.4.3 Design, construction and commissioning 114

6.4.4 Production 114

6.4.5 Project decline and closure, remediation and restoration 114

6.5 The nature of the minerals industry 115

6.5.1 Location 115

6.5.2 Hazardous nature 115

6.5.3 Size and structure 116

6.5.4 Minerals companies 117

6.5.5 Industry associations 120

6.5.6 Industry culture 120

6.5.7 Trends shaping the industry 121

6.6 The economic and social impacts of mining 122
6.6.1 Mining as a route to development 122
6.6.2 The resources curse 123
6.6.3 Artisanal and small-scale mining 124
6.7 The minerals industry and sustainable development 124
6.7.1 Industry developments and formation of the ICMM 124
6.7.2 Sustainability reporting and sustainability indicators 125
6.7.3 Status of the industry 128
6.8 References 128
6.9 Useful sources of information 130

7 Producing ores and concentrates 131

7.1 Extracting rock from the crust 131
7.1.1 Surface mining 132
7.1.2 Underground mining 134
7.1.3 Solution mining 136
7.2 Beneficiating mined material 136
7.2.1 Size reduction 137
7.2.2 Separating particles 140
7.2.3 Separating solids from water 143
7.2.4 Agglomerating particles 146
7.3 Examples of mineral beneficiation flowsheets 147
7.3.1 Mineral sand concentrates 147
7.3.2 Production of iron ore fines and lump 147
7.3.3 Base metal sulfide concentrates 152
7.4 References 152
7.5 Useful sources of information 152

8 Producing metals and manufactured mineral products 153

8.1 Theoretical considerations 153
8.2 Metals 155
8.2.1 The principles of metal extraction 156
8.2.2 Metallurgical reactors 161
8.2.3 Smelting 161
8.2.4 Leaching 167
8.2.5 The stages in the extraction of a metal 170
8.2.6 The production of some important metals 174
8.3 Cement and concrete 183
8.4 Glass 185

8.5 Mineral fertilisers 186
8.6 Commodity ceramics 187
8.7 References 188
8.8 Useful sources of information 188

9 Energy consumption in primary production 189

9.1 Direct and indirect energy and gross energy requirement 189
9.2 Embodied energy 191
9.2.1 Calculation of embodied energy 192
9.2.2 Values of embodied energy 194
9.3 Embodied energy and global warming potential 196
9.3.1 Hydrometallurgy *versus* pyrometallurgy 197
9.3.2 Global greenhouse gas production 198
9.3.3 Impact of the source of electricity used 198
9.4 The effect of declining ore grade and liberation size on energy consumption 198
9.5 The lower limits of energy consumption 200
9.5.1 Energy required for moving materials 202
9.5.2 Energy required for sorting and separating material 202
9.5.3 Energy required for chemical processing 203
9.6 Energy sustainability indicators and reporting 205
9.7 References 210

10 The role of water in primary production 213

10.1 Global water resources 213
10.2 Water in the minerals industry 215
10.3 The embodied water content of metals 216
10.4 Water sustainability indicators and reporting 218
10.5 References 219
10.6 Useful sources of information 220

11 Wastes from primary production 223

11.1 Wastes and their origin 223
11.2 Solid wastes 225
11.2.1 Calculation of the quantities of solid wastes 225
11.2.2 Quantities produced 228
11.3 Liquid wastes 228
11.3.1 Waste water 228
11.3.2 Acid and metalliferous drainage 229
11.4 Gaseous wastes 232

11.4.1 The types of gases produced in smelting 232
11.4.2 The quantities of gas produced in smelting 232
11.5 The impact of wastes on humans and the environment 235
11.5.1 Examples of the impacts of mining wastes 236
11.5.2 Toxicity 238
11.5.3 Bioavailability 241
11.6 The international regulation of wastes 242
11.6.1 The Basel Convention 242
11.6.2 REACH and the European Chemicals Agency 243
11.6.3 Implications of the Basel Convention and REACH 243
11.7 References 244
11.8 Useful sources of information 245

12 Management of wastes from primary production 247

12.1 Management of solid wastes 247
12.1.1 Waste rock 248
12.1.2 Tailings 249
12.1.3 Residues from leaching operations and water treatment 253
12.1.4 Slags 254
12.2 Management of liquid wastes 255
12.2.1 Technologies for water treatment 255
12.2.2 Management of cyanide solutions 257
12.2.3 Management of AMD 258
12.3 Gaseous wastes 262
12.3.1 Gas cooling and heat recovery 262
12.3.2 Gas cleaning 263
12.3.3 Sulfur dioxide removal 266
12.4 Waste, effluent and emission sustainability indicators 267
12.5 References 268
12.6 Useful sources of information 269

13 Secondary materials and recycling 271

13.1 Options for end-of-life products 271
13.1.1 Recycling 271
13.1.2 Reuse 272
13.1.3 Remanufacturing 272
13.2 Drivers of recycling, reuse and remanufacturing 273
13.3 The benefits and limitations of recycling 273
13.4 Recycling terminology 274

13.5 Recovery, recycling and return rates for common materials 275
13.6 The energy required for recycling 275
13.6.1 The Gross Energy Requirement for recycling 278
13.6.2 The effect of repeated recycling 279
13.7 The effect of recycling on resource life 279
13.8 Recycling materials from simple products 281
13.8.1 Construction and demolition wastes 281
13.8.2 Glass 281
13.8.3 Metals 282
13.9 Recycling materials from complex products 284
13.9.1 Cars 284
13.9.2 Waste electrical and electronic equipment 287
13.10 Design for the Environment 291
13.11 References 292
13.12 Useful sources of information 294

14 The future availability of minerals and metals 295

14.1 The determinants of long-term supply 295
14.2 Potential sources of minerals 296
14.3 Crustal resources 297
14.3.1 The distribution of the elements in the crust 297
14.3.2 The mineralogical barrier 297
14.3.3 Hubbert's curve and the concept of peak minerals 299
14.3.4 Are many mineral deposits still to be discovered? 300
14.3.5 Crustal rocks as a source of scarce elements 304
14.4 Resources in seawater 305
14.5 Resources on the seabed 308
14.5.1 Deposits originating from land sources 308
14.5.2 Deposits originating from sources in ocean basins 310
14.5.3 Deposits originating from sources on continents and in ocean basins 310
14.5.4 Recovery and processing of deep ocean deposits 311
14.5.5 Legal aspects: the Convention of the Sea 312
14.6 Summary and conclusions 313
14.7 References 313
14.8 Useful sources of information 314

15 The future demand for minerals and metals 315

15.1 The determinants of long-term demand 315
15.2 Projections of the demand for mineral commodities 316

15.3 Materials and technological substitution 318

15.3.1 Substitution limits and constraints 321

15.4 Dematerialisation 322

15.4.1 Intensity-of-use 322

15.4.2 Drivers of dematerialisation 325

15.4.3 Counters to dematerialisation 327

15.4.4 A case study 328

15.5 The IPAT equation 329

15.6 Summary and conclusions 330

15.7 References 330

15.8 Useful sources of information 331

16 Towards zero waste 333

16.1 The waste hierarchy 333

16.2 Reducing and eliminating wastes 335

16.3 Cleaner production 336

16.4 Wastes as raw materials 337

16.5 Waste reduction through process re-engineering 346

16.5.1 Examples of flowsheet simplification 346

16.5.2 Examples of novel equipment 348

16.5.3 Examples of novel processing conditions 350

16.6 Industrial ecology 352

16.7 Making it happen 359

16.8 References 363

16.9 Useful sources of information 365

17 Towards sustainability 367

17.1 Closing the materials cycle 367

17.1.1 The ICCM stewardship model 368

17.1.2 The Five Winds stewardship model 370

17.1.3 An integrated strategy for the minerals and metals sector 371

17.1.4 Drivers of stewardship 373

17.2 Market- and policy-based approaches to transitioning to sustainability 374

17.3 What does the future hold? 375

17.3.1 The 'Great Transition' scenario 375

17.3.2 The World Business Council for Sustainable Development scenario 378

17.4 Summary and conclusions 379

17.5 References 380

Appendix I: A note on units and quantities 383
International System of Units 383
Scientific notation, significant figures and order of magnitude 383
Appendix II: A review of some important scientific concepts 387
II.1 The nature of matter 387
II.2 Conservation of matter 389
II.3 Energy, heat and the laws of thermodynamics 389
II.4 Electromagnetic radiation 392
II.5 Heat transfer 393
Appendix III: GRI Sustainability Indicators 395
Appendix IV: Processing routes for extraction of common metals from their ores 401
Index 407
Elements arranged in alphabetical order 420
The Periodic Table 422

Preface

Minerals and rocks derived from the Earth provide many of the materials needed by society. This book examines the nature of minerals and mineral resources, the role that minerals and their products play in society, the nature and technology of the minerals industry, the forces shaping the supply of and demand for mineral-derived materials, and the future of the industry. It is an unabashed attempt at an all-encompassing view of an important sector of the global economy. The book takes a multi-disciplinary approach which integrates the physical and Earth sciences with the social sciences, ecology, economics and the principles of sustainability. It is not a 'how to' book or a 'best practices' book for practitioners; rather, it is a book that attempts to anchor the minerals industry and its activities in fundamental knowledge and concepts to provide the reader with a moderately deep but wide understanding on which to draw and build as required. To my knowledge, there is no other book which has attempted to do this.

The book is not polemical. It takes no particular stand and advocates no particular world view. It attempts to discuss all sides of contentious issues or those where the empirical evidence is not definitive. It does, however, build on three postulates. First, the sources of materials obtained from the Earth's crust, though very large, are finite. Second, the capacity of the Earth's ecosystems to cope with the wastes from human production and consumption is finite and by many measures is approaching (if not already exceeding) the limit. Third, the neo-liberal economic model of unrestrained capitalism, as practised by developed and many developing countries, and despite its obvious material benefits over the past three decades, has resulted in social and environmental problems on a scale that cannot be solved by market forces alone. These problems also require, at the least, major public policy solutions at national and international level. The first of these postulates is fact and the second has virtually universal support from scientists and other professionals with deep knowledge of the issues. The third postulate is a hypothesis but one that is increasingly supported by empirical evidence. The book takes a global view – it is not specific to particular mineral commodities or to any particular country or region. This is appropriate since the minerals industry is global, with many companies having operations around the world and many mineral commodities being traded internationally. No country is entirely self-sufficient in mineral resources. And the challenges of sustainability, while being implemented locally, need to be addressed as global issues.

A wide range of readers should find this book useful. Students of engineering and applied science in areas such as mining, metallurgy, civil engineering, chemical engineering, environmental science and engineering, geology and resource economics, may use it as an integrated overview of the minerals industry and/or as an introduction to specific aspects of the industry. It will also be useful to practising engineers, geologists and scientists, and to professionals in government service in areas such as resources, environment and sustainability, who need an overview of the modern minerals industry. Journalists, science teachers, commentators and others with an interest in the resources sector may also find topics of interest. Many readers will be interested in some topics more than others, so the depth of treatment has been kept at a similar level for all major topics. Extensive cross-referencing and a small amount of repetition to aid selective readers have been provided. In general, no prior knowledge of the minerals industry or of the technology used by the industry has been assumed.

The book contains much information that is very recent but, of necessity, it also contains much that is well established. What, I hope, is original is the overall approach and structure and some of the insights. For ease of reading, extensive referencing has not been used. Widely available information and data have not always been referenced as it is often difficult, if not impossible, to find the primary source. Wherever possible, seminal references have been cited and discussed in order to show something of the history of the development of ideas. References have been given for information that is not commonly available or is relatively new, and where material has been used in a form that closely follows that of another author. References have also been provided where a statement or piece of information may be controversial. Internet sites have been used extensively, and referenced, since much sustainability-related and other information is now accessible through this medium and, increasingly, only through this medium. A selected bibliography at the end of each chapter lists references and websites which provide greater detail or additional information. These are not exhaustive lists; they contain sources that are authoritative and which I have found useful in preparing the chapter.

I have attempted to keep the book concise and readable, and have tried to avoid much of the jargon and hubris associated with sustainability and the minerals industry, both of which create wide-ranging emotional responses. I have tried to be disinterested, to give facts and present ideas and concepts as I understand them with, wherever possible, substantial and quantitative information as support. Nevertheless, some subjectivity is bound to be present. Finally, while I have attempted to check all facts and statements no doubt some errors remain. Although I have received much advice and assistance in writing this book, all omissions and errors are mine. I would welcome comments and corrections from readers, for future editions.

Acknowledgements

Many people and a number of organisations have assisted in the writing of this book. Financial and in-kind support was provided by the Commonwealth Scientific and Industrial Research Organisation (CSIRO), the former Centre for Sustainable Resource Processing (CSRP), Alcoa of Australia Ltd, BHP Billiton Ltd, Hatch Associates Pty Ltd, Rio Tinto Limited, Xstrata plc, BlueScope Steel Ltd and OneSteel Ltd. Philip Bangerter of Hatch's Brisbane office was an early supporter who helped get the project off the ground. Bruce See read an early draft of the first half-dozen chapters and provided feedback that led to significant restructuring and rewriting at a relatively early stage. Joy Bear also commented on an early draft and encouraged my efforts. Doug Swinbourne read every chapter and gave advice on technical aspects, presentation and readability. David Langberg also read many of the chapters and provided helpful comments. Hal Aral, Warren Bruckard, Alex Deeve, Malisja deVries, Markus Fietz, Bart Follink, Graeme Hayes, Jonathan Law, Roy Lovel, David McCallum, Terry Norgate, Mark Pownceby, Ivan Ratchev, Shouyi Sun, Steve Wright and Marcus Zipper, all from CSIRO, reviewed one or more chapters and provided critical comments. Access to discussion papers, forums and other vehicles of the CSIRO Minerals Down Under Flagship was most helpful. The support of Sharif Jahanshahi in this, and other, respects is greatly appreciated. I particularly want to acknowledge the contribution of my long-time colleague Terry Norgate, whose work I drew on heavily in the chapters on energy, water and recycling. Terry has worked in the sustainability field for many years, quantifying the environmental impacts of mineral processing and metal production. He willingly offered advice and information on many topics. To the many others who provided references, ideas, information, advice or encouragement – thank you.

1 Introduction

The supply of the goods and services used by humans depends on access to materials obtained by exploiting the non-renewable resources of the Earth's crust, particularly metallic and non-metallic minerals, rocks, coal, oil and gas. This book is about materials made from metallic and non-metallic minerals and rocks, their production and recycling, and the environmental and social issues associated with their production and consumption. Mineral-derived materials and fossil fuels form one major group of materials, the other being those derived from living matter. The latter, while not derived from the crust, rely on the crust as the ultimate source of the nutrients they require.

The greatest challenge facing the world is to ensure that all people can have a good standard of living and quality of life without continuing to degrade the environment. The Earth's resources are finite and its land and water ecosystems have finite capacities to cope with the wastes produced by human production and consumption. Every year, the people of the United States, for example, consume more than 21 billion tonnes of resources of all kinds – about 80 tonnes per person per year, consisting of 76 tonnes of non-renewable resources and 4 tonnes of biomass (Adriaanse *et al.*, 1997). Only 19 tonnes are used as direct inputs to processing; the rest is waste. Further quantities of wastes are produced during processing of the direct inputs and during the use and ultimate disposal of the products made from them. Other developed countries have similar, though lower, patterns of consumption. The per capita consumption of materials in the European Union in the latter half of the 1990s, for example, was 49 tonnes per year (Moll *et al.*, 2005).

How to deal with the problems caused by this unprecedented level of production, the associated consumption of resources and disposal of the wastes produced, is a global challenge. This challenge will only grow as living standards in China, India, Brazil and many other developing countries continue to rise. The concept of sustainable development, usually defined as development that meets the needs of the present without compromising the ability of future generations to meet their own needs, arose as an attempt to address these issues. Sustainable development principles have had growing influence on the development of environmental and social policy in many countries in recent decades. They have been adopted and promoted by international organisations, particularly the United Nations, the International Monetary Fund and the World Bank. Sustainable development, or sustainability, is an important theme of this book.

Non-renewable resources pose a unique challenge for sustainable development, in particular how global material needs can be satisfied sustainably when so many sources of materials are non-renewable and when their production and consumption results in huge quantities of wastes, many of which are environmentally harmful. The primary aim of this book is to examine systematically the issues raised by this dilemma, and possible solutions. The focus is on the environmental aspects of mineral and metal production and use. The overarching concern is environmental sustainability and its implications for non-renewable mineral resources. A second aim of the book is to provide a comprehensive overview of the science and technology underlying the production and consumption of materials produced from minerals. It is not the intention to provide a detailed examination of the key mineral-related disciplines of geology, mining engineering and metallurgy. These are very adequately covered in more specialised texts. The aim is to give an integrated overview of the science and technology, and combine it with an overview of the socio-economic nature of the minerals industry and its environmental impact. A third aim is to examine some of the social and economic aspects of sustainable development, in particular the role of the minerals industry in wealth creation and the impact of mining on communities. These are vitally important considerations. However, they are not the main focus of this book. Understanding these, and developing solutions to the social problems caused by mining, draws more on the social sciences and requires treatment from a different perspective. Nevertheless, social and economic aspects are discussed where appropriate to provide a more balanced view.

The book is anchored firmly in the traditional sciences of chemistry, physics, geology and biology, and in engineering. There are 17 chapters, which are best read sequentially. Chapters 1 to 3 introduce the concept of materials and their sources, how materials are utilised in society (with particular focus on inorganic materials from the Earth's crust) and the environmental basis of our existence. Chapter 4 introduces the concept of sustainability and examines its interpretations and the issues it raises for the use of non-renewable resources. Chapter 5 discusses the geological basis of the minerals industry and Chapter 6 describes the structure and nature of the industry. Chapters 7 and 8 review the technologies by which mineral resources are extracted from the Earth's crust and processed to make materials for use in construction and manufacturing. Chapters 9 and 10 examine the usage of energy and water by the minerals industry, with important environmental implications. Chapters 11 and 12 survey the types and quantities of wastes resulting from the production of mineral and metal commodities, the human and environmental impacts of waste dispersion, and how wastes from mining and processing are managed. Chapter 13 examines the recycling of mineral-derived materials and the role of secondary materials in meeting material needs. Chapter 14 surveys the future sources of minerals and the factors that will determine their long-term supply. Chapter 15 surveys the socio-economic and technological factors that will determine the long-term demand for mineral-derived materials. Chapters 16 and 17 look to the future. Chapter 16 discusses how the quantities of wastes formed during the production of mineral and metal commodities can be reduced, or eliminated, through technological developments and socio-political changes. Finally, Chapter 17 addresses the concept of stewardship and the role the minerals industry should play in the ongoing transition to sustainability.

The chemical and physical basis of materials is quantifiable. We can talk about the quantities of various elements and minerals in the Earth's crust; the quantities of products made and wastes produced, and their composition (in terms of mass or volume per cent, for example); the quantities of energy and water required for different operations; the concentrations of elements in ores; the quantities of substances recovered or lost during processing; the size of mines and the capacity of various pieces of equipment; and so on. Wherever appropriate, quantitative as well as qualitative aspects of topics are examined in order to enable a better appreciation and deeper understanding of the issues. SI units are used; a brief review of these, and some useful numerical concepts, is given in Appendix I. An elementary knowledge of chemistry and physics has been assumed. Concepts which are particularly relevant are reviewed briefly in Appendix II.

REFERENCES

Adriaanse A, Bringezu S, Hammond A, Rodenburg E, Rogich D and Schütz H (1997) *Resource Flows: The Material Basis of Industrial Economies.* World Resources Institute: Washington DC.

Moll S, Bringezu S and Schütz H (2005) *Resource Use in European Countries.* Wuppertal Institute: Wuppertal, Germany.

2 Materials and the materials cycle

2.1 NATURAL RESOURCES

All the material needs of humans are met ultimately from the Earth's natural resources. *Natural resources* are the naturally occurring substances and systems that in their relatively unaltered state are useful to humans and that provide the basis for our physical existence. They include:

- the atmosphere;
- water (oceans, rivers, lakes and water in aquifers);
- forests and forest products (timber and other forms of biomass);
- land in its natural state;
- fresh and salt water fisheries and their products;
- minerals, fossil fuels (natural gas, oil and coal);
- non-mineral energy sources (wind, tidal, solar and geothermal).

The capacity of all parts of the environment to undertake the essential role of absorbing, treating and recycling wastes created by humans can also be considered a natural resource. These are also called *environmental resources* or ecosystem services. Cultivated products, while not being natural resources, are reliant on natural resources (air, water, minerals etc.) for their production.

Some natural resources are renewable by natural processes while others are not. Living renewable resources such as fish and forests can regenerate (restock themselves) whereas non-living renewable resources such as wind, tides, solar radiation and geothermal heat do not need regeneration. They are essentially an infinite resource though they are available only at a finite rate. For example, wind might be available in a particular location at an average velocity of 10 km h^{-1} but it is available, at least from a human perspective, for ever. Renewability, however, requires appropriate management of a resource, particularly in the case of living resources. If a living resource is consumed at a rate that exceeds its natural rate of replacement, the stock of the resource will decrease, become subcritical and eventually collapse. Changes to a system or related system can have irreversible effects on renewable resources. Sulfur dioxide emitted from coal-burning power stations can produce acid rain (discussed in Section 3.4.5) which causes dieback of trees in forests; heavy metals from the wastes of mining operations may enter fresh water systems and cause harm to wildlife; global warming may cause reduced rainfall in some areas, resulting in loss of forest or other biomass and wildlife. The unprecedented destruction of many renewable natural resources in recent decades poses a major challenge to achieving sustainability.

The main types of non-renewable natural resources are rocks, minerals and fossil fuels from the crust of the Earth. Rocks and minerals are used for making many useful materials such as construction products, metals and alloys, and specialty materials with vast numbers of uses. Fossil fuels are burned to produce heat to generate electricity and power machines and vehicles and, in the case of oil, to make hydrocarbon-based materials, particularly plastics. Fossil fuels are usually consumed in use (when they are burned); rocks and minerals are transformed into other solid materials. In both cases, however, when they have been extracted from the Earth's crust and used, the original resource is no longer available. While geological processes can generate new stocks of coal, oil, gas and minerals in the Earth's crust by the same processes that have occurred in the past, the time scale on which this occurs (tens to hundreds of millions of years) is far too long for these resources to be considered, in the human context, as renewable.

2.2 MATERIALS, GOODS AND SERVICES

In the technical sense, *materials* are substances which are used by humans to create goods. These consist of structures (such as buildings, dams and roads); vehicles (cars, trains, ships and aeroplanes); machines; electrical and electronic equipment and other devices (for both commercial and consumer use); works of art; and other objects (Bever, 1986). All materials are ultimately derived from the Earth's natural resources.

The variety and complexity of materials used by humans has increased throughout history. Originally, natural materials such as wood, stone, leather and bone were the main materials used; we now also use complex electronic materials, plastics, synthetic building materials, coating materials (e.g. paints, enamels and polymers) and countless others. History is often divided into periods named after materials, as shown in Table 2.1. People started using stones around 32 000 years ago for hunting, cutting, chopping, grinding seeds and building shelters – the Stone Age. This does not mean, of course, that stone replaced wood and other plant materials, which were used previously. However, the use of stone was a new technology that expanded possibilities, both by using it for building to replace wood and using it for new applications such as knives and axes to cut and chop other materials. Similarly, when copper and, later, bronze, began to be used these replaced some applications of stone, particularly those relating to cutting and chopping. However, the use of stone for building purposes and for grinding continued (and remains today). The properties of copper and bronze created opportunities to make new products, such as armour plate and pots for heating and cooking. Each new material partially supersedes one or more earlier material and creates opportunities for new products. The present period is sometimes referred to as the Information Age. The Information Age has important implications for materials – the types required (often very complex) and the quantities (often less, for example, the replacement of paper-based information with electronically stored and transmitted information). The Information Age also has created opportunities that were not feasible previously, such as personal computers and the internet, which in turn create demand for new materials.

Materials in their natural, unprocessed or minimally processed state, such as iron ore, wool or tree logs, are called *raw materials*. Materials reclaimed from used or obsolete products, such as scrap metal or printed circuit boards from electronic equipment, are called *secondary materials*. Usually, raw materials and

Table 2.1: Periods in history based on important materials and technologies

Period	Approximate date period commenced
Stone Age	30 000 BCE
Copper Age	3000 BCE
Bronze Age	2500 BCE
Iron Age	1000 BCE
Coal Age	1600
Industrial Revolution (based on coal and iron, and steam power)	1750
Oil Age	1875
Atomic Age	1945
Information Age	1960

secondary materials undergo some processing to put them in a form suitable for use in manufacturing, construction or agriculture. Such intermediate products are called *basic materials*. Examples of basic materials are steel sheet, copper wire, textiles, fertilisers, lumber, plastics, bricks and cement.

Modern societies consume vast quantities of materials to build infrastructure, manufacture machines and produce durable consumer goods (such as cameras, televisions, cars and white goods) and non-durable consumer goods (such as packaging and clothing). The materials involved in making these undergo predominantly physical changes in use. These are referred to as *engineering materials*. For example, aluminium ingots may be melted and alloyed then cast to form gearboxes or engine blocks. Timber may be cut to various lengths, widths and thicknesses or laminated or veneered or converted into particle board. In these cases the original characteristics of the raw materials (aluminium and wood) are largely retained. The most common types of engineering materials are listed in Table 2.2 according to their nature and application. While there are many other ways of classifying materials, which are useful for particular purposes (for example, according to whether they are naturally occurring or synthetic, inorganic or organic, primary or secondary, structural or non-structural), this classification is probably the most useful. The application of some materials involves major chemical or nuclear changes and the materials are transformed in the process. For example, fossil fuels are burned to produce heat and in the process are converted largely into carbon dioxide and water vapour. Other examples include uranium, artificial fertilisers and pharmaceutical and cosmetic products, all of which are consumed in the process of producing a desired effect. These are sometimes called *effect substances* since they are used to produce an effect rather than a physical product.

The terms goods and services, and related terms such as commodities and products, are used frequently throughout this book and an explanation of their meaning is appropriate at this point. *Goods* are physical (tangible) things that can be delivered to a buyer. They involve the transfer of ownership from a seller to a buyer. Goods are made from materials. A *service* is an economic activity that does not involve transfer of ownership. A service creates benefit by bringing about some change in a customer, a customer's possessions or a customer's intangible assets. A service is the non-ownership equivalent of a good. The delivery of services requires the use of infrastructure (such as

Table 2.2: The most common types of engineering materials

Types of materials, based on their nature	Types of materials, based on their application
Aggregate	Industrial minerals (both bulk and specialty)
Metals and alloys	Electrical materials
Ceramics	Electronic materials
Glasses	Superconducting materials
Cement	Nuclear materials
Other inorganic materials	Materials for other energy applications
Polymers	Magnetic materials
Elastomers	Optical materials
Fibres	Biomedical materials
Composite materials	Dental materials
Wood	Building and construction materials
Paper and paperboard	Smart materials
Nanomaterials	
Other materials of biological origin	

Source: After Bever (1986); with modifications.

Table 2.3: Common goods and services sectors in an economy

Goods-producing sectors	Services sectors
Agriculture	Wholesale and retail trade
Forestry	Transportation and warehousing
Fishing and hunting	Information and cultural industries
Mining, including oil and gas extraction	Finance services
Utilities (water, electricity, gas)	Insurance services
Construction	Real estate services
Manufacturing	Professional, scientific and technical services
	Administrative and support services
	Waste management and remediation services
	Education
	Health care and social assistance
	Arts, entertainment and recreation
	Accommodation and food services
	Public administration

Source: After Victor (2009).

transport, electricity and information technology) which requires materials for its construction and operation. Hence, services indirectly consume materials. Some goods and services sectors of a modern economy are listed in Table 2.3.

Commodities are goods that are interchangeable with similar goods. The term is used quite loosely, but generally commodities have a number of characteristics. They are physical substances; they are useful to humans; they are relatively undifferentiated - one supplier's product is interchangeable with another supplier's similar product (in economic terms, a commodity is said to be fungible); they are traded; and they are storable for a reasonable period of time. While it is generally true that commodities are physical substances, economists consider anything that trades on a commodity exchange to be a commodity even if it is not a physical substance, such as foreign currency and financial instruments. Commodities include not only natural resources but also resources produced artificially through agriculture or basic processing. Common examples are grains, sugar, beef, pork, rubber, wool, cotton, ethanol, oil, coal, ores, minerals, metals, paper pulp and timber. Gem-quality diamonds would not be considered commodities whereas industrial diamonds would be. Because of their relatively undifferentiated nature, the price of commodities is determined largely by the global market for that commodity on the basis of supply and demand. The cost of production makes up a major component of the price. Commodities are often traded through commodity exchanges which act as agents for individual producers. For example, many common metals (aluminium, copper, lead, zinc, nickel) are traded on the London Metal Exchange.

The distinction between a product and a commodity lies in the degree of differentiation. Manufactured goods, including consumer goods, are considered to be products and not commodities because there are usually considerable (actual or perceived) differences between the product of one supplier and a product with a similar function from another supplier. Hence, the price of a product is set more by its actual or perceived functionality and desirability; there is usually a much less direct relationship to the cost of production. There has been a trend in recent decades for businesses to reposition as products some things which have been considered commodities, through branding and advertising campaigns, since profit margins on products can be much greater than on commodities. The use of bottled water (a product) to replace tap water (a commodity) in cities where clean tap water is readily available is a good example. In parallel, there has been a commodification of some manufactured goods. Low-cost clothing, appliances and cars, for example, are now perceived by many people in affluent countries as commodities rather

Table 2.4: Some common materials used in house construction, and their origin (in brackets)

Drains and gutters	Formed steel or aluminium sheet; PVC pipes
Exterior walls	Brick, concrete (cement plus aggregate), timber or steel framing, steel or aluminium sheet cladding
Foundations, driveway, paving	Cement (quartz, limestone etc.), aggregate, limestone, clay
Insulation – ceiling and walls	Vermiculite, cellulose, glass wool (quartz, limestone, feldspar)
Interior walls	Plaster (gypsum)
Hardware – nails, screws, hinges and other fittings	Steel, zinc
Paint	Titania, iron oxides etc. pigments, mineral fillers
Plumbing – pipes and fittings	Brass (copper, zinc), stainless steel (iron, nickel, chromium), ceramics (clays), steel; PVC and other plastic pipes and fittings
Roof	Timber or steel frame, tile (cement or ceramic), slate, galvanised steel sheet (iron, zinc) cladding
Sewer pipes	Ceramics (clay), iron
Windows	Glass (quartz, limestone, feldspar), aluminium, steel or timber frames
Wiring	Copper, aluminium, plastic coating

than products because of the lack of differentiation between products of different brands.

2.3 THE MATERIAL GROUPS

The materials in Table 2.2 can be grouped into four broad categories according to their nature: biomass, plastics, metals and alloys, silicates and other inorganic materials. These are all derived from the crust or biosphere. All are widely used, with many common and specialist applications. Table 2.4 lists some materials commonly used in the construction of a house, and their origin.

2.3.1 Biomass

Wood, paper and textiles (such as cotton and wool) are the most important biomass materials. They originate in the biosphere. Wood consists of complex organic compounds and its structure is based on cells composed mainly of cellulose, a long chain, linear molecule $(C_6H_{10}O_5)_n$ in which the value of n can range from hundreds to tens of thousands. The cells are of different sizes and shapes according to the species of wood and are bound together by lignin, a mixture of complex organic compounds with relative molecular mass up to 15 000. This structure largely determines the properties (and hence potential applications) of wood and its products, including paper. Drywood cells may be empty or partly filled with gums, resins and other substances. Softwoods consist mainly (90–95%) of long spindle-shaped cells forming fibres typically 3–8 mm in length. Hardwoods are more complex in structure; their fibres, which make up about 50% of the wood, are shorter and typically about 1 mm in length. Most of the remaining volume of hardwood is composed of much wider cells called vessel elements. These are joined end-to-end to form tubes along the stems and branches, and appear as pores in cross-section. The complex composite nature of wood gives it its unique structural characteristics and makes it such a versatile material. Paper is made from wood pulp which retains the basic fibre structure of wood. Wood pulp is produced by digesting wood chips with chemicals to break down the lignin without degrading the fibres. The Kraft process is the dominant chemical pulping method, and uses a solution of sodium hydroxide and sodium sulfide to digest the lignin. The sulfite process, which uses sulfurous acid to digest lignin, is also employed. Wood and paper products have useful lives of a few months (newsprint, packaging) to years (documents, storage containers), decades and even centuries (furniture, buildings, books). Textiles have useful lives of months to many years according to their usage and function.

2.3.2 Plastics

Plastics are complex organic compounds manufactured from oil and gas extracted from the Earth's crust. Plastics belong to a class of compounds called polymers, which have molecules of very large relative molecular mass composed of repeating structural units, called monomers. The monomers are often linked to form a chain-like structure. The ability of single monomers to link together in a huge number of combinations makes it possible to design polymers with specific properties. The attractive forces between polymer chains play a large part in determining a polymer's bulk properties and, hence, applications. There are two broad types: thermoplastic polymers, which soften on heating; and thermosetting polymers, which do not. The main polymers used commercially are polyethylene, polypropylene, polyvinyl chloride, polyethylene terephthalate (PET), polystyrene and polycarbonate. Polymers degrade in use. They lose strength, change shape and change colour under the influence of environmental factors, particularly heat, light and chemicals. This occurs through the random breakage of bonds that hold the atoms together to form smaller molecules and through bonds breaking at points where the repeating units join to release the constituent monomer. The latter process is called depolymerisation. Plastic-containing products have useful lives of a few months (consumer products, packaging) to years (domestic appliances, auto parts, coatings) or decades (furniture).

2.3.3 Metals and alloys

Metals are chemical elements. They are differentiated from other materials by their excellent thermal and electrical conductivities and their high mechanical strength and ductility. These are the properties that make metals an important class of materials for engineering purposes. Metals occupy the bulk of the periodic table. A line drawn from boron to polonium separates the metals from the non-metals (Figure II.1, Appendix II). Elements to the lower left are metals, those to the upper right are non-metals and most elements on the line are metalloids. The latter are semi-conductors which exhibit electrical properties common to both conductors and non-conductors. Different metals have different physical and chemical properties due to the atomic-level characteristics of their metallic bonding. Metals are often used in the form of alloys. An *alloy* is a mixture of two or more elements, usually in solid solution, at least one of which is a metal. Alloys are usually produced by melting a metal, adding other elements, stirring to dissolve the constituents then cooling below the melting point. Alloys are homogeneous on a macroscopic scale but are often inhomogeneous on a microscopic scale due to the precipitation of some phases during solidification. Alloys are predominantly elemental, not molecular, in nature. The object of alloying is to make materials that are more ductile, harder or resistant to corrosion, or that have more attractive appearance (such as colour or lustre). The most common alloys are those of iron – carbon steels, stainless steels, cast iron, tool steels and alloy steels. Other important alloys are those of aluminium, titanium, copper and magnesium. Copper alloys include bronze and brass and have a long history of use. The alloys of aluminium, titanium and magnesium have been developed more recently and find use in applications requiring a high strength-to-weight ratio, such as in aeroplanes and cars. Metals can degrade through oxidation – rusting and corrosion of steel is a common example – and metals exposed to the environment are often protected by coatings to prevent or reduce oxidation. Metal-containing products have useful lives of a few months (beverage cans, consumer products) to years (computer components, car bodies, white goods) or many decades (machines, motors, structures such as bridges).

2.3.4 Silicates and other inorganic compounds

Silicates are the most common minerals in the Earth's crust and are extracted for their unique properties or as raw materials for manufacturing other substances. The major bulk products made from silicates are dimension stone, aggregate and cement (for use in construction), glass, and ceramics such as bricks and porcelain. Silicates are naturally occurring inorganic polymers built on the structural monomer SiO_4^{4-}

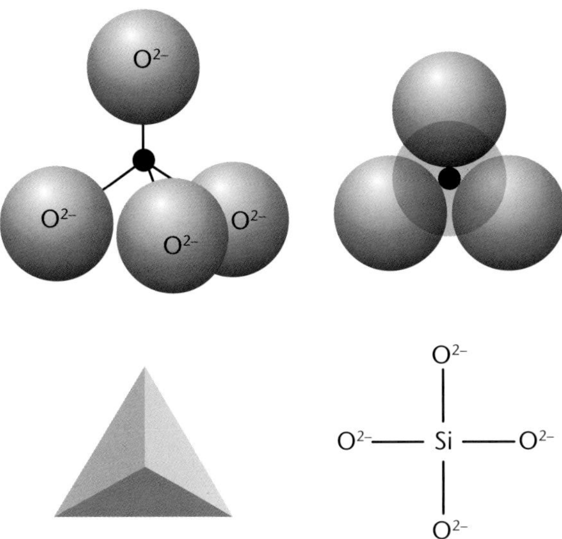

Figure 2.1: The silica tetrahedron.

which has the shape of a regular tetrahedron[1] with a silicon atom at the centre and oxygen anions at each apex (Figure 2.1). The four positive charges of the silicon ion, Si^{4+}, are balanced by four negative charges, one from each of the four oxygen ions, O^{2-}, leaving each tetrahedron with four negative charges. When the oxygen anions link with oxygen anions from adjacent tetrahedra a three-dimensional regular network of silica tetrahedra is formed. This is silica, or quartz, (SiO_2). When some of the oxygen anions are balanced electrically by cations, such as Ca^{2+}, Mg^{2+}, Fe^{2+} and Fe^{3+}, various classes of silicate minerals are formed. The structure of silica and silicate minerals is complex and is discussed further in Section 5.4.1.

The term *aggregate* describes a broad class of particulate material used in construction. It includes naturally occurring sand, gravel and crushed stone as well as secondary materials such as crushed concrete and slag. Aggregates are a major component of composite construction materials such as cement-based concrete and asphalt concrete, in which the aggregate serves as reinforcement, to add strength. *Dimension stone* is naturally occurring stone or rock that has been selected and fabricated to specific sizes or shapes. Colour, texture, pattern and surface finish are important characteristics, as are durability and ability to maintain its distinctive characteristics.

Cement, glass and ceramics are manufactured mineral products, to make which the raw materials undergo chemical reactions at high temperatures. Cement and glass are made of silicates and other naturally occurring inorganic substances, as are many ceramic materials. Cement is an ingredient of concrete. *Concrete* is made by mixing aggregate and cement (usually in the ratio 5:1) with water. The cement reacts with water to form a solid matrix which binds the materials. Commodity ceramics are primarily low-value-added materials, such as bricks, tiles and pottery, and are manufactured from naturally occurring silicate minerals. At the other extreme are engineering or fine ceramics which are low-volume, high-value-added, highly processed materials possessing carefully controlled properties. These include electronic ceramics, structural ceramics (strong, fracture-resistant materials), wear-resistant ceramics, optical ceramics and bioceramics (low-reactivity materials for use inside the body). A special class of ceramics is *refractories*. These are materials used for lining furnaces and for holding hot or molten materials, such as molten metals, alloys, glass and slags. Refractories are made from high melting point, stable, unreactive materials.

There are many other inorganic substances in the crust which are not silicates and which also find use in materials. These are too numerous to mention but some are discussed in later chapters. Some are used in large quantities; others are specialty substances with narrow, but often critical, applications. Mineral fertilisers are an important commodity in this category. Fertilisers are fed to plants to promote growth through providing essential elements, particularly nitrogen, phosphorus and potassium. They are consumed and largely dissipated in use. Some components become part of the plant and enter the food chain; some of these may ultimately be recycled through the ecosystem. The remainder is effectively lost shortly after addition, through leaching of the soil and run-off into streams and rivers. Most of this eventually enters the oceans. Mineral fertilisers are manufactured from substances found in the Earth's crust, particularly phosphate rock.

1 A tetrahedron is a four-sided solid figure in which each face is a triangle. In a regular tetrahedron, each face is an equilateral triangle.

Products containing silicates and other inorganic materials have useful lives of a few months (toothpaste, cosmetics, printer ink, mineral fertilisers) to years (fillers for plastics, pigments for paint, ceramic products, dental products) to decades or centuries (buildings, bridges, dams, roads).

2.4 THE MATERIALS CYCLE

When a product, structure or other object has reached the end of its useful life, or is no longer wanted or needed, the question arises of what to do with it. In some situations it may be reusable through repairs or modifications, or remanufactured. In other situations the individual materials or components from which it is made may be able to be separated and recycled as secondary materials to a manufacturing process. In yet other situations it may be disposed of by burning (usually to produce useful energy) if combustible, put into landfill sites if it is deemed not too polluting, or put into permanent storage if it is hazardous to the environment or humans (e.g. radioactive materials). The choice depends on many factors, including available technologies for reusing and recycling, the relative costs of recovering, recycling and disposal, and government regulations.

There is a cycle which starts with materials being obtained from the Earth, transformed in various ways, then used and finally returned to the Earth. Figure 2.2 shows the cycle from the perspective of the

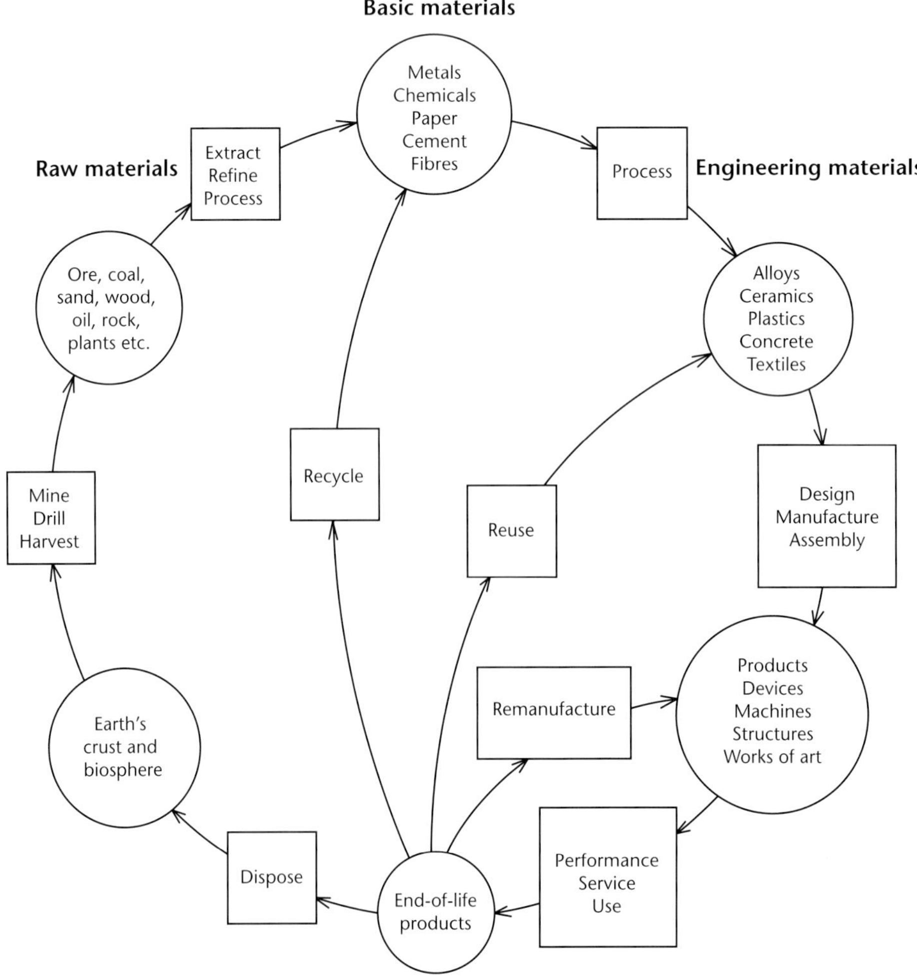

Figure 2.2: The materials cycle (after Altenpoel, 1980; with modifications).

useful components of raw materials obtained from the Earth. This is called the *materials cycle*. It is also referred to as the *life cycle* of a material. The part of the cycle during which natural resources are transformed into useful products is referred to as *value-adding*, with more value being created as the material is progressively transformed. The stages through which material moves to the point of use in a product are together called the *value chain* or *value-adding chain*. In normal circumstances, the chain of activities gives the products more added value than the sum of the costs of performing all the activities.

When a material is produced other substances are also produced as a consequence of the processes used. These are *wastes*, which can be thought of as substances for which there are no present uses, or *by-products*, which are substances of value produced as a result of manufacturing another, usually more valuable, substance. Figure 2.3 illustrates that wastes and by-products are formed at all stages in the materials cycle, and shows many of the common types. Many of the terms in Figure 2.3 may be unfamiliar to the reader at this point but they will be defined and explained progressively throughout the book. The primary production of mineral and metal commodities is the part of the materials cycle during which the greatest quantities of waste are produced. These wastes are an important theme of this book. Figure 2.4 illustrates this point by comparing the quantities of wastes produced for some important mineral and metal commodities. The wastes include those produced directly (from mining and processing) and

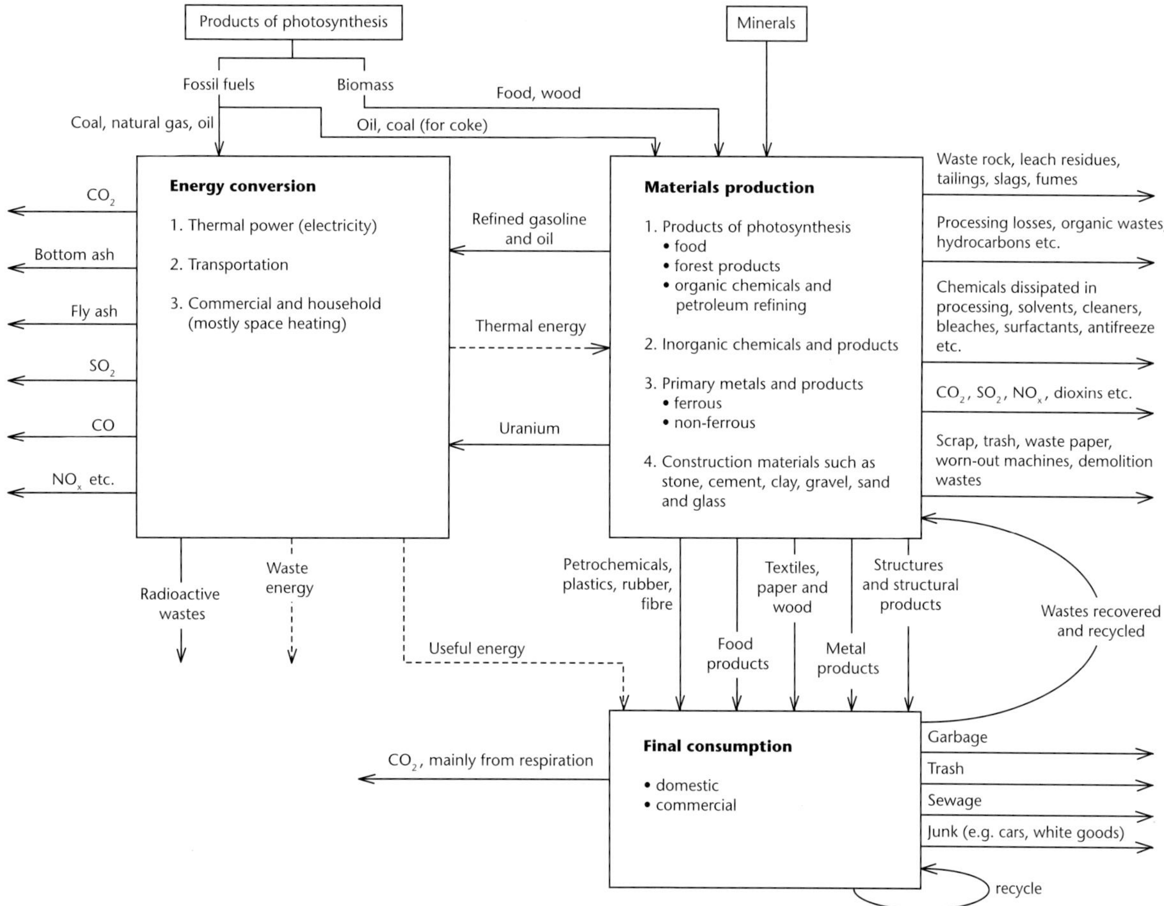

Figure 2.3: Material flows in the economy, with a focus on the wastes produced at various stages (after Ayres and Kneese, 1969; with modifications).

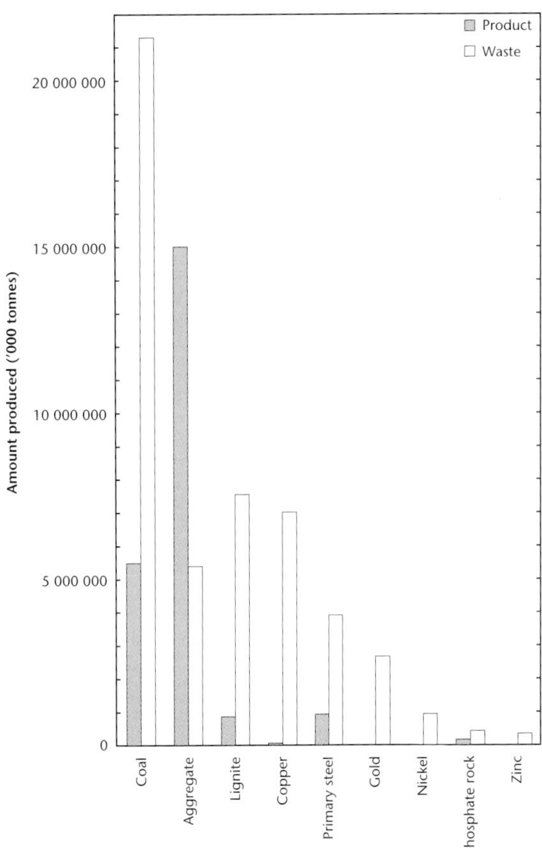

Figure 2.4: Quantities of wastes produced annually in the primary production of some common mineral and metal commodities (production data from Table 6.4; product:waste ratios values used to estimate waste production from Douglas and Lawson, 2002). The quantities of gold, nickel and zinc produced are too small to show.

those produced indirectly (for example, from mining and combustion of coal to produce the electricity consumed in the production of the commodity). In many cases the quantity of wastes produced far exceeds the quantity of valuable product. The discarding of used or unwanted products, usually into landfill, is the second-largest source of waste in the materials cycle and is an aspect of mounting concern.

When materials are returned to the environment they are usually in forms that are very different, often physically and chemically, from those in which they were originally obtained from the Earth. This may lead to harmful environmental impacts such as disturbance to land through landfill, contamination of soil, contamination of ground and surface water, contamination of air, disturbance of terrestrial and ocean ecosystems, and climate change (due to build-up of carbon dioxide in the atmosphere). In recent decades there has been increasing concern about wastes. This has resulted in increasing efforts to find ways to minimise their formation and find uses which minimise the release of harmful substances into the environment. These are important considerations in later chapters.

2.5 THE RECYCLABILITY OF MATERIALS

All materials can be recycled to some extent. However, the quality of the recycled materials and the economics of recovering the materials (collection, sorting, transportation) and of the recycling process are important factors in determining the feasibility of recycling. These are influenced by the nature of the material and how it was used, and its form in the product. The latter affects how efficiently the material can be recovered and the former affects its potential for reuse.

Wood, paper and plastics have limited recyclability since their quality declines through the process of recycling. The fibres in paper and paper products degrade during recycling due to physical breakage of the cell walls; each fibre can be recycled only a relatively few times (perhaps five to 10). New pulp must be blended with recycled material if product quality is to be maintained, or the pulp must be used in progressively lower-quality applications. The recycling characteristics of plastics vary according to their molecular constitution. Only some types can be successfully recycled and the cost is often high. In practice, clean streams of thermoplastic polymers are the most economical to recycle. Thermosetting polymers cannot be easily recycled and to be reused they must be machined into new products or ground and mixed as filler with new polymer. Some plastics can be converted into their basic constituent monomers through chemical processing and can then be used as inputs for making new plastics and other products. However, this is energy-intensive and therefore expensive, and potentially polluting. The ultimate destination for paper and plastic products is landfill or energy recovery through combustion. Combustion puts an end to any material renewal or recycling and breaks the value chain from resource to material to product.

Metals can be recycled indefinitely in principle, since elements cannot degrade, and recycling is relatively cheap since melting of scrap metal is a physical process and not particularly energy-intensive. However, collection and separation are important steps (and can be costly) and melting introduces impurities into the metal if separation is not complete. Thus recycled metals are often less pure than the metal originally used to manufacture the recycled products. They have to be refined (if that is technically and economically feasible), blended with new (virgin) metal produced from ores to dilute the impurities, or used in applications where lower performance standards are acceptable. Loss of metal occurs directly through corrosion, wear and dispersive uses and much metal is still disposed of in landfill. The interaction of metals with the environment can pose environmental and health risks due to the toxicity of some metals in solution in water. Hence, the use of metals in some applications and the disposal of metal-containing products have come under increasing scrutiny in recent decades.

Silicates and other inorganic materials are often used without chemical modification, so the compounds comprising them remain unaltered. In these cases, the intrinsic characteristics of the compound or the structure of the material give it its usefulness; for example, its size, strength, hardness or colour. If the integrity of a material can remain intact during reclamation it can be reused in the same or similar applications. If the integrity is partially compromised it may be able to be reused in a lower-grade application. Building stones and bricks, for example, can often be recovered virtually intact and reused; if broken during recovery they may be usable as a fill or aggregate material. Inorganic compounds in the form of fine powders which are used, for example, in paper, plastics, toothpastes and paint, cannot easily be recovered even though they are chemically and physically unaltered through use of the product. These are dissipated (lost) when the host material is removed from the materials cycle through landfill or combustion.

There are some important silicate materials that are chemically modified in preparation and/or in use. These include cement, glass and ceramics. Cement mixed with water and aggregate to form concrete undergoes a chemical reaction which causes the mixture to set. While concrete can be broken up and reused as rubble and aggregate, the cement component cannot be recycled for reuse in new concrete because it has been chemically transformed. Glass is produced by heating mixtures of silicate minerals and other inorganic compounds to a temperature at which they melt and fuse together. Glass containers and glass sheet, for example, can be recovered and remelted and the glass recycled but, as with metals, the melting process causes any different types of glass in a stream to be fused together. Thus the composition and colour of recycled glass depends on how well the glass was sorted. Hence, recycled glass is often used in lower-grade applications.

2.6 QUANTIFYING THE MATERIALS CYCLE

Because matter is conserved it is possible to follow the mass flows of materials or substances into and out of systems. Similarly, since energy is conserved it is possible to quantify the inputs and outputs of energy across systems. Various methodologies have been developed for materials and energy accounting according to the scale of the system, the level of detail known or desired, and the purpose for which it is being undertaken. The law of conservation of matter and the first law of thermodynamics provide the basis for the various methodologies. All are based on the simple principles that for a particular system and interval of time:

$$\begin{aligned}&\textit{Mass of material input} = \\ &\textit{Mass of material output} + \\ &\textit{Mass accumulated}\end{aligned} \qquad 2.1$$

and

$$\begin{aligned}&\textit{Energy input} = \\ &\textit{Energy output} + \textit{Energy accumulated}\end{aligned} \qquad 2.2$$

A key requirement is to define the system of interest and the boundaries so inputs and outputs can be monitored. The scale of a system may range from a single piece of equipment to an operating plant, an entire company or part of it, an industry sector, a geographical or political region, or the entire world.

2.6.1 Materials and energy balances

The balances in Equations 2.1 and 2.2 are called, respectively, the *materials balance* and *energy balance* of the system. Materials and energy balances are commonly used as engineering accounting tools for designing and optimising processes, determining the quantities of reagents and fuels required, and following the partitioning of substances into product, by-product and waste streams to better control a process and the wastes produced. They also provide direct information of environmental interest. For example, the amount of a toxic substance released into the air or water in concentrations or quantities too small or difficult to measure accurately can be determined by difference if the quantities entering the process and leaving in other streams are known. They also provide information for incorporation into higher-level material flow analyses, as discussed in the following section.

In non-reactive systems, substances do not change their nature and the material balance is simply obtained by monitoring the inputs and outputs of each substance in the system. In reactive systems, individual elements rather than compounds or other substances need to be monitored since the forms in which the elements occur change as a result of chemical reactions. In this case, material balances are performed on individual elements. Thus, for each element in a system comprising p input substances and q output substances:

$$\sum_{i=1}^{i=p} m_i f_i = \sum_{j=1}^{j=q} m_j f_j' + \Delta m \qquad 2.3$$

where m_i is the mass of input substance i, m_j is the mass of output substance j over a given period of time, f_i and f_j' are the mass fractions of the element in the input and output substance, respectively, and Δm is the mass of the element which accumulates in the system during the time period (Δm can be positive or negative). In a steady-state system, over a given period of time Δm will be zero.

2.6.2 Material flow analysis

Material flow analysis (MFA) is a methodology used in understanding the flow of a substance (often of environmental interest, such as cadmium, mercury, phosphorus or CFCs), a specific material (wooden products, plastics), a bulk material (steel, aluminium, copper) or a product (car, television set) and the associated wastes within a company, industry sector, region, country, continent or the entire world (Bringezu and Moriguchi, 2002). Material flow analyses are often undertaken to assess the environmental impact of particular substances, materials or products and to identify opportunities for improving environmental performance. There are two broad types of material flow analysis, with several variations of each (Table 2.5). Type I refers to analyses driven by concerns over specific properties of substances, materials or products which could be hazardous or critical for some reason. Type II refers to analyses with a focus on companies, sectors, regions or countries, and how their environmental performance is affected by the throughput of substances, materials or products. The following examples illustrate some MFA methodologies and demonstrate their usefulness in identifying and quantifying environmental issues.

Economy-wide material flow analysis

Figure 2.5 shows an application of the MFA methodology to the flow of all materials through an economy. An economy in this context may be a region, a country, a continent or the world. The system boundary is the interface between the environment and the economy. Materials cross the boundary when they are purchased and cross back into the environment as waste when they no longer play a role in the economy.

Inputs to the economy from the environment consist of the quantities of commodities (grains, petroleum, minerals, metals, timber) produced domestically and the corresponding quantities imported. Together these make up the *direct material input (DMI)* to the economy:

$$DMI = Domestic\ extraction + Imports \qquad 2.4$$

Hidden flows (indirect flows) refer to materials which are moved or disturbed in the environment in

Table 2.5: Types of material flow analysis

	Issues related to the impacts per unit flow of ...			Issues related to the throughput of ...		
	Substances	*Materials*	*Products*	*Companies*	*Sectors*	*Regions*
	... within specific companies, sectors, regions			... associated with substances, materials, products		
Type of analysis	**Ia**	**Ib**	**Ic**	**IIa**	**IIb**	**IIc**
	Substance flow analysis	Material flow analysis	Life cycle assessment	Business-level material flow analysis	Input–output analysis	Economy-wide material flow analysis

Source: Bringezu and Moriguchi (2002).

the course of providing commodities for economic use but which do not themselves enter the economy. Examples of hidden flows are: biomass removed from the land along with timber and grain, that is later separated and discarded; soil and rocks excavated and/or disturbed in order to provide access to an ore body; and soil eroded as a result of agricultural practices. Hidden flows are of two types – those produced locally and those associated with the production of imported commodities in the country of production. Both these flows should be included in an MFA to obtain a complete picture, but are sometimes omitted. The *total material requirement* (*TMR*) of an economy is the sum of the total material input and the hidden material flows:

$$TMR = DMI + Domestic\ hidden\ flows + Foreign\ hidden\ flows \quad 2.5$$

The *TRM* is an overall estimate of the environmental impact associated with natural resource extraction and use.

The *domestic processed output* (*DPO*) is the quantity of materials, extracted from the domestic environment and imported from other countries, that have been used in the domestic economy then flow to the domestic environment. These flows occur at the

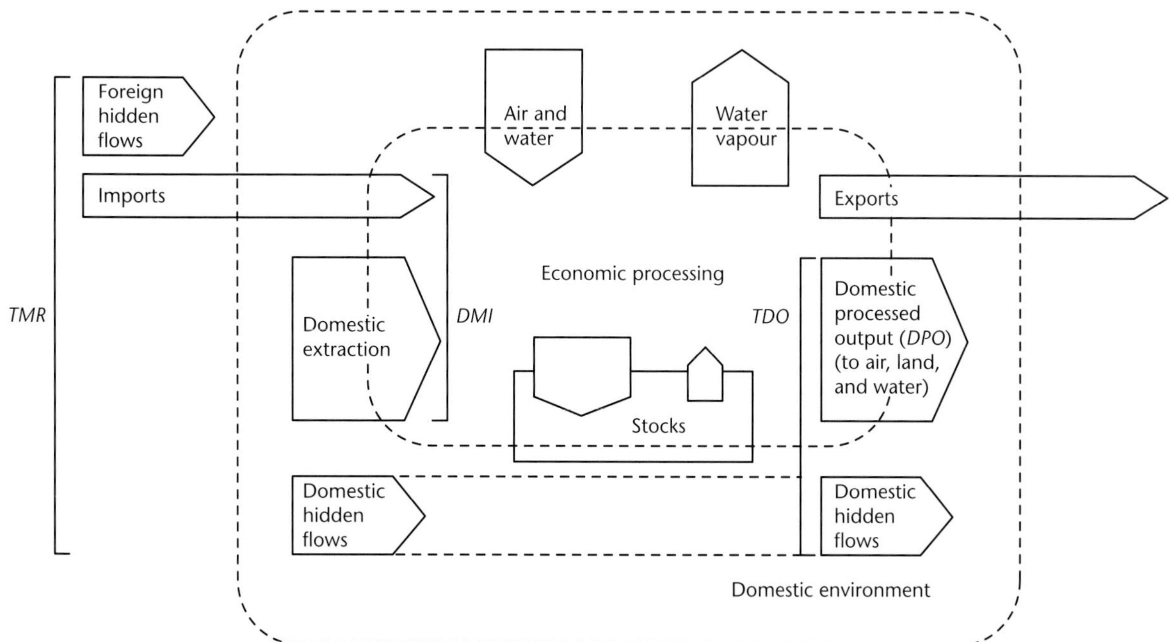

Figure 2.5: Material flows into, through and out of an economy (after Aman *et al.*, 2000).

processing, manufacturing, use and final disposal stages of the materials cycle. Exported materials are excluded because their wastes occur in other countries and are accounted for there. All wastes are included: for example, emissions to air from commercial fossil fuel combustion and other industrial processes; industrial and household wastes deposited in landfills; material loads in waste water; and materials dispersed into the environment as a result of product use. Flows of recycled materials in the economy (metals, paper, glass) are not included. *DPO* is given by the relation:

$$DPO = DMI - \text{Net additions to stock} - \text{Exports} \tag{2.6}$$

The *total domestic output* (*TDO*) of an economy is the sum of domestic processed output and domestic hidden flows:

$$TDO = DPO + \text{Domestic hidden flows} \tag{2.7}$$

TDO represents the total quantity of material outputs to the domestic environment caused directly and indirectly by human economic activity.

Table 2.6 presents a summary of total material flow for the United States, Japan and Germany as an example of a mass flow analysis. At the summary level presented, this analysis does not distinguish between different types of materials nor show the contributions of various sectors of the economy. The summary is obtained, however, in a bottom-up manner; more detailed information is normally available. This illustrates that the level of detail of an MFA can be varied according to the purpose for which the data are required. For example, the following useful observations can be made from Table 2.6 regarding material flows in the three countries.

- Around 45–85 tonnes of natural resources were required per person in 1991. About 10–20 wt% of this was from renewable resources, including agricultural products (Adriaanse *et al.*, 1997), the balance was from non-renewable resources.
- For every tonne of natural resources that entered the economy in 1991, around 2–3 tonnes of additional natural resources were moved or otherwise disturbed.

Table 2.6: Material flows for the US, Japan and Germany. Input values are for 1991, output values are for 1996. Values in italics are calculated values

		US	Japan	Germany	Year
Domestic extraction	(Mt)	4581	1424	1367	1991
Imports	(Mt)	568	710	406	"
Direct material input (DMI)	(Mt)	*5149*	*2134*	*1773*	"
Domestic hidden flows	(Mt)	15 494	1143	2961	"
Domestic hidden flows/domestic extraction		*3.4*	*0.8*	*2.2*	"
Foreign hidden flows	(Mt)	594	2439	2030	"
Total hidden flows	(Mt)	*16 091*	*3583*	*4993*	"
Total hidden flows/DMI		*3.1*	*1.7*	*2.8*	"
TMR	(Mt)	*21 240*	*5716*	*6764*	"
Domestic processed output (DPO)	(Mt)	6774	1407	1075	1996
Total domestic output (TDO)	(Mt)	*22 268*	*2550*	*4036*	"
Total hidden flows per capita	(t)	*64*	*29*	*63*	1991
DMI per capita	(t)	*20*	*17*	*22*	"
TMR per capita	(t)	*84*	*46*	*86*	"
DPO per capita	(t)	*25*	*11*	*13*	1996
TDO per capita	(t)	*83*	*20*	*49*	"

Source: Adriaanse *et al.* (1997); Aman *et al.* (2000).

Table 2.7: Breakdown of domestic extraction for the US in 1991. Values in italics are calculated values

Material category		Domestic extraction	Domestic hidden flows	Domestic hidden flows/ domestic extraction
Fossil fuels	(Mt)	1684	5846	*3.5*
Metals	(Mt)	185	1750	*9.5*
Industrial minerals	(Mt)	105	312	*3.0*
Construction materials	(Mt)	1730	159	*0.1*
Infrastructure	(Mt)		3473	
Soil erosion	(Mt)		3710	
Renewable resources	(Mt)	878	244	*0.3*
Domestic extraction	(Mt)	*4581*		*3.4*
Domestic hidden flows	(Mt)		*15 494*	

Source: Adriaanse *et al.* (1997).

- Around 10–25 tonnes of waste per person from the production, manufacture, use and disposal of goods were released to the land (as landfill), water or the atmosphere in 1996.

These figures (and data for other developed countries; e.g. Moll *et al.*, 2005) quantify some of the issues raised in Chapter 1 concerning the level of consumption of many finite resources and the quantities of wastes associated with their production, use and disposal. Table 2.7 breaks the value of domestic extraction for the United States down to the industry sector level. This provides an additional level of understanding. It is apparent that mining-related activities account for the bulk of hidden domestic flows and that fossil fuel production accounts for the single largest amount. Preparation and excavation for infrastructure projects and soil erosion are also major contributors. More detailed material flows for the United States and the world, and trends over the past century, have been provided by Rogich and Matos (2002).

Substance flow analysis

A substance flow analysis (SFA) is a variation of MFA which follows the flow of a specific substance through a company, industry sector or region. As an example, the flow of copper in North America (the US, Canada and Mexico) is summarised in Figure 2.6. This figure provides a quantitative overview of both the absolute and relative flows of copper in North America, from which a coherent understanding can be developed. It shows that the majority of copper used in North America in 1994 was primary copper, mined, smelted and refined within the continent. About 3000 kt of copper were mined from ores or obtained from the reworking of tailings. Additionally, around 400 kt of copper were imported in the form of semi-manufactures and finished products. Based on a population of 412 million, this equates to nearly 8 kg of copper per person per year. The copper was used mainly in infrastructure, buildings, industry and transportation. Of the 8 kg per person, about 5 kg was added to stock. Over 80% was used in the form of pure copper and the remainder was in the form of alloy. About 60% of waste copper was recycled, the vast majority from production rather than from end-of-life products. In total, about 1400 kt of domestic copper waste entered the waste management system. Of this, 190 kt of scrap was exported, around 500 kt was recycled domestically and 700 kt was disposed of in landfill. Thus, approximately 5 kg of copper per person was discarded. Waste electronic and electrical equipment and end-of-life vehicles made up only 7% of the overall waste stream but these streams contained about 80% of the total copper entering the waste management system. The losses of copper from North American production in 1994 were 370 kt in tailings and 41 kt in slag.

Life cycle assessment

Life cycle assessment (LCA) is a methodology for assessing the environmental impacts associated with a

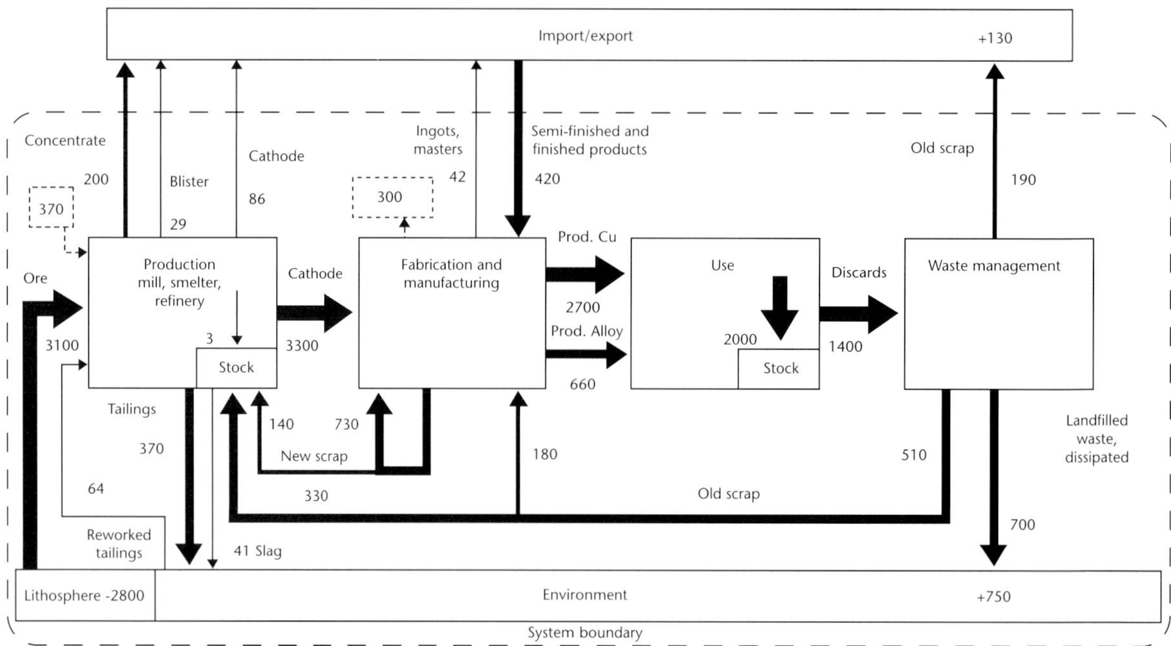

Figure 2.6: The copper cycle in North America, 1994. All quantities are expressed in kilotonnes. The dotted boxes and associated arrows indicate the distance to closure of the mass balance, i.e. the quantity by which the various statistical sources do not agree (after Lifset *et al.*, 2002).

specific material (steel, copper, aluminium, plastic) or product (car, personal computer, drink can) over its life cycle. The concept is illustrated in Figure 2.7. When the entire life cycle from raw material acquisition to material manufacture to product manufacture, use and disposal is considered, it is referred to as cradle-to-grave LCA. When it is applied to the production of primary materials, the life cycle stops at the material production step. This is referred to as cradle-to-gate LCA.

An LCA involves compiling an inventory of all relevant environmental exchanges of the material or product through the life cycle and evaluating the potential environmental impacts of those exchanges. The methodology has been progressively refined since the 1970s when it was first used and it is now the subject of a series of international standards: ISO14040 to 14044 (International Standards Organization). These describe the principles and framework for carrying out LCAs. There are four stages in performing an LCA.

- *Definition of the goal and scope.* The goal and scope of the assessment are determined according to the intended application of the results. Selecting the functional unit (for example, 1 tonne of material, 1 unit of a product) as the reference point and setting the system boundary are important aspects.
- *Inventory analysis.* The material and energy inputs and outputs are identified and quantified. This is usually a difficult and time-consuming part of the assessment and results in a materials and energy balance for environmentally relevant flows in the system.
- *Impact assessment.* The inventory data are grouped into impact categories according to the environmental problems to which they contribute. These include global warming, acidification, eutrification, photochemical smog, ozone depletion, ecotoxicity, human toxicity and resource depletion. The inventory data are weighted using equivalency factors which indicate how much a substance contributes to a particular environmental impact, compared to a reference substance. For example, the global warming potential of a substance is measured relative to the effect of 1 kg of CO_2, acidification potential is measured relative to the effect of 1 kg of SO_2, eutrification is measured relative to 1 kg of phosphate ions, toxicity is

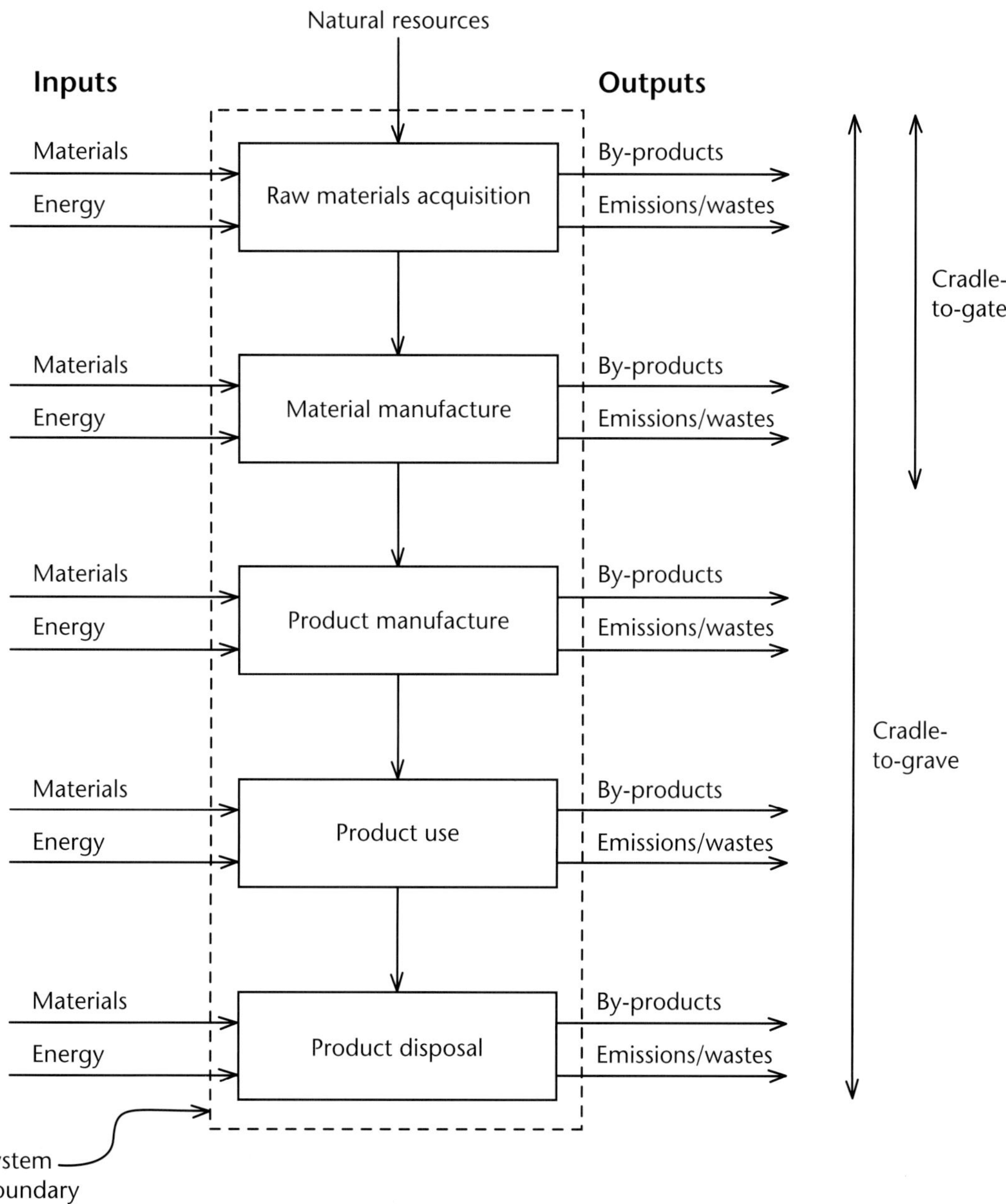

Figure 2.7: Inputs and outputs at various stages of a product's life cycle and the system boundary for cradle-to-gate and cradle-to-grave LCAs.

measured relative to 1 kg of 1,4-dichlorobenzene, and resource depletion is measured relative to world reserves. Each inventory amount is multiplied by its equivalency factor and an aggregated score for each impact category is obtained. Examples of equivalency factors for substances that contribute to global warming and acidification are given in Table 2.8.

- *Interpretation, reporting and critical review.* In this stage, the point(s) in the life cycle where particular environmental impacts occur are identified and, from a design or optimisation point of view, opportunities to reduce them are sought. LCAs are an important tool in improving the design of a product or process from an environmental perspective.

Table 2.8: Equivalency factors of some common substances contributing to global warming and acidification

Global warming		Acidification	
CO_2	1	SO_2	1
CH_4	21	NOx	0.7
N_2O	310	HCL	0.88
		HF	1.6

Source: de Bruijn *et al.* (2002).

Table 2.9 illustrates the application of the LCA methodology to the primary production of some common metals. It lists values of global warming potential and acidification potential for the cradle-to-gate production of the metals. The values illustrate the range of potential environmental impact for two important impact categories. Note the high global warming potential of aluminium and nickel and the low global warming potential of steel. The reasons for this are discussed in Chapter 9. A problem that often arises in the inventory analysis phase is how to allocate inventory data when there is more than one useful product. This may be done on mass, volume, energy content or economic value. The most common approach is to allocate on the basis of mass (Norgate and Rankin, 2002). This is how the impacts for lead and zinc in Table 2.9 were allocated in the case of the Imperial Smelting Process, which produces lead and zinc as co-products.

The Ecological Rucksack

The Ecological Rucksack is the total quantity (in kilograms or tonnes) of materials moved from the environment to create a product or service. The Material Input per Service unit (MIPS) is a closely related concept (Ritthoff *et al.*, 2002). The Ecological Rucksack and MIPS provide an indication of the environmental impact or resource use efficiency of a product or service over its life cycle. The concept was proposed by Friedrich Schmidt-Bleek of the Wuppertal Institute for Climate, Environment and Energy in Germany. The Ecological Rucksack of a product containing n different materials is given by:

$$MI = \sum_{i=1}^{i=n} m_i R_i \qquad 2.8$$

where m_i is mass of material i in the product and R_i is its Rucksack factor. Values of Rucksack factor have been estimated for common materials; some are listed in

Table 2.9: Cradle-to-gate life cycle global warming and acidification potentials for the production of 1 kg of some common primary metals. Required electrical energy is assumed to be supplied from black coal-fired power stations (with 35% efficiency)

Metal	Process	Global warming potential (kg CO_2-e/kg)	Acidification potential (kg SO_2-e/kg)
Steel	BF/BOF[1]	2.3	0.02
Aluminium	Electrolytic[2]	22.4	0.13
Copper	Pyromet[3]	3.3	0.04
	Hydromet[4]	6.2	0.05
Lead	BF[5]	2.1	0.02
	ISP[6]	3.2	0.02
Zinc	Electrolytic[7]	4.6	0.06
	ISP[6]	3.3	0.03
Nickel	Pyromet[8]	11.4	0.13
	Hydromet[9]	16.1	0.07

1. Blast furnace and basic oxygen furnace (iron ore, 64 wt% Fe, 50 wt% lump, 50 wt% fines, open-cut mine).
2. Bayer and Hall-Héroult processes (bauxite ore, 17.4 wt% Al, open-cut mine).
3. Matte smelting, converting and electro-refining (sulfide ore, 3.0 wt% Cu, underground mine).
4. Heap leaching, solvent extraction and electro-winning (sulfide ore, 2.0 wt% Cu, underground mine).
5. Blast furnace (ore 5.5 wt% Pb, underground mine).
6. Imperial Smelting Process (ore 8.6 wt% Zn, 5.5 wt% Pb, underground mine).
7. Roasting and electrolysis (ore 8.6 wt% Zn, underground mine).
8. Flash furnace smelting and Sherritt-Gordon refining (sulfide ore, 2.3 wt% Ni, underground mine).
9. Pressure acid leaching, solvent extraction and electro-winning (laterite ore, 1.0 wt% Ni, underground mine).

Source: Norgate and Rankin (2002).

Table 2.10: The Rucksack factor of some common materials according to impact category

Material	Abiotic material (tonne per tonne)	Biotic material (tonne per tonne)	Earth movements (tonne per tonne)	Water (tonne per tonne)	Air (tonne per tonne)
Steel, primary	6.97			44.6	1.3
Steel, secondary	3.36			57.5	0.56
Aluminium, primary	85.4			1378.6	9.78
Aluminium, secondary	3.45			60.9	0.37
Copper, primary	500			260.0	2.0
Copper, secondary	9.66			105.6	0.72
Plastics (PVC)	8.02			117.7	0.69
Plastics (PEP)	5.4			64.9	2.1
Cotton (USA, west)	8.6	2.9	5.01	6814.0	2.74
Paper (primary)	1.2	5.0		14.7	0.24

Source: Ritthoff *et al.* (2002).

Table 2.10. It is common to separate these into five categories so the environmental impact can be better seen.

- *Abiotic raw materials.* These are minerals and fossil fuels used in production of the material, plus all unused materials, overburden, tailings and other wastes.
- *Biotic raw materials.* These are the various forms of plant biomass used in production of the material, from cultivated and uncultivated areas.
- *Earth movement.* This is the amount of Earth moved due to agriculture and erosion.
- *Water.* This is the amount of water from all sources used in the production of the material.
- *Air.* This is the amount of air used in combustion and chemical and physical processes in the production of the material.

2.7 REFERENCES

Adriaanse A, Bringezu S, Hammond A, Rodenburg E, Rogich D and Schütz H (1997) *Resource Flows: The Material Basis of Industrial Economies.* World Resources Institute: Washington DC.

Altenpoel DG (1980) *Materials in World Perspective.* Springer-Verlag: Berlin.

Aman C, Bringezu S, Fischer-Kowalski M, Hütler W, Kleijn R, Moriguchi Y, Ottke C, Rodenburg E, Rogich D, Schandl H, Schütz H, Van der Voet E and Weisz H (2000) *The Weight of Nations: Material Outflows from Industrial Economies.* World Resources Institute: Washington DC.

Ayres RU and Kneese AV (1969) Production, consumption and externalities. *American Economics Review* **59**(3): 282–289.

Bever MB (1986) Introduction. In *Encyclopedia of Materials Science and Engineering.* Vol. 1. (Ed. MB Bever) pp. xi–xxiii. Pergamon Press: Oxford, UK.

Bringezu S and Moriguchi Y (2002) Material flow analysis. In *A Handbook of Industrial Ecology.* (Eds RU Ayres and LW Ayres) pp. 79–90. Edward Elgar: Cheltenham, UK.

De Bruijn H, van Duin R and Huijbregts E (2002) *Handbook on Lifecycle Assessment: Operational Guide to the ISO Standards.* Kluwer Academic Publishers: Dordrecht, The Netherlands.

Douglas I and Lawson N (2002) Material flows due to mining and urbanization. In *A Handbook of Industrial Ecology.* (Eds RU Ayres and LW Ayres) pp. 351–364. Edward Elgar: Cheltenham, UK.

Lifset RJ, Gordon TE, Graedel TE, Spatari S and Bertram M (2002) Where has all the copper gone: the stocks and flow project, Part 1. *JOM* **54**(10): 21–26.

Moll S, Bringezu S and Schütz H (2005) *Resource Use in European Countries.* Wuppertal Institute: Wuppertal, Germany.

Norgate TE and Rankin WJ (2002) The role of metals in sustainable development. In *Green Processing*

2002. *Proceedings of International Conference on Sustainable Processing of Minerals*. pp. 49–55. Australasian Institute of Mining and Metallurgy: Melbourne.

Ritthoff M, Rohn H and Liedtke C (2002) *Calculating MIPS: Resource Productivity of Products and Services*. Wuppertal Institute for Climate, Environment and Energy: Wuppertal, Germany.

Rogich DG and Matos GR (2002) Material flow accounts: the USA and the world. In *A Handbook of Industrial Ecology*. (Eds RU Ayres and LW Ayres) pp. 260–287. Edward Elgar: Cheltenham, UK.

Victor PA (2009) Scale, composition and technology. *Bulletin of Science, Technology and Society* **29**(5): 383–396.

2.8 USEFUL SOURCES OF INFORMATION

Baumann H and Tillman A-M (2004) *The Hitch Hiker's Guide to LCA: An Orientation in Life Cycle Assessment Methodology and Application*. Studentlitteratur: Lind, Sweden.

Bringezu S and Bleischwitz R (Eds) (2009) *Sustainable Resource Management: Global Trends, Visions and Policies*. Greenleaf Publishing: Sheffield, UK.

Hertwich E, van der Voet E, Suh S, Tukker A, Huijbregts M, Kazmierczyk P, Lenzen M, McNeely J and Moriguchi Y (2010) 'Assessing the environmental impacts of consumption and production: priority products and materials. A report of the Working Group on the Environmental Impacts of Products and Materials to the International Panel for Sustainable Resource Management'. United Nations Environment Programme: Paris, France.

Horne R, Grant T and Verghese K (2009) *Life Cycle Assessment: Principles, Practice and Prospects*. CSIRO Publishing: Melbourne.

3 An introduction to Earth

Earth is the third planet from the Sun and orbits at an average distance of 1.50×10^8 km. The atmosphere extends about 150 km above the surface of the Earth and consists of mainly nitrogen (78% by volume), oxygen (21% by volume) and argon (0.9% by volume). Other components include water vapour, carbon dioxide, hydrogen and the inert (noble) gases. Together, the atmosphere and the Earth constitute the *Earth system*. The Earth has a diameter of about 12 750 km and a mass of 6.0×10^{24} kg. The Earth consists of the biosphere, the hydrosphere, the lithosphere, the mantle and the core. Its structure is illustrated in Figure 3.1. The biosphere and hydrosphere form a thin layer over the lithosphere and are not shown in the figure. From a utilitarian point of view, the biosphere, hydrosphere and the upper part of the lithosphere, called the *crust*, are of most interest. The crust is the ultimate source of the inorganic materials and fossil fuels used by humans, and the source of the nutrients required by living matter. The biosphere and hydrosphere are the sources of materials made from living matter and of our food.

This chapter examines the nature of the crust, hydrosphere and biosphere and considers some implications of the law of conservation of matter and the laws of thermodynamics for the Earth system. The emphasis on the biosphere and hydrosphere provides the background and context required for an understanding of the principles of environmental sustainability discussed in Chapter 4. The crust, as the source of minerals and rocks, is examined in greater detail in Chapter 5.

3.1 THE CRUST

The crust is the outermost layer of the Earth and is broadly of two types. The crust under the oceans (*oceanic crust*) is typically 6–7 km thick and has an average density of around 3.0 g mL^{-1}. The crust making up the continents (*continental crust*) is typically 30–50 km thick and has an average density of about 2.7 g mL^{-1}. The temperature within the lithosphere increases with depth to about 500–1000°C at the boundary with the underlying mantle. The temperature continues to increase through the mantle and reaches around 4000–6000°C in the core. The high temperatures and pressures within the mantle cause the mantle material to become plastic (a solid which is capable of being shaped or formed without fracturing).

The upper portion of the mantle has an average density of about 4.0 g mL^{-1}. The crust has lower density

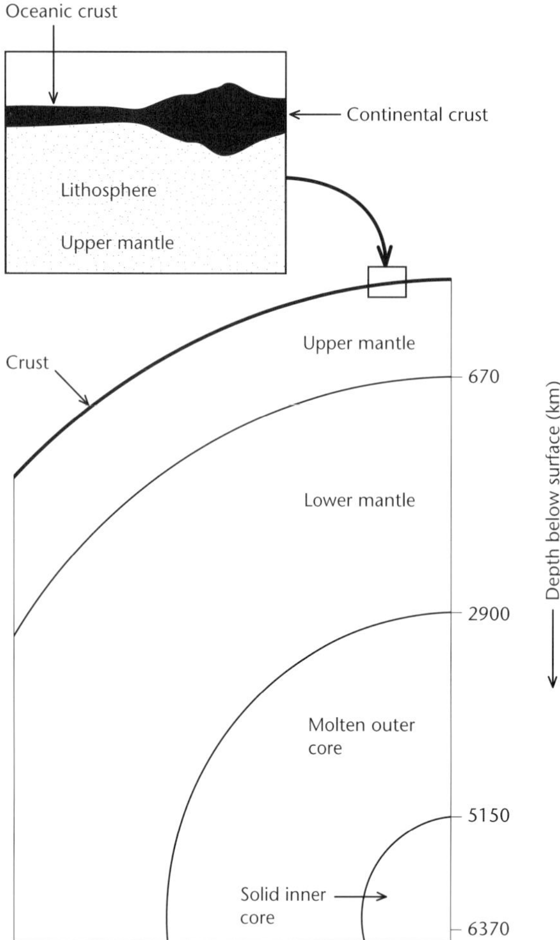

Figure 3.1: The structure of the Earth, showing the major zones.

and floats on the mantle in the same way that a body with a density less than that of water will float on water. The less-dense continental crust floats much higher than the oceanic crust (proportionately less is submerged in the mantle) and has an average elevation above sea level of about 850 m, compared to an average depth below sea level of around 3800 m for oceanic crust. Because of slow convection currents in the underlying plastic upper mantle, driven by the temperature gradient across the mantle, the crust is broken into sections (*tectonic plates*), which move very slowly in relation to one another at rates of around 50–100 mm per year. The mechanism is illustrated in Figure 3.2. In some regions, descending slabs of crust are heated, soften and lose form, and eventually become part of the mantle, while in other regions fresh material from the mantle is brought to the surface. There are seven major and about 20 smaller plates (Figure 3.3). Earthquakes, volcanic activity, mountain-building and formation of mid-ocean ridges occur along plate boundaries. These are examined in more detail in Chapter 5.

3.2 THE HYDROSPHERE AND BIOSPHERE

The *hydrosphere* consists of water in all its forms on Earth:

- the oceans;
- other surface waters, including inland seas, lakes, and rivers;

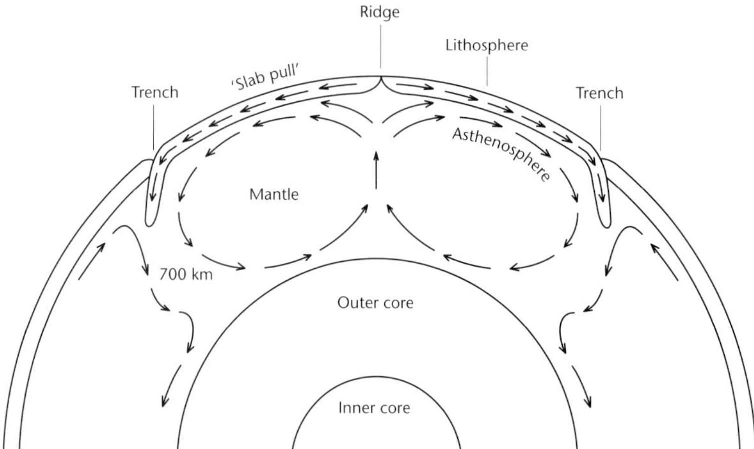

Figure 3.2: Schematic diagram illustrating the concept of convection cells in the mantle causing movement of the crustal plates (courtesy United States Geological Survey).

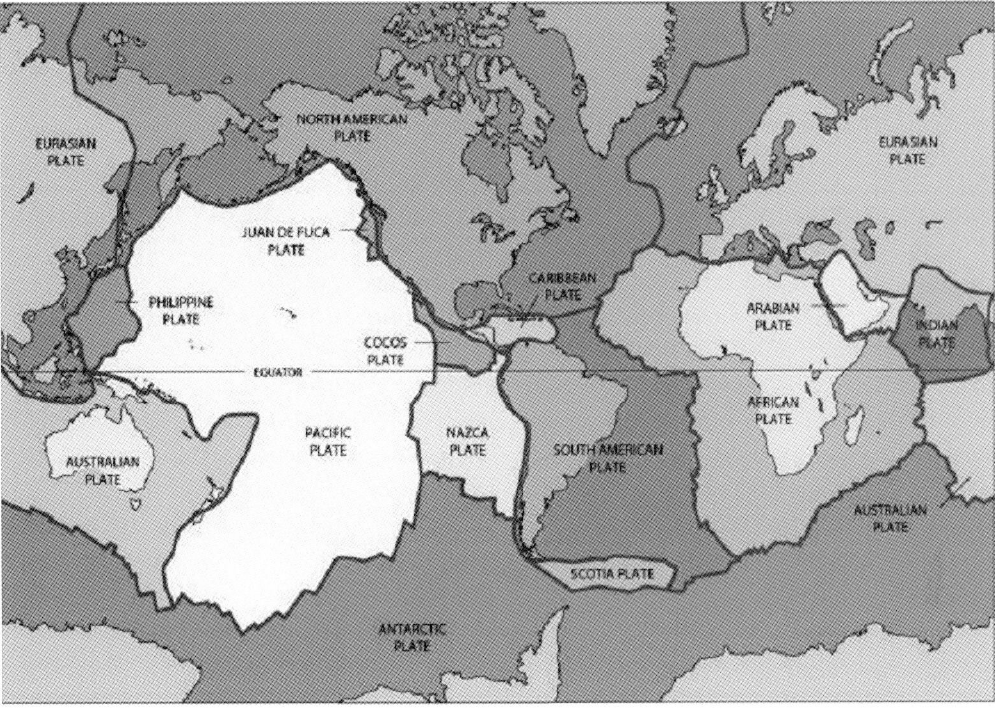

Figure 3.3: Map of the Earth showing the major tectonic plates (courtesy United States Geological Survey).

- rain;
- underground water;
- ice (at the poles, in glaciers and as snow);
- atmospheric water vapour.

The average depth of the oceans is 3800 m, more than four times the average height of the continents, and about 71% of Earth's surface is covered by water (69% by seawater and 2% by fresh water). Thus, since the radius of the Earth is about 6.375×10^3 km, the volume of water in the oceans ($4\pi r^2 h$, where r is the radius of the Earth and h is the depth of the oceans) is approximately:

$$0.69 \times 4 \times \pi \times (6.375 \times 10^3)^2 \times 3.800$$
$$= 1.34 \times 10^9 \text{ km}^3$$

Since the density of seawater is about 1020 kg m^{-3}, the mass of water in the oceans is approximately 1.37×10^{18} tonnes. This abundance of water is a unique feature that distinguishes the Earth from other planets in the solar system. Earth's orbit around the Sun is outside the band of orbits within which planets are warm enough for water to be present in liquid form (0–100°C) and it is only the natural greenhouse effect created by the presence of carbon dioxide and water vapour in the atmosphere which keeps the mean surface temperature of Earth above the freezing point of water.

The *biosphere* consists of the air, land, surface rocks, soil and water, within which life occurs. All living organisms exist within *ecosystems*, which are dynamic entities consisting of plants, animals and microorganisms functioning together in a defined area with the non-living matter (soil, air and water). In its broadest sense the biosphere is the global ecosystem. The mean temperature of the biosphere is around 15°C.

3.2.1 Life on Earth

Nearly every part of the Earth's surface supports life of some kind. There are two broad types of biological organisms – producers and consumers. *Producers* store energy from the environment in carbon-based compounds. Plants are the most familiar producers but algae and cyanobacteria are also important producers. Plants absorb energy from the electromagnetic radiation from the Sun and use it to convert carbon dioxide from the atmosphere into carbohydrates by the process of photosynthesis:

$$nCO_2 + nH_2O = (CH_2O)_n + nO_2$$

where n is any number greater than three. *Consumers* obtain the energy they need from the carbon compounds made by producers. Animals are consumers; they obtain energy by consuming producers (in which case the animals are called herbivores) or by consuming other animals (in which case the animals are called carnivores). Omnivores are animals that consume both plants and animals. *Decomposers* are consumers that breakdown (or decompose) the waste products of animals and the dead remains of plants and animals and release matter back into the ecosystem. The primary decomposers are bacteria and fungi.

Figure 3.4 shows the flow of energy from the Sun through the biosphere, with it all ultimately forming thermal energy. The dotted lines show the flow of nutrients through the environment. These are inorganic substances (phosphorus, nitrogen, iron etc.) which are essential to life, but often in relatively small quantities. Producers obtain nutrients from the soil or water in which they grow and the nutrients are passed along the food chain by organisms consuming them. These organisms are in turn consumed by other organisms, and so on. Waste matter and dead organisms are decomposed; at this stage the last of the energy is extracted (and lost as heat) and the inorganic nutrients are returned to the soil or water to be reused. While inorganic nutrients are recycled, energy is not.

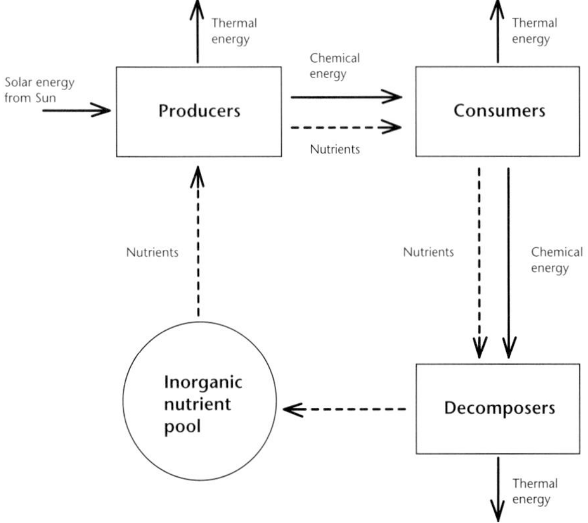

Figure 3.4: Energy and nutrient flows in the biosphere.

3.2.2 The Earth's biomes

The biosphere can be conveniently divided into biomes, each inhabited by broadly similar types of flora and fauna (Olson *et al.*, 2001). *Biomes* are regions which have similar climatic conditions but are geographically distinct and which, as a result of natural selection, have similar types of ecosystems. The major biomes are listed in Table 3.1. The terrestrial biomes tend to be separated geographically, primarily by latitude. The term *biomass* is used to describe the total amount of living matter of all kinds in a particular area. It is usually reported as dry mass per square metre or square kilometre. Since, on average, water makes up about 70 wt% of biomass, reported biomass values represent only about 30% of the actual mass of living matter.

Global production of biomass

Table 3.1 summarises some important quantitative data for biomes. Column 2 lists the approximate areas occupied by each biome type. Columns 3 and 4 list the estimated respective quantities of biomass produced annually per square metre, called the *productivity*. Column 5, the product of columns 2 and 4, lists the approximate annual net production of biomass for each biome. These data show that biomes within the Arctic and Antarctic circles are relatively barren and that most of the more populous biomes are near the equator.

Terrestrial ecosystems produce around 120 billion tonnes of biomass annually (on a dry basis) whereas marine ecosystems, which occupy about 69% of the Earth's surface, produce only about 55 billion tonnes (on a dry basis). The average net productivity of cultivated land is about 650 g km^{-2} of biomass annually, which is low among terrestrial biomes. Hence, agriculture does not increase net production of biomass, and can actually decrease it in some areas. Open oceans, which make up about 66% of the Earth's surface, have a productivity very similar to that of deserts; all the ocean areas, except for estuaries, algal beds and reefs, have productivities less than that of

Table 3.1. The world biomes and world biomass production

		Net primary production (dry basis)			Stocks of biomass (dry basis)			Average lifespan
(1)	(2)	(3)	(4)	(5)	(6)	(7)	(8)	(9)
Biome	Area (10^6 km^2)	Range (g m^{-2} year^{-1})	Mean (g m^{-2} year^{-1})	Total (10^9 t year^{-1})	Range (kg m^{-2})	Mean (kg m^{-2})	Total (10^9 t)	Years
Tropical rainforest	17.0	1000–3500	2200	37.4	6–80	45	765	20.5
Tropical seasonal forest	7.5	1000–2500	1600	12.0	6–60	35	263	21.9
Temperate forest								
– evergreen	5.0	600–2500	1300	6.5	6–200	35	175	26.9
– deciduous	7.0	600–2500	1200	8.4	6–60	30	210	25.0
Boreal forest	12.0	400–2000	800	9.6	6–40	20	240	25.0
Woodland and shrubland	8.5	250–1200	700	6.0	2–20	6	51	8.6
Savannah	15.0	200–2000	900	13.5	0.2–15	4	60	4.4
Temperate grassland	9.0	200–1500	600	5.4	0.2–5	1.6	14	2.7
Tundra and alpine	8.0	10–400	140	1.1	0.1–3	0.6	5	4.3
Desert and semi-desert scrub	18.0	10–250	90	1.6	0.1–4	0.7	13	7.8
Extreme desert: rock, sand, ice	24.0	0–10	3	0.1	0–0.2	0.02	0.5	6.7
Cultivated land	14.0	100–4000	650	9.1	0.4–2	1	14	1.5
Swamp and marsh	2.0	800–6000	3000	6.0	3–50	15	30	5.0
Lake and stream	2.0	100–1500	400	0.8	0–0.1	0.02	0.04	0.050
Total continental	*149*		*782*	*117.5*		*12.2*	*1840*	15.7
Open ocean	332.0	2–400	125	41.5	0–0.005	0.003	1.0	0.024
Upwelling zones	0.4	400–1000	500	0.2	0.005–0.1	0.02	0.008	0.040
Continental shelf	26.6	200–600	360	9.6	0.001–0.04	0.01	0.27	0.028
Algal beds and reefs	0.6	500–4000	2500	1.5	0.04–4	2	1.2	0.800
Estuaries (excluding marsh)	1.4	200–4000	1500	2.1	0.01–4	1	1.4	0.667
Total marine	*361.0*		*155*	*54.9*		*0.01*	*3.9*	0.071
Global total	*510*		*336*	*172.3*		*3.6*	*1844*	10.7

Note: small inconsistencies are due to rounding errors.
Source: Whittaker and Likens (1975).

agriculture. The main reason is that nutrients pass quickly through the upper layer of ocean waters by settling under the action of gravity whereas nutrients in terrestrial ecosystems remain in the system for much longer periods. Thus, plankton in the upper, lighted zones of the oceans, where photosynthesis occurs, are starved of nutrients relative to their plant counterparts in terrestrial ecosystems.

Humans appropriate some of the world's biomass either deliberately for their own purposes or unintentionally through their activities. Appropriated biomass is the sum of:

- the quantity of biomass consumed directly by people and domestic animals;
- the quantity of biomass consumed in human-dominated ecosystems by organisms that are different from those in ecosystems not dominated by humans;
- the quantity of biomass lost as a consequence of human activities (such as by clearing of forest).

With this definition, Vitousek *et al.* (1986) estimated that humans appropriate about 40% (38% for land and 2% for aquatic) of the net primary production of the Earth.

The energy stored in biomass

The quantity of energy captured from the Sun annually in biomass can be estimated from the total annual net production of biomass and the combustion value (heat of combustion) of biomass. The combustion value is the energy released by biomass when its complex hydrocarbon molecules are oxidised to produce carbon dioxide and water. It is the quantity of heat released when an organism decomposes naturally or is combusted (in both cases, the final products are the same). This value is equal to the quantity of energy in sunlight that was stored in complex molecules in the biomass through photosynthesis. The average combustion values for terrestrial biomass and ocean biomass are about 17.8 kJ g^{-1} and 19.9 kJ g^{-1} dry biomass, respectively (Leith, 1973). Applying these to the net annual production of biomass in Table 3.1 yields 2.09×10^{21} and 1.09×10^{21} J per year of energy stored by photosynthesis in biomass on the continents and in the oceans, respectively, giving a total of 3.18×10^{21} J per year. This value can be compared with the quantity of energy consumed by humans through combustion of fossil fuels. In 2005, global consumption of energy generated by humans was about 4.79×10^{20} J, of which 81% was from fossil fuels and the remainder from nuclear and renewable sources (IEA, 2007). Thus the energy produced annually from combustion of fossil fuels is about 15% of the energy extracted annually from sunlight by biomass.

The global stock of biomass

The values in column 5 of Table 3.1 are the net quantities of biomass produced each year; they are not the total quantities of biomass present in each biome at a particular time. The latter are referred to as the stock of biomass; values of these are listed in columns 6 and 7 as the kilograms of biomass per square metre. These values are constant over time unless a biome is being degraded or enhanced through natural or artificial processes. The total quantity of biomass in each biome is listed in column 8, which is the product of columns 2 and 7. The values in column 8 show that the stock of biomass is very much greater in terrestrial biomes than in aquatic biomes and that the global stock of biomass is about an order of magnitude greater than the net quantity of biomass produced annually. The ratio of total biomass (column 8) to the annual quantity produced (column 5) represents the average lifespan (in years) of biomass in the ecosystem (column 9). The total energy stored in biomass at any time is given by:

$$\begin{aligned}&\textit{Energy stored in continental biomass} + \\ &\textit{Energy stored in marine biomass} = \\ &(1840 \times 10^9 \times 17.8 \times 10^9) + \\ &(3.9 \times 10^9 \times 19.9 \times 10^9) = 3.28 \times 10^{22}\ \text{J}\end{aligned}$$

As expected, this is about 10 times the quantity of energy stored in biomass each year.

3.2.3 Ecosystem services

Humans depend on the biosphere to provide essential life-supporting services. These can be grouped as follows (UNEP, 2005).

- *Provisioning services.* These include food, water, timber and fibre.

- *Regulating services.* These affect climate, flood control, disease, waste and water quality.
- *Cultural services.* These provide recreational, aesthetic and spiritual benefits.
- *Supporting services.* These include soil formation, photosynthesis and nutrient recycling.

Collectively, these are called *ecosystem services.* Ecosystem services are distinct from other ecosystem functions because there is a human need for them. The United Nations Millennium Ecosystem Assessment report identifies the essential requirements for human well-being as having access to the basic materials required for a good life, good health, good social relations, security and freedom of choice and action. All of these depend, to a greater or lesser extent, on ecosystem services (illustrated in Figure 3.5). Throughout history many ecosystem services have been freely available, not rationed nor paid for directly. However, as population has increased, the demand on ecosystems has increased. At the same time the area of the Earth occupied by natural ecosystems has decreased due to the rising use of land for agriculture and urban living. As discussed in Section 4.1, there is strong evidence that ecosystems can no longer cope with this increased demand and are degrading.

3.3 SOME IMPLICATIONS OF THE BASIC LAWS OF SCIENCE

There is a small transfer of matter between the Earth's surface and outer space. A small quantity falls on the Earth every day from space, from meteors and dust, and a small quantity of very light gases in the

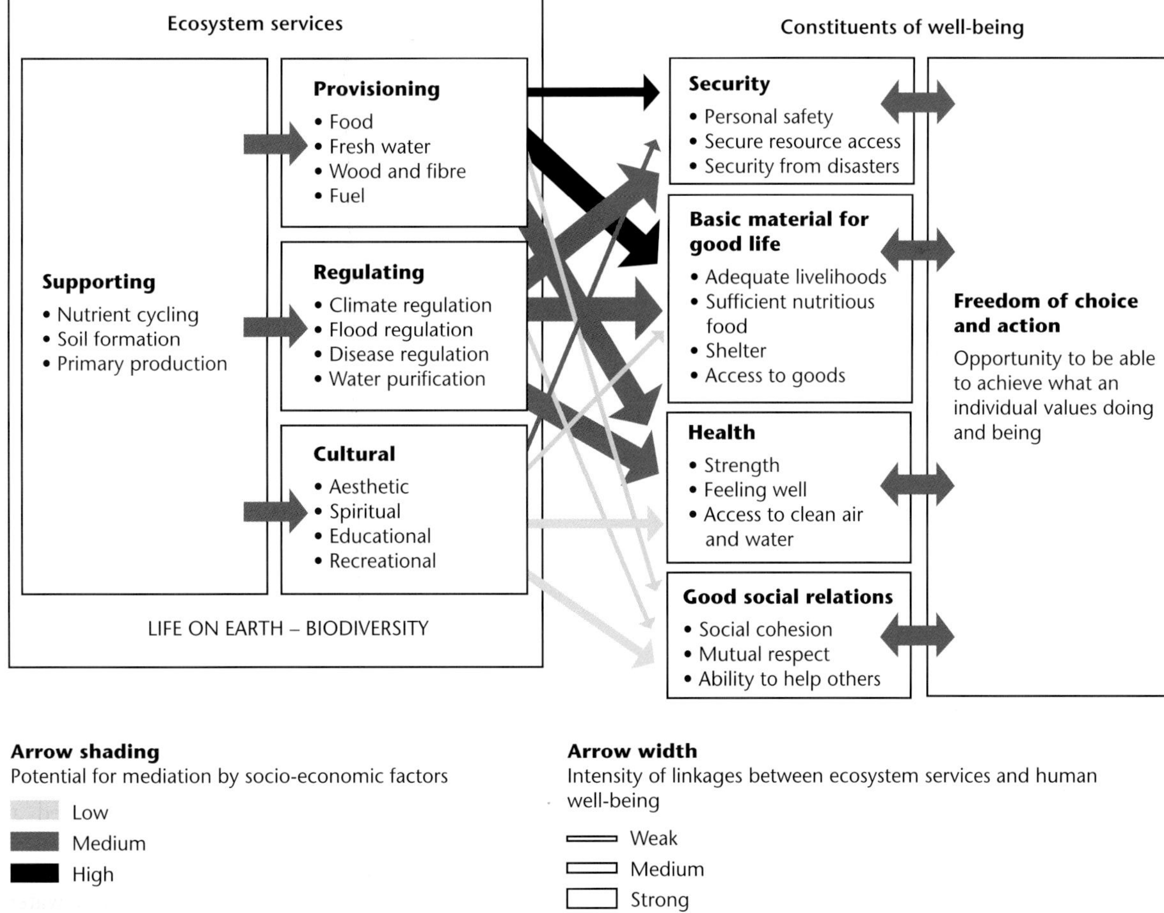

Figure 3.5: The links between ecosystem services and human well-being (after UNEP, 2005).

atmosphere, mainly hydrogen and helium, escape the gravitational pull of the Earth. However, given the mass of the Earth, these exchanges are insignificant. There is a large exchange of matter between the crust and the mantle through the action of plate tectonics but, in human time frames, this is also insignificant. It follows that the crust and the biosphere together can be considered a closed system with respect to matter. This matter can change form (physically and chemically) but the total amount must remain constant (the law of conservation of matter).

3.3.1 Thermal energy flows to the biosphere and hydrosphere

While the crust and biosphere is a closed system with respect to matter, it is an open system with respect to energy since it receives energy from the Sun and from the Earth's interior. Heat flows from hot to cold (second law of thermodynamics), i.e. from the Sun to the Earth and from inside the Earth to the surface. The mean energy flux from the Sun to the Earth, as measured at the outer edge of the atmosphere in a plane perpendicular to the radiation, is about 1366 W m^{-2} (Mizutani, 1995). The total quantity of energy received by Earth each year from the Sun is obtained by multiplying the solar flux per unit area by the cross-sectional area of the Earth (since the solar flux is measured perpendicularly to the radiation). The cross-sectional area of the Earth is:

$$\pi r^2 = \pi \times 6375^2 = 1.277 \times 10^{14}\ m^2$$

Hence, the mean solar flux to the Earth is $1.277 \times 10^{14} \times 1366 = 1.744 \times 10^{17}$ W, or 5.50×10^{24} J per year. The heat flux to the Earth's surface from the interior of the Earth is on average about 0.08 W m^{-2} (Furlong, 1992). The surface area of the Earth is:

$$4\pi r^2 = 4 \times \pi \times 6375^2 = 5.107 \times 10^{14}\ m^2$$

Therefore, the total heat flux from the Earth's interior to the surface is 4.1×10^{13} W. This is nearly three orders of magnitude smaller than the energy received from the Sun. Thus, the main source of energy in the biosphere and hydrosphere is radiation from the Sun. This makes possible all life and many non-biological processes in the biosphere and hydrosphere, and the flow of thermal energy from the Earth's interior drives major tectonic processes within the mantle and crust. Convection plays an important role in heat and mass transfer in the mantle region of the Earth; conduction is the major mechanism of heat transfer through the Earth's crust from the hotter interior. While radiation is the mechanism by which heat is transferred from the Sun to the Earth, convection, driven by the energy from the Sun, is the major mechanism of heat and mass transfer in the Earth's atmosphere and oceans. It drives most aspects of the climate and weather.

About half of the solar flux to the Earth reaches the surface; the remainder is reflected back into space or absorbed in the upper layers of the atmosphere (ASDC, 2010). About half of the energy reaching the surface is at a wavelength suitable for photosynthesis (Mizutani, 1995). Therefore, about 1.38×10^{24} J per year of solar energy is potentially available for capture by photosynthesis. The energy captured by biomass through photosynthesis was previously estimated to be about 3.18×10^{21} J per year. Therefore, only about 0.25% of solar energy potentially available for photosynthesis is actually used for photosynthesis. The remaining energy that reaches the Earth, about 2.25×10^{24} J per year, is absorbed in the lower atmosphere or at the Earth's surface and is converted directly to thermal energy. This is equivalent to about 6000 times the quantity of energy produced annually by combustion of fossil fuels. The energy stored through photosynthesis as chemical energy is ultimately released as thermal energy at various stages along the food chain (Figure 3.4).

3.3.2 The greenhouse effect

Some objects have nearly perfect ability to absorb and emit electromagnetic radiation. These are called black bodies. The quantity of electromagnetic radiation emitted by a black body is related to its temperature by the Stefan-Boltzmann Law:

$$J = \sigma T^4 \qquad 3.1$$

where J is the quantity of radiation, or flux, emitted by the body (W m^{-2}), T is the temperature of the body (K) and σ is a constant (5.67×10^{-8} $Wm^{-2}K^{-4}$). Because of the effect of the fourth power, a small increase in the temperature of a body results in a large quantity of additional radiation. The Earth is not a black body; about 30% of the radiation reaching Earth is reflected back into space, the balance being absorbed in the atmosphere or at the surface (ASDC, 2010). If it is assumed that this energy is ultimately re-radiated back into space, the mean surface temperature of the Earth can be calculated as follows.

$$\textit{Energy flux absorbed by Earth} = 0.7 \times 1.744 \times 10^{17} = 1.23 \times 10^{17}\ \text{W}$$

Therefore:

$$\textit{Energy flux from the Earth's surface} = \frac{1.23 \times 10^{17}}{4\pi r^2} = 2.305\ \text{Wm}^{-2}$$

and, from Equation 3.1:

$$\textit{Mean surface temperature of Earth} = \left(\frac{2.305}{5.67 \times 10^{-8}}\right)^{0.25} = 254.8\ \text{K} = -18.3°\text{C}$$

That the mean surface temperature of the Earth is actually about 15°C is due to the natural greenhouse effect (discussed below).

The wavelength of the radiation emitted from a body follows a distribution about a mean value given by Wien's law:

$$\lambda_{\text{max}} = \frac{b}{T} \qquad 3.2$$

where λ_{max} is the peak wavelength (the wavelength carrying the maximum energy) of the radiation emitted by the body (m), T is the temperature of the body (K) and b is a constant (2.897×10^{-3} m K). Wien's law indicates that as the temperature of a body increases, the wavelength of maximum emission becomes smaller. According to Equation 3.2, the wavelength of maximum emission from the Sun (at 5500°C) is about 0.5×10^{-6} m (in the visible light range) while the wavelength of maximum emission from the Earth (with a mean temperature of 15°C) is about 10×10^{-6} m (in the infrared wavelength range). Some gases, particularly CO_2, H_2O, CH_4 and N_2, have strong absorption bands in the infrared wavelength range and therefore absorb the energy radiated from the surface or lower atmosphere to a greater extent than the energy received directly from the Sun. These gases then re-radiate the energy in all directions, both towards the surface and to outer space. The result is that the Earth's surface is about 30°C warmer than the temperature predicted by the Stefan-Boltzmann equation.

The atmosphere creates a greenhouse effect within the biosphere that is similar to that caused by the glass walls and roof of a greenhouse. If small variations in the intensity of radiation from the Sun are ignored, then the steady-state mean temperature at the surface of the Earth will be the temperature at which the quantity of energy radiated from Earth into space, over any period of time, is equal to the quantity of energy received by the biosphere from the Sun over the same period of time. If this were not the case, the biosphere would heat or cool, according to the Stefan-Boltzmann law, until a temperature was reached at which the resulting increased or decreased rate of radiation of energy from the Earth matched the rate of incident radiation. In other words, the Earth would move to a new stable surface temperature.

If the concentration of greenhouse gases increases or decreases, the quantity of infrared radiation absorbed and re-radiated will also increase or decrease and the surface temperature of Earth will increase or decrease. Most greenhouse gases are formed by natural processes though all, and some others such as chlorofluorocarbons, are also produced by human activities, particularly the burning of fossil fuels. The change in carbon dioxide concentration in the Earth's atmosphere over the past 250 years is shown in Figure 3.6. The increase in the quantities of greenhouse gases in the atmosphere due to human activity is believed to be the cause of global warming.

3.3.3 The Sun as driver of both change and order

It might be expected that energy and matter on Earth would become dissipated or dispersed over time through natural processes since, according to the second law of thermodynamics, entropy increases in isolated systems undergoing spontaneous (or natural) changes. However, the biosphere, together with the crust, is not an isolated system – it is open with respect to energy. Energy from the Earth's interior and from the Sun is responsible for the dynamic nature of the

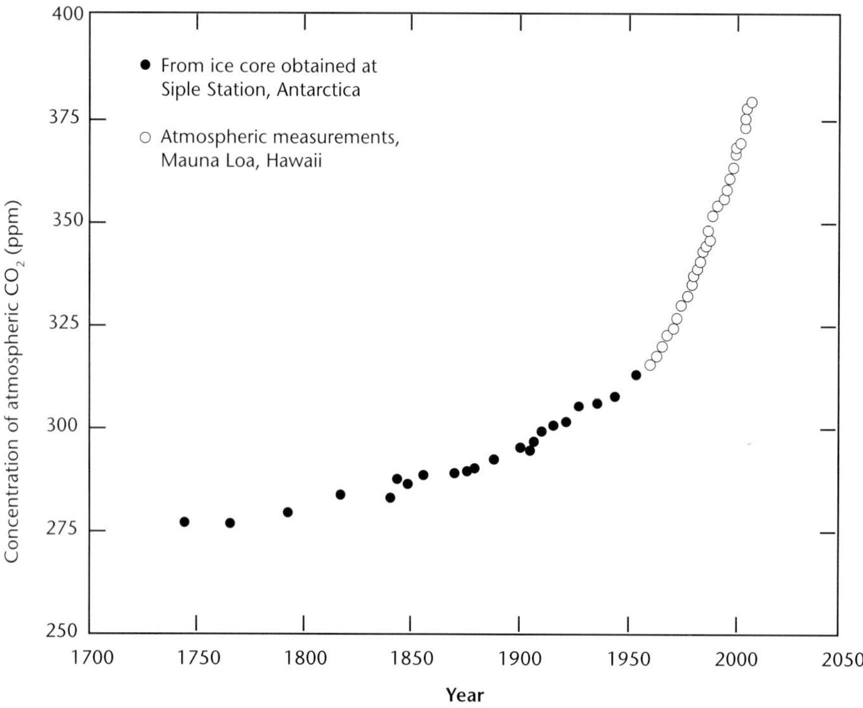

Figure 3.6: The concentration of carbon dioxide in the atmosphere from 1744 to 2005 (Pidwirny, 2006a).

Earth system. Without them, Earth would be a cold, rigid, lifeless body. Energy from the Sun drives all life on Earth and many non-biological processes, particularly ocean currents and the weather system. It does this through photosynthesis in plants and by solar warming, which creates convection currents in the oceans and atmosphere. Plants take energy from sunlight and store the energy in chemical forms; plants, in turn, provide energy for animals. The flow of energy from the Sun continually creates structure and order from the natural tendency of the biosphere to become more disordered. The energy from the Sun also drives the great biogeochemical cycles (discussed in Section 3.4). Evaporation of water from the oceans by solar heating produces most of the Earth's fresh water and thermal warming causes winds which distribute evaporated water around the world, importantly over the continents where it can fall as rain and provide water for biomass. Solar energy also drives some geological processes on the surface. Weathering and erosion by water, wind and ice are dissipative processes; depositions of sediments to form sedimentary rocks and fossil fuels are ordering processes. Energy from within the Earth drives other, larger-scale geological processes, particularly plate tectonics which formed the continents and mountain ranges. The time scales are very different for the weather, erosion and deposition, and tectonic processes, being around 10^{-3}–10^{-1} years for weather events and 10^{6}–10^{9} years for major tectonic processes.

3.4 THE BIOGEOCHEMICAL CYCLES

While energy does not cycle through the biosphere, instead entering it from the Sun and leaving as infrared radiation radiated back into space, many substances, including water and nutrients, do cycle through the biosphere driven by the energy from the Sun (Figure 3.4). Each inorganic substance has its unique and often complex cycle. The major cycles are: carbon and oxygen; water; nitrogen; phosphorus; and sulfur. Only these are considered here. More details are available elsewhere (for example, Ayres, 1999). Cycles are characterised by reservoirs (or sinks) in which the substance is held in large quantities for long periods, and exchange pools where the substance is held for short periods. The oceans are a reservoir for water, for example, while a cloud is an exchange pool. Water may reside in an ocean for thousands of years, but in a cloud for only a few days. The biomass in an ecosystem may serve as a reservoir or exchange pool and also serve to move substances from one stage of a cycle to another. For example, trees may store particular elements for weeks or months (in leaves) and for years or decades (in the trunk and branches) but may also continuously move water from below ground into the atmosphere through transpiration via the leaves.

3.4.1 The carbon and oxygen cycles

Although all life is based on the element carbon, it makes up only a very small fraction of the Earth's crust. The major reservoirs of carbon are listed in Table 3.2, with estimates of the quantities in each. Carbon is exchanged between the reservoirs by biological and chemical processes as illustrated in Figure 3.7. Carbon contained in limestone deposits ($CaCO_3$) and fossil fuels are listed separately because they are three to seven orders of magnitude greater than the other reservoirs and because this carbon is effectively locked up and does not participate in the carbon cycle on human time scales. The dissolved inorganic carbon in the oceans, mainly bicarbonate ions (HCO_3^-), is the source of limestone; bicarbonate ions react with

Table 3.2: Estimates of the magnitude of major carbon reservoirs

Reservoir	Quantity (10^9 tonnes)	Wt%
Atmosphere	750	1.8
Soils	1500	3.6
Deep ocean	38 100	91.0
Sediments	150	0.4
Dissolved organic carbon (in deep oceans)	740	1.8
Terrestrial biota	610	1.5
Marine biota	3	0.0
Subtotal	*41 853*	*100.0*
Fossil fuels	10^6	
Carbonates in crust	10^7	

Source: Ayres (1999).

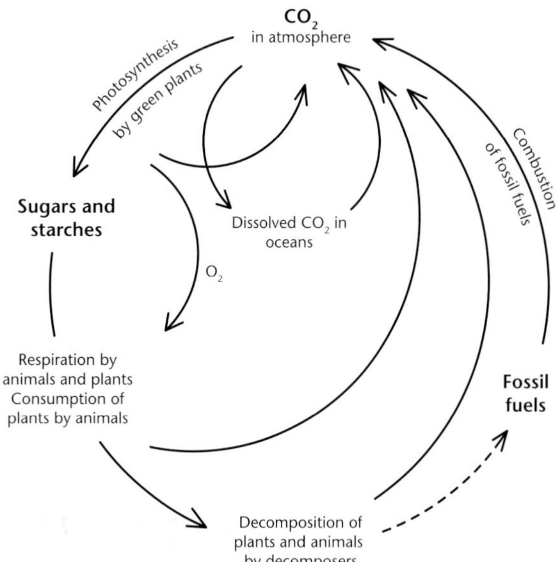

Figure 3.7: The carbon cycle.

calcium cations released by the weathering of calcium silicate minerals (which are plentiful in the crust) to form $CaCO_3$ which precipitates and settles to the ocean floor. This is an important natural, but slow, mechanism for regulating the CO_2 content of the atmosphere.

The main biological processes in the carbon cycle are respiration and photosynthesis. In respiration, carbohydrates and oxygen combine to produce carbon dioxide and water and to release energy. In photosynthesis, carbon dioxide and water combine and absorb energy from the Sun to produce carbohydrates and oxygen. The outputs of respiration are the inputs of photosynthesis, and *vice versa*. Both plants and animals respire (breathe) but only plants carry out photosynthesis. Carbon dioxide dissolves readily in water and the major reservoir for carbon is the oceans. Carbon in plants has three possible fates. It can be transferred to the atmosphere through respiration, it can be consumed by an animal, or it can remain in the plant when it dies. Animals obtain carbon from their food. Carbon in animals has the same three possible fates. Carbon dioxide released to the atmosphere through respiration is either taken up by plants in photosynthesis or it dissolves in the oceans. When an animal or plant dies the carbon is released as CO_2 to the atmosphere by decomposers, or the animal or plant is buried and ultimately forms coal, oil or natural gas.

Humans have a significant impact on the carbon cycle through burning of fossil fuels, which puts additional carbon dioxide into the atmosphere and oceans at a rate greater than natural processes can absorb (Figure 3.6). The build-up of CO_2 in the atmosphere is believed to be the cause of global warming. The increase in dissolved CO_2 content of the oceans results in increased acidity, due to the formation of bicarbonate ions, and this can adversely affect organisms living in the oceans.

3.4.2 The water cycle

The major reservoirs of water are listed in Table 3.3. The water cycle (hydrological cycle) is relatively simple and is illustrated in Figure 3.8. The oceans are the major reservoir and provide most of the water in the atmosphere. Energy from the Sun causes evaporation from water surfaces and biomass. It also provides the energy which drives the weather systems which move water vapour from one place to another. About 425 000 km^3 of water evaporate per year from the oceans and about 71 000 km^3 per year from terrestrial biomass, rivers and lakes. About 111 000 km^3 per year of water is precipitated as rain and snow over the continents (about 22% of total precipitation), the remainder falling into the oceans (Speidel and Agnew, 1982). Rain and melting snow soak into the ground to form *groundwater* or flow across the surface into streams, rivers, lakes and wetlands (Figure 3.9). Groundwater is either artesian or subartesian. *Artesian water* is confined and pressurised within a porous geological formation called an aquifer, a layer of porous rock, such as sandstone confined between impermeable layers of rock (aquicludes) or semi-impermeable layers (aquitards). Aquifers absorb water from precipitation and convey it deep underground. When an artesian aquifer is tapped by a bore, water will rise up the bore to the surface under its own pressure. *Sub-artesian water* occurs in porous rocks resting on bedrock or other impervious layer. It is thus not under pressure and, when tapped by a bore, must be pumped to the surface. Under the influence of gravity, surface and groundwater progressively moves to lower levels and ultimately finds its way to the ocean, unless it encounters a local depression effectively cut off from the ocean (e.g. the Caspian Sea, Lake Eyre in Australia). Frozen water may be trapped in cooler parts of the Earth (the poles, glaciers, on mountain tops etc.) as snow and ice and may remain there for very long periods before re-entering the water cycle. As water moves towards the oceans, soluble minerals dissolve from the rocks exposed to it and are transported to the sea. The oceans are salty because, when water leaves the oceans by evaporation, dissolved salts are left behind.

Table 3.3: Estimates of the magnitude of major water reservoirs

Reservoir	Volume (10^6 km^3)	% of total
Oceans	1370	97.25
Ice caps and glaciers	29	2.06
Groundwater	9.5	0.67
Lakes	0.125	0.01
Soil moisture	0.065	0.005
Atmosphere	0.013	0.001
Streams and rivers	0.0017	0.0001
Biomass	0.0006	0.00004
Total	*1409*	*100.00*

Source: Pidwirny (2006b).

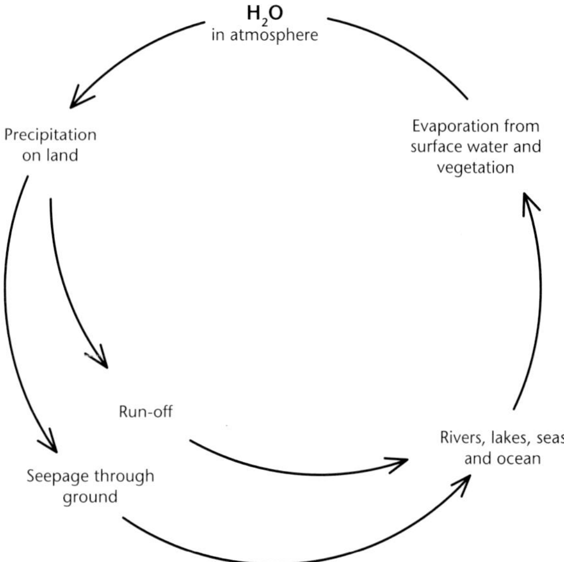

Figure 3.8: The water cycle.

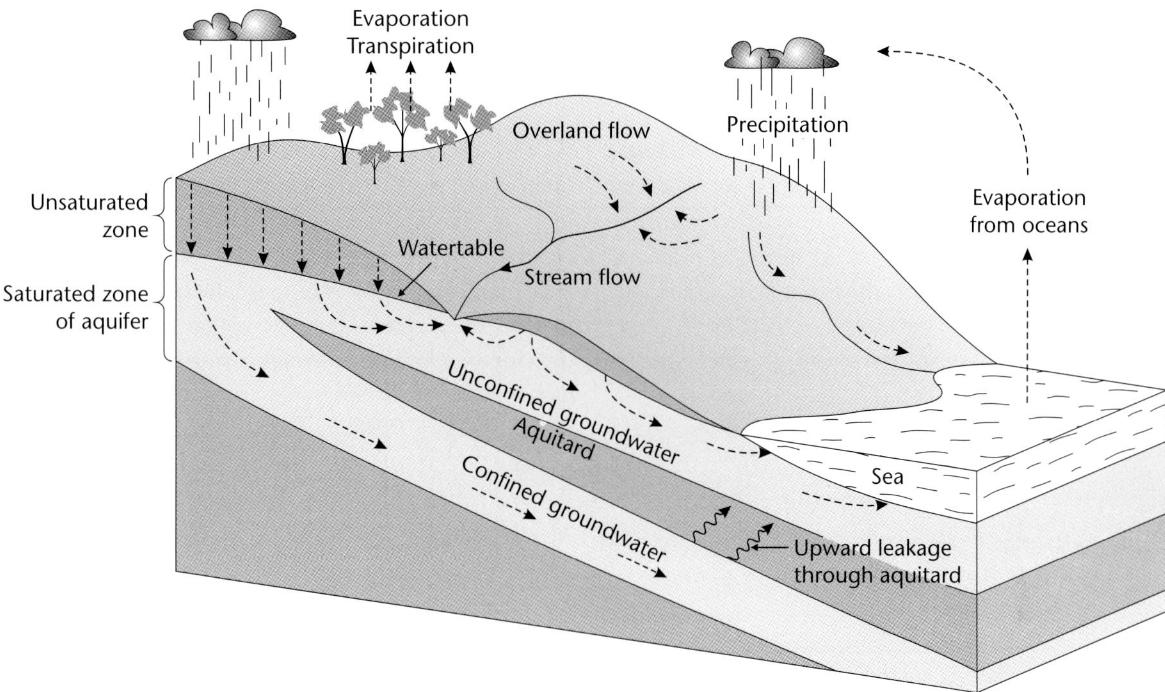

Figure 3.9: The water cycle, showing the role played by groundwater (after Shand *et al.*, 2007).

3.4.3 The nitrogen cycle

Nitrogen is an essential element for all life and occurs in compounds such as proteins and nucleic acids. The atmosphere contains about 78% by volume of nitrogen as diatomic N_2 and is the major reservoir of nitrogen. Diatomic nitrogen is relatively unreactive and most organisms cannot use nitrogen in this form. Plants obtain nitrogen in chemically combined, or fixed, forms such as nitrate ions (NO_3^-), ammonia (NH_3) and urea, $CO(NH_2)_2$. Animals obtain nitrogen from plants or from other animals that have fed on plants. The cycle is illustrated in Figure 3.10.

Fixation occurs in two ways. Energy from lightning can break nitrogen molecules, which enables nitrogen atoms to combine with oxygen in the air to form nitrogen oxides. These dissolve in rain and are carried to the Earth. Fixation by lightning probably contributes around 5–8% of the total nitrogen fixed. Nitrogen is also fixed by bacteria and archaea which form symbiotic relationships with legumes (such as alfalfa, clover, peas, beans, lentils, lupins and soybeans), other plants and some animals, or which live freely in soil and water. Many legumes also directly convert some of their organic nitrogen to soluble nitrites and nitrates.

The nitrogen in plants enters and passes through the food chain and at each step metabolism produces organic nitrogen compounds that return to the

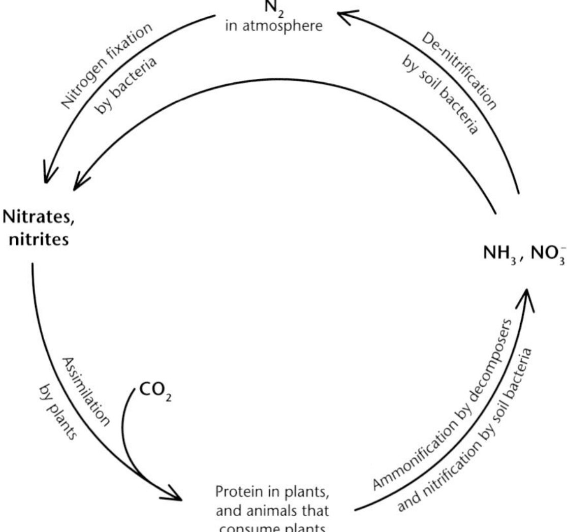

Figure 3.10: The nitrogen cycle.

environment as waste products and in dead organisms. These are broken down into ammonia by decomposers in soil and water. Ammonia can be taken up directly by plants through their roots but most is converted into nitrites by bacteria, which are then converted to nitrates by other bacteria. This process is called nitrification. Nitrates are readily taken up by plants. In de-nitrification nitrates are reduced back to nitrogen, which is released to the atmosphere, thus completing the cycle. This is also performed by bacteria which, in this case, are deep in soil and in aquatic sediments where there is no free oxygen.

Nitrogen can be artificially fixed as ammonia in the Haber process by reacting nitrogen (obtained from the atmosphere) and hydrogen (usually obtained from natural gas, CH_4) at high temperatures and pressures:

$$N_2 + 3H_2 = 2NH_3$$

Most ammonia produced in this way is further reacted to produce urea ($CO(NH_2)_2$) and ammonium nitrate (NH_4NO_3), both of which are used as fertilisers. Agriculture is now responsible for about half of all nitrogen fixation through the manufacture of fertilisers and the cultivation of legumes. De-nitrification processes often cannot keep up with the additional load of nitrates and there are many cases of nitrogen build-up in ecosystems. This can cause algal blooms in lakes and rivers as nitrogen leaches from the soil of nearby farms and provides nutrient to the algae. The algae consume oxygen dissolved in the water and can kill other aquatic life by depleting the oxygen supply. The accumulation of dissolved nutrients in a body of water is called *eutrophication*.

3.4.4 The phosphorus cycle

Phosphorus is an essential nutrient for plants and animals and is an important component of bones and teeth. It is present in DNA molecules, in molecules that store energy and in fats of cell membranes. Phosphorus most commonly occurs naturally as the phosphate ion (PO_4^{3-}); the major reservoirs are phosphate salts in mineral deposits on the seafloor and in crustal rocks. Unlike the elements of the other major cycles, phosphorus is not present in the atmosphere. Producers absorb phosphates from soil and water. These may be consumed by herbivores who in turn may be consumed by carnivores, thus carrying phosphorus through the food chain. Phosphates are released by decomposers from the waste products of consumers and from dead animals and plants, and are returned to the soil or water. Run-off from land may carry phosphate ions to the ocean, where some may be deposited in sediments on the ocean floor and eventually be incorporated into new sedimentary rocks. Marine birds play a role by removing fish from the ocean. Their waste (guano) contains high levels of phosphorus and in this way marine birds help return phosphorus from the ocean to the land. Phosphates move quickly through plants and animals but the processes that move them through the soil and ocean are very slow; the phosphorus cycle is one of the slowest natural cycles. In some locations guano is mined for use as fertiliser, though most deposits have by now been exhausted. Phosphorus-rich rocks are also mined as a raw material for making fertiliser, for example superphosphate. Use of phosphorus-containing fertilisers can cause a local overabundance of phosphorus (eutrophication), particularly in coastal regions and at the mouths of rivers and this can result in excessive growth of algae. The release of sewage into bodies of water can also create algal blooms due to its high phosphorus content.

3.4.5 The sulfur cycle

Sulfur is a constituent of many proteins, vitamins and hormones. It is important for the functioning of proteins and enzymes in plants, and in animals that depend upon plants for sulfur. The sulfur cycle is illustrated in Figure 3.11. Most sulfur occurs naturally as sulfide minerals in the crust and in oceanic sediments. Sulfur enters the atmosphere through natural and human processes. The cycle begins with the weathering of rocks in the presence of oxygen and water and the formation of sulfates, which are taken up by plants and microorganisms. Animals consume these and move the sulfur through the food chain. When organisms die, sulfur-containing proteins are degraded into their constituent amino acids by the action of soil organisms. Some of the sulfur is released as sulfates and some enters the tissues of the microorganisms. Sulfur in the form of sulfates can be transported by water and eventually reaches the oceans. The sulfur of amino acids is

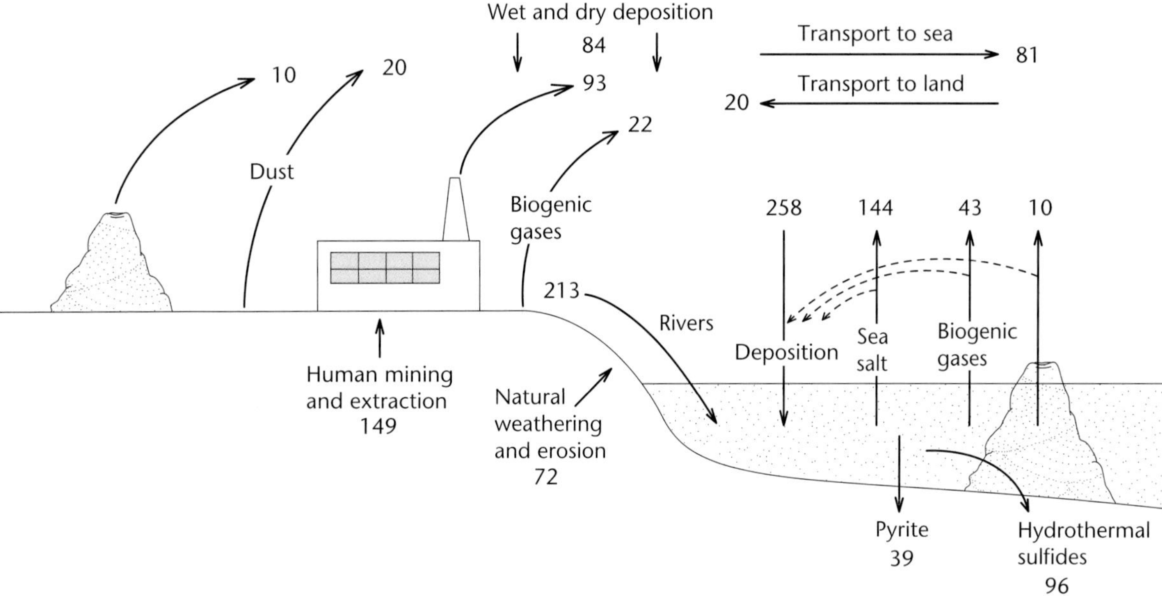

Figure 3.11: The sulfur cycle. Numerical values are the flux of sulfur, in millions of tonnes per year (after Schlesinger, 1991).

converted to hydrogen sulfide (H_2S) by other soil microbes and is released to the atmosphere. Sulfur is also emitted directly into the atmosphere from natural sources, including volcanic eruptions, dust, the breakdown of organic matter in swamps and tidal flats, and sea spray. Sea spray is the largest natural source of sulfur entering the atmosphere, but about 90% of this re-enters the oceans very quickly. Volcanoes are a large but variable source; dust (mainly gypsum from desert areas) is an important land-based source. Sulfur emissions to the atmosphere by human activities are significant, being greater than the combined quantities from volcanic and biological sources. Ninety per cent of sulfur emissions due to human activities is from combustion of fossil fuels, the remainder being due to smelting of sulfide minerals to produce metals such as copper, nickel, lead and zinc. Non-oxide forms of sulfur are rapidly oxidised in the atmosphere to sulfates. All atmospheric forms of sulfur are deposited back into the oceans or onto land within a few days. Hence, sulfur species in the atmosphere do not travel large distances and there is no global cycle for sulfur as there is for water, carbon and nitrogen. About 80% of sulfur deposition from the atmosphere is by precipitation, the balance being by dry deposition and absorption of SO_2 by soil and plants.

Acid rain

Pure water has a pH[2] of 7. Natural rain is slightly acidic, with a pH of about 5.2, because CO_2, SO_3, NO and NO_2 in the atmosphere dissolve in it to form dilute acids. These acids ionise to form hydrogen ions (H^+), and carbonate (CO_3^{2-}), sulfate (SO_4^{2-}) and nitrate (NO_3^-) ions, respectively. *Acid rain* is rain or any other form of precipitation that is unusually acidic. Acid rain is a phenomenon closely related to the carbon, sulfur and nitrogen cycles; the acceleration of the sulfur cycle by human activity has greatly increased the level of acidification. Acid rain is harmful to ecosystems. The hydrogen ions increase soil and water acidity, which is harmful to plants and animals. They can release metallic ions bound in the soil, including potentially toxic elements. Acid rain is also detrimental to infrastructure, causing corrosion and other

2 The pH of a solution is defined as the negative logarithm of the thermodynamic activity of the hydronium ions (H_3O^+) in the solution. The pH of a solution is a measure of how acidic or basic it is. The pH scale ranges from 0 to 14. A pH value of 7 is neutral, a pH value of less than 7 is acidic and a pH value greater than 7 is basic. Since the scale is logarithmic, each integer value of pH below 7 is 10 times more acidic than the next higher value, and each integer value above 7 is 10 times more basic than the next lower integer value.

forms of chemical attack on steel, concrete and other construction materials.

3.5 REFERENCES

ASDC (2010) Earth's Radiation Budget Facts. NASA Langley Atmospheric Science Data Center; <http://eosweb.larc.nasa.gov/EDDOCS/radiation_facts.html>.

Ayres RU (1999) Industrial metabolism and the grand nutrient cycles. In *Handbook of Environmental Economics*. (Ed. JCJM van den Bergh) pp. 912–945. Edward Elgar: Cheltenham, UK.

Furlong KP (1992) Heat flow through the Earth. In *Encyclopedia of Earth System Science*. Vol. 2. (Ed. WA Nierenberg) pp. 491–505. Academic Press: San Diego.

IEA (2007) *Key World Energy Statistics*. International Energy Agency: Paris.

Leith H (1973) Primary production: terrestrial ecosystems. *Human Ecology* **1**(4): 303–332.

Mizutani H (1995) Primary production and humanosphere – is energy sufficient for sustainable humanosphere? *Biological Sciences in Space* **9**(4): 303–313.

Olson DM *et al.* (2001) Terrestrial ecoregions of the world: a new map of life on Earth. *BioScience* **51**(11): 933–938.

Pidwirny M (2006a) Atmospheric composition: carbon dioxide. In *Fundamentals of Physical Geography*. 2nd edn. <http://www.physicalgeography.net/fundamentals/7aCO2.html>.

Pidwirny M (2006b) The hydrologic cycle. In *Fundamentals of Physical Geography*. 2nd edn. <http://www.physicalgeography.net/fundamentals/8b.html>.

Schlesinger WH (1991) *Biogeochemistry: An Analysis of Global Change*. Academic Press: San Diego.

Shand P, Edmunds WM, Lawrence AR, Smedley PL and Burke S (2007) 'The natural (baseline) quality of aquifers in England and Wales'. Research Report RR/07/06. British Geological Survey: London.

Speidel DH and Agnew AF (1982) *The Natural Geochemistry of Our Environment*. Westview Press: Boulder, CO.

UNEP (2005) 'Millennium ecosystem assessment, synthesis report: ecosystems and human well-being: general synthesis (2005)'. Island Press: Washington DC.

Vitousek PM, Ehrlich PR, Ehrlich AH and Matson PA (1986) Human appropriation of the products of photosynthesis. *BioScience* **36**: 368–373.

Whittaker RH and Likens GE (1975) The biosphere and man. In *Primary Productivity of the Biosphere*. (Eds H Leith and RH Whittaker) pp. 305–328. Springer-Verlag: Berlin.

3.6 USEFUL SOURCES OF INFORMATION

Bever MB (Ed.) (1986) *Enclyclopedia of Materials Science and Engineering*. Vols 1–8. Pergamon Press: Oxford, UK.

Cleveland CJ (Ed.) *Encyclopedia of Earth*. Environmental Information Coalition, National Council for Science and the Environment: Washington DC. <http://www.eoEarth.org>.

Cotterill R (1985) *The Cambridge Guide to the Material World*. Cambridge University Press: Cambridge.

Hertwich E, van der Voet E, Suh S, Tukker A, Huijbregts M, Kazmierczyk P, Lenzen M, McNeely J and Moriguchi Y (2010) 'Assessing the environmental impacts of consumption and production: priority products and materials. A report of the Working Group on the Environmental Impacts of Products and Materials to the International Panel for Sustainable Resource Management'. United Nations Environment Programme: Paris, France.

PhysicalGeography.net; Created by Dr Michael Pidwirny, University of British Columbia: Okanagan. <http://www.physicalgeography.net/home.html>.

Schlesinger WH (1991) *Biogeochemistry: An Analysis of Global Change*. Academic Press: San Diego.

4 An introduction to sustainability

Earth provides enough to satisfy every man's need but not every man's greed.

Mahatma Gandhi (1869–1948)

It was proposed in Chapter 1 that the greatest long-term challenge facing the world is to ensure that all people can have a good standard of living and quality of life without continuing to degrade the environment. The Earth's natural resources are finite, though some are renewable, and the biosphere has a finite capacity to cope with the wastes resulting from production and consumption and to provide the air and water that life relies on (ecosystem services). The concept of sustainability, or sustainable development, as a means of addressing these issues has had growing influence in recent decades and has been adopted by several major bodies, particularly the United Nations, the International Monetary Fund (IMF), the World Bank and the European Union. It has become the platform of many non-governmental organisations with a concern for the environment and social equity, such as WWF and Greenpeace. Companies in all sectors of the economy are adapting, often reluctantly, to this changing situation.

The supply of goods and services within society largely depends on the exploitation of non-renewable resources, including metallic and non-metallic minerals, coal, oil and gas. This raises a number of important questions.

- Can global material needs continue to be satisfied indefinitely into the future?
- Can future energy needs be met without continuing to increase the levels of CO_2 levels in the atmosphere?
- Can the environment cope with the emissions and wastes caused by meeting the seemingly ever-increasing material and energy needs of society?
- Can resource-rich societies and local communities benefit from the exploitation of non-renewable resources and prosper beyond the time of resource exhaustion?

These questions are not answered in this chapter but are addressed in later chapters. This chapter sets the scene by reviewing the complex concept of sustainability from a historical and environmental perspective, then develops a deeper understanding of the meaning of sustainability. It then examines various frameworks that have been proposed to help organisations and communities contribute to transitioning towards sustainability, and the ways developed to measure their progress. The chapter concludes with a description of a conceptual model which provides a more concrete basis for assessing where an organisation is in its implementation of sustainability principles. The model can also serve as a tool to help organisations integrate sustainability principles into business practices and decision-making processes.

4.1 THE ENVIRONMENTAL CONTEXT

Throughout history, there have been indigenous cultures and pre-industrial civilisations that persisted for many hundreds and sometimes thousands of years. There are also many that didn't. Some collapsed as a result of invasion or the spread of infectious diseases but in other cases the cause is less obvious. Many explanations have been offered and these have most recently been summarised by Diamond (2005), who identified eight factors which have historically contributed to the collapse of societies:

- deforestation and habitat destruction;
- soil problems, including erosion, salinity and loss of fertility;
- water management problems;
- overhunting;
- overfishing;
- effects of introduced species on native species;
- overpopulation;
- increased per capita impact of people on the environment.

These are all environmentally related and clearly point to key issues which the world faces today.

4.1.1 The state of the environment

There is now a wide consensus that all key indicators of the health of the environment are in decline. An excellent overview of the state of the environment is provided in the United Nations millennium ecosystem assessment report (UNEP, 2005), based on a study coordinated by the United Nations Environment Programme to which over 2000 authors and reviewers contributed. The key findings are summarised below.

- Approximately 60% of the ecosystem services examined in the study are being degraded. These include fresh water, fisheries, air and water purification, and the regulation of regional and local climate, natural hazards, and pests. The full costs of the loss and degradation of these services are substantial and growing.
- Many ecosystem services have been degraded as a consequence of actions taken to increase the supply of other services, such as food. These trade-offs often resulted in the costs of degradation being shifted from one group of people to another, or deferred the costs to future generations.
- Changes being made in ecosystems are increasing the likelihood of non-linear changes occurring in ecosystems (including accelerating, abrupt and potentially irreversible changes), with important consequences for human well-being. Such changes include emergence of diseases, abrupt alterations in water quality, the creation of dead zones in coastal waters, the collapse of fisheries and changes in regional climate.
- The harmful effects of the degradation of ecosystem services are being borne disproportionately by the poor, are contributing to growing inequities and disparities across groups of people and are sometimes the principal factor causing poverty and social conflict.
- The consumption of ecosystem services, which the study found is unsustainable in many cases, will continue to grow as a consequence of a likely three- to six-fold increase in global Gross Domestic Product[3] (GDP) by 2050 even though global population growth is expected to slow and level off in mid-century. Most of the important direct drivers of ecosystem change are unlikely to diminish in the first half of the century and two drivers – climate change and excessive nutrient loading – will become more severe.

Ironically, the green revolution in agriculture since World War II has caused environmental degradation, as well as obvious benefits. Land has been cleared and intensively cultivated, usually with single commodities (monocultures). These require heavy use of pesti-

3 The Gross Domestic Product of a country is the total market value of all final goods and services produced in the country in a particular year. It is equal to total consumer, investment and government spending, plus the value of exports, minus the value of imports.

cides and artificial fertilisers which, through run-off, enter and damage ecosystems. Modern agriculture is typically water- and energy-intensive, requiring manufacture of artificial fertilisers, transport of fertilisers and produce over long distances, and mechanised farming practices. Increasingly, modern agricultural techniques, while greatly increasing the quantity of food produced, are contributing to the general degradation of ecosystems.

4.1.2 The ecological footprint

Various attempts have been made to quantify the human demand on ecosystems, for example Vitousek *et al.* (1986) as discussed in Section 3.2.2. In recent years the ecological footprint concept, developed by Rees (1992) and Wackernagel (1994), has become increasingly popular as an easily understood measure (Global Footprint Network). Ecological footprint analysis compares human demand on nature with the biosphere's ability to regenerate resources and provide services. It does this by assessing the biologically productive land and marine area required to produce the resources a population consumes and to absorb the resulting waste, using prevailing technology.

The *ecological footprint* is defined as the area of biologically productive land and water needed to regenerate the renewable resources a human population consumes and to absorb and render harmless the wastes produced by humans (Hails, 2008). It is expressed in units of global hectares (gha) to distinguish it from total area, which includes non-productive land and water. The total ecological footprint is calculated as the sum of:

- the area of cropland, grazing land, forest and fishing grounds required to produce the food, fibre and timber consumed by humans and to provide space for infrastructure;
- the area of productive land needed to absorb the CO_2 emissions from fossil fuel use and land disturbance, other than the portion absorbed by the oceans.

Methods for measuring the ecological footprint vary, but emerging standards will make results more comparable and consistent (Global Footprint Network, 2009). Table 4.1 lists the estimated ecological footprint for the world's major geographical regions and for a selection of countries.

It is evident from Table 4.1 that the wealthier the country, the greater the ecological footprint. In 2005, it ranged from an average of 6.4 gha per person for high-income countries to an average of 1.0 gha per person for low-income countries; the actual range was 9.4 (United States) to 0.6 (Bangladesh) gha per person. The United States and China had the largest total footprints, each using about 21% of the planet's productive area. China had a much smaller footprint per capita than the United States, as its population was more than four times as large. India's footprint was the next largest, occupying about 7% of the Earth's total productive area. The global ecological footprint in 2005 was 17.5 billion gha, or 2.7 gha per person. On the supply side, the total productive area available was 13.6 billion gha, or 2.1 gha per person. Thus demand for productive area exceeded supply by about 30%.

The ecological footprint first exceeded the Earth's total productive area in the 1980s and the magnitude of the overshoot has been increasing since. To achieve a sustainable relationship with the environment, it is necessary to reduce the impact of humans and their activities on the environment through reducing the quantities of resources consumed and the quantities of wastes produced. It has been claimed that, to meet projected human material needs within the carrying capacity of the Earth, a decrease in resource consumption of an order of magnitude is required over the next 30–50 years (Schmidt-Bleek, 1997). The term *Factor 10* has been coined to describe this. Based on European patterns of consumption, this would mean that per capita consumption of non-renewable resources should not exceed around 5 tonnes per year, compared with around 45 tonnes per year at present (Schmidt-Bleek, 2008). There appears to be no empirical or theoretical basis for the factor of 10; the term Factor 10 is best thought of as a 'metaphor for the strategic economic goal of approaching sustainability by increasing the overall resource productivity' (Factor 10 Institute, 2009). Perhaps more usefully, the concept *Factor X* has been proposed as a measure of the degree of dematerialisation achieved by a particular industry or product.

Table 4.1: The estimated ecological footprint and water footprint of the major world regions and for a selection of countries

		Ecological footprint, 2005 (global hectares per person)							Water footprint, 1997–2001 (m^3 per person per year)		
Country/region	Population (millions)	Total	Carbon	Cropland	Grazing land	Forest	Fishing ground	Built-up land	Total	Internal	External
World	6476	2.7	1.41	0.64	0.26	0.23	0.09	0.07	1243	1043	199
High-income countries	972	6.4	4.04	1.15	0.28	0.61	0.17	0.13	–	–	–
Middle-income countries	3098	2.2	1.00	0.62	0.22	0.18	0.09	0.08	–	–	–
Low-income countries	2371	1.0	0.26	0.44	0.09	0.15	0.02	0.05	–	–	–
Africa	902.0	1.4	0.26	0.54	0.25	0.24	0.03	0.05	–	–	–
Middle East and Central Asia	365.6	2.3	1.34	0.69	0.08	0.08	0.04	0.08	–	–	–
Asia-Pacific	3562.0	1.6	0.78	0.49	0.08	0.13	0.07	0.06	–	–	–
Latin America and the Caribbean	553.2	2.4	0.65	0.57	0.72	0.32	0.10	0.08	–	–	–
North America	330.5	9.2	6.21	1.42	0.32	1.02	0.11	0.10	–	–	–
Europe (EU)	487.3	4.7	2.58	1.17	0.19	0.48	0.10	0.17	–	–	–
Europe (non-EU)	239.6	3.5	2.00	0.94	0.04	0.29	0.17	0.07	–	–	–
United States	298.2	9.4	6.51	1.38	0.30	1.02	0.10	0.10	2483	2018	464
Australia	20.2	7.8	1.98	1.93	2.82	0.94	0.08	0.06	1393	1141	252
Canada	32.3	7.1	3.44	1.83	0.50	1.00	0.21	0.09	2049	1631	418
United Kingdom	59.9	5.3	3.51	0.87	0.21	0.46	0.08	0.20	1245	369	876
Finland	5.2	5.2	1.68	1.24	0.06	1.96	0.15	0.16	1727	1026	701
Sweden	9.0	5.1	0.95	0.95	0.31	2.59	0.10	0.20	1621	759	861
Japan	128.1	4.9	3.68	0.58	0.04	0.24	0.28	0.08	1153	409	743
France	60.5	4.9	2.52	1.28	0.32	0.39	0.17	0.25	1875	1176	699
Italy	58.1	4.8	2.77	1.19	0.22	0.43	0.06	0.10	2332	1142	1190
Germany	82.7	4.2	2.31	1.21	0.09	0.36	0.04	0.21	1545	728	816
Netherlands	16.3	4.0	2.29	1.22	–0.03	0.36	0.00	0.18	1223	220	1003

Country/region	Population (millions)	Ecological footprint, 2005 (global hectares per person)							Water footprint, 1997–2001 (m^3 per person per year)		
		Total	Carbon	Cropland	Grazing land	Forest	Fishing ground	Built-up land	Total	Internal	External
Poland	38.5	4.0	2.06	1.10	0.16	0.52	0.04	0.08	1103	785	317
South Korea	47.8	3.7	2.47	0.66	0.04	0.19	0.31	0.06	1179	449	730
Russia	143.2	3.7	2.24	0.92	0.03	0.34	0.15	0.06	1858	1569	289
Mexico	107.0	3.4	1.92	0.77	0.31	0.23	0.07	0.08	1441	1007	433
Chile	16.3	3.0	0.56	0.52	0.41	0.77	0.60	0.13	803	486	317
Iran	69.5	2.7	1.66	0.69	0.11	0.04	0.09	0.09	1624	1333	291
Turkey	73.2	2.7	1.37	1.00	0.04	0.17	0.05	0.08	1615	1379	236
Brazil	186.4	2.4	0.04	0.61	1.11	0.49	0.02	0.08	1381	1276	106
South Africa	47.4	2.1	1.03	0.44	0.23	0.27	0.04	0.07	931	728	203
China	1323.3	2.1	1.13	0.56	0.15	0.12	0.07	0.07	702	657	46
Thailand	64.2	2.1	0.89	0.64	0.01	0.16	0.37	0.06	2223	2037	185
Egypt	74.0	1.7	0.71	0.72	0.02	0.11	0.01	0.10	1097	889	207
Nigeria	131.5	1.3	0.12	0.95	0.00	0.19	0.02	0.06	1979	1932	47
Vietnam	84.2	1.3	0.46	0.56	0.00	0.15	0.03	0.07	1324	1284	40
India	1103.4	0.9	0.33	0.40	0.01	0.10	0.01	0.04	980	964	16
Indonesia	222.8	0.9	0.09	0.50	0.00	0.12	0.16	0.08	1317	1182	135
Philippines	83.1	0.9	0.07	0.42	0.01	0.08	0.25	0.04	1543	1378	164
Pakistan	157.9	0.8	0.30	0.39	0.01	0.07	0.02	0.05	1218	1153	65
Bangladesh	141.8	0.6	0.13	0.33	0.00	0.07	0.01	0.04	896	865	31

Source: Hails (2008).

4.1.3 The tragedy of the commons

The answer to the question 'Why is the natural environment being allowed to degrade?' lies mainly in our system of economics, which places little value on the environment, environmental resources and ecosystem services. The benefits from the exploitation of environmental resources are usually captured privately (by individuals or companies). The costs, particularly degradation of the environment and social disruption, are usually not borne by the party who is responsible and who benefits most, but by society at large. The benefits are internalised while the costs are externalised. Economists refer to this as market failure. Degradation occurs because the benefits of exploitation accrue to individuals (or organisations) who therefore try to maximise their use of the resource; the costs of exploitation, however, are borne by all to whom the resource is potentially available. The demand for the resource increases because others also attempt to gain private benefit and the problem compounds until, eventually, the resource is destroyed or becomes exhausted. William Forster Lloyd (1795–1852) first described this phenomenon (Lloyd, 1833) using the example of herders sharing a common pasture, on which they all graze cows. It is in an individual herder's interest to put as many cows as possible onto the land to maximise the individual benefit. The herder receives all the benefits from the additional cows while any damage to the pasture is shared by the group. If all herders do this, however, the pasture will be overgrazed and ultimately destroyed, and all will suffer. This concept is usually referred to as 'the tragedy of the commons', a phrase coined by Hardin (1968).

The tragedy of the commons has been interpreted as a justification for privatising resources held in common since, it is argued, property rights give the owner an incentive to conserve the resource. This may be true in some situations, but many resources are difficult to privatise (the air, rivers and oceans) and others perhaps should not be (mineral resources, national parks, state-owned forests and fisheries). The most common approach to address market failure in these cases has been regulation by government or government agencies to limit the quantity of goods that can be extracted, through permit systems (mining, fishing and hunting), and to prescribe the types and quantities of wastes that can be released into the environment. Increasingly, the instruments of taxation on undesirable wastes (landfill, carbon emissions) and trading schemes are being adopted. These mechanisms are difficult to establish and may not stop, let alone reverse, the degradation of the environment.

Elinor Ostrom has argued that, in addition to governmental regulation or privatisation of a common resource, there is a third approach to resolving the problem of the commons; namely, the design of durable cooperative institutions that are organised and governed by the resource-users themselves (Ostrom, 1990).[4] Ostrom claims the three dominant models – the tragedy of the commons, the prisoner's dilemma (Tucker, 1950; Poundstone, 1992) and the logic of collective action (Mancur, 1965) – are inadequate because they are based on the 'free-rider' problem in which individual, rational resource-users act against the best interest of the users collectively. She claims these models apply only when the many, independently acting individuals involved want or need quick returns, have little mutual trust and no capacity to communicate or to enter into binding agreements, and do not arrange for monitoring and enforcing mechanisms to avoid over-investment and over-use. Ostrom's analysis of some long-standing and viable common property regimes, including Swiss grazing pastures, Japanese forests and irrigation systems in Spain and the Philippines, identified a number of common themes. These include:

- they had clearly defined boundaries;
- they had monitors who were either resource users or were accountable to them;
- there were graduated sanctions imposed on users breaching agreed rules;
- the mechanisms to resolve conflicts and to alter the rules were dominated by the users;
- higher authorities recognised the self-determination of the community.

Ostrom proposed that these can be used as design principles for managing common property.

4 For this work, Ostrom received the Sveriges Riksbank Prize in Economic Sciences in Memory of Alfred Nobel in 2009.

4.2 A BRIEF HISTORY OF THE IDEA OF SUSTAINABILITY

4.2.1 The rising public awareness

Since the 1960s there has been a growing scientific understanding of the environment's importance to our well-being and of its interconnected but fragile nature. There has been a parallel growth in public awareness, particularly in the developed world, of environmental issues as a result of the impacts of local environmental problems and internationally reported extreme events, for example:

- the Bhopal disaster in India, in which a Union Carbide pesticide manufacturing plant released a mixture of poisonous gases on the night of 2–3 December 1984, resulting in the worst industrial disaster in history and the loss of thousands of lives;
- the grounding of the *Exxon Valdez* oil tanker off the coast of Alaska on 24 March 1989 and the resulting oil spill;
- the off-shore failure of the BP oil well in the Gulf of Mexico in 2010.

Silent Spring (Carson, 1962), often regarded as the book that began the public's concern with the environment, drew attention to the environmental and health problems created by the unregulated use of DDT and other pesticides in agriculture. Growing public awareness led governments to introduce laws regulating the use of pesticides and the release of toxic substances to land, air and water during the 1960s; the laws have become progressively more stringent. Other books, articles in popular journals and magazines, and documentaries drew public attention to global development issues and further raised concern for the environment. Former United States Vice-President Al Gore's documentary film *An Inconvenient Truth* and book *Our Choice: A Plan to Solve the Climate Crisis* (Gore, 2009) are recent examples.

The Population Bomb (Ehrlich, 1968) was another early book. Ehrlich predicted that hundreds of millions of people would starve to death during the 1970s and 1980s, that nothing could be done to avoid this and that radical action was needed to limit overpopulation. The book was a modern restatement of the catastrophe hypothesis of Thomas Malthus (1766–1834), who argued that population increases geometrically with time but agricultural output increases only arithmetically and, therefore, demand for food will always outstrip production of food (Malthus, 1826). Fortunately, the predictions failed to materialise[5] but *The Population Bomb* raised serious issues and was a highly influential book at the time.

Another important book was *The Limits to Growth* (Donella *et al.*, 1972), which presented the results of a study commissioned by the Club of Rome, a not-for-profit organisation which studies problems facing humanity. The study involved modelling the implications of the impact of finite resources in a world with rapidly growing population. Five variables were examined: population, industrialisation, pollution, food production and resource depletion. The study found that continued growth of the global economy would lead to planetary limits being exceeded during the 21st century, most likely resulting in a decrease in population and collapse of the economic system. Equally importantly, though often overlooked by critics, the Club of Rome study found that collapse could be avoided through a combination of early changes in behaviour, policy and technology. The Club of Rome predictions have stood the test of time. A recent study found that data for the period 1970–2000 closely matched the simulated results for the standard case scenario (Turner, 2008), which predicts global collapse by the middle of the present century.

4.2.2 International developments

International forums convened by the United Nations raised environmental and related social issues at national and international levels. The UN Conference on the Human Environment in Stockholm in 1972 debated the need for a change in approach to development. It achieved a level of consensus and

5 It is easy to see why Ehlich's predictions were wrong. World population growth has in fact slowed during the past several decades and food production has increased dramatically since 1945 as a result of the green revolution. Malthus (and Ehrlich!) failed to understand the role of technological progress in improving the productivity of agriculture. They also failed to understand that the availability of land for agriculture is an issue of relative scarcity rather than absolute scarcity, since land of lower productivity can progressively be used as demand for food increases.

understanding among the participating nations, in the Stockholm Declaration. The Declaration raised the key challenges facing the world and set out the principles for addressing these (http://www.unep.org). The next major development was the World Commission on Environment and Development, convened in 1983 to address the growing concern 'about the accelerating deterioration of the human environment and natural resources and the consequences of that deterioration for economic and social development'. It has become known as the Brundtland Commission, after its chairperson Gro Harlem Brundtland. The report issued by the Brundtland Commission (UN, 1987) first popularised the term *sustainable development* and defined it as development that meets the needs of the present without compromising the ability of future generations to meet their needs.

A decade later, in 1992, the United Nations Conference on Environment and Development (commonly referred to as the Earth Summit) in Rio de Janeiro resulted in the Rio Declaration (http://www.unep.org) and produced *Agenda 21*, a major publication that provided 26 principles for sustainable activity across most areas of human endeavour (UN, 1992). Principle 15 of the Rio Declaration contains a statement that has become known as the *precautionary principle*:

> *Where there are threats of serious or irreversible damage, lack of full scientific certainty shall not be used as a reason for postponing cost-effective measures to prevent environmental degradation.*

The precautionary principle implies that preventative measures should be undertaken before there are scientific results proving that protection of the environment is necessary, and the burden of proof should be on those who believe that an economic activity has only negligible detrimental consequences on the environment (Neumayer, 2003). Principle 15 further states that, in order to protect the environment, the precautionary approach should be widely applied by states according to their capabilities.

A decade after the Earth Summit, the United Nations World Summit on Sustainable Development, Earth Summit 2002, was held in Johannesburg. It brought together leaders from business and non-governmental organisations. The Johannesburg Declaration built on the earlier declarations but was a more general statement and focused on social, economic and political issues rather than the environment. It was an agreement to focus particularly on 'the worldwide conditions that pose severe threats to the sustainable development of our people, which include: chronic hunger; malnutrition; foreign occupation; armed conflict; illicit drug problems; organized crime; corruption; natural disasters; illicit arms trafficking; trafficking in persons; terrorism; intolerance and incitement to racial, ethnic, religious and other hatreds; xenophobia; and endemic, communicable and chronic diseases, in particular HIV/AIDS, malaria and tuberculosis'.

4.2.3 Corporate developments

The World Business Council for Sustainable Development (WBCSD) was formed in 1995, shortly after the Earth Summit. It is a CEO-led, global association with headquarters in Geneva, formed to provide 'a platform for companies to explore sustainable development, share knowledge, experiences and best practices, and to advocate business positions on these issues in a variety of forums, working with governments, non-governmental and intergovernmental organizations' (http://www.wbcsd.org). It has around 200 member companies. Many companies and other organisations have adopted the language and, to varying degrees, the principles of sustainability. The reasons often cited for doing this are varied:

- to enhance their reputation;
- to secure a 'social licence to operate' (i.e. approval, or at least acquiescence, by the community to carry out a business);
- to attract and retain employees;
- to improve access to investors;
- to identify potential cost savings;
- to increase the scope for innovation;
- to align stakeholders' needs with management objectives, or create a basis for stakeholder engagement.

These are often referred to as the *business case for sustainable development*. The business case for sustainability has grown out of a recognition that companies can cause environmental and social harm and have responsibility in these areas which goes beyond maximising shareholder value and obeying the letter of the law. Many companies now measure and report on activities that they consider contribute to sustainability.

In recent years, some companies have adopted the principles of *corporate social responsibility* (CSR), in response to growing community expectations. The World Business Council for Sustainable Development gives the following definition:

> *Corporate social responsibility is the continuing commitment by business to behave ethically and contribute to economic development while improving the quality of life of the workforce and their families as well as the local community and society at large.*

CSR implies the voluntary acceptance by companies of social responsibilities, sustainability standards and codes of ethics above and beyond legal requirements. Ideally, adoption of CSR principles means that a company would monitor and modify its operations to ensure it adheres not only to the law but to high ethical, social and environmental standards and to international norms. The company would accept responsibility for the impact of its activities and products on the environment, consumers, employees, community and other stakeholders, and promote the public interest by encouraging and supporting community development and eliminating practices that are harmful, even if those practices are legal. Supporters of CSR argue that there is a good business case for CSR, in that companies benefit in many tangible and intangible ways by operating with a broad long-term perspective. Critics argue that CSR distracts from the primary role of business – to maximise shareholder value. Others argue that it is nothing more than window-dressing, or 'greenwash', to permit business-as-usual and retain a so-called 'licence-to-operate'. Examples of each can easily be found.

4.3 THE CONCEPTS OF SUSTAINABLE DEVELOPMENT AND SUSTAINABILITY

The widely accepted definition of sustainable development is that used in the Brundtland Report (UN, 1987):

> *Sustainable development is development that meets the needs of the present without compromising the ability of future generations to meet their own needs.*

This definition contains two concepts – the concept of needs and the concept of limits to the environment's ability to meet present and future needs. Although critics argue that it is a vague statement open to interpretation, it has nevertheless proved durable and provided a goal to which many people aspire. The term *sustainability* is often preferred to the phrase 'sustainable development'. The word 'development' may be perceived as implying growth and, therefore, that sustainable development means ameliorating the problems caused by, but not challenging, continued economic growth. In this book, the term 'sustainability' is preferred as being the more value-neutral.

The term 'sustainability' is frequently used in different ways. Evironmentalists often mean ecological sustainability when they speak of sustainability. Many business people really mean economic sustainability when they speak of sustainability. However, sustainability has three aspects – ecological, social and economic – and it is not possible to achieve a particular level of ecological, social or economic sustainability independently without achieving at least a basic level of all three forms simultaneously. It is not possible for subsystems to be sustainable within an unsustainable global system – sustainability is a property of the Earth system as a whole, particularly the hydrosphere, biosphere and lithosphere. For example, the sustainability of a community can only be defined in terms of the global system. Similarly, a firm or organisation is most unlikely to be sustainable if the society at large is not sustainable. The term 'sustainable mining' is an oxymoron, since mining exploits a non-renewable resource. A more useful way of thinking about sustainability and mining is to attempt to answer the question: How can mining contribute to the transition to sustainability?

4.3.1 Alternative definitions of sustainability

Attempts to define sustainability more rigorously than in the Brundtland definition are usually couched in economic or ecological terms. The economic definition can be stated as:

> *Development is sustainable if it does not decrease the capacity of a system to provide non-declining per capita utility (Neumayer, 2003) or, more simply, development is sustainable if consumption or production does not decline over time.*

Ecological definitions of sustainability include:

> *Development is sustainable if the stock of natural capital does not decline over time.*
>
> *Development is sustainable if resources are managed so as to maintain a sustainable yield of ecosystem services.*
>
> *Development is sustainable if minimum conditions of ecosystem stability and resilience are maintained over time.*

The economic definitions define sustainability in terms of the economy's ability to maintain material production or consumption indefinitely. Since this is not possible without ongoing use of environmental resources, economic interpretations imply that there must be at least some degree of environmental sustainability. The ecological interpretations define sustainability directly in terms of the environment and its capacity to continue to respond to demands on it.

Saying that the productive capacity or level of consumption or production is to be maintained indefinitely (the economic definition) is the same as saying that the stock of capital assets must be preserved indefinitely since that is the stock which will ensure sufficient productive capacity for a non-declining flow of output or consumption. The stock of capital assets consists of several types; an important issue in sustainability theory is the extent to which various forms of capital can substitute for one another, in particular the possibility of substituting other forms of productive capital for declining stocks of environmental resources. The following forms of capital can be readily identified (Porritt, 2005).

- *Natural capital (environmental resources).* This is the stock of natural resources (energy and matter) and processes that produce valuable goods and services. It consists of resources (renewable and non-renewable) and ecosystem services.
- *Human capital.* This is the health, knowledge, skills and motivation required for productive work, and the individual's emotional and spiritual capacity. In this context, human capital is understood to include intellectual capital (intellectual property), which is often considered a separate form of capital. Intellectual capital is derived from the stocks of useful knowledge societies accumulate over time. It includes the skills, knowledge, secrets and know-how, and the relationships that an individual, organisation or society as a whole has developed. It resides collectively in individuals and groups of individuals, in patents and in documents about processes, customers, research results etc. that have value to an organisation or institution.
- *Social capital.* This is the structures, institutions and relationships which enable individuals to maintain and develop their human capital in partnership with others, and to be more productive working together than in isolation. It includes networks, communication channels, families, communities, businesses, trade unions, schools, voluntary organisations, legal and political systems and educational and health bodies, as well as social norms, values and trust.
- *Reproducible (manufactured) capital.* This consists of material goods and infrastructure owned, leased or controlled by an organisation that contribute to production or provision of services but are not part of its output (tools, machines, buildings, roads, dams).
- *Financial capital.* This consists of an organisation's assets that exist in a form of currency that can be owned or traded, including shares, bonds and banknotes. It has no intrinsic value; its value is representative of the other forms of capital.

Human-made capital is the sum of reproducible, human, social and financial capital assets. The sum of human-made capital and natural capital at any time, therefore, is the stock of productive assets. The distinction between human-made capital and natural capital is important because the issue of whether productive capacity can be maintained indefinitely depends on the degree to which human-made capital can substitute for natural capital. This leads to the concepts of weak and strong sustainability.

4.3.2 Interpretations of sustainability

Weak sustainability

The concept of weak sustainability is based on the assumption that human-made capital can substitute for natural capital both as an input for production of goods and services for consumption and directly as a provider of ecosystem services. This means that natural capital can be allowed to degrade as long as enough human-made capital is built up to compensate. In this view, some parts of the total stock of assets, including non-renewable natural resources, can be allowed to decline – the decline is not important provided other types of capital substitute for declining natural capital. The weak sustainability model takes an optimistic view of resources, which can be summarised in four propositions (Neumayer, 2003). If a resource *A* is becoming scarce its price will rise. This will lead to a number of responses, which are not mutually exclusive:

- Demand could shift from resource *A* and another resource *B* would become economical as a substitute for *A*.
- It could become economical to explore for, extract and recycle more of resource *A*. As a consequence, the price of resource *A* would decline, indicating an ease in scarcity.
- Human-made capital could substitute for resource *A*.
- More effort could be put into research and development to reduce the quantity of *A* required per unit of output, thus easing any resource constraint.
- More effort could be put into research and development to make resource extraction cheaper and more economical, again easing any resource constraint.

The assumption that human-made capital can substitute for natural capital in a virtually unlimited way occurs frequently in economic modelling, possibly as a mathematical convenience (Ayres, 2005). The classic mathematical works of Solow (1974) and Stiglitz[6] (1974a, 1974b) showed that perpetual economic growth of consumption is possible provided the elasticity of substitution[7] of human-made capital for natural capital is 1 or greater. They argued that the elasticity of substitution was likely to be at least 1, in practice. Solow went so far as to say that the world could, under some circumstances, get along without natural resources (Solow, 1974)! There is no theoretical justification for assuming that human-made capital can substitute for natural capital in a significant way, and empirical evidence indicates that little substitution is possible. While considerable substitution between the various forms of human-made capital is clearly possible, natural capital has characteristics that distinguish it from human-made capital (Neumayer, 2003) and that make substitution of natural capital with human-made capital problematical. Some forms of natural capital provide basic life-support functions that no other form of capital can provide. These are the ecosystem services that make human life on Earth possible. Also, some forms of natural capital are unique and cannot be rebuilt once they have been destroyed. In general, this is not the case for human-made capital – reconstruction may be expensive or slow but, in principle, it is possible. It is most unlikely, therefore, that human-made capital can substitute to any great extent for most forms of natural capital. This conclusion renders the works of Solow and Stiglitz of little practical value and leads to the strong sustainability interpretation.

Strong sustainability

The concept of strong sustainability is based on the assumption that human-made capital cannot substitute for natural capital either as an input for production of goods and services for consumption or directly as a provider of ecosystem services. There are two interpretations of strong sustainability (Neumayer,

6 Solow and Stiglitz won the Sveriges Riksbank Prize in Economic Sciences in Memory of Alfred Nobel, in 1987 and 2001, respectively.

7 Elasticity of substitution refers to the responsiveness of buyers to price changes in substitutes for a good or service. It is defined as the ratio of the change in the relative demand for two goods to the change in their relative prices. When the elasticity of substitution is greater than 1, a small change in relative prices will cause a large change in the inputs used.

2003). In one, the total value of natural capital should be preserved. This interpretation implies that the scarcity rents[8] from non-renewable resource extraction should be invested in the development of alternatives to keep the total value of natural capital constant. For example, royalties from coal mining would go into developing renewable energy sources. In the second interpretation, the physical stock of those forms of natural capital that are considered to be non-substitutable for human capital (critical natural capital) should be preserved; there should be no substitution between different forms of critical natural capital. This interpretation implies that renewable resources should be used only to the extent that their stock does not deteriorate and that the environment should be used as a sink for wastes only to the extent that its natural absorptive capacity does not deteriorate.

4.3.3 Responses to the challenge of sustainability

The modern environmental movement has its origins in industrial society and industrialism. Industrialism can be defined as an overarching commitment to growth in the quantity of goods and services produced and the material well-being that growth brings (Dryzek, 2005). Industrial societies have produced many competing ideologies, particularly liberalism, conservatism, socialism, communism and fascism – whatever their differences, these ideologies are all committed to industrialism. They have a similar attitude to the environment, which involves largely discounting or suppressing environmental concerns. Possible responses to the sustainability challenge do not have to accept industrialism and should include alternatives. Responses can be radical or reformist. Reformist responses are those that seek to work within the present understanding of industrialism. Radical responses are those that seek to break the nexus between growth, consumption and well-being. These form one dimension of a matrix of possible responses. A second dimension takes account of the fact that responses can be prosaic (unimaginative) or imaginative, to use Dryzek's terminology (Dryzek, 2005). Prosaic responses are those that accommodate the politico-economic arrangements of industrial society and imaginative responses are those that seek to redefine politico-economic arrangements. Putting the two dimensions together results in the four responses to the challenge of sustainability shown in Table 4.2.

- *Environmental problem-solving.* This response accepts the politico-economic *status quo* and assumes environmental problems can be solved by public policy through use of market-type incentives (cap-and-trade schemes, carbon tax etc.) or regulations, or a combination.
- *Survivalism.* This response assumes that population growth will eventually, if it hasn't already, reach limits set by the Earth's stock of finite resources and ecosystem services. It is radical because it seeks a redistribution of power within the industrial-political economy and reorientation away from continued economic growth. It is prosaic because it sees solutions within the context of industrialism, particularly greater control of existing systems by administrators, scientists and other 'responsible' elites.
- *Sustainable development.*[9] This response seeks ways to remove the conflicts between environmental and economic values and to redefine growth and development in ways which make the 'limits' argument meaningless. It sees economic growth and environmental protection as complementary.
- *Green radicalism.* This response is both radical and imaginative. It rejects the basic structure of industrial society in favour of a variety of alternative interpretations of society and the place of humans in the world. Because of its radical and imaginative nature, green radicalism is subject to widely varying approaches and deep internal divisions.

The proponents of these responses engage with industrialism, even if in the case of green radicalism it is only to distance themselves from it. Their engagement with industrialism and its defenders is often

8 Scarcity rent is the term used by economists to refer to that part of the income that accrues to the owner of a resource due to the scarcity of the resource. This is discussed further in Section 6.3.

9 Dryzek uses the term sustainability rather than sustainable development. However, because of the more encompassing meaning of sustainability, the term sustainable development, as defined by Brundtland, is used here.

Table 4.2: A classification of the departures from industrialism

	Reformist	Radical
Prosaic	Environmental problem-solving	Survivalism
Imaginative	Sustainable development	Green radicalism

Source: Dryzek (2005).

more pronounced than their engagement with each other (Dryzek, 2005). Their difficulty in engaging with each other is due largely to their very differing world views. We will return to these responses later, in Section 4.5 and in Chapter 17.

Government and corporate responses

Sustainability is a complex and subjective concept. As a relatively new concept, the theory of, and how to transition to, sustainability is not well developed and approaches change as understanding and awareness improve. As demonstrated above, the various world views make it difficult, if not impossible, to achieve a common discourse let alone consensus. At the international, national and industry sector levels, decision-makers in government, business and other institutions often exploit this to keep debate at bay and to delay taking needed actions by pointing to the lack of agreement on the issues and appropriate responses. This has been particularly apparent in the field of global warming and climate change – debate over the proof of climate change and the best approaches to tackling it have been used by governments to delay taking major action. Coal companies have promoted the concepts of carbon-capture and geosequestration as the solution to global warming due to coal combustion, to justify continued growth of the industry, despite lack of evidence that the technology can be widely deployed. Many governments have been willing participants and have found that providing relatively small amounts of funding for research into capture and sequestration and other mitigation technologies is a cheap way of demonstrating commitment to solving the problem, while allowing business as usual. At individual company level, these strategies used for short-term advantage have often weakened the organisational capacity to make practical sense of sustainability concepts and to create the new knowledge and culture necessary to shape business practices.

4.4 SUSTAINABILITY FRAMEWORKS

Given the vagueness of the Brundtland definition of sustainable development and the interpretations and responses discussed in Sections 4.3.2 and 4.3.3, it is not surprising there are many views on how best to transition to sustainability. Many approaches have been proposed; some of the better-known are listed in Table 4.3. These are often expressed as sets of guidelines, rules or principles which can be regarded as constituting sustainable development or sustainability frameworks. They have often been promoted through books aimed at a general audience and by not-for-profit organisations which advocate sustainability concepts to the public and to governments. All approaches embody assumptions, sometimes explicitly stated but often implicit, about the meaning of sustainability and what a sustainable world would look like, and all are values-based to some extent. They reflect the range of responses discussed in Section 4.3.3. Some attempt to anchor their approach on scientific principles, others more on social or economic principles. Sustainability frameworks have been developed for a variety of purposes. They may be

Table 4.3: Some important sustainability frameworks

Framework	Basis	Reference
Triple bottom line	Social and economic	Elkington (1997)
Eco-efficiency	Science and engineering	WBCSD (2000)
The Natural Step	Science	The Natural Step (2009)
Natural capitalism	Social and economic	Hawken *et al.* (1999)
Biomimicry	Biology and ecology	Benyus (2002)
The five capitals model	Social and economic	Forum for the Future (2009)
Green chemistry	Physical sciences	Anastas and Warner (1998)
Green engineering	Physical sciences	USEPA (2009)

intended for use by individual companies or institutions, by groups of companies or institutions in a particular economic sector, by communities, or by governments at local, regional or national levels. They vary in level of detail and prescriptiveness and are not necessarily mutually exclusive.

4.4.1 Triple bottom line

An early attempt to come to terms with the abstract nature of sustainable development was through the triple bottom line (TBL) concept of social–environmental–financial accountability (Elkington, 1997), also known as people–planet–prosperity. TBL was adopted by many companies during the 1990s as a simple way of demonstrating their environmental and social credentials. It was used to varying degrees as a management tool for planning and integrating the company's environmental and social responsibilities to the community with its financial responsibility to shareholders. The TBL concept played an important role in raising sustainable development consciousness in the business community during the 1990s and is now widely used as a reporting framework. In 2006, the South African government adopted TBL as a framework for sustainable development (DEAT, 2008).

4.4.2 Eco-efficiency

The concept of eco-efficiency can be summarised as 'doing more with less'. The general principles are:

- reduce material intensity;
- reduce energy intensity;
- reduce dispersion of toxic substances;
- enhance recyclability;
- maximise use of renewables;
- extend product durability;
- increase service intensity.

These principles may be thought of as being concerned with three broad objectives (WBCSD, 2000).

- *Reduce the consumption of resources.* This includes minimising the use of energy, materials, water and land, enhancing recyclability and product durability, and closing material loops.
- *Reduce the impact on nature.* This includes minimising air emissions, water discharges, waste disposal and the dispersion of toxic substances, as well as fostering the sustainable use of renewable resources.
- *Increase product or service value.* This means providing more benefits to customers through product functionality, flexibility and modularity, providing additional services (such as maintenance, upgrading and exchange services) and focusing on selling the functional needs that customers actually want.

Selling a service instead of the product itself raises the possibility of the customer receiving the same functional need with fewer materials and less resources. It also improves the prospects of closing material loops because responsibility and ownership, and therefore concern for efficient use, remain with the service provider.

Eco-efficiency is targeted at environmental sustainability; it does not consider equity and other social aspects of sustainability. Nevertheless, it is a useful framework for driving innovation, both technological and organisational, and for breaking down the seemingly insurmountable challenges of Factor 10 type reductions in environmental impact into more manageable tasks. The World Business Council for Sustainable Development adopted the term eco-efficiency in 1991; it is now the central business concept of the WBCSD for bringing about corporate progress towards sustainability.

4.4.3 The Natural Step

The sustainability framework of The Natural Step, a not-for-profit organisation founded in Sweden in 1989 by Karl-Hendrik Robèrt, consists of a number of system conditions and principles. The system conditions are as follows (The Natural Step, 2009).

1. In a sustainable society, nature is not subjected to systematically increasing concentrations of substances extracted from the Earth's crust, increasing concentrations of substances produced by society, and increasing degradation by physical means.
2. In a sustainable society, people are not subjected to conditions that systematically undermine their capacity to meet their needs.

These conditions lead to the following principles:

- eliminate human contribution to the progressive build-up of substances extracted from the Earth's crust;
- eliminate human contribution to the progressive build-up of chemicals and compounds produced by society;
- eliminate human contribution to the progressive physical degradation and destruction of nature and the natural environment;
- eliminate human contribution to conditions that undermine people's capacity to meet their basic needs.

4.4.4 Natural Capitalism

Natural Capitalism is the title of a book by Hawken *et al.* (1999). More a polemical text with numerous case studies and examples than a set of principles or guidelines, the book argued that the next industrial revolution will result from the adoption of four central strategies:

- the conservation of resources through more effective manufacturing processes;
- the reuse and recycling of materials in ways that mimic natural systems;
- a change in social values from quantity to quality;
- investing in natural capital, or restoring and sustaining natural resources.

Implicit in the argument is the assumption that these can and will be achieved within the existing economic system and that companies that adopt these ideas early will become more successful.

4.4.5 Biomimicry

Biomimicry is the imitation of natural (often biological) processes and products in industrial situations. A study of the structure and chemistry of leaves, for example, might lead to the development of better solar cells. The central idea is that nature, through natural selection over several billion years, has solved many of the problems we are confronted with. Benyus (2002) gives nine principles underpinning the concept of biomimicry: nature runs on sunlight; nature uses only the energy it needs; nature fits form to function; nature recycles everything; nature rewards cooperation; nature banks on diversity; nature demands local expertise; nature curbs excesses from within; and nature taps the power of limits. The framework proposed by Benyus has three components.

- *Nature as model.* Biomimicry studies nature's models then emulates these forms, processes, systems and strategies to solve human problems.
- *Nature as measure.* Biomimicry uses an ecological standard to judge the sustainability of human innovations since nature has learned what works and what lasts.
- *Nature as mentor.* Biomimicry introduces an era based not on what can be extracted from the natural world, but what can be learned from it.

4.4.6 The five capitals model

The five capitals model, developed by the Forum for the Future, a not-for-profit organisation established in 1996, is based on five forms of capital – natural, human, social, manufactured and financial. The model uses 12 features to describe what a sustainable society should look like in terms of these forms of capital. Its aim is to help organisations evaluate projects. The underlying assumption is that by investing appropriately in all capital stocks, the following statements will be true and will lead to a successful capital investment strategy for sustainable development (Forum for the Future, 2009). The features of a sustainable society are as follows.

Natural capital.

- In their extraction and use, substances taken from the Earth do not exceed the environment's capacity to disperse, absorb, recycle or otherwise neutralise their harmful effects on humans and/or the environment.
- In their manufacture and use, artificial substances do not exceed the environment's capacity to disperse, absorb, recycle or otherwise neutralise their harmful effects on humans and/or the environment.
- The capacity of the environment to provide ecological system integrity, biological diversity and productivity is protected or enhanced.

Human capital.

- At all ages, individuals enjoy a high standard of health.
- Individuals are adept at relationships and social participation, and throughout life set and achieve high personal standards of development and learning.
- There is access to varied and satisfying opportunities for work, personal creativity and recreation.

Social capital.

- There are trusted and accessible systems of governance and justice.
- Communities and society at large share key positive values and a sense of purpose.
- The structures and institutions of society promote stewardship of natural resources and development of people.
- Homes, communities and society at large provide safe, supportive living and working environments of manufactured capital.
- All infrastructure, technologies and processes make minimum use of natural resources and maximum use of human innovation and skills.

Financial capital.

- Financial capital accurately represents the value of natural, human, social and manufactured capital.

4.4.7 Green chemistry and green engineering

Two technically based frameworks are green chemistry and green engineering. These have relatively limited objectives. Green chemistry, also known as sustainable chemistry, is the design of chemical products and processes that reduce or eliminate the use or generation of hazardous substances. Anastas and Warner (1998) developed some principles of green chemistry that are now widely accepted (Table 4.4). Most of these principles are more applicable to the manufacture of organic than inorganic substances, and hence are not directly relevant to minerals and metal production. Green engineering is not yet a well-developed concept. The United States Environmental Protection Agency (USEPA) defines green engineering as the design, commercialisation and use of processes and products that are feasible and economical while reducing the generation of pollution at the source and minimising the risk to human health and the environment. Green engineering embraces the idea that decisions to protect human health and the environment can have the greatest impact and cost-effectiveness when applied early to the design and development phase of a process or product. The USEPA proposed nine principles of green engineering (USEPA, 2009), as listed in Table 4.4. These seem sufficiently generic to be applicable to all industry sectors. Anastas and Zimmerman (2003) proposed an alternative list,[10] also given in Table 4.4.

4.4.8 Putting the frameworks into context

Robèrt (2000) described a five-level hierarchy of interrelated principles, activities and metrics applicable to sustainability, based on the approach used for strategic planning: describe the system; describe favourable outcomes; describe the principles for how to reach a favourable outcome; describe the activities that must be aligned with those principles; and describe ways of measuring and monitoring those activities to ensure they align with the principles.

Level 1 – Principles of the biosphere/society system. These describe how the system is constituted, for example the ecological and social principles.

Level 2 – Principles for sustainability. These define a stage or a favourable outcome in the system.

Level 3 – Principles for sustainable development. These describe a process for meeting principles for sustainability (both the transition to sustainability and the safe development thereafter).

Level 4 – Activities. These are tasks that align with the principles for sustainable development, for example to change from non-renewable energy to renewable energy or to recycle material. Activities for sustainable development should not be confused with principles for sustainable development.

10 Some of the jargon has been removed to make the list more comprehensible.

Table 4.4: Principles of green chemistry and green engineering

Principles of green chemistry	Principles of green engineering	
Anastas and Warner (1998)	**US Environmental Agency**	**Anastas and Zimmerman (2003)**
Design chemical syntheses to prevent waste.	Engineer processes and products holistically, use systems analysis and integrate environmental impact assessment tools.	Ensure that all material and energy inputs and outputs are as non-hazardous as possible.
Design chemical products to be fully effective, yet have little or no toxicity.	Conserve and improve natural ecosystems while protecting human health and well-being.	It is better to prevent waste than to treat or clean up waste after it is formed.
Design syntheses to use and generate substances with little or no toxicity to humans and the environment.	Use life cycle thinking in all engineering activities.	Separation and purification operations should be designed to minimise energy consumption and materials use.
Use raw materials and feedstocks that are renewable.	Ensure that all material and energy inputs and outputs are as inherently safe and benign as possible.	Products, processes and systems should be designed to most efficiently utilise mass, energy, space and time.
Minimise waste by using catalytic reactions.	Minimise depletion of natural resources.	Products, processes and systems should be 'output pulled' rather than 'input pushed'.
Avoid chemical derivatives since derivatives use additional reagents and generate waste.	Strive to prevent waste.	Embedded entropy and complexity should be viewed as an investment when making design choices on recycle, reuse or beneficial disposition.
Design syntheses so that the final product contains the maximum proportion of the starting materials.	Develop and apply engineering solutions, while being cognisant of local geography, aspirations and cultures.	Targeted durability, not immortality, should be a design goal.
Avoid using solvents, separation agents or other auxiliary chemicals or, if these are necessary, use innocuous chemicals.	Create engineering solutions beyond current or dominant technologies; improve, innovate and invent (technologies) to achieve sustainability.	Design for unnecessary capacity or capability (e.g. one size fits all) solutions should be considered a design flaw.
Carry out chemical reactions at ambient temperature and pressure whenever possible.	Actively engage communities and stakeholders in development of engineering solutions.	Material diversity in multi-component products should be minimised to promote disassembly and value retention.
Design chemical products to break down to innocuous substances after use so that they do not accumulate in the environment.		Design of products, processes and systems must include integration and interconnectivity with existing and available energy and materials flows.
Include in-process real-time monitoring and control during syntheses to minimise or eliminate the formation of by-products.		Products, processes and systems should be designed for a commercial afterlife.
Design chemicals to minimise the potential for accidents including explosions, fires and releases to the environment.		Material and energy inputs should be renewable rather than non-renewable.

For example, it is possible to violate principles for sustainability and sustainable development with renewable energy or recycling: changing from using non-renewable to renewable energy could lead to the impoverishment of forests and other parts of ecosystems, or to increased concentrations of scarce metals in ecosystems from poorly recycled photovoltaics.

Table 4.5: Sustainability frameworks classified according to Robèrt's hierarchy

Robèrt's hierarchy		Frameworks
1	Principles of the biosphere/ society system	
2	Principles for sustainability	The Natural Step; biomimicry; natural capitalism
3	Principles for sustainable development	Green chemistry; green engineering
4	Activities	Eco-efficiency
5	Metrics	Triple bottom line; five capitals model; ecological footprint; ecological rucksack; Factor X

Level 5 – Metrics. These are measures used to assess the relevance, quality and quantity of activities to ensure they are aligned with the principles for sustainable development, for example measurements to determine that material flows are really decreased to levels that are sustainable.

The hierarchy provides a useful tool for better understanding the role and contribution of different sustainability frameworks. Table 4.5 shows various frameworks classified according to Robèrt's hierarchy. It also includes, for completeness, some measures of environmental impact discussed previously. This classification should be considered as indicative only since there is considerable overlap between frameworks and some frameworks cover more than one level of the hierarchy. Nevertheless, the table is helpful in highlighting the major areas of applicability of the different frameworks.

4.5 A MODEL OF SUSTAINABILITY

Medveçka and Bangerter (2007), writing specifically about the minerals industry, claimed that key questions that remain unanswered with respect to sustainability are 'How to integrate sustainability into business practices?' and, more specifically, 'What are the factors that influence integration of sustainable development in the decision making process?' To address these questions, they developed a two-dimensional model (Table 4.6) as a tool by which organisations can engage at a practical level with the challenges of sustainability.

The sustainability dimension of the model (expressed vertically in Table 4.6) follows the logic of conventional strategic planning and is based on Robèrt's hierarchy. The first level, the context, concerns the nature of the biosphere – understanding the world we live in and what makes life possible, and social and ecological principles. The second level, the goal, consists of the conditions required for a state of sustainability to be achieved, the principles for a

Table 4.6. A two-dimensional model of sustainability

	Interpretations of sustainability			
Hierarchy	**Compliers**	**Innovators**	**Marketeers**	**Localisers**
Context	Risk reduction Weak sustainability	Innovation	Ecosystem limitation Strong sustainability	Ecosystem limitation Strong sustainability
Goal	Environmentalism; pollution prevention	Ecological footprint reduction; eco-efficiency	Decoupling of impacts from GDP; industrial ecology	Development that allows humankind to be sustained
Business strategy	Sustaining development	Investment in social capital	Use market- (and policy-) based instruments	GDP growth must be limited
Tasks	Based on corporate social responsibility	Based on enterprise development		Local production of food and distributed electricity generation (for example)
Monitoring and reporting	Triple bottom line		Five capitals model	Changes in social values

Source: After Medveçka and Bangerter (2007); with modifications.

favourable outcome and the desired end point. The third level, the strategies, consists of the guidelines for reaching a sustainable outcome and the technical strategies and process principles that underpin the transition to sustainability. The fourth level, the tasks, consists of the specific activities, projects and initiatives that should contribute to the goal of sustainability. The fifth level, the tools, consists of the frameworks for monitoring and reporting progress.

The levels are the points of entry into the field of sustainability. They indicate at which level – from performance reporting (level five), project and operational work (level four), company, industry or government strategies (level three), to institutional driving goal (level two) – decisions and actions are taking place in relation to sustainability. As in conventional strategic planning, tasks and monitoring activities are the result of decisions made in levels two and three where goals, strategies, principles, guidelines, systems and the like are developed and defined. This does not mean that what happens in level four and five cannot influence what happens at levels two and three, or that the same people may act at different levels of the hierarchy.

The different responses to sustainability are summarised in the second dimension (expressed horizontally in Table 4.6) in four categories: *Compliers,*[11] *Innovators, Marketeers* and *Localisers*. These cover the range industrialism to radicalism, as proposed by Dryzek (2005), and weak to strong sustainability. *Compliers*, *Innovators* and *Marketeers* are adherents of industrialism, though of different forms. They are committed to material well-being and to growth in the production of material goods. *Localisers* reject traditional industrialism and have a world view based on different values. Their interpretations of the relationships between the ecosystem and human needs differ radically and demand different institutions.

Compliers are people and organisations committed to risk reduction. Their historical perspective is that of sustaining their business in a constantly changing set of community expectations, including those regulated by legislation. They believe in the ideas of weak sustainability. For Compliers, the business case for sustainable development is clear, as it is a necessity to mitigate risks, both environmental and social. Managing reputation and ensuring optimum conditions to secure licence to operate, access to land, finance and market are prime concerns. Therefore, Compliers are strong proponents of the concept of corporate social responsibility. They may exceed legal requirements and receive acknowledgement for good husbandry of the environment and the communities where they operate. Meeting triple bottom line targets is an example of a compliance approach.

Innovators are committed to searching for and pioneering inventive solutions that drive sustainability into the heart of their activities. They see a reduction in ecological footprint as a business imperative and they believe in improving social equity by investing in social capital. They are less risk-averse than Compliers. Eco-efficiency is a key focus of Innovators in the manufacturing and metals sectors. They invest in research and development and are open to innovative ways of designing their facilities.

Marketeers appreciate there are limits imposed by the finite nature of the biosphere and recognise that current human activities endanger ecosystems and deplete natural resources to a level that requires attention by society. They generally believe in strong sustainability. If the critical elements that sustain life are imperilled and the capacity to develop, let alone sustain, current business activities is jeopardised then the marketplace needs to drive a reversal. They perceive the market as a powerful force that can generate good results for all. They want to adapt the economic system and rely on market-driven approaches to decouple economic growth from adverse ecological and social impacts. They claim that the right price signals through trading schemes, for example, are all that is needed to move industry towards sustainability. They promote the concept of no waste and whole systems thinking, and focus on radical technological innovation within the current economic paradigm.

Localisers, like Marketeers, are concerned about the limits imposed by nature and believe in the ideas of strong sustainability. They understand and value the relationships and complex interactions between the biosphere, human activities, social equity and the institutions humans create to manage their affairs, including the economic system. Localisers differ from

11 Medveçka and Bangerter use the term *Compliance* but *Compliers* is more consistent with the terminology of the other columns.

the other three categories in that they question the institutions. They see the current marketplace as the problem, not the solution, and aim to change it. Although Localisers accept the notion of progress, they refute economic growth as the primary indicator of human progress and advocate replacing it with improvement of human well-being. Characteristically, they propose a return to more localised food production, more community-based activity, less work aimed at producing consumer goods, and more equity. They do not necessarily want a low-tech world, contrary to what their critics often claim, but advocate a shift in societal values from consumer-centred growth to collaborative and equitable human development.

Medveçka and Bangerter's model helps understand the difficulties encountered when different groups or individuals discuss sustainability. When mineral industry representatives speak about sustainability they are almost invariably speaking from the Compliers perspective. When 'green' politicians or activists speak about sustainability they are usually speaking from the Localiser perspective. There is almost no common ground between these perspectives. Those who speak from the Innovator or Marketeer perspective have some common ground but tend to think that Compliers and Localisers have extreme and unrealistic views. These ideas are pursued further in Chapter 6, where the mineral industry's response to sustainability is discussed. Later chapters explore opportunities for further contribution by the industry to transitioning to a sustainable world.

4.6 REFERENCES

Anastas PT and Warner JC (1998) *Green Chemistry: Theory and Practice*. Oxford University Press: New York.

Anastas PT and Zimmerman JB (2003) Design through the 12 principles of green engineering. *Environmental Science and Technology* **37**(5): 94A–101A.

Ayres RU (2005) 'On the practical limits to substitution'. Interim Report IR-05-036, 28 July 2005. International Institute for Applied Systems Analysis: Laxenburg, Austria.

Benyus JM (2002) *Biomimicry: Innovation Inspired by Nature*. Perennial: New York.

Carson R (1962) *Silent Spring*. Houghton Mifflin: New York.

DEAT (2008) *People – Planet – Prosperity: A Strategic Framework for Sustainable Development in South Africa*. Department of Environmental Affairs and Tourism, South Africa, July 2008. <http://www.participation.org.za/docs/NFSD%20July%202008.doc>.

Diamond JM (2005) *Collapse: How Societies Choose to Fail or Succeed*. Viking Books: New York.

Donella H, Meadows DH, Meadows DL, Randers J and Behrens WW (1972) *The Limits to Growth*. Universe Books: New York.

Dryzek J (2005) *The Politics of the Earth: Environmental Discourses*. Oxford University Press: New York.

Ehrlich P (1968) *The Population Bomb*. Ballantine Books: New York.

Elkington J (1997) *Cannibals with Forks: The Triple Bottom Line of 21st Century Business*. Capstone: Oxford, UK.

Factor 10 Institute (2009) <http://www.factor10-institute.org/index.html>.

Forum for the Future (2009) The five capitals model. <http://www.forumforthefuture.org/projects/the-five-capitals>.

Global Footprint Network (2009) *Ecological Footprint Standards 2009*. Global Footprint Network: Oakland, CA. <www.footprintstandards.org>.

Global Footprint Network. <http://www.footprintnetwork.org>.

Gore A (2009) *Our Choice: A Plan to Solve the Climate Crisis*. Rodale: Emmaus, PA.

Hails C (Ed.) (2008) 'Living planet report 2008'. WWF International: Gland, Switzerland. <http://assets.pnda.org/downloads/living_planet_report_2008.pdf>.

Hardin G (1968) The tragedy of the commons. *Science* **162**(3859): 1243–1248.

Hawken P, Lovins AB and Lovins LH (1999) *Natural Capitalism: Creating the Next Industrial Revolution*. Little, Brown and Co.: Boston.

Lloyd WF (1833) *Two Lectures on the Checks to Population*. Oxford University Press: Oxford, UK.

Malthus TR (1826) *An Essay on the Principle of Population*. 6th edn. <http://www.econlib.org/library/Malthus/malPlong.html>.

Mancur O (1965) *The Logic of Collective Action: Public Goods and the Theory of Groups*. Harvard University Press: Cambridge, Mass.

Medveçka J and Bangerter P (2007) Engineering sustainable development into industry: unlocking institutional barriers. In *Cu2007. Vol. 6 – Sustainable Development, HS&E and Recycling*. (Eds D Rodier and W Adams) pp. 13–26. Canadian Institute of Mining, Metallurgy and Petroleum: Toronto, Canada.

Neumayer E (2003) *Weak versus Strong Sustainability*. 2nd edn. Edward Elgar: Cheltenham, UK.

Ostrom E (1990) *Governing the Commons*. Cambridge University Press: Cambridge, UK.

Porritt J (2005) *Capitalism as if the World Matters*. Earthscan: London.

Poundstone W (1992) *Prisoner's Dilemma*. Doubleday: New York.

Power TM (2002) *Digging to Development? A Historical Look at Mining and Economic Development*. Oxfam America: Washington DC.

Rees WE (1992) Ecological footprints and appropriated carrying capacity: what urban economics leaves out. *Environment and Urbanisation* **4**(2): 121–130.

Robèrt K-H (2000) Tools and concepts for sustainable development, how do they relate to a general framework for sustainable development, and to each other? *Journal of Cleaner Production* **8**: 243–254.

Schmidt-Bleek F (1997) *Statement to Government and Business Leaders*. Wuppertal Institute: Wuppertal, Germany.

Schmidt-Bleek F (2008) Future beyond climate change. *Position Paper 08/01*. Factor 10 Institute: Carnoules, France.

Solow RM (1974) The economics of resources or the resources of economics. *American Economic Review* **64**(2): 1–14.

Stiglitz J (1974a) Growth with exhaustible natural resources: efficient and optimal growth paths. *Review of Economic Studies* **41**: 123–137.

Stiglitz J (1974b) Growth with exhaustible natural resources: the competitive economy. *Review of Economic Studies* **41**: 139–152.

The Natural Step (2009) <http://www.naturalstep.org>.

Tucker A (1950) *A Two-Person Dilemma*. Stanford University Press: Palo Alto, CA.

Turner GM (2008) A comparison of 'The Limits to Growth' with 30 years of reality. *Global Environmental Change* **18**(3): 397–411.

UN (1987) *Our Common Future*. Oxford University Press: Oxford, UK.

UN (1992) *Agenda 21*. United Nations: New York. <http://www.un.org/esa/sustdev/documents/agenda21/english/agenda21toc.htm>.

UNEP (2005) 'Millennium ecosystem assessment, synthesis report: ecosystems and human well-being: general synthesis'. Island Press: Washington DC.

USEPA (2009) Green Engineering. United States Environmental Protection Agency. <http://www.epa.gov/oppt/greenengineering>.

Vitousek PM, Ehrlich PR, Ehrlich AH and Matson PA (1986) Human appropriation of the products of photosynthesis. *BioScience* **36**: 368–373.

Wackernagel M (1994) Ecological footprint and appropriated carrying capacity: a tool for planning toward sustainability. PhD thesis, School of Community and Regional Planning, University of British Columbia, Vancouver, Canada.

WBCSD (2000) *Eco-efficiency: Creating More Value for Less*. World Business Council for Sustainable Development. <http://www.wbcsd.org/web/publications/eco_efficiency_creating_more_value.pdf>.

4.7 USEFUL SOURCES OF INFORMATION

International Council on Mining and Metals (ICMM). <http://www.icmm.com>.

World Business Council for Sustainable Development (WBCSD). <http://www.wbcsd.org>.

5 Mineral resources

Although it makes up only about 0.5% of the Earth's mass, the crust is the most important part of the Earth from a human perspective because it is the ultimate source of the inorganic and organic materials we use and is the source of the nutrients required by living organisms. This chapter describes the formation of the Earth and the crust, then examines the composition of the crust and the form and distribution of the various materials in it, particularly those of value to humans. Of particular interest are geochemical anomalies, since these produce higher than average concentrations of some elements and hence are potential sources of those elements or their compounds. The chapter concludes with a discussion of the implications for extracting substances of value.

5.1 FORMATION OF THE EARTH

The model of the Universe which best explains observable data is the Big Bang theory, originally postulated in 1927 by Georges Lemaître and subsequently extensively modified and elaborated.[12] According to this theory, the Universe began about 13.7 billion years ago as an infinitesimally small, infinitely hot and infinitely dense singularity. The singularity began to expand and cool (the Big Bang) and it continues to expand. Though a major scientific achievement and the best explanation to date, the Big Bang theory is rather unsatisfactory. Apart from the issue of the nature of the singularity, the theory accounts for only about 10% of the mass of the Universe, the remainder being described as dark matter that is as yet undetected. It has been necessary to introduce the concept of dark energy, which exerts a negative force, to account for the observation that the Universe's rate of expansion is increasing rather than decreasing as would be expected due to gravitational attraction. The nature of this dark energy is not known and it is presently undetectable.

The expansion resulting from the Big Bang produced protons, neutrons and electrons which very quickly became organised into atoms, predominantly of hydrogen and helium. As the mass expanded, its slightly uneven distribution resulted in gravitational instabilities which caused it to coalesce into separate huge clouds that coalesced further and eventually became galaxies and clusters of galaxies. Uneven distribution of mass within each galaxy and further gravitational collapse caused denser bodies to form

12 The term Big Bang to describe the expanding Universe model was first used by the cosmologist Fred Hoyle in 1949 in a radio broadcast.

within the galaxies. These eventually became suns, or stars, as they were heated by the potential energy released by the gravitational collapse. They eventually reached temperatures at which thermonuclear fusion reactions commenced. These are called first-generation stars. Thermonuclear fusion reactions proceeded in stages, producing successively heavier elements. The larger stars, which reached the stage at which silicon atoms fused to form iron atoms, expanded explosively due to the heat generated by the very fast rate of the reaction. Such an explosion is called a supernova.

Around 4.5–5 billion years ago our solar system began to coalesce from the remnants of one or more supernova. Thus, the Sun is a second-generation star. In its form today, the solar system consists predominantly of the Sun and the eight planets that orbit around it. All the planets move in the same direction, in elliptical orbits that lie almost in the same plane. The moons that revolve around some of the planets, planetoids (including Pluto), asteroids, comets and meteors are also part of the solar system. The inner planets – Mercury, Venus, Earth and Mars – are relatively small and have densities consistent with being composed of rock and metal. These are called the *terrestrial planets*. The outer planets – Jupiter, Saturn, Uranus and Neptune – are composed largely of hydrogen and helium and have lower densities. They are called the *gas planets* and are much larger than the terrestrial planets.

As the gas that formed the solar system collapsed under gravity it formed a rotating disc and heated up due to the release of gravitational potential energy. Eventually it started to cool and elements with the highest boiling points and the most stable compounds began to condense as solid particles. These included the platinum-group elements, aluminium oxide, metallic nickel and iron, calcium aluminates, and calcium-, iron- and magnesium silicates. These were followed by less stable compounds, particularly the more complex silicates and the sulfides of the heavy metals. This precipitation sequence has been observed in primitive meteorites found on Earth. Before the temperature cooled in the inner regions of the disc sufficiently for the most volatile elements and compounds to condense, thermonuclear fusion of hydrogen commenced in the core. The solar wind produced as a result (which consists mainly of electrons and protons from the fusion reaction) drove hydrogen, helium, nitrogen, carbon and their compounds, such as methane (CH_4) and ammonia (NH_3), to the outer regions of the disc where ultimately they concentrated to form the gas planets.

The particles began to coalesce to form planetesimals (which may have grown to about 10–100 km in diameter), which coalesced further. The Earth was formed from the early solid particles; accordingly, the elements contained in them are enriched in the Earth relative to their overall abundance in the solar system. As the proto-Earth grew from collision with planetesimals it heated up due to the release of gravitational potential energy. It eventually became entirely molten. The denser iron, and elements soluble in it, particularly nickel, sank towards the centre under the influence of gravity and formed the core, while the less dense compounds, mainly silicates, rose to the surface and formed the mantle. Droplets of metal moved to the core, slowly at first then increasingly quickly, between around 100 million and 500 million years after the Earth first formed. This period is called the Iron Catastrophe. This event created the present structure of the Earth (Figure 3.1). The mantle is solid, and has been throughout most of Earth's history; the core consists of a liquid metal outer core and a solid metal inner core. The inner core is solid because at the immense pressure at the centre of the Earth the melting point of the iron-nickel alloy is higher than the temperature of the inner core. At an early stage of formation, a very large planetesimal struck the Earth a glancing blow and a large quantity of mantle material was ejected into space. Most of this eventually fell back to Earth, under the action of gravity, but some collected into a second planetesimal circling Earth – the Moon. Since the Moon is composed mainly of mantle material, it has a lower average density than Earth.

There are three main sources of heat in the Earth: the heat from when the planet formed by accretion, which has not yet been lost; heat due to the release of gravitational potential energy of the settling core material during the Iron Catastrophe; and heat from the decay of radioactive elements. Some heat may also be generated by tidal forces on the Earth as it rotates, causing cycles of compression and elongation. Earth has been losing heat by radiation into space since its formation, but it is producing almost as much heat as it is losing. When the core starts to cool, a little more of the molten iron of the outer core solidifies and the released latent heat maintains the temperature of the core. The magnitude of radioactive heating is uncertain since the abundances of the radioactive elements (particularly potassium, uranium and thorium) in the core and mantle are not known accurately, but it may be around 2.7×10^{13} W (Rybach, 2007), a substantial component of the 4.1×10^{13} W total heat flux from the Earth's interior (Section 3.3.1).

5.2 THE GEOLOGICAL TIME SCALE

Geologists have established a time scale which relates geological and biological changes on Earth to time. It has developed progressively over several hundred years as knowledge of the Earth and of evolution has increased. A modern version is shown in Figure 5.1. The scale divides the Earth's past into Eons, Eras, Periods and Epochs of unequal intervals of time; the demarcations correspond to geological events or biological events such as mass extinctions.

Eon	Era	Period	Epoch		Millions of years
Phanerozoic	Cenozoic	Quaternary	Holocene		0.01
			Pleistocene	Late	
				Early	1.8
		Tertiary	Pliocene	Late	
				Early	5.3
			Miocene	Late	
				Middle	
				Early	23
			Oligocene	Late	
				Early	33.9
			Eocene	Late	
				Middle	
				Early	55.8
			Paleocene	Late	
				Early	65.5
	Mesozoic	Cretaceous	Late		
			Early		145
		Jurassic	Late		
			Middle		
			Early		200
		Triassic	Late		
			Middle		
			Early		251
	Paleozoic	Permian	Late		
			Early		299
		Carboniferous	Late		
			Early		359
		Devonian	Late		
			Middle		
			Early		416
		Silurian	Late		
			Early		443
		Ordovician	Late		
			Middle		
			Early		490
		Cambrian	D		
			C		
			B		
			A		543
Precambrian	Protero-zoic	Late			
		Middle			
		Early			2500
	Archean	Late			
		Middle			
		Early			~4000

Figure 5.1: The geological time scale. Note that the scale at the right does not represent equal intervals of time, but rather the boundaries of the Epochs.

5.3 FORMATION OF THE CRUST

It is likely that the Earth had a thin, continuous, solid crust from a very early stage because the outermost layer would have cooled rapidly by radiation of heat into space. Nevertheless, the thin primitive crust was probably broken up regularly by the action of molten material from below, which in turn solidified as it cooled at the surface. This resulted in extensive masses being progressively built up. Large, thick, solid blocks, which ultimately formed the early continents, were present about 4 billion years ago. The lower areas between the blocks ultimately formed the oceanic crust. The early crust was also subjected to impact from planetoids and small asteroids, which were relicts of the formation of the planets and much more common then than they are today.

Convection currents gradually became established within the mantle as the Earth cooled progressively from the surface inwards and a temperature gradient was established. As a result, hotter and therefore less dense material from the upper mantle is continually carried upwards to the surface by buoyancy. It is replaced by cooler and therefore more dense material from near the surface. It is thought that convection within the mantle was not significant before about 2.5 billion years ago, therefore plate tectonics (as discussed in Section 2.1) commenced around then. Through the action of tectonic processes, which continue to the present time, slabs of crust are recycled at subduction zones and taken down to the mantle; new crust is formed from magma at mid-ocean ridges, as illustrated in Figure 5.2. Subduction is not an even process; ocean crust is subducted very rapidly but continental crust, having lower density and greater height, is less amenable to subduction. No known oceanic crust is older than about 200 million years while the continental crust has been well preserved, with parts in Canada and Australia over 4 billion years old.

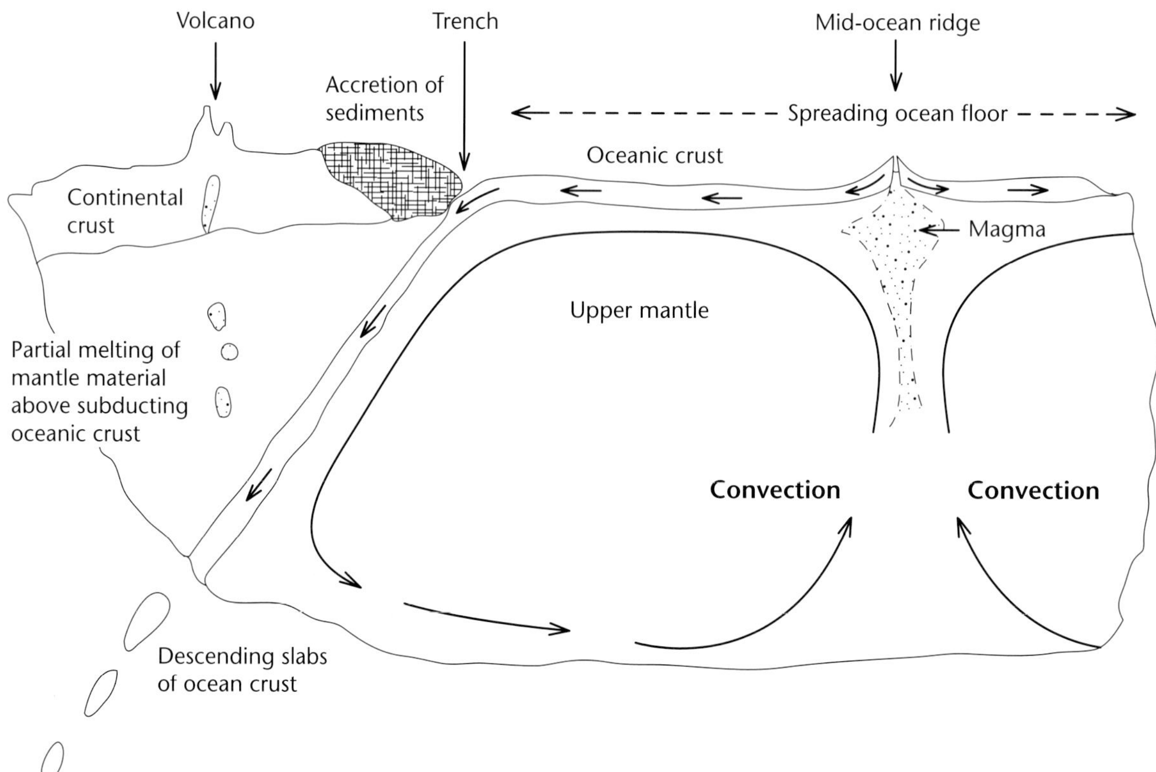

Figure 5.2: The action of convection currents in the Earth's mantle showing the formation of subduction zones, trenches and mid-ocean ridges (after Selley *et al.*, 2004).

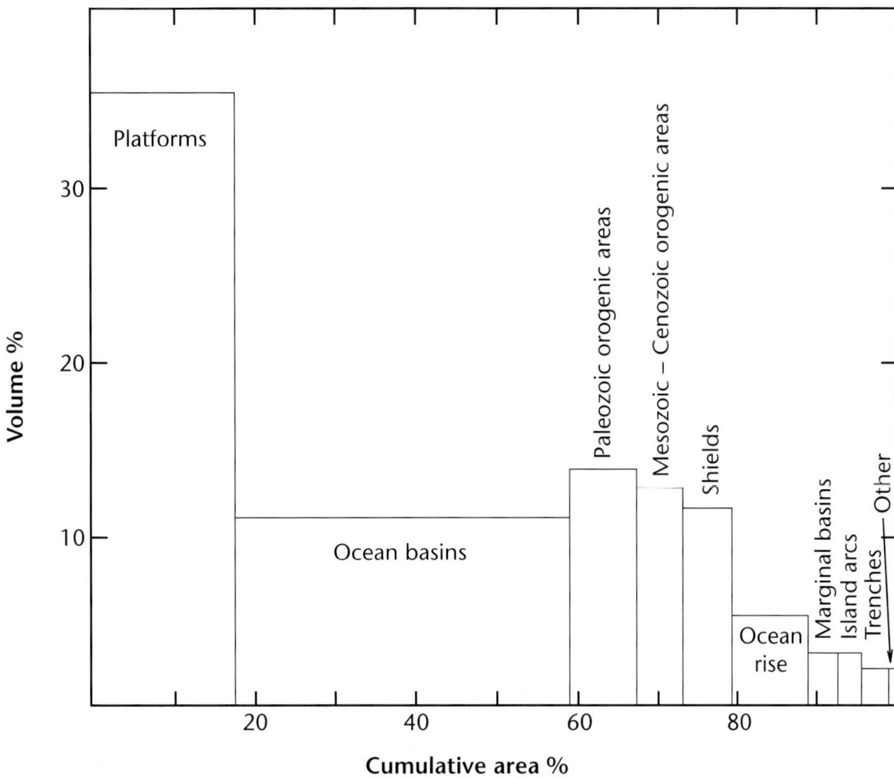

Figure 5.3: The proportions of the major crustal structures, by area and volume (after Condie, 1976).

The present continental and oceanic crusts are not geologically uniform, comprising distinct types of structures resulting from plate tectonics, weathering and other geological processes over billions of years. The major structural types and their relative areas and volumes are shown in Figure 5.3.

5.3.1 Continental crust

The continental crust consists largely of shields, platforms, orogens and rifts.

Shields are expanses of low profile which have been stable since the Precambrian and therefore subjected to extensive weathering and erosion. The Canadian Shield, which extends from Lake Superior to the Arctic Islands and from western Canada to Greenland, and the Australian Shield, which occupies most of the western half of the Australian continent, are examples of large shields.

Platforms are similar to shields. They are composed of Precambrian basement rocks but are covered in thick layers of sedimentary material which may date entirely or in part from the Phanerozoic. The loose, unconsolidated and re-cemented fragments of rock, sand, alluvium and soil (products of weathering and erosion) that cover the underlying, more coherent bedrock are collectively called *regolith*. The term *craton* is frequently used to describe the old and stable parts of the continental crust that have survived the merging and splitting of continents. Cratons are remnants of earlier continents. They have a thick crust which extends deep into the mantle and they form the core of the present day continents, which themselves are large aggregations of various tectonic elements brought together at different times and in different ways. Cratons usually make up the central portion of continents and consist of both shields and the basement of platforms.

Orogens are long, curved belts of folded rocks usually formed where tectonic plates collide. They vary in width from several hundred to several thousand kilometres and may extend in length for thousands of kilometres. They may form present mountain ranges or may have been almost completely eroded away. The oldest orogenic belts were formed about 2 billion years ago.

Rifts are areas where the crust is being pulled apart by the action of plate tectonics. Most rifts occur along the axis of mid-ocean ridges where new oceanic crust is created along the boundary between two plates. They also, more rarely, occur on continents where they are typically 30–70 km wide and from tens of kilometres to thousands of kilometres long. For example, the East African Rift, which is part of the larger Great Rift Valley, is a narrow zone where the African Plate is in the process of splitting into two new plates.

5.3.2 Oceanic crust

The oceans and bodies of fresh water formed when the Earth cooled sufficiently for water vapour to condense. This water occupied the lowest-lying areas, thus oceanic crust is largely covered by water. Oceanic crust consists largely of volcanic islands, island arcs, mid-ocean ridges, trenches, ocean basins, marginal basins and inland sea basins.

Volcanic islands are scatterings or chains of islands which mark hot spots in the mantle where mantle plumes reach the surface. Examples include the Galapagos Islands and the Hawaiian Islands.

Island arcs lie above subduction zones and are formed by melting of the subducted plate as it penetrates the mantle. Island arcs are sites of vulcanism. Examples include the Aleutian Islands and the Kuril Islands.

Mid-ocean ridges are raised, wide linear belts, which form at the boundary between tectonic plates and usually have a rift valley running along their axis. They form from magma rising from the mantle through convection and emerging through weaknesses at plate boundaries. The mid-ocean ridges, which are covered today by water, are interconnected and form a single global system with a total length of about 65 000 km. Mid-ocean ridges are geologically active, with new magma constantly emerging onto the ocean floor.

Trenches are the outer edges of subduction zones and are usually parallel to, and several hundred kilometres from, a volcanic island arc. They are the deepest parts of the ocean and are typically 3–4 km below the level of the surrounding ocean floor. The Mariana Trench, in the western Pacific Ocean, is the deepest trench. It is about 2550 km in length and 69 km wide and has a maximum depth of about 11 km.

Ocean basins are the large, flat, deep ocean areas beyond the continental shelves. They occupy most of the Earth's surface. They are tectonically stable and typically have a coating of sediment around 0.3 km thick.

Marginal basins are segments of the oceanic crust between island arcs or between an island arc and a continent. They are often coated with lake or river deposits or with marine sediments. They are located on the overriding tectonic plate behind island arcs at a subduction zone.

Inland sea basins are seas within continents (for example, the Caspian Sea, Black Sea, the Great Lakes of North America, Lake Eyre in Australia) or seas which are nearly surrounded by continents (for example, the Gulf of Mexico).

5.3.3 The distribution of elements

Relics of the accretion process that formed the Earth are still present in the solar system and occasionally fall to Earth as meteorites. The compositions of these have been used to deduce the average composition of the Earth by mathematically combining portions of the various types of meteorites, according to their relative abundance, to produce mixtures that have densities equal to that of the Earth (about 5.3 g mL^{-1}). When these calculations are performed, values such as those in Table 5.1 are obtained; these should be viewed as approximate estimates. Clearly, the overall composition of the Earth is dominated by the elements iron, oxygen, silicon, magnesium, nickel and sulfur. All other elements can be considered minor constituents of the Earth as a whole. However, the elements are not uniformly distributed. The mantle makes up most of the Earth's mass and is composed largely of iron- and magnesium silicates, the core is composed largely of metallic iron and nickel, while the crust (which is only about 0.5% of Earth's total mass) is composed mainly of silicates consisting of oxygen, silicon, aluminium, iron, calcium, sodium and potassium.

Table 5.1: Estimated average composition of the entire Earth

Element	Wt%	Goldschmidt classification
Iron	30–35	Siderophile
Oxygen	30–31	Lithophile
Silicon	15–17	Lithophile
Magnesium	13–16	Lithophile
Sulfur	1.9–4.7	Chalcophile
Nickel	1.7–2.4	Siderophile
Calcium	1.1–1.9	Lithophile
Aluminium	1.0–1.4	Lithophile
Sodium	0.7	Lithophile

Source: Condie (1976).

V.M. Goldschmidt (1888–1947), often called the father of geochemistry, classified the elements into four groups according to their chemical preferences for associating with other elements (Goldschmidt, 1954): atmophile, siderophile, lithophile and chalcophile. Figure 5.4 shows the periodic table divided into these four groups and Table 5.1 shows the major Earth-forming elements classified accordingly. Goldschmidt's classification provides a qualitative explanation of the distribution of the elements within the Earth. The *atmophile elements* (literally, gas-loving) are the elements most commonly occurring in the Earth's atmosphere, particularly H, C, N and the inert gases. They may occur in an uncombined state (e.g. N_2, Ar) or combined, for example, H_2O, CO_2, CH_4, NH_3 and H_2S. These substances are gases at ambient temperature. Although oxygen is a major component of the atmosphere it is not considered to be an atmophile element: most of Earth's oxygen is present in the crust as oxide and silicate minerals and it is present in the atmosphere only because of the action of photosynthesis. The *siderophile elements* (iron-loving) are the metals near iron in the periodic table that exhibit metallic bonding and are soluble in molten iron. The *lithophile elements* (oxygen-loving) are the elements which form ionic bonds and have a strong affinity for oxygen. These elements have a tendency to form oxides and silicates. The *chalcophile elements* (sulfur-loving) are the elements that bond, usually covalently, with S, Se, Te, Sb and As.

The atmophile elements and related compounds remained in the gaseous state as the Earth cooled but had not grown to a size to have sufficient gravity for

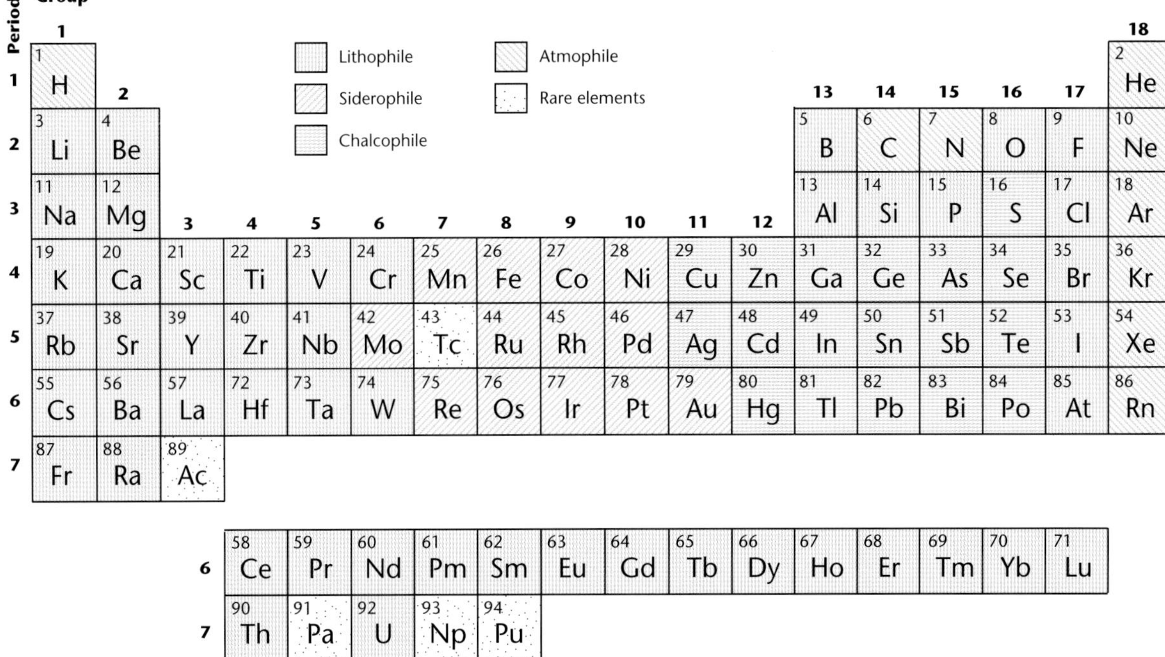

Figure 5.4: The periodic table showing the elements classified as atmophile, siderophile, lithophile and chalcophile.

easy retention. As a result, they are strongly depleted in Earth as a whole relative to their overall abundance in the solar system. Where they do occur, they are concentrated in the atmosphere. The siderophile elements, being dense, segregated into a metallic phase in the Earth's core when the Earth was molten. Thus, they are strongly depleted in the Earth's crust relative to their solar system abundance but are probably present in the Earth as a whole in approximately their solar system abundance. The lithophile elements are present mainly as silicates and oxides which, being of low density, concentrated in the mantle and crust where they are enriched relative to their solar system abundance. Chalcophile elements, being dense, separated along with the siderophile elements into the core. They are depleted in the Earth's crust relative to their solar system abundance, but to a lesser degree than siderophiles.

Composition of the crust: abundant and scarce elements

Table 5.2 lists the average elemental composition of the crust. Oxygen makes up nearly half of the crust's mass and 11 elements – oxygen, silicon, aluminium, iron, calcium, magnesium, sodium, potassium, titanium, phosphorus and manganese – together make up nearly

Table 5.2: The mean elemental composition of the Earth's crust

Geochemically abundant elements					Geochemically scarce elements											
	>0.1 wt%		Cumulative wt%			>10 ppm		Cumulative wt%			>1 ppm			<1 ppm		
O	45.5	wt%	45.5	wt%	F	544	ppm	99.63	wt%	Pb	13	ppm	Tl	0.70	ppm	
Si	27.2	wt%	72.7	wt%	Ba	390	ppm	99.67	wt%	Pr	9.1	ppm	Tm	0.50	ppm	
Al	8.3	wt%	81.0	wt%	Sr	384	ppm	99.71	wt%	B	9.0	ppm	I	0.46	ppm	
Fe	6.2	wt%	87.2	wt%	S	340	ppm	99.75	wt%	Th	8.1	ppm	Y	0.31	ppm	
Ca	4.66	wt%	91.86	wt%	C	180	ppm	99.76	wt%	Sm	7.0	ppm	In	0.24	ppm	
Mg	2.76	wt%	94.62	wt%	Zr	162	ppm	99.78	wt%	Gd	6.1	ppm	Sb	0.20	ppm	
Na	2.27	wt%	96.89	wt%	V	136	ppm	99.79	wt%	Er	3.5	ppm	Cd	0.16	ppm	
K	1.84	wt%	98.73	wt%	Cl	126	ppm	99.81	wt%	Yb	3.1	ppm	Hg	0.08	ppm	
Ti	0.632	wt%	99.36	wt%	Cr	122	ppm	99.82	wt%	Hf	3.0	ppm	Ag	0.08	ppm	
P	0.112	wt%	99.47	wt%	Ni	99	ppm	99.82	wt%	Cs	2.6	ppm	Se	0.05	ppm	
Mn	0.106	wt%	99.58	wt%	Rb	78	ppm	99.84	wt%	Br	2.5	ppm	Pd	0.015	ppm	
					Zn	76	ppm	99.84	wt%	U	2.3	ppm	Pt	0.010	ppm	
					Cu	68	ppm	99.85	wt%	Eu	2.1	ppm	Bi	0.008	ppm	
					Ce	66	ppm	99.86	wt%	Sn	2.1	ppm	Os	0.005	ppm	
					Nd	40	ppm	99.86	wt%	Be	2.0	ppm	Au	0.004	ppm	
					La	35	ppm	99.86	wt%	As	1.8	ppm	Ir	0.001	ppm	
					Co	29	ppm	99.87	wt%	Ta	1.7	ppm	Te	0.001	ppm	
					Sc	25	ppm	99.87	wt%	Ge	1.5	ppm	Re	0.0007	ppm	
					Nb	20	ppm	99.87	wt%	Ho	1.3	ppm	Rh	0.0001	ppm	
					Ga	19	ppm	99.87	wt%	W	1.2	ppm				
					N	19	ppm	99.87	wt%	Tb	1.2	ppm				
					Li	18	ppm	99.88	wt%	Mo	1.2	ppm				

Source: Greenwood and Earnshaw (1984).

99.6% of the mass of the crust. These are called the geochemically abundant elements. All the other elements are geochemically scarce. With only a few exceptions, such as gold, silver and the platinum group metals, elements do not occur in the uncombined state in the crust but as naturally occurring compounds (minerals). Since oxygen is the major constituent of the crust and it is relatively reactive, it is not surprising that most elements in the crust occur as compounds of oxygen. Fluorine, sulfur and chlorine are the next most abundant anions which form stable compounds but these are present in very much lower concentrations. Chlorides, fluorides and sulfides are present in the crust in much lesser quantities than oxides. Nevertheless, these compounds, particularly sulfides, are important sources of some metals.

5.4 MINERALS AND ROCKS

The chemical elements in the crust and mantle are present predominantly in the form of minerals. *Minerals* are naturally occurring, homogeneous, solid crystalline chemical elements or compounds. Minerals are usually formed by inorganic processes. The term 'homogeneous' means that, chemically and physically, a mineral has the same characteristics at all scales down to the basic atomic unit cell. *Mineraloids* are mineral-like substances that do not demonstrate crystallinity (for example, opal and obsidian). Many common natural substances, such as granite and basalt, are composites (mixtures) of several minerals. These are called *rocks*. Oil, natural gas and coal are not considered to be minerals since they are mixtures of compounds, are not crystalline and are organic.

Oxygen and silicon make up 73% by mass of the crust and nine elements (oxygen, silicon, aluminium, iron, calcium, magnesium, sodium, potassium and titanium) make up 99.4% by mass (Table 5.2). The remaining elements make up only about 0.6% and are present in trace quantities. Oxygen and silicon readily combine to form the silicate anion, SiO_4^{4-} (Figure 2.1), which readily bonds with Al, Fe, Ca, Mg, Na, K and Ti cations to form silicates. As a result, 99% by mass of the crust is made up of silicate and oxide (and hydroxide) minerals. The next most important mineral classes, based on the next most abundant anions – fluorine (0.05%), sulfur (0.03%) and carbon (0.02%) – are the sulfides and carbonates. Fluorine, while more abundant than sulfur and carbon, tends not to form simple compounds, other than fluorite (CaF_2); it occurs most commonly as a component in multi-anionic minerals. The most common sulfides are compounds of iron and the most common carbonates are compounds of Ca, Mg and Fe. In summary, the crust is essentially composed of silicates with small quantities of a few oxides and hydroxides, the carbonates of calcium, magnesium and iron, and the sulfides of iron.

Although the 94 naturally occurring elements could, in principle, combine to produce an enormous number of compounds, there are only about 4500 known minerals. Goldschmidt (1954) provided a qualitative explanation for this. When two cations have similar charges, one can substitute for the other in a mineral without fundamentally changing the crystal structure (atomic arrangement) and thereby producing a different mineral (though distortion of the crystal lattice will occur). The extent of substitution depends on the temperature and pressure at which a mineral is formed and on the relative size of the substituting cation. If the ionic radii differ by less than about 15%, complete substitution is usually possible. When the difference is greater than 15%, some substitution is possible. However, there are limits beyond which addition of the larger or smaller cation will cause the crystal structure to change fundamentally. Either the new cation becomes accommodated within the structure, which then becomes a different mineral, or, more commonly, it forms a separate mineral (in addition to the original mineral) in which the substituting element becomes the major cation. Because most elements occur in such small quantities in the crust they can usually be accommodated by substitutions within the atomic structures of common oxide and silicate minerals without forming separate minerals. The only way minerals of scarce elements can form, therefore, is for some geological process to cause a local enrichment of a scarce element so that substitution limits are exceeded. This rarely occurs in the common rock-forming processes; these rare occurrences are referred to as a *geochemical anomaly*. Geochemical anomalies are very important as they are the source of the metals and many of the industrial minerals that humans use.

5.4.1 Mineral classes

Minerals are grouped according to their anionic element or the nature of their anionic polyhedron. This is because minerals usually contain only one anion or polyhedral group but may contain many different cations. Linus Pauling (1901–94) developed five rules which determine the structure of crystalline materials at the atomic level (Pauling, 1929). Because anions are larger than cations, the anions of a crystalline material form into a polyhedron around each cation (Rule 1). These polyhedra form the building blocks of crystalline, inorganic materials. A polyhedron is a geometric three-dimensional form with flat faces and straight edges where the faces meet. Polyhedra can take many forms with varying numbers of faces. The simplest polyhedra are the tetrahedron (four triangular faces), the cube (six square faces) and the octahedron (eight triangular faces). The distance between the cations and anion in a polyhedron is equal to the sum of the ionic radii, and the number of anions in the polyhedron (the coordination number) is determined by the ratio of the radii – the larger the ratio, the larger the coordination number. An ionic structure is stable if the sum of the strengths of the electrostatic bonds that reach an anion equals the charge on that anion (Rule 2). This means that in order to be stable, the polyhedra in an ionic substance must be arranged in such a way as to preserve local electroneutrality. The sharing of edges and, particularly, faces by two polyhedra decreases the stability of an ionic structure (Rule 3). Silica tetrahedra, for example, never share faces in silicate compounds because this would not leave any free oxygen ions on the shared face to provide bonds for the rest of the structure. In crystalline substances which contain different cations, cations with high charge and small coordination number tend not to share parts of their polyhedron (Rule 4). The number of different kinds of structural arrangements within a crystal is always small (Rule 5). This is known as the rule of parsimony: it means that elements that are chemically similar usually have similar environments within crystal structures. Thus, mineral structures tend to be simple and contain at most only a few types of cation sites. In minerals of complex composition, therefore, a number of different ions may occupy the same structural position. This is consistent with Goldschmidt's explanation.

The major classes of minerals are native elements, sulfides, halides, oxides, hydroxides, carbonates, sulfates, phosphates, borates and silicates.[13]

Native elements are elements which occur naturally in elemental form, not as compounds. They include gold, silver, copper and platinum (all of which are metals), graphite and diamond (which are forms of carbon), and sulfur.

The *sulfides* contain sulfur as the major anion in combination with chalcophile cations. Common examples are pyrite (FeS_2), pyrrhotite (FeS), sphalerite (ZnS), galena (PbS) and chalcopyrite ($CuFeS_2$). Sulfides are predominantly covalent.

The *halides* contain the halogen elements as the dominant anion. They are ionically bonded and typically contain cations of alkali and alkaline Earth elements. Common examples are halite or rock salt (NaCl) and fluorite (CaF_2).

The *oxides* contain various cations and are ionically bonded. Common examples are hematite (Fe_2O_3), magnetite (Fe_3O_4), corundum (Al_2O_3), rutile (TiO_2) and cassiterite (SnO_2). Quartz (SiO_2) is considered with the silicates not with the oxides, since its structure is based on the SiO_4^{4-} tetrahedron.

The *hydroxides* contain the polyhedral group OH^- as the dominant anion. Examples include limonite, $FeO(OH).nH_2O$, and gibbsite, $Al(OH)_3$.

The *carbonates* contain the CO_3^{2-} polyhedron in which the C^{4+} ion is surrounded by three O^{2-} ions in a planar triangular arrangement. Common examples are calcite ($CaCO_3$) and dolomite ($MgCO_3$).

The *sulfates* contain SO_4^{2-} as the major polyhedron in which the S^{6+} ion is surrounded by four oxygen anions in a tetrahedron. Common examples are barite ($BaSO_4$) and gypsum ($CaSO_4.2H_2O$).

The *phosphates* contain the PO_4^{3-} polyhedral group which has the form of a tetrahedron. Common examples are apatite, $Ca_5(PO_4)_3OH$, and monazite, $(Ce,La,Y,Th)PO_4$.

The *borates* contain triangular BO_3^{3-} or tetrahedral BO_4^{5-} polyhedral groups. Both may occur in the

13 There are numerous internet sites which show photographs of minerals and give accounts of their properties and uses. The interested reader is referred to these for further information. A list of some of these sites is given at the end of the chapter.

same mineral. A common example is borax, $Na_2B^{III}B_2{}^{IV}O_5(OH)_4.8H_2O$.

The *silicates* are by far the largest group of minerals, both in terms of the number of minerals in the group and the quantity of silicate minerals relative to other minerals in the crust. Silicate minerals contain SiO_4^{4-} as the dominant polyhedral group in which the Si^{4+} cation is surrounded by four oxygen anions (O^{2-}) in the form of a regular tetrahedron, as described in Section 2.3.4. The silicates are divided into subgroups based on the degree of polymerisation of the SiO_4^{4-} tetrahedra (Figure 5.5). There are six subgroups of silicate minerals.

1 *Nesosilicates* (orthosilicates) contain isolated SiO_4^{4-} groups in which the oxygen ions of the tetrahedron are bound to one Si atom only, i.e. they are not polymerised. Examples are olivine, $(Mg,Fe)_2SiO_4$, and pyrope, $Mg_3Al_2SiO_{12}$.
2 *Sorosilicates* contain double silicate tetrahedra in which one of the oxygen atoms is shared with an adjacent tetrahedron so that the polyhedron has the formula $(Si_2O_7)^{6-}$. An example is epidote, $Ca_2(Fe^{3+},Al)Al_2(SiO_4)(Si_2O_7)O(OH)$.
3 *Cyclosilicates* contain six-membered rings of silicate tetrahedra with the formula $(Si_6O_{17})^{10-}$. Examples are tourmaline:
$(Na,Ca)(Al,Li,Mg)_3(Al,Fe,Mn)_6(Si_6O_{18})(BO_3)_3(OH)_4$
and beryl, $Al_2Be_2Si_6O_{18}$.
4 *Inosilicates* contain silica tetrahedra that are polymerised in one dimension to form single chains or double chains. In single-chain inosilicates, two of the four oxygen atoms are shared with adjacent tetrahedra. In the double-chain inosilicates only half the tetrahedra share two oxygen ions while the other half share three. This leads to silicon:oxygen ratios of 1:3 for single chains and 4:11 for double chains. The single-chain silicates include the pyroxene (e.g. diopside, $CaMgSi_2O_6$) and pyroxenoid minerals, and the double-chain silicates include amphibole minerals (e.g. tremoliote, $Ca_2Mg_5Si_8O_{22}(OH)_2$).
5 *Phyllosilicates* contain silica tetrahedra that are polymerised in two dimensions to form sheets with an overall silicon:oxygen ratio of 2:5. Common examples are the micas – muscovite:

$$KAl_2(AlSi_3O_{10})(F, OH)_2$$

and biotite:

$$K(Mg,Fe)_3(AlSi_3O_{10})(F,OH)_2)$$

and talc, $Mg_3Si_4O_{10}(OH)_2$, in which the cleavage reflects the sheet structure of the mineral.
6 *Tectosilicates* contain silica tetrahedra that are polymerised in three dimensions to form a framework in ways that neutralise the four negative charges. This is achieved by sharing the four oxygen ions with neighbouring tetrahedra (e.g. quartz and the feldspars), or by sharing one, two or three of the oxygen ions with neighbours, leaving the remaining negative charges to be neutralised by metal ions in appropriate quantities. The feldspars consist of three end-members – orthoclase ($KAlSi_3O_8$), albite ($NaAlSi_3O_8$) and anorthite ($CaAl_2Si_2O_8$). $KAlSi_3O_8$ and $NaAlSi_3O_8$ form a solid solution series known as the alkali feldspars, and $NaAlSi_3O_8$ and $CaAl_2Si_2O_8$ form a solid solution series known as the plagioclase feldspars. In the alkali feldspars every fourth Si^{4+} ion is replaced by Al^{3+} and in the plagioclase feldspars every second to every fourth Si^{4+} ion is replaced by Al^{3+}. This allows K^+, Na^+ and Ca^{2+} cations to occupy void spaces to maintain electrical neutrality.

Other groupings and classifications of minerals are possible. An important group is the *rock-forming minerals*, the most common and widely distributed of which are listed in Table 5.3. Mineralogically, the continental crust consists of approximately 60–65 wt% feldspar, 10 wt% quartz, 25 wt% iron and magnesium silicates, 0.5 wt% apatite and a few per cent of clays and carbonates (Ronov and Yaroshevsky, 1969). *Ore minerals* form another important group. These are minerals which are mined for the purpose of extracting one or more element, usually a metal, and are valuable because of the usefulness of the element. Some of the important ore minerals, and common applications of the metals produced from them, are listed in Table 5.4. Another economically important group is *industrial minerals*. These are mined for their intrinsic properties, not as fuels or sources of metals. They are often used in their natural state or after beneficiation as raw materials or as additives in a wide range of applications. A list of

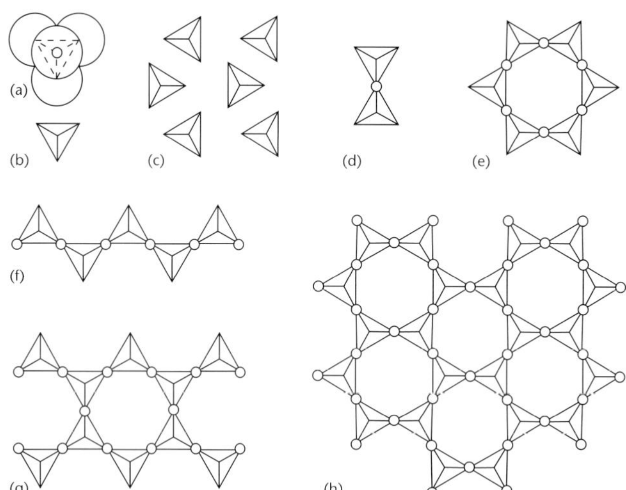

Figure 5.5: The major structural arrangements of the silica tetrahedron in silicate minerals (after Holmes, 1965). (a) The silica tetrahedron. (b) Conventional representation of the silica tetrahedron (in subsequent figures, shared oxygen ions are represented by open circles). (c) No shared oxygen ions (neosilicates). (d) A pair of tetrahedra sharing one oxygen ion (sorosilicates). (e) A ring of six tetrahedra, each sharing two oxygen ions (cyclosilicates). (f) A single chain of tetrahedra, each sharing two oxygen ions (inosilicates). (g) A double chain of tetrahedra (inosilicates). (h) A sheet of tetrahedra (phyllosilicates).

Table 5.3: The common rock-forming minerals

Primary mineral	Products of weathering
Feldspars:	
Orthoclase ..... Albite .... Anorthite	Clay minerals; e.g. $Al_2Si_2O_5(OH)$
$KAlSi_3O_8$ $NaAlSi_3O_8$ $CaAl_2SiO_8$	Bauxite (in tropics), $AlO(OH) + Al(OH)_3$
Quartz, SiO_2	Does not weather
Pyroxenes:	
Enstatite .... Hypersthene	Serpentine, $Mg_3SiO_5(OH)_4$
$MgSiO_3$ $(Mg,Fe)SiO_3$	
Micas and other sheet silicates:	
e.g., Muscovite .... Biotite	Chlorite, $(Mg,Fe,Al)_6(Al,Si)_4O_{10}(OH)_8$
$KAl_2(AlSi_3O_{10})(OH,F)_2$ $K(Mg,Fe)_3(AlSi_3O_{10})(OH,F)_2$	
Amphiboles:	
Hornblende, $NaAlCa_2(Mg,Fe)_4(Al_2Si_6)O_{22}(OH)_2$	Chlorite, $(Mg,Fe,Al)_6(Al,Si)_4O_{10}(OH)_8$
Olivines:	
Forsterite .... Fayalite	Serpentine (from forsterite), $Mg_3Si_2O_5(OH)_4$
Mg_2SiO_4 Fe_2SiO_4	
Calcite, $CaCO_3$	
Magnetite, Fe_3O_4	
Dolomite, $(Ca,Mg)CO_3$	
Chromite, $FeCr_2O_4$	
Hematite, Fe_2O_3	
Ilmenite, $FeTiO_3$	
Siderite, $FeCO_3$	Limonite, $FeO.H_2O$

Note dots between minerals indicates a solid solution series; the end member compositions are shown. (Mg,Fe) indicates that Mg and Fe substitute for one another in variable quantities.
Source: Holmes (1965).

important industrial minerals, and some of their applications, is given in Table 5.5. Some minerals have many uses and may be both an industrial mineral and an ore mineral; for example, ilmenite, chromite and magnesite.

5.4.2 Rock classes

Rocks belong to three broad classes according to their mode of formation: igneous, sedimentary and metamorphic.[14] *Igneous rocks* are formed by solidification and crystallisation of molten silicate material within the crust. *Sedimentary rocks* are formed from sediments created when rocks exposed at the Earth's surface are weathered and eroded by the action of water (and, to a lesser extent, wind and ice). Usually, the sediments are deposited in layers and may be buried and compressed. Chemical reactions may also occur, in which minerals are deposited from solution within the sediments. *Metamorphic rocks* are formed when existing rocks are physically or chemically altered by the action of heat or pressure, or both. Sedimentary rocks are formed at or near the Earth's surface, while metamorphic rocks are generally formed within the crust where temperatures and pressures are high. Igneous rocks can be formed either on the surface by lava flows or within the crust. The approximate abundance of the major rock types present in the crust is listed in Table 5.6. Many common rocks are mined or quarried in large quantities and used in their relatively natural state for building and construction purposes. Granite, sandstone and limestone, for example, are used in buildings as construction materials or, more commonly, as cladding for their visual attractiveness and resistance to weathering. Other rocks, particularly basalt, are crushed to produce aggregate for concrete, road base or ballast for railway tracks.

Igneous rocks

The nature of the igneous rocks that form from magma depends on the composition of the magma and the rate of cooling. Magma is of four broad compositions: *felsic* (acid), *intermediate*, *mafic* (basic) and *ultramafic* (ultrabasic) depending on the silica content. The major types of rock arising from these magmas are listed in Table 5.7. Extrusive igneous rocks form on the Earth's surface by lava flows and are fine-grained because the lava solidifies quickly and the mineral crystals do not have time to grow. Intrusive igneous rocks form below the Earth's surface and are coarser-grained because the magma, insulated by the surrounding crustal material, cools slowly, allowing the mineral crystals to become much larger. Large igneous intrusions are called *plutons*, or *batholiths* when many plutons converge. Smaller, tabular intrusions are called *dykes* when they intrude across layers and *sills* when they intrude between layers and run parallel to them.

Felsic magma is rich in sodium, aluminium and potassium and contains more than 65 wt% silica. Felsic rocks are light in colour because of the dominance of quartz and feldspars. Rhyolite and dacite form from continental lava flows and are fine-grained. Granite and granodiorite form during periods of mountain-building on the continents and are coarse-grained because they form below the Earth's surface. Granite is the most common rock in continental crust. Pegmatite is a very coarse-grained igneous rock with grain sizes of 20 mm or more; it occurs near the edges of igneous intrusions and often has a layered structure. Mafic magma is rich in calcium, iron and magnesium and contains 45–52 wt% silica. Some common mafic igneous rocks are basalt (fine-grained) and gabbro (coarse-grained). Mafic igneous rocks are dark in colour because they contain large quantities of pyroxene, amphiboles and olivine. Basalt is more common than gabbro; it occurs in the upper layer of the oceanic crust and in large continental lava flows. Gabbro occurs in the lower layer of oceanic crust and in small intrusions in the continental crust. Intermediate igneous rocks have compositions between mafic and felsic and are composed largely of plagioclase, amphibole and pyroxene. Examples are andesite and diorite. Andesite is a common fine-grained extrusive igneous rock formed from lavas from volcanoes along continental margins. Coarser-grained diorite occurs in intrusive igneous bodies in continental crust. Ultramafic magma contains less than 45% silica and ultramafic rocks are

14 There are numerous internet sites which show photographs of rocks and give accounts of their characteristics and uses. The interested reader is referred to these for further information. A list of some of these sites is given at the end of the chapter.

Table 5.4: Some important ore minerals and common uses of the metals

Metal	Oxide minerals	Sulfide minerals	Common bulk uses of metal
Aluminium	Bauxite, $AlO(OH) + Al(OH)_3$ Alunite, $K_2O.3Al_2O_3.4SO_4.6H_2O$		As a lightweight, corrosion-resistant, general-purpose structural material; high thermal and electrical conductivity applications
Cobalt		Cobaltite, $CoAsS$ Smaltite, $CoAs_2$	In superalloys for turbine blades (gas turbines and jet engines); in cobalt-chromium-molybdenum alloys for prosthetic parts; as an alloying agent in special steels; alloyed with aluminium, nickel and iron for permanent magnets; in batteries; as a catalyst for chemical reactions
Chromium	Chromite, $(Fe,Mg)(Cr,Al)_2O_4$		As an alloying element in steels, particularly stainless steels, to impart toughness and corrosion and thermal resistance
Copper	Cuprite, Cu_2O Azurite, $2CuCO_3.Cu(OH)_2$ Chrysocolla, $CuSiO_3.2H_2O$	Chalcopyrite, $CuFeS_2$ Chalcocite, Cu_2S Bornite, $Cu_2S.CuS.FeS$	As thermal and electrical conductors; the major component of brass (with zinc) and bronze (with tin)
Iron/steel	Hematite, Fe_2O_3 Magnetite, Fe_3O_4 Limonite, $FeO.H_2O$	Pyrite, FeS_2 Pyrrhotite, FeS	As a general-purpose structural material
Mercury		Cinnabar, HgS	Dental amalgam (with silver and small quantities of tin, copper and zinc); in drugs and chemicals
Magnesium	Dolomite, $(Ca,Mg)CO_3$		In magnesium alloys used for light weight and strength; desulfurising agent in iron-making
Manganese	Pyrolusite, MnO_2		As an alloying element in steels; as MnO_2 in dry-cell batteries
Molybdenum		Molybdenite, MoS_2	As an alloying element in steels; as MoS_2 as a solid lubricant and high-pressure, high-temperature anti-wear agent
Nickel	Garnierite, $(Mg,Ni)_6Si_4O_{10}(OH)_8$	Pentlandite, $(Fe,Ni)S$	As an alloying element in steels (especially stainless steels), nickel brasses and bronzes, and alloys with copper, chromium, aluminium, lead, cobalt; in batteries
Lead	Cerussite, $PbCO_3$	Galena, PbS	In lead-acid batteries; corrosion-resistant, pliable applications (in roofing and plumbing); as ballast material; in ammunition; previously used as tetraethyl lead in gasoline to reduce engine 'knocking'

Metal	Oxide minerals	Sulfide minerals	Common bulk uses of metal
Silicon	Quartz, SiO_2		In aluminium-silicon alloys, in silicone compounds (sealants and lubricants), as ferrosilicon used as a heavy medium in mineral separation processes and as a de-oxidising agent in steel-making
Tin	Cassiterite, SnO_2		Corrosion-resistant coatings on steel (tin plate); as a component of pewter and bronze alloys; with lead in solder
Titanium	Rutile, TiO_2 Ilmenite, $FeTiO_3$		In titanium lightweight alloys for structural applications (particularly aerospace and military) and for high-temperature and corrosion-resistant applications in petrochemical and chemical industries; as ferro-titanium as an alloying agent in steel
Uranium	Pitchblende, U_3O_8		As fuel in nuclear power plants; as depleted uranium in high-density penetrators in military applications and as a shielding material for storing and transporting radioactive materials
Vanadium	Carnotite, $K_2O.2UO_3.V_2O_5.3H_2O$	Patronite, V_2S_5	As ferrovanadium as an alloying element in steel for strength
Tungsten	Wolframite, $FeWO_4$ Scheelite, $CaWO_4$		In high-temperature applications such as light bulbs, cathode ray tube filaments, heating elements, rocket engine nozzles, welding; alloying element in special steels; as tungsten carbide in high-speed cutting tools for metals
Zinc		Sphalerite, ZnS	In corrosion-resistant applications, particularly as a coating on steel (galvanised); diecast products; in form of ZnO as white pigment, filler for plastic and rubber to confer UV protection, in sunscreens
Zirconium	Zircon, $ZrSiO_4$		In corrosion-resistant and/or high-temperature applications (high-performance pumps and valves); in nuclear reactors; as an alloying agent in steel for surgical equipment

Table 5.5: Some important industrial minerals and their uses

Mineral	Composition	Common uses
Apatite	$Ca_5(PO_4)_3(OH,F,Cl)$,	For production of fertiliser (superphosphate)
Asbestos	Hydroxy-silicates of Mg and Fe: chrysotile, crocidolite and amosite	Previously used for insulation, as fibre reinforcing in cement sheet for construction, in brake pads and disc linings in vehicles; now banned in most countries for health reasons
Barite	$BaSO_4$	In muds used in rotary drilling of oil and gas wells and in mining for lubricating and cooling; small quantities in electronics, CRTs, rubber production, glass ceramics, paint, radiation shielding and medical applications
Bentonite	Clay consisting mainly of montmorillonite, $(Na,Ca)_{0.33}(Al,Mg)_2(Si_4O_{10})(OH)_2 \cdot nH_2O$	A wide range of uses, including drilling muds; binder for foundry sands; ingredient of cement, adhesives, ceramics; as a binder in agglomeration processes; as an absorbent (in cat litter and wine-making, decolourising oils)
Chromite	$FeCr_2O_4$	As a constituent of refractories for lining furnace walls
Dolomite	$(Ca,Mg)CO_3$	As an ornamental stone and construction aggregate; as a source of magnesium oxide; as a flux in smelting of iron and steel; an ingredient in glass; soil additive
Fluorite or fluorspar	CaF_2	As flux in steel-making and aluminium smelting; ingredient in ceramics and enamels; for production of hydrofluoric acid; for making lenses for high-performance telescopes, cameras and microscopes.
Fuller's earth	A form of clay often containing montmorillonite, attapulgite and kaolinite	As a decolourising agent for oils; filler for drilling muds
Ilmenite	$FeTiO_3$	Most is used as a source of pure TiO_2 for use as a white pigment in paints, paper and other products; as sand-blasting agent
Kaolinite	$Al_2Si_2O_5(OH)_4$	Major use as coating and white pigment in paper; in paint to extend TiO_2; in rubber for strength; in toothpaste and cosmetics; in ceramics (porcelain ware)
Limestone	$CaCO_3$	Raw material for cement and glass manufacture; as dimension stone; for lime production; smelting flux; ingredient in manufacture of paper, plastics, paint, tiles (as white pigment and cheap filler); toothpaste and cosmetics
Magnesite	$MgCO_3$	As a constituent of refractories for lining furnace walls; ingredient in the production of synthetic rubber, magnesium chemicals, fertilisers
Mica	A complex phyllosilicate	As a filler in plasterboard, paints and plastics; in capacitors for radio frequency applications; as thermal insulation; soil conditioner
Perlite	An amorphous volcanic glass	In lightweight plasters, mortars and ceiling tiles; for building insulation; as a filter aid; soil additive
Pumice	Solidified frothy lava	As aggregate in lightweight concrete and low-density construction blocks; as an abrasive (especially in polishes, pencil erasers, cosmetic exfoliants and some soaps)

Mineral	Composition	Common uses
Quartz	SiO_2	As a flux in smelting; as a raw material for glass and cement manufacture; as sand for aggregate; as an abrasive
Rutile	TiO_2	As a source of pure TiO_2 for use as a white pigment in paints, paper, plastics and other products and in sunscreens; in manufacture of refractory ceramics
Sillimanite	Al_2SiO_5	As a constituent of refractories for lining furnace walls
Sulfur	S	In vulcanising rubber; as a soil additive; main use is as a precursor to other chemicals – most is converted to sulfuric acid (most of which is used to make fertiliser from phosphate rock); sulfur compounds are also used in manufacture of cellophane and rayon, detergents, fungicides, dyestuffs and agrichemicals; sulfites are used as bleaching agents
Talc	Hydrated magnesium silicate	As a filler in paper, plastics and paint; in cosmetics and pharmaceuticals; in ceramics
Vermiculite	A complex phyllosilicate; $(MgFe,Al)_3(Al,Si)_4O_{10}(OH)_2 \cdot 4H_2O$	In high-temperature insulation applications; fireproofing of structural steel and pipes; as loose-fill insulation; soil conditioner and ingredient of potting mixes and slow-release fertilisers; lightweight aggregate for plaster and concrete; as an additive to fireproof wallboard; soundproofing
Wollastonite	$CaSiO_3$	In ceramics; as a filler in paints, paper and vinyl tiles
Zeolites	Complex Al-containing tectosilicates	Largest use is in production of laundry detergents; as ion exchange materials in domestic and commercial water softening; as catalysts in the petrochemical industry; for removing and trapping fission products from nuclear waste

Table 5.6: The approximate abundance of the major rock types in the crust

Rock type	Abundance (vol.%)
Basalts, gabbros etc.	42.5
Gneisses	21.4
Granodiorite and quartz diorite	11.2
Granites	10.4
Schists	5.1
Clays and shales	4.2
Carbonates	2.0
Sandstone	1.7
Marbles	0.9
Ultramafics	0.2
Other	0.4
Total	*100.0*

Source: Ronov and Yaroshevsky (1969).

dominated by olivine, plagioclase and pyroxene. Peridotite is the most common ultramafic rock.

Porphyry is an igneous rock consisting of large-grained crystals of feldspar or quartz within a fine-grained matrix of feldspars. Porphyry deposits are formed when magma cools in two stages, initially deep in the crust, which allows large crystals to grow, then more quickly at or near the surface, which results in fine crystals. The cooling leads to a separation of dissolved metals into distinct zones and local enrichment of metals such as gold, copper, molybdenum, lead, tin, zinc and tungsten.

Sedimentary rocks

Sedimentary rocks are classified according to the source of sediment from which they form. Clastic sedimentary rocks are composed of fragments of material derived from other rocks which have been transported then deposited. They are converted into rock by a combination of compaction and cementing by physical and chemical processes. During transportation, particles often become rounded due to abrasion and can become highly sorted into size fractions. Clastic sedimentary rocks are classified according to particle type and size; the common groups are listed in Table 5.8. Non-clastic sedimentary rocks form by chemical precipitation from water and crystallisation or, less commonly, by compaction and cementing of the skeletons of dead organisms. The major non-clastic sedimentary rocks are listed in Table 5.9. Calcium, sodium, potassium and magnesium are commonly released into the environment by chemical processes. They can then dissolve in water and be transported. If this solution enters a basin environment where the rate of evaporation exceeds the rate of precipitation and in-flow, evaporite minerals can form from the loss of water from the solution. The oceans are nearly saturated with dissolved calcium carbonate from the shells of marine organisms; under appropriate conditions, the dissolved calcium carbonate can precipitate to form limestone. Dolomite may form subsequently by the action of magnesium-rich solutions on the limestone.

Metamorphic rocks

Metamorphism is the alteration of existing rocks by heat and pressure which cause chemical or physical changes to the minerals composing the rock. Physical changes may involve the reorganisation of minerals into layers or the segregation of minerals into specific areas within the rock. At temperatures above about 200°C, and below their melting point (usually greater than 600°C), minerals in rocks can change form and react with other minerals. In this way new minerals form and the rocks are transformed. Rocks can be subjected to heat in two main ways. Rocks at the

Table 5.7: The common igneous rocks

	Felsic (or acid) (>65 wt% SiO_2)	Intermediate	Mafic (or basic) (45–52 wt% SiO_2)	Ultramafic (or ultrabasic) <45 wt% SiO_2
Intrusive (coarse-grained)	Granite, granodiorite, pegmatite	Diorite	Gabbro	Peridotite
Extrusive (fine-grained)	Rhyolite, dacite	Andesite	Basalt	Komatiite

Table 5.8: The common clastic sedimentary rocks

	Particle type	Typical size of particles (mm)
Limestone	Broken shells, coral and other marine skeletons	
Mudstone	Clay and silt	<0.002
Shale	Clay	~0.002
Siltstone	Silt	0.002–0.063
Sandstone	Sand consisting predominantly of quartz	0.063–2.00
Conglomerate	Rounded gravel	2.00–263
Breccia	Angular gravel	2.00–263

surface can be carried into the crust at subduction zones and heated as they move downwards. Alternatively, magma can migrate up and through the crust, particularly at continental boundaries where subduction is taking place, and heat the rocks in contact with it. Rocks that are buried are subjected to pressure because of the mass of overlying material. Pressure can also be exerted on rocks by the forces involved in tectonic processes. The most obvious effect of pressure on rocks is the reorientation of mineral grains (foliation). Pressure never acts in isolation – the temperature in the crust increases with depth, as does the mass of overlying material. Water present in small quantities in rocks, either weakly bound chemically or as free water in pores, can enhance chemical reactions during metamorphism by providing a medium in which ions from minerals can move and react with other ions.

The most common metamorphic rocks are slate, schist, gneiss, marble and quartzite. Slate is a fine-grained foliated rock formed by mild metamorphism of shale or mudstone which, because of the foliation, cleaves easily into thin plates. Schist, which is a medium- to coarse-grained foliated rock, results from more intense metamorphism. Gneiss is a coarse-grained metamorphised igneous rock which exhibits both recrystallisation and foliation of quartz, feldspars, micas and amphiboles into alternating light- and dark-coloured bands. Marble is non-foliated metamorphised limestone or dolomite, and quartzite is recrystallised quartz formed from sandstone.

Table 5.9: The common non-clastic sedimentary rocks

Halite	$NaCl$
Gypsum	$CaSO_4$
Silcretes	SiO_2
Limestone	$CaCO_3$
Dolomite	$(Ca,Mg)CO_3$

5.4.3 The rock cycle

The rock cycle, which shows how the formation of all rock types is interlinked, is illustrated in Figure 5.6. All rocks can ultimately be traced back to *magma*, which is a partially melted mixture of compounds and elements with overall composition similar to that of igneous rocks. Magma forms where the local temperature exceeds the melting point of the rock. This can occur, for example, at subduction zones, continental rift zones and mid-ocean ridges, and where there are local concentrations of radioactive elements. All types of rocks (igneous, sedimentary and metamorphic) can be degraded by weathering processes. In *physical weathering*, rocks are broken down into particles, mainly by the action of water. Water in cracks may freeze; the force exerted as a result of the increase in volume can be sufficient to break off pieces of rock. Heating and cooling of rock, and the resulting expansion and contraction, also contribute to breakage, particularly in hot, dry regions. Physical weathering contributes to further breakdown of rock by increasing the surface area exposed to chemical agents. In *chemical weathering*, rocks are broken down by reaction of the minerals with air and water. Rock particles in the form of clay, silt, sand and gravel formed by weathering may be transported by water and, less frequently, by ice and wind to new locations and re-deposited, often in layers and usually at a lower elevation. This process is called *erosion*. Sediments are deposited from streams and rivers when the flow rate falls; for example, in alluvial fans, on floodplains, in river deltas and on the bottom of lakes and the seafloor. Wind may also move large quantities of sand and other smaller particles. Glaciers may transport and deposit great quantities of unsorted rock material. Particles deposited from water, wind and ice may eventually become compacted and cemented together to form sedimentary rocks. These rocks contain mainly minerals which are resistant to further mechanical and chemical breakdown, such as quartz, zircon, rutile and magnetite. Quartz is one of the most physically and chemically resistant minerals. Geological processes resulting from the movement of tectonic plates, such as folding, faulting and deep burial, can act on igneous and sedimentary rocks to form metamorphic rocks. Finally, rocks of all types can be returned to the mantle at subduction zones. These processes, and the processes discussed subsequently which form mineral deposits, have occurred throughout geological time. They exist today and will continue into the future. They are, of course, immensely slow by human time scales – apart from catastrophic events such as earthquakes, volcanic eruptions and soil erosion, the Earth appears to us to be unchanging. This is far from being the case.

5.5 MINERAL DEPOSITS

Oxygen, silicon, aluminium, iron, calcium, magnesium, sodium, potassium and titanium make up

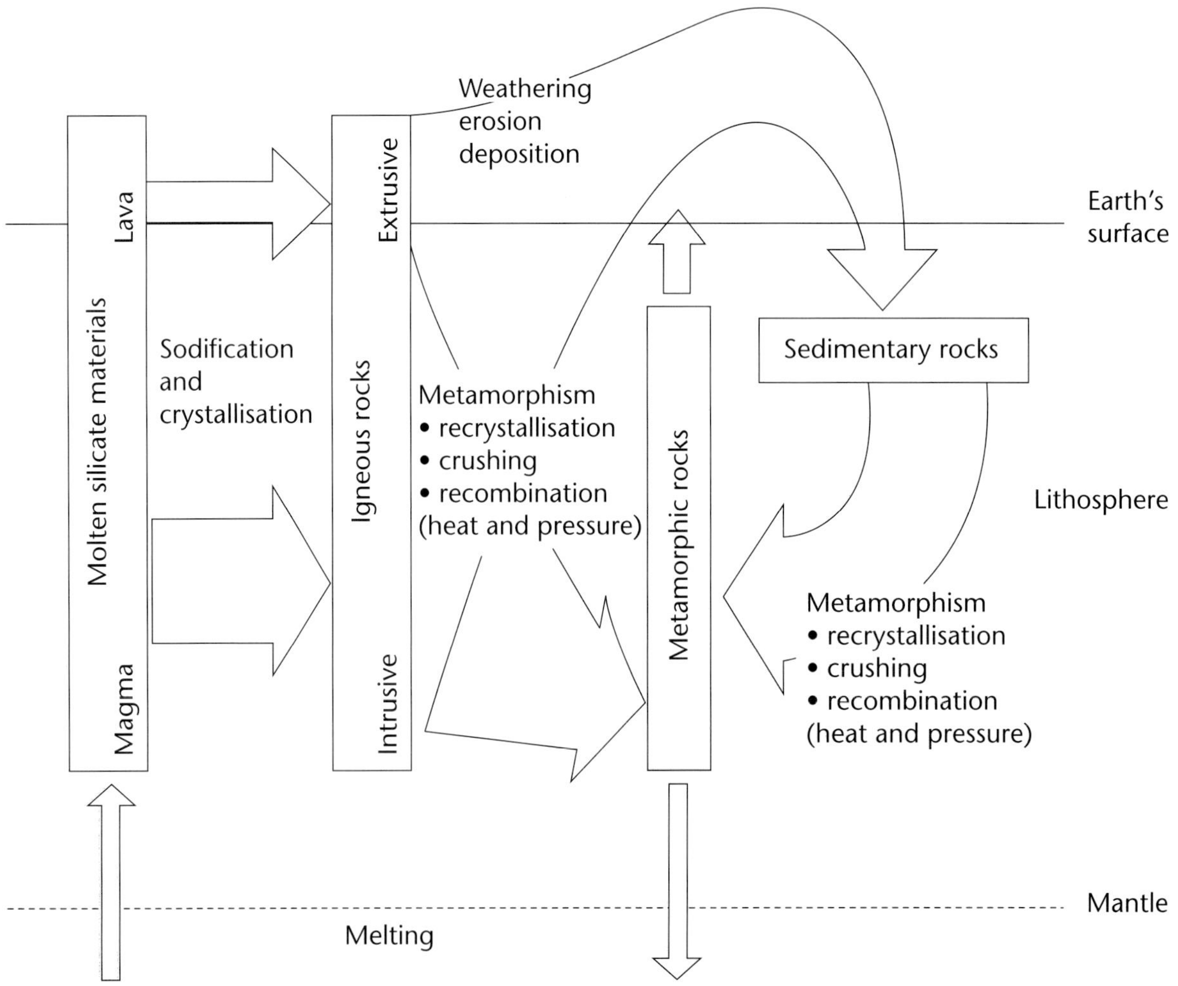

Figure 5.6: The rock cycle.

99.4 wt% of the crust (Table 5.2). The next two most abundant elements are phosphorus and manganese. These 11 elements alone make up nearly 99.6 wt% of the crust and are the *geochemically abundant elements*. Of these, silicon, iron, aluminium, magnesium, titanium, phosphorus and manganese are used in large quantities either in elemental form or as compounds. All the other elements, including many commercially and industrially important elements such as copper, nickel, lead, zinc, lead, chromium, silver, gold and platinum, together make up less than 0.5 wt% of the Earth's crust. These are the *geochemically scarce elements*. Fortunately, many of these occur in localised concentrations within the crust as *mineral deposits* or *geochemical anomalies*. If the elements of interest are concentrated enough that they can be mined profitably, we refer to the deposit as an ore body.

5.5.1 Formation of mineral deposits

Many of the geochemically scarce elements occur in substitution within the crystal structure of major rock-forming minerals for reasons discussed previously. However, the less reactive elements such as copper, nickel, lead, zinc, arsenic, bismuth, antimony, gold, silver and platinum, which do not have a strong affinity for other elements, have only a limited ability to substitute in rock-forming minerals (Candela, 2007). During igneous processes, these tend to concentrate into liquid phases. In intrusive igneous processes these liquids permeate into host rocks through cracks and pores, and fracture rocks under the pressure of the liquids. As the liquid cools, the less reactive elements crystallise to form their own minerals, usually oxides and sulfides, within *veins* (*lodes*) in the host rocks. Minerals that form in this manner are referred to as vein minerals. Veins are an important source of many metals. In extrusive igneous processes, the fluids are ejected at the surface and spread out, precipitating minerals as they cool.

Mineral deposits occur in three main types, according to their mode of formation (Polya, 1992). Some form as precipitates from liquids, as described above. The liquids can be aqueous solutions, in which case the minerals form by hydrothermal processes, or silicate melts, in which the minerals form by magmatic processes. These are the most common type of deposit for gold, copper, nickel, lead and zinc. They are called *primary deposits*. Other deposits form *in situ* by selective removal of the more soluble geochemically abundant non-ore elements (particularly silicon, calcium, magnesium, sodium, potassium and iron) from bedrock through weathering by water percolating through bedrock, particularly in hot, wet tropical areas. The remaining elements thus become more concentrated in the residual material. A third type of deposit forms from clastic sediments at or near the Earth's surface. This involves the physical separation of chemically unreactive, usually high-density minerals from low-density rock-forming minerals through a combination of weathering of host rock and transportation and deposition of sediments by water. These are referred to as *placer deposits*. Mineral deposits formed through the weathering of primary rocks are called *secondary deposits*.

Many primary minerals, particularly sulfides, are stable only in dry, oxygen-free environments. The region of a mineral deposit which is above the watertable is generally accessible to atmospheric oxygen. With the rise and fall of the watertable and downward-percolating rainwater (containing dissolved oxygen), some minerals dissolve and new minerals are precipitated. In the dissolution of sulfide minerals, the water becomes acidic, thereby enhancing the dissolution of other minerals. This process results in the formation of a gossan, which consists primarily of iron oxides and hydroxides and quartz. When a gossan is well developed and the secondary minerals are sufficiently concentrated, it might be profitable to mine it because it is near the surface (and often outcrops at the surface) and is rich in valuable minerals, for example, malachite, azurite and chrysocolla (in copper deposits), goethite and hematite (in iron deposits), anglesite and cerussite (in lead deposits), pyrolusite and rhodochrosite (in manganese deposits), garnierite (in nickel deposits) and smithsonite (in zinc deposits). There is often a narrow zone just below the watertable called the supergene zone, in which metals have been precipitated from liquids percolating downwards from the oxidised zone above. This is often the richest part of a mineral deposit, though frequently a relatively small part of it. Common minerals found in supergene zones are chalcocite and bornite (copper deposits), galena (lead deposits), violarite (sulfide nickel deposits), sphalerite and wurtzite (zinc deposits).

5.5.2 Common forms of mineral deposits

Table 5.10 summarises the main ways in which mineral deposits occur. Clearly, some minerals and metals can occur in deposits formed by a variety of geological processes. Iron and aluminium are two of the most abundant elements and they are also the two metals most used in modern society. The largest and richest iron deposits are Precambrian banded iron formations, known as BIF, or taconites in the United States, which consist of alternating silica- and iron-rich layers. Important examples include those of the Pilbara region of Western Australia, Lake Superior in the United States and Canada, and Carajas in Brazil. These deposits were formed by the weathering of igneous rocks, the dissolution of iron, the transport of the solution containing dissolved iron and its precipitation as magnetite onto the surface of ocean basins

Table 5.10: Classification of ore and mineral deposits according to their common modes of occurrence

1 Deposits associated with igneous intrusions	
a) Primary deposits	b) Secondary deposits formed by *in situ* weathering
Chromium	Aluminium (as bauxite)
Manganese	Nickel (as laterite)
Copper, nickel, lead, zinc, tin	
Precious metals (gold, silver, platinum group metals)	
Uranium	
Ilmenite, rutile	
Feldspar (pegmatite deposits)	
Quartz (pegmatite deposits)	
Rare earth minerals (pegmatite deposits)	
Beryllium minerals (pegmatite deposits)	
Lithium minerals (pegmatite deposits)	
2 Deposits associated with sedimentary rocks (concentration by weathering and/or precipitation from solution)	
Iron	
Manganese	
Uranium	
Copper	
3 Placer deposits	
Monazite	
Zircon	
Ilmenite, rutile	
Zirconia	
Tin (as cassiterite)	
Gold	
Quartz sand	
4 Evaporitic and chemical deposits	
Phosphorus (as phosphate rock)	
Barium (as barite)	
Strontium (as celestite)	

under appropriate conditions. These deposits were buried under sediment and eventually became sedimentary rock. Later weathering of some deposits converted magnetite to hematite. In some cases, as in the Pilbara, placer deposits were formed through weathering, erosion and concentration of hematite. Bauxite, the main ore for making aluminium, forms from the *in situ* weathering of aluminium-rich igneous rocks through processes that selectively remove silica and iron. Important deposits occur around the northern and western Australian coastline, China, Brazil, Guinea and Jamaica.

Copper, nickel, lead, zinc and platinum are chalcophile elements and often occur together as sulfide minerals, sometimes with gold and silver. Primary copper occurs predominantly in porphyry deposits in which copper minerals, mainly chalcopyrite, and other minerals are concentrated in a two-stage crystallisation process. The deposits of South America (e.g. at Chuquicamata and Escondida) are important examples. Sedimentary-hosted copper deposits, such as those of the Central African copper belt, were formed by hydrothermal processes. Aqueous solutions in buried sediments discharged onto the surface of ocean basins and precipitated copper, and often lead and zinc, sulfides. These were then buried under sediment and eventually formed sedimentary rocks. Copper occurs in these deposits in more oxidised form, particularly as chalcocite. Primary nickel occurs in mafic and ultramafic igneous rocks derived from mantle magmas which had sufficient sulfur, nickel and iron to form sulfide minerals such as pentlandite, pyrite, chalcopyrite and pyrrhotite during crystallisation. The large nickel deposit at Voisey's Bay in Canada was formed this way. Secondary nickel occurs in laterite deposits, which make up about 80% of known nickel resources. These were formed *in situ* by the weathering of ultramafic or mafic rocks. Lead and zinc usually occur with silver. Most are hydrothermal deposits formed in a similar way as sedimentary copper deposits, or by hydrothermal replacement of limestone in so-called skarn deposits and Mississippi Valley-type deposits. Platinum and palladium are generally found in ultramafic igneous rocks which had sufficient sulfur to form sulfide minerals during crystallisation. The Bushveld Igneous Complex in South Africa is a large layered igneous intrusion which has eroded and now outcrops around what appears to be the edge of a large basin. The complex contains some of the richest ore deposits on Earth. Other important deposits occur in Ontario, Canada and Russia.

Most primary gold occurs in hydrothermal lode deposits which consist primarily of quartz veins or reefs; for example, the famous Golden Mile at Kalgoorlie in Western Australia. Placer gold deposits form from weathering and erosion of gold-bearing lodes. The gold may be deposited on the bed of streams and rivers where the flow rate falls. Some gold is recovered from streams and rivers but most is present in old sedimentary rocks formed from ancient placer deposits. The latter are known as leads or deep leads; an important example is the Witwatersrand deposits in South Africa. Secondary gold also occurs in iron oxides in laterite deposits formed from the weathering of pre-existing gold deposits, including placer deposits.

Titanium and zirconium occur most commonly as mineral sand deposits consisting of the heavy minerals rutile, ilmenite, leucoxene and zircon. Monazite often occurs in mineral sand deposits. These minerals are relatively unreactive. They were originally present in granites which were weathered to release the minerals. The mineral sand deposits form by accumulation of these dense minerals in placer deposits in beach systems; for example, along the coastline of parts of south-west Australia and the south-east coast of Africa.

5.5.3 The distribution of base and precious metal deposits

Based on a global survey of known gold, silver, copper, zinc and lead deposits, several observations concerning the distribution and size of base and precious metal deposits can be made (Singer, 1995).

- *Base and precious metals tend to be concentrated geographically and geologically.* Over 50% of the reserves of gold, silver, copper, zinc and lead occur in just four countries. These are not the same countries for each metal, but the United States and Canada are in the top four for each metal. Australia is one of the top four countries for gold, silver and lead. China is one of the top four countries for zinc and lead. This is partly explained by the large size of some countries but is also due to the geological environment, particularly the action of plate tectonics. Each metal occurs predominantly in a small number of types of deposit. For example, about 60% of gold occurs in placer deposits (modern or ancient) and 19% occurs in igneous deposits (porphyries, skarns, epithermals, massive sulfides). At least 65% of copper occurs in igneous deposits and about 23% in sediment-hosted deposits.

- *Base and precious metals tend to be concentrated into higher-grade deposits.* Over 74% of gold, silver, zinc and lead occurs in deposits having grades above the median grade of all deposits and 44% of copper occurs in deposits with grades above the median grade of all copper deposits. Thus, the common belief that the total quantity of metal contained in lower-grade deposits is greater than the total quantity of metal contained in high-grade deposits may not be correct.
- *Base and precious metals tend to be concentrated into large deposits.* The quantity of metal contained in deposits smaller than the median size of mineral deposits is quite small. The upper 10% of mineral deposits, ranked according to size, account for over 70% of the total metal. The upper 1%, the supergiants, account for over 25% of total metal.

5.6 RESOURCES AND RESERVES

In describing the quantity and quality of mineral deposits from a utilitarian perspective, rather than a geological perspective, we distinguish between resources and reserves. A *mineral resource* is a concentration of minerals of intrinsic economic interest in or on the Earth's crust in such form, quality and quantity that there are reasonable prospects for eventual economic extraction (JORC, 2004). An *ore reserve* (*ore body*) is the economically mineable part of a mineral resource. Thus, for a deposit to be called an ore body, it must be able to be profitably mined and processed with current technology and at current market prices for the products. Mineral resources and ore reserves are characterised by location (geography) and depth, the nature of the composing rocks and minerals (their geology and mineralogy) and by the quantity (mass) and concentration (grade) of economically valuable element or elements.

The *grade* of a mineral deposit is the average concentration of the valuable element or compound in the deposit. It is expressed in appropriate units according to the relative quantity, such as mass (or weight) per cent or gram per tonne (g t^{-1}). The cut-off grade is the minimum grade above which it is economical to extract an element or compound. Grade is the most important, but not the only, factor determining the cost to produce a marketable product from a mineral deposit because grade determines the quantity of rock that must be mined and processed to recover a specific quantity of valuable mineral or element. For example, 200 tonnes of ore must be mined and processed to produce 1 kg of gold from a typical gold ore containing 5 g t^{-1} of gold, whereas only 0.1 tonne ore needs to be mined and processed to produce the same quantity of copper from a typical copper ore containing 1 wt% of copper. It is possible to mine gold profitably at 5 g t^{-1} and copper at 1 wt% because the market price of gold is over three orders of magnitude greater than the price of copper.

As illustrated in Figure 5.7, mineral resources can be divided into inferred, indicated and measured (JORC, 2004). An *inferred mineral resource* is that part of a resource for which tonnage, grade and mineral content can be estimated with only a low level of confidence, inferred from geological evidence. An *indicated mineral resource* is that part of a resource for which tonnage, densities, shape, physical characteristics, grade and mineral content can be estimated with a reasonable level of confidence. It is based on geological exploration, sampling of the deposit and testing of samples. A *measured mineral resource* is that part of a resource for which tonnage, densities, shape, physical characteristics, grade and mineral content can be estimated with a high level of confidence. It is based on detailed and reliable geological exploration, sampling of the deposit and testing of samples. The *reserve base* is the sum of the measured and indicated resources

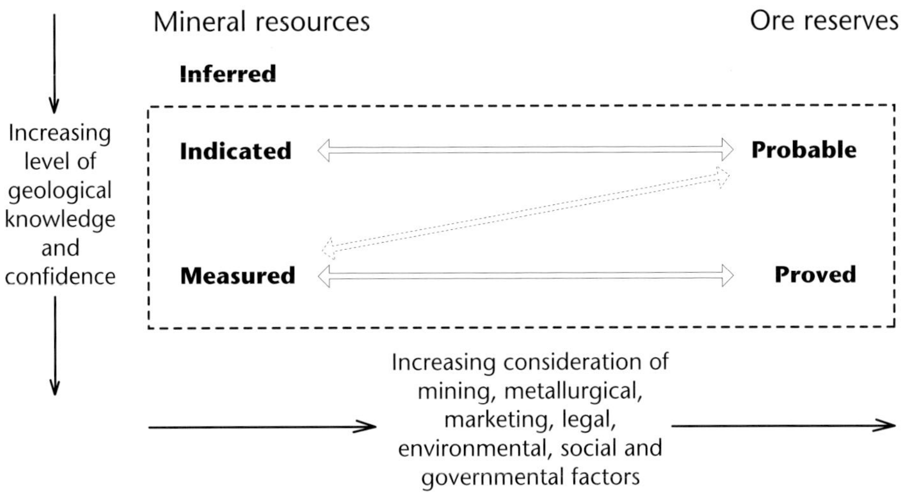

Figure 5.7: The general relationship between mineral resources and ore resources (JORC, 2004; with modifications).

and is the resource from which reserves are estimated. It includes resources that are currently economic and marginally economic, and some of those that are currently subeconomic. Ore reserves can be subdivided into probable and proved. A probable ore reserve is the economically mineable part of an indicated mineral resource. A proved mineral resource is the economically mineable part of a measured mineral resource.

Mineral deposits vary greatly in size. The largest copper deposits have total reserves and resources of 10^9–10^{10} tonnes of ore. Bauxite and iron ore deposits with total reserves and resources of over 10^8 tonnes of ore are common. At the other end of the scale, there are innumerable silver- and gold-bearing veins which contain only 10–100 tonnes of ore. The concentrations of the valuable element in ore bodies also varies greatly, from about 0.0001 wt% to greater than 50 wt%. To avoid the need to use small numbers like 0.0001 wt%, the units parts per million (ppm) and grams per tonne (g t^{-1}) are usually used for precious metal grades. Note: 0.0001 wt% is equal to 1 ppm or 1 g t^{-1}. Many gold deposits are mined profitably at 1–5 ppm Au, copper is commonly mined at 0.5–1.5 wt% Cu and aluminium deposits at grades above about 30 wt% Al. Many iron ores contain over 60 wt% Fe.

The feature that is common to all ore bodies is the considerable enrichment of at least one element above its average concentration in the crust. The ratio of the concentration of the element in the deposit to its average concentration in the crust is called the *enrichment factor* (or Clarke value). For many of the geochemically scarce elements, enrichment factors of 10^2–10^4 are common. The enrichment factors for important geochemically abundant elements, such as silicon, iron, aluminium, magnesium, titanium and manganese, are very much smaller. Typical values of minimum ore grades and enrichment factors for some common ore elements are given in Table 5.11.

In practice, for any particular ore element, there is a range of minimum concentrations at which deposits can be mined economically. This is because ore grade is not the only factor that influences production costs. The size of the deposit is also important since the cost of mining, per tonne of ore, usually decreases as the mine size increases. A large deposit can justify a larger scale of mining (often open cut) and hence a lower mining cost. The location of the deposit is also important. A deposit in a difficult area (mountainous, very dry or very wet, remote from infrastructure, politically unstable etc.) or far from the market for its products has higher cost and might require a higher grade to be mined profitably.

The minimum grade at which it is profitable to mine a particular element (the *cut-off grade*), has varied throughout history. There is no upper concentration

Table 5.11: Typical ore grades and enrichment factors for important metals

	Typical minimum ore grade		Crustal abundance	
	(ppm)	(wt%)	(ppm)	Enrichment factor
Geochemically abundant elements				
Aluminium	320 000	32	80 000	4
Iron	250 000	25	50 000	5
Magnesium	150 000	15	25 000	6
Titanium	100 000	10	5000	20
Manganese	300 000	30	1500	200
Geochemically scarce elements				
Nickel	10 000	1	100	100
Copper	5000	0.5	50	100
Zinc	40 000	4	80	500
Platinum	4	0.0004	0.004	1000
Silver	100	0.01	0.065	1500
Gold	4	0.0004	0.002	2000
Chromium	250 000	25	100	2500
Lead	40 000	4	10	4000
Tin	50 000	5	2	25 000

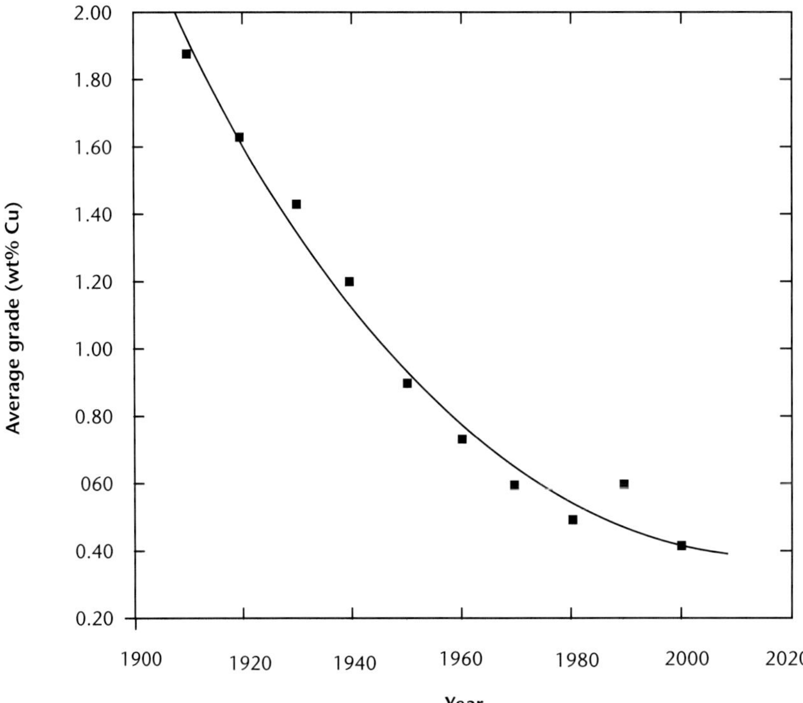

Figure 5.8: The trend in the average grade of ore mined for copper in the United States during the 20th century. Data for 1919–90 from Hiskey (1995) and for 2000 from the United States Geological Survey.

Source: Data from Polya (1992) and various public domain sources.

limit at which an element can be profitably mined and, of course, the richest deposits discovered were always the first to be exploited. As richer deposits become depleted it becomes necessary to utilise progressively lower-grade deposits. This is illustrated in Figure 5.8, which shows the trend in the average grade of copper ores mined in the United States since 1900. In earlier centuries, the cut-off grades were even higher; for example, in the 17th century most copper ores had grades in excess of 10-15 wt%. The trend in Figure 5.8 is typical of the other geochemically scarce elements in common use (nickel, lead, zinc and tin) and of the precious metals. However, the grades of deposits mined for the geochemically abundant elements used in large quantities (iron, aluminium, titanium, manganese and magnesium) have not decreased to the same extent, if at all, because they occur in such vast quantities that human consumption has not yet depleted the richest sources of these elements.

5.7 EXTRACTING VALUE FROM THE CRUST

The approaches adopted for extracting value from mineral deposits are summarised in Figure 5.9. Some deposits of valuable minerals or rocks occur in relatively large bodies and in fairly pure form; they are not intermingled with other unwanted minerals or rocks. In these cases, usually the entire mineral assemblage has value. For example, large deposits of granite, quartz, basalt, limestone, marble, sand and clay for building, construction and other purposes are relatively common. These can be quarried and the material obtained used with relatively little further preparation beyond, perhaps, washing, shaping or crushing and sizing (Figure 5.9a). Most ore minerals and industrial minerals occur at low concentrations and recovering the valuable mineral or its constituent valuable element(s) poses a problem. Miners would like to do the least amount of work (use the least amount of energy) and disturb the land around the deposit as little as possible (do the least environmental harm). Ideally, we would like to pick the valuable minerals out of the host rock, which would remain undisturbed. However, the only way known to do this is by *in situ* leaching, in which a solvent is circulated through an ore body to selectively dissolve the valuable element. The element is then recovered from the solution. This method, which for technical reasons has quite limited and specific application, is discussed further in Section 7.1.3. Generally, all the rock containing the valuable mineral has to be mined (broken

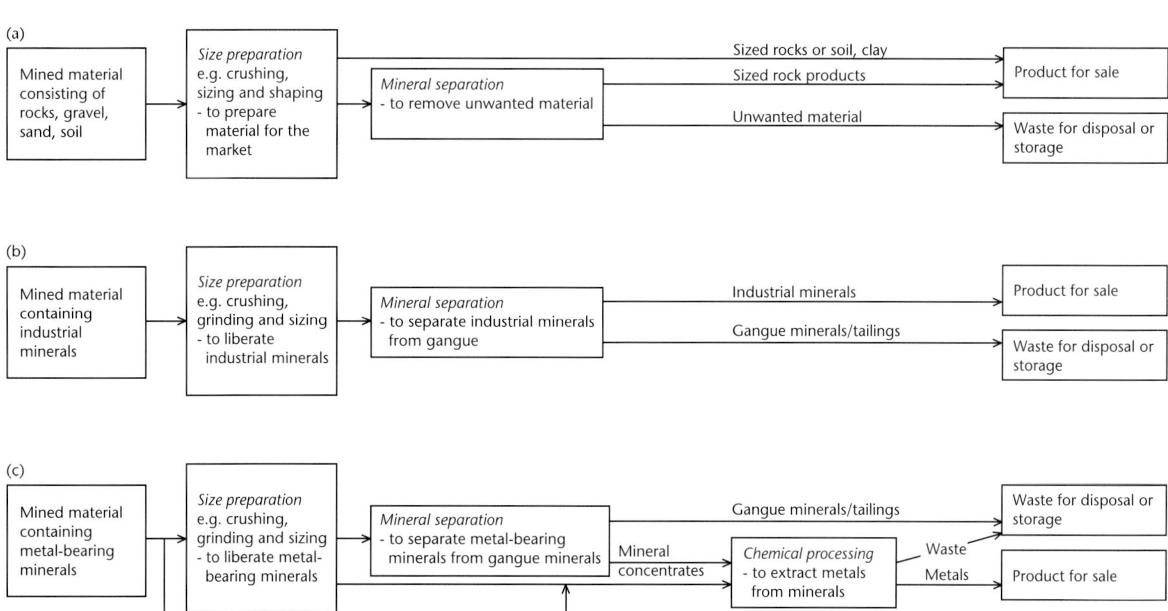

Figure 5.9: The broad approaches to the processing of mined mineral material.

Geometric Classification of Basic Intergrowth Patterns of Minerals

Type 1a: Simple intergrowth or locking type; rectilinear or gently curved boundaries. Most common type, many examples.

Type 1b: Mottled, spotty, or amoeba-type locking or intergrowth. Simple, common pattern; many examples.

Type 1c: Graphic, myrmekitic, or "eutectic" type. Common; examples: chalcopyrite and stannite; quartz and feldspars; etc.

Type 1d: Disseminated, emulsion-like, drop-like, buckshot or peppered type. Common; examples: chalcopyrite in sphalerite or stannite; sericite, etc. in feldspars; tetrahedrite in galena; etc.

Type 2a: Coated, mantled, enveloped, corona-, rim-, ring-, shell-, or atoll-like. Common; examples: chalcocite or covellite around pyrite, sphalerite, galena; etc.; kelyphite rim, and other rims.

Type 2b: Concentric-spherulitic, or multiple shell-type. Fairly common; examples: uraninite with galena, chalcopyrite, bornite; cerussite-limonite; Mn- and Fe-oxides; etc.

Type 3a: Vein-like, stringer-like, or sandwich-type. Common; examples: molybdenite-pyrite; silicates; carbonates; phosphates; etc.

Type 3b: Lamellae-layered, or polysynthetic type. Less common; examples: pyrrhotite-pentlandite; chlorite-clays; etc.

Type 3c: Network, boxwork, or Widmanstätter-type. Less common; examples: hematite-ilmenite-magnetite; bornite or cubanite in chalcopyrite; millerite-linneite; metals; etc.

Between most of these nine common locking types there are naturally gradational transitions with regard to both, pattern and size. Particle or grain size data are a prerequisite of any accurate study of rocks and mineral deposits and enhance the value of this chart.

Figure 5.10: Common mineral intergrowth patterns (Amstutz, 1962). Source: *Mining World*; reproduced courtesy E&MJ.

into pieces small enough to be physically brought up to the Earth's surface) then processed further. The rocks are composites consisting of different minerals strongly locked together, often in complex geometrical arrangements. The sizes and shapes of the constituent minerals and their distribution and intergrowth constitute the *texture* of a rock. Amstutz (1962) identified nine common rock textures (Figure 5.10). These patterns give an idea of the complexity of intergrowths in mineral deposits and illustrate the problems of extracting the desired mineral or element. There is no length scale in Figure 5.10 because the figures represent geometrical arrangements which can occur on many scales under different circumstances, from micrometre scale to millimetre or larger scale. Several approaches are possible for extracting the valuable components; all involve the use of energy to act selectively on the desired minerals or on the unwanted minerals to achieve a separation of the minerals or the extraction of the valuable element from the minerals (Figure 5.9b and c). With present technology, there are two ways of applying energy for this purpose: physically and chemically.

5.7.1 Physical separation

The most common approach is to apply physical force to progressively break the rock into smaller and smaller pieces until the minerals making up the rock are individually discrete, or *liberated*. The mineral mixture can then be separated into fractions – a *concentrate* which contains the valuable mineral(s), *tailings* which contain unwanted minerals, and *middlings* which contain rock particles in which the valuable and unwanted minerals were not completely liberated. The process is illustrated in Figure 5.11. There are many techniques for breaking rocks and grinding particles on large scale and for carrying out separation of liberated minerals based on differences in their physical properties. These are discussed in Chapter 7. A concentrate is often an intermediate product. It may be treated in a subsequent chemical process to extract the valuable element(s), in elemental or compound form (Figure 5.9c). This is most commonly done by leaching or smelting, as discussed in Chapter 8.

There is a complication with this approach. Liberation of a mineral is never perfect because of the way rocks break. Rocks fracture along planes of greatest weakness, which are not usually the surfaces of the individual grains making up the rock. This can be illustrated using a simple three-dimensional model

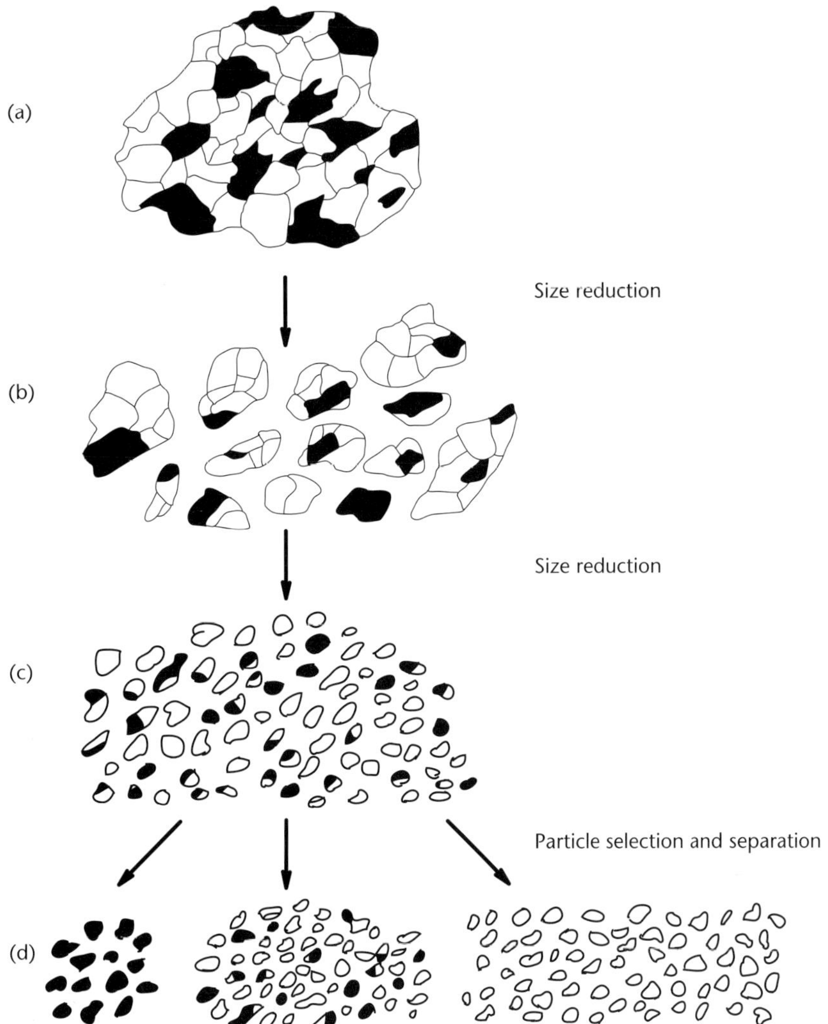

Figure 5.11: Size reduction and particle selection and separation for a simple rock consisting of two minerals (after Hayes, 1993).

(Gaudin, 1939). Suppose Figure 5.12a represents a section cut through an ore-bearing rock. For simplicity, assume that the ore mineral (dark) and the unwanted mineral (light) both occur as cubic particles of side 1 mm. If the rock could be broken into discrete dark and light particles, by breaking it along surfaces separating the particles, a reduction in particle size to 1 mm would be all that is necessary to achieve complete liberation. However, if the rock broke along fracture planes that run through the centres of the particles, in directions at right angles (Figure 5.12b) not one particle would be fully liberated even though the rock would be broken into 1 mm size particles. Assume the rock is broken into 0.5 mm cubes, again offsetting the fracture surface from the mineral surfaces (Figure 5.12c). Inspection of the particles produced shows that one-eighth of them are all black (completely liberated), one-eighth are completely white (again, completely liberated) but three-quarters contain both black and white. This can be repeated using increasingly smaller cubes and, while the proportion of fully liberated minerals will increase, it is not possible to completely liberate the minerals. In reality, rocks usually contain more than two different minerals, mineral grains are not uniform in size or regularly shaped, and textures are much more complex. Also, particles do not fracture uniformly but break along defects in the structure.

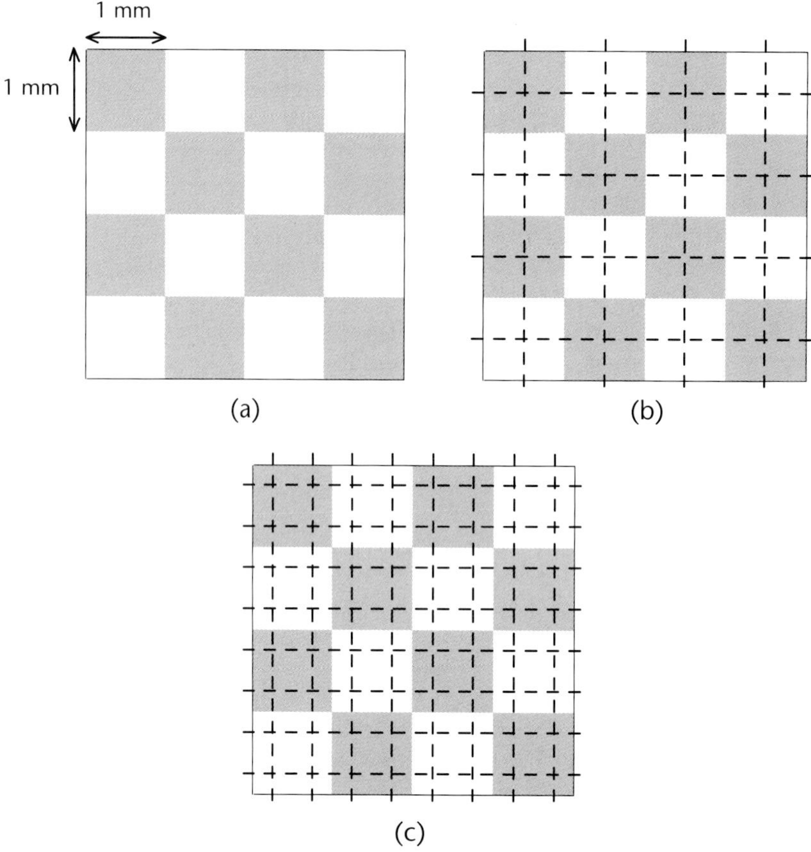

Figure 5.12: A two-dimensional representation of a rock containing ore mineral (dark) and unwanted mineral (light) particles. For simplicity, the minerals are equal-sized squares arranged in a rectangular array.

Nevertheless, the general principle illustrated by the example is valid.

An inspection of Figure 5.10 suggests that good liberation could be achieved relatively easily with rocks of Type 1a, 1b and 3a texture but it would be more difficult for rocks with the other textures, particularly Type 1c. In addition to the increasing technical difficulty of separating finer and finer minerals, the grinding of rocks uses large amounts of energy, the quantity increasing rapidly as particle size decreases using available technologies. Grinding also produces large quantities of very fine, and therefore very chemically reactive, waste minerals which have to be disposed of carefully to prevent environmental damage. In practice, a compromise must always be made between the size to which the particles are reduced and the percentage of valuable mineral recovered in the concentrate.

5.7.2 Chemical separation

The second approach to separation is to use a chemical which selectively attacks the valuable mineral. This is most commonly done in a process called leaching, usually using aqueous acidic or basic solutions that dissolve the value-bearing mineral(s) but leave the unwanted minerals relatively unaltered. The extent to which the valuable minerals will be attacked by a reagent depends on how effectively they are exposed to the solution. Any mineral surrounded by insoluble waste minerals will not be leached. While complete physical liberation is not required in this case, pathways for the solution to reach the desired minerals are essential. For example, particles with Type 1b, 1c, 2a and 3c textures (Figure 5.10) would be amenable to leaching if the wanted mineral is represented by the dark phase since it would be readily accessible to the leaching solution.

If the light phase is the wanted mineral, these particles would not suitable for leaching. The leach solution is treated in subsequent operations to recover the valuable element or compound.

Even if a texture is suitable for leaching, leaching may not be an option. A relatively selective leaching agent must be available, otherwise other minerals will also be dissolved, increasing the quantity and cost of the reagent. There would also be technical problems and additional costs in recovering the valuable element from the leach solution. The reaction products formed by leaching must be soluble, otherwise a layer can form over the target mineral and prevent further reaction. In rare cases, it may be necessary to subject the rock to thermal treatment in the presence of solid or gaseous reactants to change the nature of a mineral, usually the valuable mineral, to make it easier to liberate and separate or to leach. This is performed at temperatures at which the materials comprising the rock are still in the solid state.

Sometimes, effective liberation or selective leaching cannot be achieved. The texture of the rock may be too fine or complex for liberation, the desired elements may occur in substitution in minerals and liberation is not possible, or no effective leaching system or thermal pretreatment can be devised for the particular combination of minerals. The only option in these cases is to chemically process the entire rock. The rock could be dissolved in a suitable leaching solution, which would then be purified and the desired element recovered. Alternatively, the rock could be smelted. In smelting, the valuable mineral is reduced and the desired element is collected in a molten phase (usually as a metal or alloy) and unwanted minerals in a second molten, immiscible phase (slag). These approaches are energy-intensive and produce large quantities of waste materials that are potentially environmentally harmful.

5.7.3 The effect of breakage on the surface area of materials

As the particle size of a quantity of rock or ore is reduced the number of particles increases and, therefore, the surface area of the mass of material also increases. This means the material becomes more reactive the smaller the particle size, since more surface is available for reaction. This has advantages in leaching operations when ore is ground to increase the rate of reaction. However, it also means that waste rock from mining operations and tailings from mineral beneficiation processes are more reactive than the host rock from which they were produced. This poses problems when these materials are disposed of as wastes. Metals and other potentially toxic elements can leach from them due to the action of air and water (from rain or groundwater). This is a major problem for waste materials containing sulfide minerals which react with oxygen and water to form sulfuric acid, which in turn leaches metals from the ore. This phenomenon, called acid and metalliferous drainage, or acid rock drainage, is discussed further in Section 11.3.2.

The effect of size reduction on the surface area of a mass of material can be demonstrated by the following calculation. Assume we start with a rock in the shape of a cube with a side of 100 mm. Now, assume this is broken into successively smaller cubes of uniform size. The number of cubes will increase each time the material is broken and the total surface area will be the sum of the surface areas of all the cubes at each stage of breakage (Table 5.12). Thus, as the particle size is reduced from 100 mm to 0.1 mm (100 μm), a common size for the liberation of metallic sulfide minerals, the number of particles increases from 1 to 10^9 and the surface area of the material increases from 0.06 m^2 to 60 m^2. Even though in practice a rock would not break into perfect cubes, this simple exercise demonstrates

Table 5.12: The effect of breakage on the surface area of a material

Size of particle, S (mm)	100	10	1	0.1
No. of particles created from original particle = $(100/S)^3$	1	10^3	10^6	10^9
Surface area of 1 particle (m^2) = $6 \times (S/1000)^2$	6×10^{-2}	6×10^{-4}	6×10^{-6}	6×10^{-8}
Total surface area (m^2)	0.06	0.6	6	60

Table 5.13: Common by-products that occur during the production of copper, nickel, lead, zinc and platinum from their ores

Major metal	Copper	Nickel	Lead	Zinc	Platinum
By-products	Cobalt	Copper	Zinc	Lead	Rhodium
	Silver	Cobalt	Antimony	Cadmium	Palladium
	Gold	Manganese	Bismuth	Silver	Iridium
	Antimony		Silver	Germanium	Osmium
	Arsenic		Selenium	Gallium	Ruthenium
	Selenium		Tellurium	Indium	
	Tellurium				

the rapid increase in surface area that occurs when a rock is ground to produce smaller particles.

5.7.4 By-products and co-products

A *by-product* is a secondary or incidental product from a production process. A by-product may be useful and marketable or it can be considered a waste, with no value and needing disposal or safe storage. A by-product of value is often called a *co-product*. Tailings are usually considered a waste but they could be considered a co-product since many tailings contain valuable minerals and elements that could be extracted if it were economical to do so. Thinking of wastes as co-products is useful as it focuses on their potential value. Concentrates also contain elements other than the main element of interest and many of these are often valuable. This is particularly the case for base metals, which rarely occur in isolation in mineral deposits. They are present in the concentrate as a result of incomplete liberation of minerals or incomplete separation of liberated minerals, or because they are present in the crystal structure of the main minerals as minor elements. A number of by-products are usually produced during the extraction and refining of base and other metals (Table 5.13).

5.7.5 The efficiency of extraction

The proportion of valuable material in the input materials that passes into the product is called the recovery efficiency or, more commonly, the *recovery*. It may be applied to the entire process or to any stage. Recovery is expressed as a fraction or percentage. For example, 0.9 or 90% recovery means that 90 wt% of the valuable product is recovered in the product stream (such as the concentrate from a mill or the metal from a smelter) and 10% is lost in waste streams (such as the tailings from a mill or the slag from a smelter). Recoveries of less than one are normal due to technological limitations, such as the need to leave some ore in mines as support material, incomplete liberation and separation during beneficiation, bound or inaccessible minerals which are not amenable to leaching, and loss of some elements in slags in smelting processes.

5.8 REFERENCES

Amstutz GC (1962) How microscopy can increase recovery in your milling circuit. *Mining World* **24**(13): 19–23.

Candela PA (2007) Ores in the Earth's crust. In *Treatise on Geochemistry*. Vol. 3. (Eds HD Holland and KK Turekian) pp. 411–431. Elsevier-Pergamon: Oxford, UK.

Condie KC (1976) *Plate Tectonics and Crustal Evolution*. Pergamon Press: New York.

Gaudin AM (1939) *Principles of Mineral Dressing*. McGraw Hill: New York.

Goldschmidt VM (1954) (Ed. A Muir) *Geochemistry*. Clarendon Press: Oxford, UK.

Greenwood NN and Earnshaw A (1984) Appendix 4: Abundance of elements in crustal rock. In *Chemistry of the Elements*. p. 1496. Pergamon Press: Oxford, UK.

Hayes PC (1993) *Process Principles in Minerals and Materials Production*. Hayes Publishing: Sherwood, Qld.

Hiskey B (1995) Metallurgy: survey. In *Kirk-Othmer Encyclopedia of Chemical Technology.* 4th edn, Vol. 16. (Eds J Kroschwitz and M Howe-Grant) p. 317. John Wiley and Sons: New York.

Holmes A (1965) *Principles of Physical Geology.* Thomas Nelson and Sons: London.

JORC (2004) *The JORC Code.* Joint Ore Reserves Committee of the Australasian Institute of Mining and Metallurgy, Australian Institute of Geoscientists and Minerals Council of Australia, December 2004. <http://www.jorc.org>.

Pauling L (1929) The principles determining the structure of complex ionic crystals. *Journal of the American Chemical Society* **51**(4): 1010–1026.

Polya DA (1992) Ore formation. In *Encyclopedia of Earth System Science.* Vol. 3. (Eds L Brekhovskikh, K Turekian, K Emery and C Tseng) pp. 481–491. Academic Press: San Diego.

Ronov AB and Yaroshevsky AS (1969) Chemical composition of the Earth's crust. In *Earth's Crust and Upper Mantle.* Monograph No.13. (Ed. JP Hart) pp. 37–57. American Geophysical Union: Washington DC.

Rybach L (2007) Geothermal sustainability. *Geo-Heat Center Quarterly Bulletin*, Oregon Institute of Technology **28**(3): 2–7.

Selley RC, Cocks LRM and Plimer IR (Eds) (2004) *Encyclopedia of Geology.* Elsevier: Amsterdam.

Singer DA (1995) World class base and precious metal deposits – a quantitative analysis. *Economic Geology* **90**: 88–104.

5.9 USEFUL SOURCES OF INFORMATION

Condie KC (1976) *Plate Tectonics and Crustal Evolution.* Pergamon Press: New York.

Deer WA, Howie RA and Zussman J (1992) *An Introduction to the Rock-forming Minerals.* Longman: Burnt Mill, Harlow, Essex, UK.

Earth Science Australia. <http://Earthsci.org>.

Geoscience, National Museum of Australia. <http://www.amonline.net.au/geoscience/index.htm>.

Harben PW and Bates R (1990) *Industrial Minerals Geology and World Deposits.* Industrial Minerals Division, Metals Bulletin Plc: London.

Holland HD and Turekian KK (Eds) (2007) *Treatise on Geochemistry.* Elsevier-Pergamon: Oxford, UK.

Kious WJ and Tilling RI (1996) *This Dynamic Earth: The Story of Plate Tectonics.* United States Geological Survey. Reston, Virginia. Available as a downloadable electronic book at <http://pubs.usgs.gov/gip/dynamic/dynamic.pdf>.

Mason B (1966) *Principles of Geochemistry.* 3rd edn. John Wiley and Sons: New York.

Mine-engineer website <http://www.mine-engineer.com/mining/mineral/mineralindx.htm> has a good display of minerals with descriptions.

Misra KC (2000) *Understanding Mineral Deposits.* Kluwer Academic Publishers: Dordrecht, The Netherlands.

Park CF and MacDiarmid RA (1970) *Ore Deposits.* 2nd edn. WH Freeman: San Francisco.

Pidwirny M *PhysicalGeography.net.* <http://www.physicalgeography.net>. University of British Columbia: Okanagan.

Robb L (2005) *Introduction to Ore-Forming Processes.* Blackwell Publishing: Malden, MA.

The Minerals Information Institute. <http://www.mii.org/index.html>.

The United States Geological Survey. <http://www.usgs.gov>.

6 The minerals industry

This chapter provides an overview of the minerals industry. It commences with a description of the most commonly traded mineral commodities, the quantities produced and their prices. This leads to a discussion of the marketing of mineral commodities and factors that determine their price. The life cycle of a typical mining operation is described and the types of companies involved in the minerals industry, and their role and mode of operation, are discussed. Next, the social and economic impact of mining on countries and regions which are endowed with mineral resources is examined. Finally, the industry's past and present responses to the challenges of sustainability are discussed.

6.1 MINERAL COMMODITIES

6.1.1 Traded commodities

For trading purposes, commodities derived from mineral resources are broadly classified as metallic, non-metallic (or industrial minerals and industrial rocks) and fuels. This classification has been developed for commercial reasons and bears no relationship to the geological classification of mineral commodities in Table 5.10. The specification of traded mineral commodities, in particular the elemental and mineralogical composition or physical form (size, shape, porosity, strength), or both, are usually specified to be within given ranges. Often these are quite narrow.

Metallic commodities

Metallic commodities include metals and alloys, at a purity and composition appropriate for use in manufacturing, and metal precursors, such as mineral concentrates. Table 6.1 lists commonly traded metallic commodities. Because they are relatively high-value materials, metallic commodities are actively traded over long distances and internationally. Traded metals are often grouped into *ferrous metals* (pig iron, steel, steel scrap and ferroalloys such as ferromanganese and ferrochromium) and *non-ferrous metals*. Non-ferrous metals are subdivided into *base metals* (copper, nickel, zinc, lead, tin), *light metals* (aluminium, magnesium, titanium, lithium), *precious metals* (gold, silver, platinum, rhodium, palladium) and *rare metals*. Rare metals are metals that are both geochemically scarce and have commercial value. The list of rare metals changes over time since, although the scarcity of metals is unchanging, the value of metals changes according to demand for them. Many scarce metals were not considered rare in the past but developments in technology, particularly in electronics and electrical

Table 6.1: Commonly traded mineral commodities and typical composition ranges (major elements only)

Traded commodity	Typical composition range (wt%)
Bauxite	Al_2O_3 33.2–76.9, H_2O 8.6–31.4 (as bound water), Fe_2O_3 0.1–48.8, SiO_2 0.3–37.8, TiO_2 <4.0
Aluminium metal	London Metal Exchange, primary aluminium: Al ≥99.7, Fe ≤0.2, Si ≤0.1
Copper (cathode)	London Metal Exchange, Grade A copper: Ag ≤0.0025, As ≤0.0005, Bi ≤0.00020, Fe ≤0.0010, Pb ≤0.0005, S ≤0.0015, Sb ≤0.0004, Se ≤0.00020, Te ≤0.00020 (sum of these elements ≤0.0065; (As + Cd + Cr + Mn + P + Sb) ≤0.0015; (Bi + Se + Te) ≤0.0003, of which (Se + Te) ≤0.00030; (Co + Fe + Ni + Si + Sn + Zn) ≤0.0020
Copper concentrate[a]	Cu 14–47 (mainly as sulfide minerals), Fe 8.5–35 (as sulfide minerals), S ≤15–37, SiO_2 ≤17, Al_2O_3 ≤5, MgO ≤2, CaO ≤2, small quantities of Bi ≤0.1, Sb 0.002–0.5, As 0.01–8.0, plus small quantities of Ni, Cr, Sn, Pb, Se, Te, Ag, Au etc.
Iron ore/concentrate	Fe 60–68, as oxide minerals; balance mainly SiO_2, Al_2O_3,CaO, MgO; small quantities of P, S
Lead concentrate[b]	Pb 55–75 (mainly as sulfide minerals), Zn 3–15, Fe 2–12, S 14–25, SiO_2 2–10, Ag 100–2000 g t^{-1}, As ≤0.5, Bi ≤10 g t^{-1}
Lead metal	London Metal Exchange, standard lead: Pb 99.970, Ag ≤0.0050, As ≤0.0010, Bi ≤0.030, Cd ≤0.0010, Cu ≤0.0030, Ni ≤0.0010, Sb ≤0.0010, Sn ≤0.0010, Zn ≤0.0005 (total ≤0.030)
Nickel concentrate[c]	Ni 6–15 (mainly as sulfide minerals), Cu <5, Fe 24–40, Co <0.7, S 22–36 (as sulfide minerals), SiO_2 <10, MgO <10
Nickel metal	London Metal Exchange: Ni ≥99.80, Co ≤0.15, Cu ≤0.02, C ≤0.03, Fe ≤0.02, S ≤0.01, P <0.005, Mn <0.005, Si <0.005, As <0.005, Pb <0.005, Sb 0.005, Bi <0.005, Sn <0.005, Zn <0.005
Steel	Plain carbon steels: iron containing 0.1–1.0 C, 0.3–1.0 Mn, <0.04 P, <0.05 S (actual composition is specified by the grade of steel)
Synthetic rutile[d] (Becher process)	TiO_2 93.0–95.0, Fe (total) 1.0–1.5, ZrO_2 0.05–0.5, MnO 0.7–1.2, Al_2O_3 0.7–1.3, SiO_2 0.7–2.0, V_2O_5 0.2–0.3, Cr_2O_3 0.06–0.40, Nb_2O_3 0.25–0.30, CaO 0.04–0.10, MgO 0.39–0.70, P_2O_5 0.01–0.04, S 0.02–0.44
Zinc concentrate[e]	Zn 45–56 (mainly as sulfide minerals), Fe 5–10, S 30–32, Pb 1–3, Cu ≤2 wt, Cd ~0.2, As ≤0.5, Sb ≤0.2
Zinc metal	London Metal Exchange, Special High Grade Zinc: nominal Zn content 99.995, Pb ≤0.003, Cd ≤0.003, Fe ≤0.002, Sn ≤0.001, Cu ≤0.001, Al ≤0.001 (total of these elements ≤0.005) Prime Western Grade[e]: Zn ≥98.0, Pb ≤1.6, Cd ≤0.5, Fe ≤0.05 (<2.0 total impurities)

a) Broadhurst (2006).
b) Sinclair (2009).
c) Mäkinen and Taskinen (2008).
d) TZ Minerals (2008).
e) Sinclair (2005).

energy storage devices that require scarce metals, have led to an increase in the number of scarce metals considered rare. For example, the electric motor in the Toyota Prius hybrid car requires 1 kg of neodymium and the battery requires 10–15 kg of lanthanum. Table 6.2 lists metals that are commonly regarded as rare. Most are used in relatively small quantities and in very specific applications. Some of these metals also belong to other groups – base metals or precious metals.

Metal precursors are most commonly concentrates of metallic minerals prepared by physical processes involving liberation and separation of minerals (discussed in Section 5.7), and still require chemical processing to extract the metal. They also include materials that have been chemically concentrated, for example alumina prepared from bauxite and synthetic rutile prepared from ilmenite. These intermediates are a necessary product on the way to the ultimate extraction of the metal from the ore but, because the metal of value is in a more concentrated form, they can be shipped more cheaply (per unit mass of metal) than mined, unprocessed ore. Some

Table 6.2: Metals that are commonly considered to be rare

Beryllium	Holmium	Palladium	Tantalum
Cerium metals	Indium	Platinum	Tellurium
Caesium	Iridium	Praseodymium	Terbium
Chromium	Lanthanum	Promethium	Thorium
Dysprosium	Lithium	Rhenium	Thulium
Erbium	Lutetium	Rhodium	Tin
Europium	Manganese	Rubidium	Vanadium
Gadolinium	Molybdenum	Ruthenium	Ytterbium
Gallium	Neodymium	Samarium	Yttrium
Germanium	Niobium	Scandium	Zirconium
Hafnium	Osmium	Selenium	

metallic ores are traded without significant concentration because they are rich in valuable metal. Of particular significance are iron ore (which can contain up to around 65 wt% iron) and bauxite (which can contain up to around 35 wt% aluminium).

Traded ores and concentrates always contain unwanted components and are purchased on the basis of contained valuable metal content. Where a constituent of an ore or concentrate could cause processing difficulties (e.g. in smelting or leaching) or a disposal problem (e.g. arsenic), penalties are applied and the price is reduced. For example, copper smelters generally accept concentrates with arsenic levels up to about 0.2 wt% but apply a penalty above that level. They usually will not accept arsenic levels greater than about 0.5 wt%. Conversely, when a concentrate contains a valuable element other than the main element, a premium may be paid. This is the case, for example, for copper concentrates containing small quantities of gold and lead concentrates containing silver. When an ore or concentrate is processed at the mine site, non-standard compositions and higher than normal impurity levels may be acceptable since it is then possible to optimise the mine–concentrator–smelter combination to achieve optimum recovery, and to select the process stage and process streams which are most appropriate (from an economic and environmental perspective) for removing particular impurity elements. Such optimisation is usually not possible when ores or concentrates are traded and processed at a distance from the mine site or are sold to a third party for processing.

Non-metallic commodities

Industrial minerals and rocks fall broadly into two classes, though there are many exceptions. Some are produced in very large quantities but have relatively low value per unit mass, while others are produced in lower volumes but have high unit values. The former are usually construction and building materials and the latter are usually materials used for their effect, due to some intrinsic property, rather than as an engineering material. These are often consumed or dissipated during their use. The term *dissipated* means that the material is lost because its form is changed somehow or it becomes diluted with waste material, thereby making it difficult to recover and recycle. Building materials, on the other hand, can in principle be reclaimed after a structure is no longer wanted or needed, and reused or recycled; the decision on doing so is determined by the cost and by regulations concerning the disposal of waste materials. Characteristics of these two classes of industrial minerals and rocks, and examples of each type, are given in Table 6.3. Many industrial minerals are used essentially in raw form, in which much of their intrinsic characteristics remain. Users of industrial minerals and rocks usually demand tight specifications, which are often unique to particular applications or products. Most industrial mineral and rock products have multiple

Table 6.3: The two broad classes of industrial minerals

High-volume/low-value materials	Low-volume/high-value materials
Usually not transported long distances	Often transported long distances and internationally
Have wide occurrence	Occurrence is restricted
Simple geology	Complex geology
Simple processing	Complex processing
Often used as construction materials	Often used for their 'effect' rather than as an engineering material
Examples: granite, slate, aggregate, clay, sandstone	Examples: fluorspar, mica, beryl, graphite, industrial diamond

uses; for example a limestone deposit could supply material for lime, dimension stone, aggregate and cement production, flue-gas desulfurisation, soil stabilisation and other agricultural uses (Jeffrey, 2006).

Fuel commodities

The third group of mineral commodities is fuels. It includes the fossil fuels coal, oil and natural gas as well as uranium and its precursor, yellowcake (mainly U_3O_8). Fossil fuels, which are composed largely of organic compounds, are said to be consumed in producing heat but they are actually converted into other, mostly gaseous products (particularly CO_2 and water vapour). Uranium is converted into other solid products. Detailed consideration of fuels falls outside the scope of this book.

Manufactured mineral commodities

Cement, mineral fertilisers, glass and ceramics are manufactured from mineral commodities and could be considered a fourth class of mineral commodity. These could be called manufactured mineral commodities. Although they are made from industrial minerals, they are distinct from industrial minerals as they do not retain the characteristics, mineralogically or structurally, of the raw materials used in their manufacture; the raw materials undergo chemical transformations during the manufacturing process.

6.1.2 Mineral commodity statistics

Production and consumption

Mineral commodities are produced in widely differing quantities, a reflection of both their relative abundance and their value in use. Table 6.4 illustrates this point for a number of important traded mineral commodities. The values are for 2007 and are representative of production just before the advent of the global financial crisis. For ease of comparison, all masses are expressed in terms of tonnes of contained metal or tonnes of ore. In practice, other units of mass are used for some commodities; the more common are listed in Box 6.1. The production rate of mineral and metal commodities has increased rapidly over the past 100 years due to industrialisation (Figure 6.1).

The consumption rate of mineral commodities over a sufficiently long period will be equal to the production rate over the same period but at any particular instant the rate of production and consumption will not be equal because of the delay in response of production to demand. The stock or inventory of each commodity represents the quantity of the commodity held in long-term and short-term storage. Companies maintain stocks of the commodities they use, so they can draw a constant quantity for their operations and smooth out fluctuations in production and/or shipments. Commodity traders maintain stocks of the commodities they trade. Governments may, for strategic reasons, maintain stocks of specific commodities. The size of the inventory is allowed to increase and decrease according to supply and demand conditions and the security of supply.

Table 6.4 shows that aggregate materials (gravel, sand, crushed rock) are by far the largest volumes mined, with annual global production exceeding 15 billion tonnes, and that cement is the second largest volume mineral commodity (excluding fossil fuels). Most aggregate is used in concrete, made by mixing aggregate and cement (in an average ratio of about

Box 6.1: Some non-SI units of mass still in use in the minerals industry

The troy ounce is the customary unit for precious metals. The troy ounce is heavier than the avoirdupois ounce, which is still commonly used in the United States as a general unit of mass.

1 troy ounce = 1.09714 avoirdupois ounces = 31.103 g

1 troy pound = 12 troy ounces

1 tonne = 32 159.7 troy ounces

The term caret is the customary unit of mass for diamonds.

1 caret = 200 mg

1 tonne = 5 × 10^6 carets

The purity of gold alloys used in jewellery is customarily expressed in terms of the carat (spelled karat in the United States and Canada). One carat is one twenty-fourth part by mass. Thus, for example, 18 carat gold is an alloy of gold which contains 18/24 or 75 wt% gold. Pure gold is 24 carat.

5 tonnes of aggregate per tonne of cement) with water. Iron ore is the third largest mined commodity. Most iron ore is used to make pig iron, an impure form of metallic iron produced in blast furnaces. Primary steel is produced from pig iron in a separate refining step to remove impurities, mainly carbon and silicon, dissolved in the iron. Scrap steel is also recycled, hence the total production of steel exceeds that of pig iron. The quantity of steel produced far exceeds that of aluminium, the next largest volume metal, by a factor of greater than 30. Aluminium metal is produced from bauxite, which is also a traded commodity. Of the base metals (copper, nickel, zinc, lead and tin), copper has the largest volume but it is less than half that of aluminium. At the other extremes of metal production, about 2500 tonnes of gold and about 230 tonnes each of platinum and palladium are produced annually.

Price

Just as the quantities of mineral and metal commodities produced vary widely, so do their prices, again reflecting their relative scarcity or abundance, the cost of extraction and their value in use. This is illustrated in Table 6.4 where, for comparative purposes, prices are expressed in US dollars per tonne (of contained metal or ore) even though prices for some commodities are usually expressed in terms of other units of mass. Prices range from around US$40 million per tonne for platinum to about US$7 per tonne for sand and gravel. Steel is the largest traded commodity in terms of sale value (nearly $1000 billion in 2007) because of the combination of relatively high price and very large volume. Cement is the next largest traded mineral commodity in terms of sale value, being about a quarter that of steel. Iron ore is the third largest traded commodity, having a value about one-tenth that of steel. This is followed very closely by copper, aluminium and sand and gravel, the last due to the very great quantities used which compensates for the very low unit value.

6.1.3 Reserves and resources of mineral commodities

Table 6.4 shows the global reserves and resources of the commodities for which data are available. These values are necessarily approximate, as they are based on assessments of data from many different sources. The terms reserves, reserve base and resources were defined in Section 5.6. The number of years of supply of a commodity can be estimated by dividing the quantity of reserves or resources by the annual rate of production; these values are shown in the last two columns. For most commodities there is at least 50 years and, in the case of the geochemically abundant elements, several centuries supply. The calculation is based on two assumptions: over time, the consumption rate of each commodity will remain constant, and the reserve base will remain constant. However, historical experience has shown that consumption rate increases with time, particularly as countries become more developed, and the reserve base increases as new discoveries are made and technological developments make it economical to mine and process lower-grade and more complex mineral deposits. Clearly, neither of these trends can continue indefinitely and there is much debate about an appropriate rate of consumption of mineral commodities that will allow future generations to also meet their

Table 6.4: Annual global production of commonly traded commodities, their indicative price and estimated global reserves and resources. Data are for 2007.

	Quantity produced	Indicative price	Reserves	Reserve base	Resources	Life of reserve base	Life of resources
	('000 t metal)	(US$/t metal)	('000 t contained metal)			(years)	(years)
Steel	1 320 000	600–860[a]	See reserves and resources for iron ore, below				
Pig iron	940 000						
Aluminium	38 000	2700	See reserves and resources for bauxite, below				
Copper	15 600	7400	490 000	940 000	>3 000 000	60	>192
Manganese	11 600		460 000	5 200 000		448	
Zinc	10 500	3500	180 000	480 000	1 900 000	46	181
Lead	3550	2400	79 000	170 000	>1 500 000	48	>423
Nickel	1660	37 700	67 000	150 000	130 000 at ≥1 wt% Ni; plus Ni in Mn nodules	90	
Tin	302	13 800	6100	11 000		37	
Cobalt	62.3	66 600	7000	13 000	15 000	209	241
Silver	20.5	431 000	270	570		28	
Gold	2.50	21 700 000	42.0	90.0		36	
Palladium	0.232	11 600 000	71[g]	80[g]	100[g]		435[g]
Platinum	0.230	40 500 000					

	Quantity produced	Indicative price	Reserves	Reserve base	Resources	Life of reserve base	Life of resources
	('000 t product)	(US$/t product)	('000 t product)				
Iron ore	1 900 000	63	150 000 000	340 000 000	>800 000 000 with >320 billion t contained iron	179	>421
Bauxite	190 000	27	25 000 000	32 000 000	~65 000 000	168	342
Phosphate rock	147 000	39	18 000 000	50 000 000	Not known. Large quantities on continental shelves	340	
Kaolin	37 800	139			Extremely large quantities		
Chromite	20 000	175			12 000 000		600
Bentonite	11 800	48			Extremely large quantities		
Rutile	6100[b]	488	7 300 002	15 000 002	>2 000 000[b] combined resources of rutile, ilmenite and anastase	246	>357[f]
Ilmenite	5600[b]	80	6 800 002	14 000 002		250	
Fluorspar	5310	111	240 000	480 000	500 000	90	94
Magnesite	4600		2 200 000		12 000 000	783	2609
Fuller's earth	4020	99			Extremely large quantities		
Crushed stone, sand and gravel	>15 000 000[c]	6.80			Virtually unlimited		
Coal	5 506 394[d]						
Lignite/ brown coal	868 412[d]						
Cement	2 600 000				Virtually unlimited raw materials		
Glass	142 000[e]				Virtually unlimited raw materials		

a) Price varies widely according to type and grade of steel.
b) Expressed as TiO_2.
c) Starke (2002), p. 36. Note: a total of 3 600 000 tonnes of aggregate was produced in 21 European countries in 2006 (Delgado *et al.*, 2006).
d) IEA (2009).
e) Consisting of 29% flat glass, 35% container glass, 28% glass fibre and 6% other forms of glass (T. Green, British Glass, private communication, 27/5/09).
f) Includes rutile, ilmenite and anastase.
g) Total platinum group metals (PGM).
Source: Data from United States Geological Survey unless otherwise indicated.

needs. This is explored further in later chapters. It is clear that there will not be a general shortage of mineral commodities in the foreseeable future, though there are likely to be specific commodities that become increasingly difficult to obtain. There may also be difficulties in securing particular commodities at a national or regional level, since they are not uniformly distributed around the Earth.

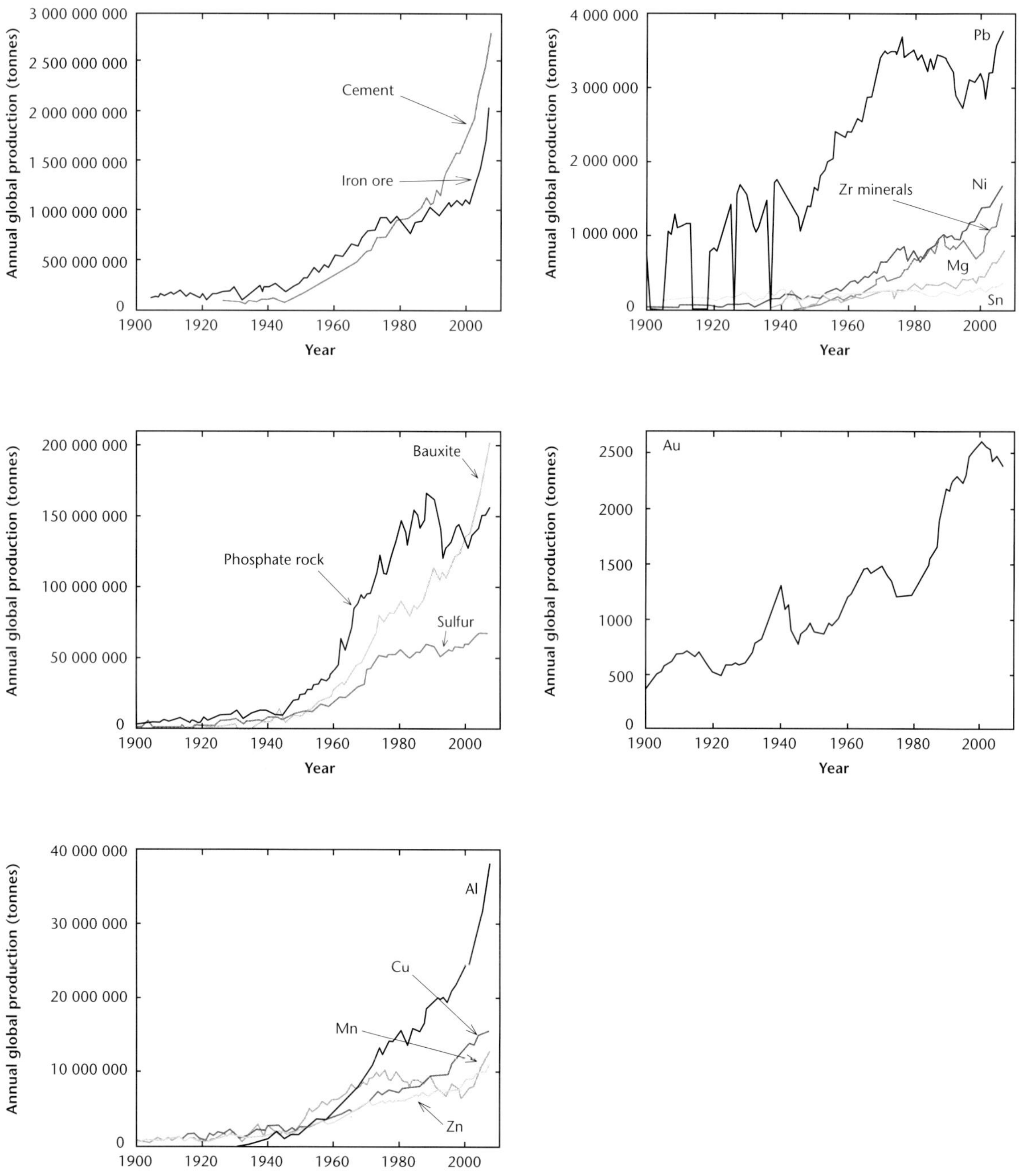

Figure 6.1: Trends in annual global production of some common mineral and metal commodities since 1900 (data from Kelly and Matos, 2009).

6.2 HOW MINERAL COMMODITIES ARE TRADED

In general, the higher the value of a commodity the greater the distance it is economic to transport. Some commodities are valuable enough that they can be traded in the global market (metals, alumina, and diamonds), some are valuable enough to be traded in broad geographical regions (coal and steel) and others have very low value and are traded mainly locally. This is particularly the case for sand and aggregate. Commodities that are traded globally are also sold regionally and domestically. Domestic output satisfies a substantial share of domestic demand for many mineral commodities in some of the large economies, particularly China, Brazil, India, the United States and Russia, and in some smaller mining-intensive economies such as Canada, Australia and South Africa. Mineral-derived materials may be sold at different stages of the value chain. Aluminium may be sold as bauxite ore, alumina, aluminium metal (as ingot or extruded forms for use in manufacturing) and aluminium scrap. Copper can be sold as concentrate, blister copper (an impure form of copper), cathode copper (a refined form), rod, wire, plate, etc. (for use in manufacturing) and scrap. Mineral commodities are sold through commodity exchanges or directly by the producer to the user.

Base and precious metals are highly homogeneous – one producer's product can readily substitute for that of another. There are many producers of these, there are many customers and there are many end uses of the products. In this situation individual producers can have little influence on the market. Prices for these commodities are uniform, determined by supply and demand in *terminal markets*. These are institutions in which dealers bid for quantities of commodities and establish a daily price. The leading terminal market for base metals is the London Metal Exchange, which accounts for 90% of global exchange business for aluminium, copper, nickel, zinc, lead and tin. The New York Mercantile Exchange is also an important terminal market.

Producers have more ability to influence the price when there are only a few producers of a commodity but many end uses and/or customers. They can also influence price when there is a degree of product differentiation. Metallic copper is not differentiated but iron ore, for example, is because each deposit has a unique chemical composition and mineralogy. In these cases, a *producer price* is set by a dominant producer or price-leader and followed closely by other producers. This form of pricing was more widespread in the past when various international producer associations attempted to influence the price of specific commodities by controlling the supply. The International Bauxite Association was the most successful. The former International Tin Council regulated the tin market for several decades in the 1950s, 1960s and 1970s. However, with the advent of aluminium containers and the increase in metal recycling, the demand for tin had decreased considerably by the early 1980s and by the mid-1980s the price could no longer be sustained. Producer pricing is still common for some industrial minerals and minor metals. Where there are only a few large producers and a few customers, the major participants know each other well and price is determined largely by negotiation. In these cases, trade is based on contracts between individual producers and customers. This is now common practice for trading in iron ore and alumina.

6.2.1 Mineral and metal markets

In the decades to the 1950s demand for mineral commodities was largely driven by the then developed nations, particularly the United States, United Kingdom and Europe. The rapid growth in the demand for commodities since the end of World War II has been driven largely by the industrialisation of Japan, South Korea and Taiwan (in the 1960s to 1980s) and of China, India and Brazil (in the 1990s and 2000s). Developing countries have a greater influence on the demand for commodities as they industrialise than do the developed nations. This is because developing countries invest heavily in infrastructure such as buildings, roads and railways, and establish export manufacturing industries, all of which consume large quantities of mineral and metal commodities. Developed countries have skilled workforces and service-oriented economies; higher personal incomes allow them to import manufactured goods from countries where labour costs are lower and there may be less concern with the environmental impact of industry.

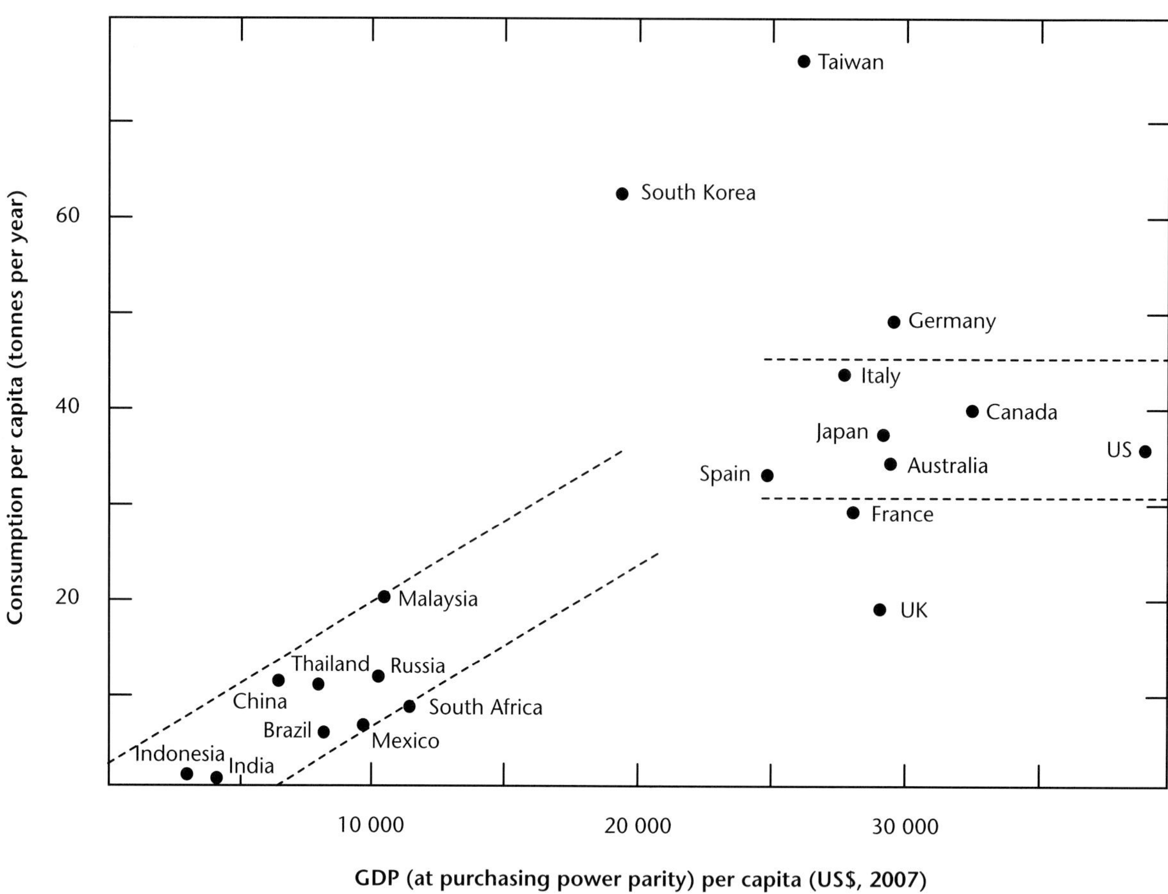

Figure 6.2: Per capita consumption of base metals as a function of per capita GDP (after Hooke, 2008).

This generally leads to a contraction of manufacturing and reduced infrastructure development as a country's gross domestic product (GDP) increases and, therefore, to decreasing demand for commodities. Figure 6.2 shows that, for developing countries, the consumption of base metals initially increases as the per capita GDP increases but that it is relatively independent of per capita GDP in developed countries. Germany has the highest consumption and the United Kingdom the lowest of the developed countries shown, presumably because Germany retains a strong export-oriented manufacturing base whereas the United Kingdom has a much more service-based economy. Taiwan and South Korea are outliers – they have relatively high GDP but remain major manufacturing countries. Similar trends occur for most other mineral commodities. Historically, per capita consumption of energy (which is largely fossil fuel based) has risen with increasing per capita GDP but, as illustrated in Figure 6.3, it may not flatten out to the extent shown by non-fuel mineral commodities as countries become more developed.

The production of mineral commodities is an important economic activity in many parts of the world. By and large, the regions which consume large quantities of mineral commodities are not major producers of them, hence there is considerable international trade in mineral commodities. An exception is the United States, which has the largest minerals sector of all countries, though it contributes less than 0.5% to its GDP. European countries produce large quantities of aggregate materials, about 20% of global production. Countries for which mineral commodities make up 25% or more of total exports are listed in

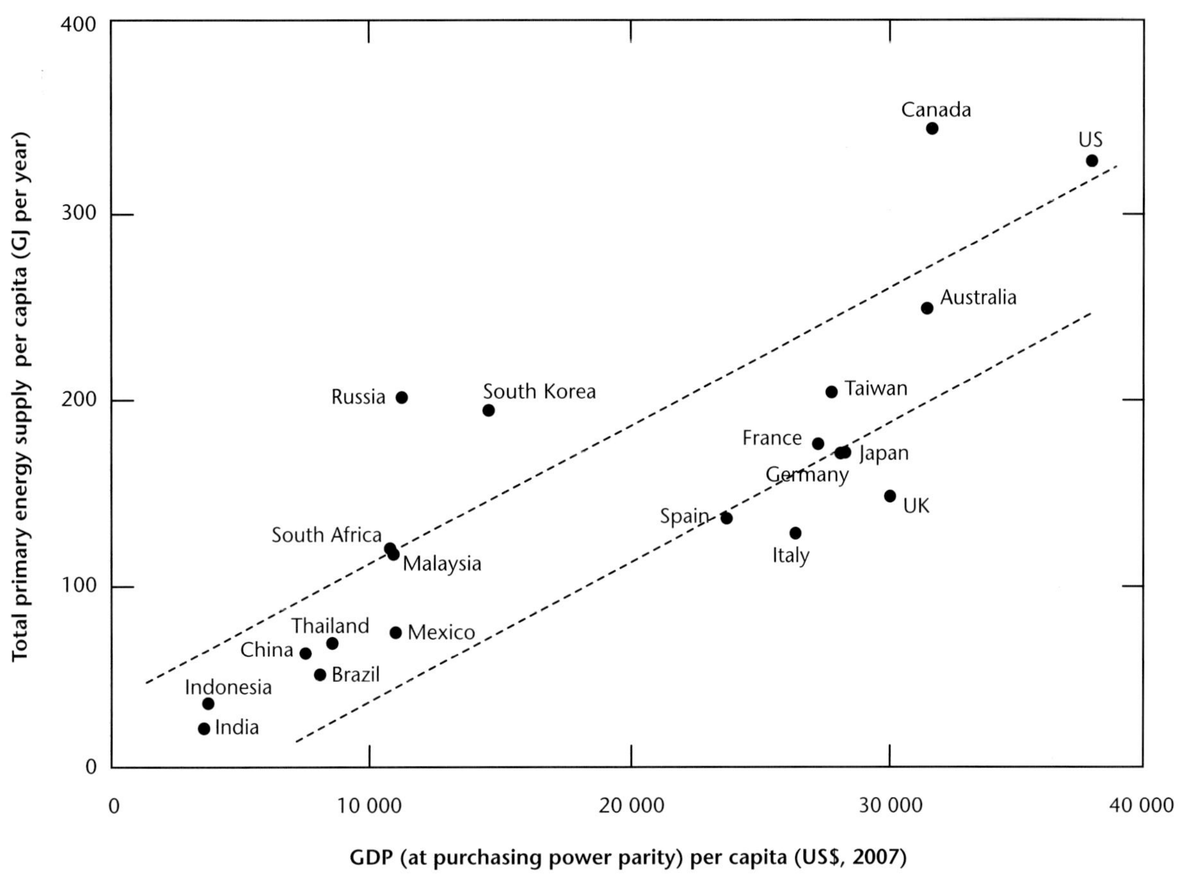

Figure 6.3: Per capita supply of primary energy as a function of per capita GDP. Total primary energy supply is equal to production plus imports minus exports minus international marine bunkers minus international aviation bunkers plus or minus stock changes. (Source of data: International Energy Agency, Statistics by country/region; <http://www.iea.org/stats/index.asp>).

Table 6.5. Many are in Africa, South America and Oceania, and many are underdeveloped. Note that the balance between export of ores and metals and of fuels varies considerably, and that a number of countries are almost totally reliant on production of fuels for export income.

6.2.2 The complexities of trading mineral commodities

The trading of mineral commodities is complex, due to a number of interrelated factors. The short-term demand for most commodities is cyclical and it responds quickly to changes in the economic climate, even though the long-term trend is for increasing levels of consumption. If prices become too high alternative materials, or lesser quantities of particular mineral commodities, are used. The short-term demand for mineral commodities is said to be elastic[15]

15 Elasticity of supply (or demand), E, refers to the responsiveness of the quantity of a product supplied (or demanded) to a change in the price of the product:

$$E = \frac{\text{\% change in quantity supplied (or demanded)}}{\text{\% change in price}}$$

If, in response to a 10% rise in the price of a product, the quantity supplied (or demanded) decreased by 20%, the price elasticity of supply is 20/10 = 2. In this case, supply (or demand) is relatively elastic (E = 0, perfectly inelastic; $1 > E > 0$, relatively inelastic; $E = 1$, elastic; $E > 1$, relatively elastic; $E = \infty$, perfectly elastic). If, however, the quantity supplied (or demanded) decreased by only 5%, $E = 0.5$ and the supply (or demand) is relatively inelastic.

Table 6.5: Countries for which export of mineral commodities is greater than 25% of total export income

Country	% of total export income: Ores and metals	Fuels	Total
Nigeria	0	99	99
Algeria	0	96	96
Libya	0	95	95
Yemen	0	93	93
Saudi Arabia	1	85	86
Venezuela	4	81	85
Kuwait	0	79	79
Oman	1	77	78
Guinea	71	0	71
Azerbaijan	1	69	70
Syria	1	68	69
Niger	67	0	67
Zambia	66	0	66
Kazakhstan	22	42	64
Mongolia	60	0	60
Norway	7	50	57
Trinidad and Tobago	0	54	54
Russia	11	41	52
Peru	40	5	45
Chile	43	0	43
Colombia	1	40	41
Egypt	4	37	41
Congo	40	0	40
Mauritania	40	0	40
Australia	17	19	36
Papua New Guinea	35	0	35
Tajikistan	35	0	35
Ecuador	0	33	33
South Africa	21	10	31
Bolivia	23	6	29
Indonesia	5	23	28
Jordan	27	0	27
Senegal	10	17	27
Togo	27	0	27

Source: Starke (2002).

since it rapidly responds to changes in the overall state of the economy and price. Production rates can be reduced quickly if circumstances require, but inventories take time to run down and increasing the production rate back to normal levels is expensive and time-consuming – mines and processing plants need to be reopened, additional staff employed and inventories re-established. Increasing production beyond the capacity of existing mines and processing plants is even more expensive and time-consuming because of the cost of developing new mines and building new plants, and the time to do the planning, gain approvals, design, construct and commission the new operation. The long-term supply of mineral commodities is said to be inelastic, since production cannot easily and quickly respond to increasing demand.

A complicating factor is that, because substitution for some materials is relatively easy, the pattern of usage of minerals and metals can change quite quickly. The invention of semi-conductors, integrated circuits and optical fibres reduced the demand for copper wire in the electronics industry. Substitution of plastics and aluminium for steel in some applications in automobiles to reduce their weight, and hence their fuel consumption, had a direct impact on the demand for steel and petroleum. These changes result partly from rising prices (in response to perceived or actual shortages) and partly from technological developments which allow the changes.

The geochemical nature of mineral deposits is another complicating factor. Many metals and industrial minerals and rocks occur together, in a variety of mineral deposit types. It is rarely possible to mine for a single commodity, even if one commodity is dominant (Table 5.13). Exceptions are iron ore, bauxite and some industrial minerals and rocks. Thus, although each commodity is traded individually, its production is linked to the production, and thus the trading, of other commodities that occur with it in the mineral deposit.

Governments have sometimes maintained stockpiles of minerals and metals they consider to be strategic, to guard against shortages in case of war, natural disaster or economic crisis. The practice became common in the late 1930s as World War II approached and increased sharply in the 1950s as a result of Cold

War tensions. The stockpiles could be sold when a government deemed it necessary and therefore potentially acted to stabilise prices. In recent decades most governments have reduced or abolished stockpiles of mineral and metal commodities.

6.3 THE ECONOMIC VALUE OF MINERAL COMMODITIES

Mineral resources, like other natural resources such as land, command a rent (also called scarcity rent or net price), which is the term used by economists to refer to that part of the income that accrues to the owner of a resource due to the scarcity of the resource. The economic concept of rent was introduced by David Ricardo (1772–1823) in relation to land for agricultural purposes. Ricardo (1817) argued that as more land is cultivated, farmers start to use less-productive land but, because a unit of produce from less-productive land sells for the same price as a unit of produce from highly productive land, farmers are willing to pay more for highly productive land. He defined rent as the difference between the quantity of produce obtained by the employment of two equal quantities of capital and labour. Thus, if only one grade of land is used for cultivation there is no rent payable; if different grades of land are used rent will be charged on the higher grades and the rent will be higher the higher the quality of the land. The market price of a commodity can be thought as the sum of the scarcity rent and the marginal cost of production. The marginal cost of production is the increase in total cost of production as a result of producing one extra unit of the commodity. It includes the fixed and variable costs, salaries and wages, taxes, dividends and a fair market return.

6.3.1 Hotelling's rule

Since mineral resources are exhaustible, the scarcity rent for mineral resources should increase in real terms as a resource is consumed because the commodity is becoming scarcer and therefore more valuable. This is recognised in Hotelling's rule, which states that in a competitive market environment the scarcity rent, or net price, of exhaustible resources should rise at the same rate as other financial assets, i.e. at the rate of interest (Hotelling, 1931). This rule was derived mathematically from a number of assumptions (discussed later). The essential ideas that come from Hotelling's work are:

- the current price of non-renewable commodities should reflect both the marginal cost of production and the scarcity rent;
- real prices (scarcity rent plus marginal cost) should rise over time to reflect increasing scarcity and the rising marginal extraction cost.

Scarcity rents should increase at the rate of interest. The real marginal cost of mining should also increase during the life of a mine as the ore becomes more difficult to extract. In underground mining, costs increase as the mine becomes deeper and mining occurs further from the shaft where the ore is brought to the surface. In open pit mining, costs increase as the haulage roads become longer, pits become larger and deeper, and more waste rock has to be removed to keep the walls stable. The implications of Hotelling's rule extend to aspects other than the price of commodities. If scarcity rent increases at the rate of interest, then its discounted value will be the same at any time. Also, if the scarcity rent of the resource grows at the rate of interest there can be no benefit from increasing or decreasing the rate of ore extraction and, therefore, no reason for the owner to vary the rate of extraction over time.

Hotelling's rule was largely ignored for 40 years but interest in it revived during the 1970s. Much work was carried out in the 1970s and 1980s to test the law against empirical data. Typical studies are those of Slade (1982) and Schmidt (1988). These found that, while prices were often highly volatile (and cyclical), real prices between 1870 and 1990 for iron, lead, zinc, coal, aggregate and petroleum remained approximately constant, decreased for copper and aluminium, and increased only for tin. More recent data, showing the trends in prices of mineral and metal commodities, are shown in Figure 6.4. There was a large increase in real prices of mineral commodities in the first decade of this century due to rapidly increasing demand, particularly from China, and the inelastic nature of production. However, real prices dropped dramatically as a result of the global financial crisis though they started to increase again during 2009. It is clear that, contrary to Hotelling's predictions, real prices of commodities have not risen over the past century (with, perhaps, a few exceptions).

6.3.2 Limitations of Hotelling's rule

Even though it is accepted that Hotelling's rule does not apply to mineral commodities, it is instructive to analyse the assumptions on which the rule is based in

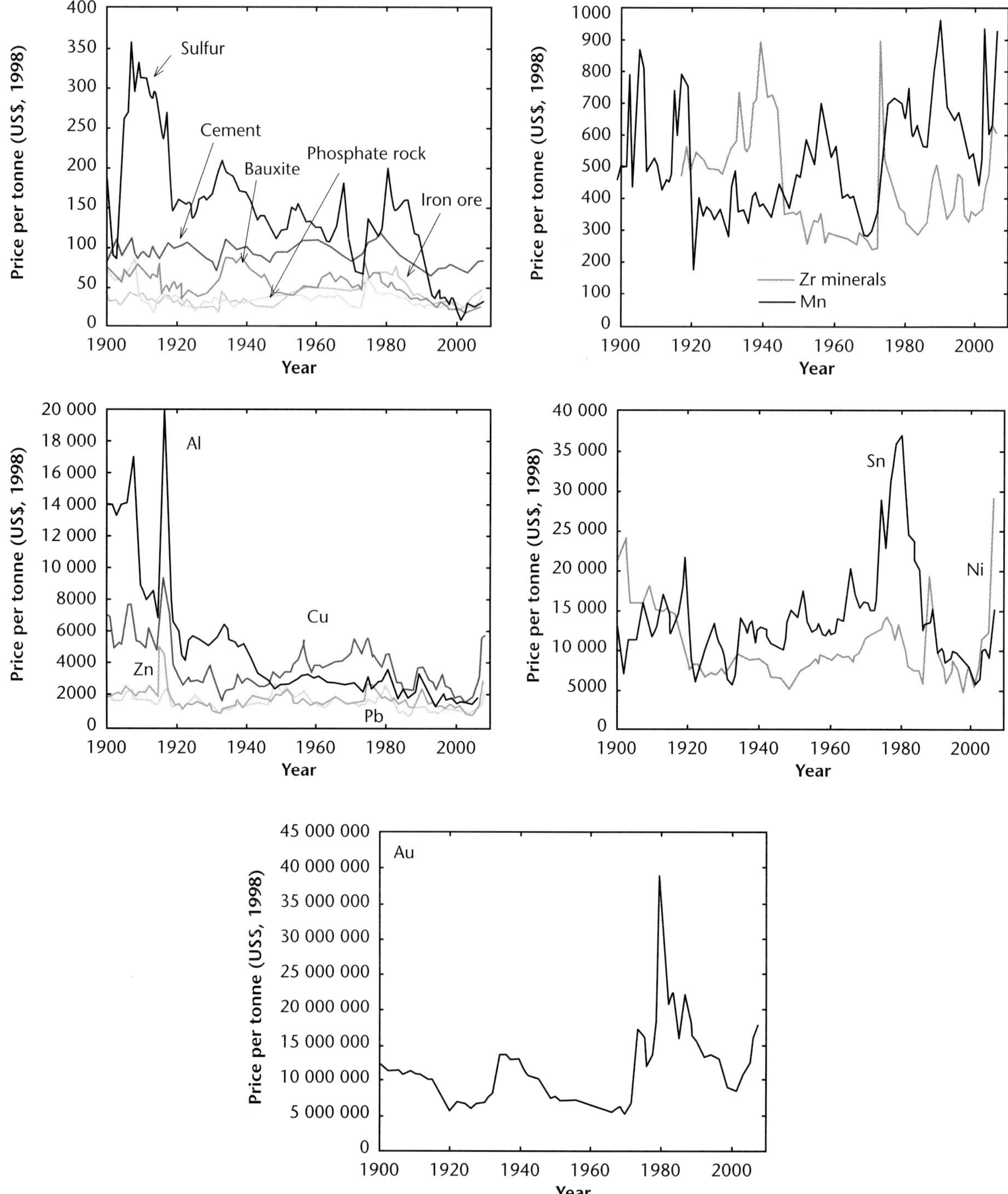

Figure 6.4: Trends in the real price of commodities, 1950–2007 (US$, 1998). Data from Kelly and Matos (2009).

order to better understand why the real long-term price of commodities has remained relatively stable. As with all economic models, Hotelling's rule requires a series of assumptions, each of which bear some relationship to reality. The appropriateness (or otherwise) of the assumptions provides useful insights.

1 Hotelling assumed that the size and grade of reserves was known from the beginning. In reality, continuing exploration, and developments in exploration, mining and processing technologies, affect the size of proven reserves (Section 5.6). As Figure 6.5 shows, the reserves of most commodities have increased, rather than decreased as assumed by Hotelling.
2 Hotelling assumed that the demand for a commodity is known and predictable. In reality, there can be dramatic changes in the patterns and intensity of usage of particular mineral commodities and in the availability of alternatives. In the short term, the supply of commodities is inelastic but in the long term substitutes tend to be developed and adopted, often becoming preferred to the original.
3 Hotelling assumed a competitive market structure. In reality, no market is completely competitive.

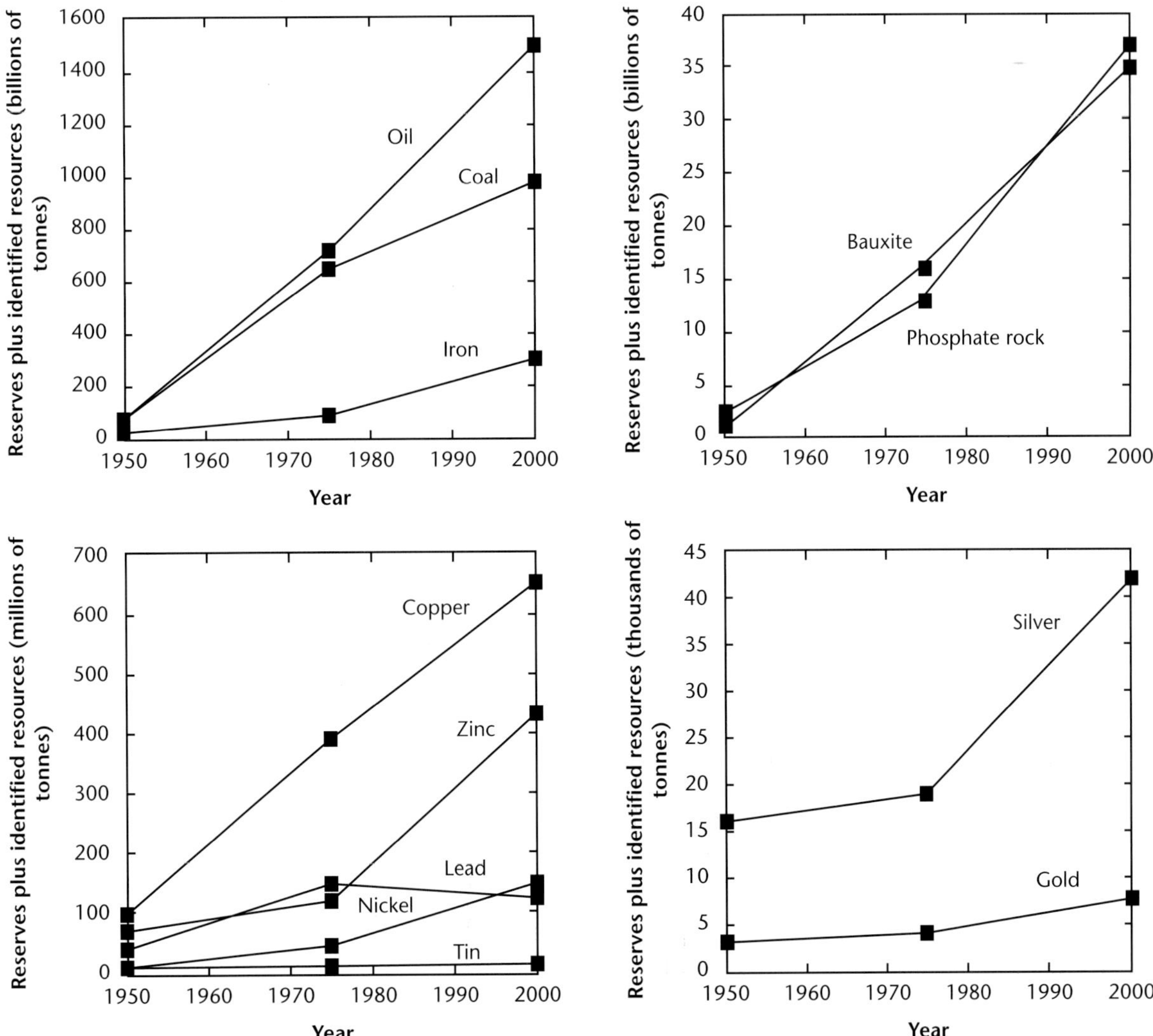

Figure 6.5: Trends in the quantity of reserves and identified resources for common mineral and metal commodities, 1950–2000 (Suslick and Machado, 2003).

Some commodities are produced by a relatively small number of producers who can, to varying degrees, influence the price. Some commodities are produced by state-owned enterprises (or by companies that are highly regulated by the state); in these cases the government may have different objectives from those assumed by Hotelling.

4 Hotelling ignored the impact of technological developments. Though the cost of mining should increase during the life of a mine due to increased distances and depths, decreasing grade and so on, development of new technologies and improvements in existing technologies have a counteracting effect on cost. Technological innovation can occur in all stages of the mining project (exploration, development, design and construction, production and closure, remediation and restoration) and the net effect over all stages is potentially large even if each individual improvement is small. Technological improvements reduce both current costs and the scarcity rent, the latter because such developments can result in new discoveries of mineral deposits and therefore increase the size of reserves.

Some economists (e.g. Slade, 1982) have argued that the real price of all commodities will eventually increase because new reserves will become increasingly difficult to find. When discovery of new deposits ceases, the rising scarcity rent will become dominant and will not be offset by technological developments. This scenario may occur for the geochemically scarce elements but it seems unlikely for the abundant elements (Table 5.2).

Livernois (2009) extended and modified Hotelling's model to take into account many of the limitations noted above. He included mathematical terms that allowed for technological changes that reduce marginal cost over time (at a rate that decreased over time); unanticipated increases in recoverable reserves (at a rate that decreased with time); and a randomness term to allow for short-term supply and demand cycles. The model predicted that real commodity prices would fall over time, but eventually increase gradually as the resource base started to decline. The dominant factors affecting the price of non-renewable resources were technological developments and upward revisions of the resource base. The structure of the market for commodities was also an important factor.

6.4 THE MINING PROJECT CYCLE

An ore body is a finite resource. Once the valuable mineral has been extracted to the extent that is economically feasible, the deposit is of no further commercial interest. All mining operations therefore have a beginning and an end and companies establish a mine in the knowledge that it will have a finite life. Mines that are closed for economic reasons may become economical to reopen if the price of the valuable commodity (or a by-product) rises sufficiently, or if new technology makes extraction and processing cheaper. Many mines have been closed and reopened on a number of occasions. Mining operations are often referred to as projects, since they have a beginning and an end. This is different from many other commercial activities, which have beginnings but not necessarily a known or planned end. For example, farming commences with the clearing of land and an expectation to continue farming the same land indefinitely. Manufacturing operations also usually anticipate ongoing activities, providing they remain competitive and make products that are wanted and needed.

The minimum life of a mine is an important factor when deciding whether to establish an operation. The financing of a minerals operation depends on being able to recover the cost of the investment and gain an acceptable return within the expected lifespan of the mine. This decision is made more difficult by the cyclical nature of the industry and the long lead times in establishing a new operation. There are countless examples of mines and processing operations that have begun operation just as the economic cycle swings down! The knowledge that a mine will be closed in the future also influences how the mine is developed, since closure is an expensive part of the project cycle.

All jurisdictions have laws which allow access to a site for exploration and potential exploitation of any deposit discovered. The intent of most mining law is to uphold the right to explore, provide secure and exclusive right for persons and companies to pursue development of a mineral discovery, protect the public interest, prevent owners tying up land without exploiting it, and provide means of dispute resolution. Once a prospective deposit has been identified, most jurisdictions require a detailed technical, environmental and socio-economic study of the proposed development. Such a study often takes several years. Public hearings may be held, and objections considered. The relevant authority will then decide, or in the case of major developments make a recommendation to the government, on whether the project should proceed. If development is approved it may be subject to certain conditions specific to that site or operation.

There are five major stages in the cycle of a minerals project: exploration; evaluation and development; design, construction and commissioning; production; and project decline and closure, remediation and restoration. The time to production can be very long, depending on the scale of the project. Exploration is an ongoing activity but it may take 20 years or more from discovery for a major ore body to be delineated. Typical times for developing a major mine are one to three years for evaluation, two to five years for mine development, two to three years for plant construction, and six months to a year for commissioning.

6.4.1 Exploration

Mineral deposits at or near the Earth's surface are discovered by visual examination of the deposit outcrop at the surface, or examination of fragments that have been eroded from the outcrop and carried away. Geological studies of the area then allow the mineral deposit to be located. This is how most deposits were discovered. Mineral deposits below the surface are much harder to detect. Advanced physical sensing techniques can probe below the surface to measure gravitational, seismic, magnetic, electrical, electromagnetic and radiometric variations in the Earth. These methods are usually applied from above ground, using aircraft and satellites to detect anomalies which may indicate mineral deposits. Geochemical anomalies at the surface may be used to deduce the existence of a mineral deposit below the surface.

Having discovered and secured a deposit, the next step is to more fully delineate it – its size, shape, depth and mineralogy. Representative samples are taken across the deposit using methods including chipping from outcrops, digging trenches and tunnels, and drilling to obtain cores of rock which reveal the vertical structure and composition. The samples are analysed using a variety of techniques such as chemical analysis (to determine the elements present and their concentrations), X-ray diffraction (to determine the minerals present), microscopic examination (to study the ore texture) and metallurgical analysis (to assess how best to process the mined material). All these enhance the knowledge of the mineral deposit. Finally, an assessment is made of the size of the deposit and its grade.

6.4.2 Evaluation and development

An evaluation is undertaken to determine the technical and economic viability of developing the deposit. At the conclusion of this stage, a decision is made to develop the project, sell it to another party or abandon it. The evaluation is done in a series of iterations with increasing levels of detail. The project will be abandoned at any stage if the evaluation indicates one or more reasons not to go ahead. These could be, for example, insurmountable technical difficulties in mining, processing or waste disposal; insufficient financial returns to justify the investment; approval for mining may be unobtainable, for political or environmental reasons; the conditions likely to be imposed by regulators are too onerous; or there is too much public opposition to the project. The stages of evaluation often include a scoping study, a feasibility study (including a bankable feasibility study which can be presented to financial institutions and other potential investors) and a final estimate, which would aim to determine costs and returns during the life of the project to within about 10%. During the evaluation stage, work will continue on exploration and on mining and processing options to help refine the study and provide more definitive information.

Some preliminary development work may be undertaken, including acquiring water and mineral rights, buying land, arranging for financing, and preparing permit applications and an environmental impact statement. Discussions and negotiations with the local community and non-governmental organisations with an interest in environmental and social issues may be undertaken. When access to land and the various approvals have been obtained, consideration will be given to infrastructure needs such as access roads, power, transport systems, mining method (underground or open cut), mineral processing facilities, waste disposal areas and surface facilities (offices, laboratories, workshops, housing for workers). Consideration of how the mine will be closed and the area rehabilitated will also be done. Finally, some surface preparation work may commence in anticipation of full development.

6.4.3 Design, construction and commissioning

Once a decision has been taken to develop a mine, funding has been obtained and approvals are in place, detailed design of the mine and processing operations and associated facilities will commence. Tenders will be let for construction. Often, different companies will be awarded tenders for different parts of the project. Overall management of the design and construction may be done in-house by an internal team with the necessary expertise, or it may be subcontracted to a service company. Individual parts of the operations are started up independently as they are constructed, to test them and train operators. Modifications will be made if necessary, then separate parts will be progressively operated together until the entire system or part of it is functioning. Commissioning is often a slow process, as this is when unanticipated technical problems usually arise. It may take months or even years before a mine or processing plant is operating at its design capacity.

6.4.4 Production

As the name implies, during the production phase the focus is on producing material from the mine and processing plants. Modern mining operations are very complex and integrated. A mechanical or electrical failure, an accident or an environmental incident, such as an unplanned release of waste into the air or water, can result in a shut-down of the entire operation or a large part of it, with the consequent loss of production and cost of restarting. Thus, equipment maintenance is important, as is continually improving operating procedures to minimise lost production. Technical development of the mining and processing operations continues throughout this stage with the aim of one or more of the following: increasing the throughput; decreasing operating costs; increasing reliability of equipment; increasing occupational health and safety for the staff; and reducing emissions of waste materials. Exploration around the ore body may continue in an attempt to find more ore to extend the life of the mine.

6.4.5 Project decline and closure, remediation and restoration

Ultimately, the mine will start to become exhausted and the rate at which ore is produced will begin to decline. Eventually, mining will cease. Not all the valuable mineral will have been extracted, but the cost of extracting the ore or of processing it because of its lower grade, or both, will have become too great using available methods. However, as technology develops, or if commodity prices increase, it may become possible to again mine and process the material economically. Thus, when a mine is closed it may be done in a way that will allow it to reopen in the future. Processing plants associated with the mine will also be closed, unless they can be used to economically treat ore mined elsewhere and shipped to the site.

In a good operation, planning for closure is undertaken in parallel with other activities and integrated at all stages of the project cycle. Mine closure is expensive; the cost must be anticipated at project commencement and incorporated into the financial modelling. Many legacy issues relate to an abandoned mine. Future employment of local people and the future of local businesses that relied heavily on the operation need to be considered. Land around the mine and surface operations needs to be remediated and, if possible, restored to its original state. Waste dumps need to be stabilised to prevent contamination of water, air and land. Unfortunately, many mining companies have not dealt well with legacy issues. This is a major reason for the poor reputation of the minerals industry in most countries – both developed and undeveloped. In response, non-binding guidelines on planning for mine closure have been developed for use by mining companies (Planning for Integrated Mine Closure: Toolkit, 2008).

6.5 THE NATURE OF THE MINERALS INDUSTRY

6.5.1 Location

The location of mining operations is determined by geology and history. Some regions or countries have an abundance of useful minerals, while others are poor in mineral resources. No country has all the minerals it needs. Historically, most mining operations were in the more developed countries in the northern hemisphere. There was a marked increase in mining in developing countries and in the southern hemisphere during the 20th century, particularly after the end of World War II, most notably in southern Africa, Australia and South America. This was due mainly to the discovery of large high-grade deposits of major commodities in previously geologically unexplored areas and to the decline in transportation costs. Figure 6.6 shows how the dominant share of global production of minerals since 1850 shifted from Europe to the United States to, more recently, Australia, Canada and a number of developing countries (including Brazil, Chile and South Africa). Figure 6.7 shows the present global distribution of mineral production. Though mining remains a major industry in North America (Canada, the United States and Mexico), the countries of the former Soviet Union and parts of Western Europe (particularly Scandinavia), there has been a decline in mining activity in many developed countries due to the depletion of high-grade ores and the increased difficulty of exploration and mining because of social factors including public opposition to mining, high population densities and high labour costs.

6.5.2 Hazardous nature

The minerals industry is inherently hazardous because of the nature and scale of its operations and its types of products and wastes. Globally, each year thousands of workers are killed and tens of thousands are injured in mining-related accidents. Most of these occur in mining operations as a result of leakage of poisonous gases (such as hydrogen sulfide) or flammable gases (especially methane), dust explosions, collapse of excavated spaces, flooding, improper use of equipment or

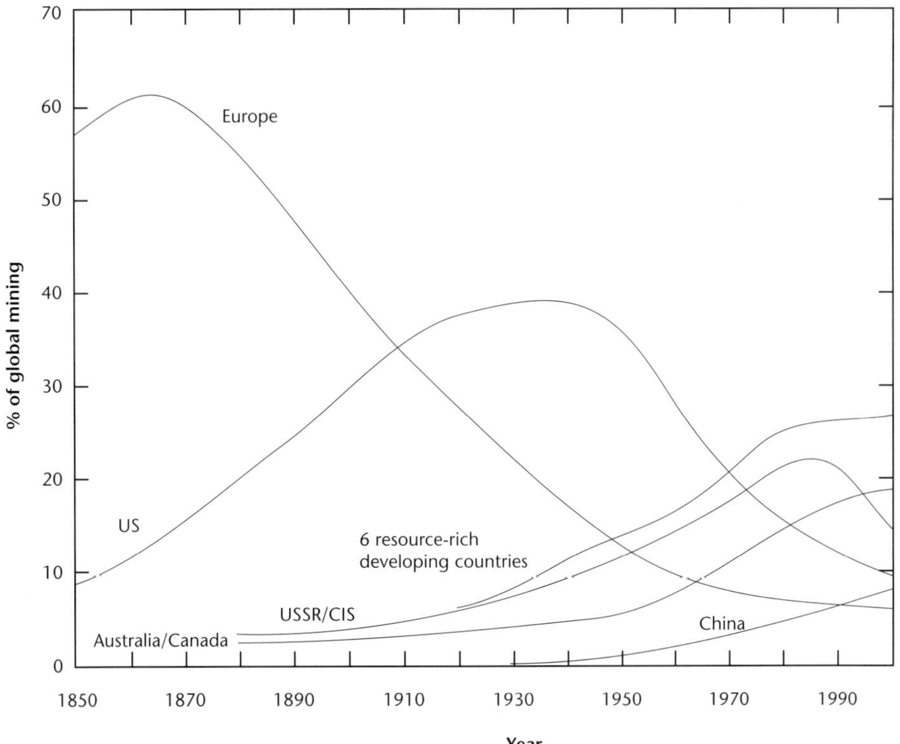

Figure 6.6: Trends in the geographical location of mineral production since 1850 (after Brett, 2006).

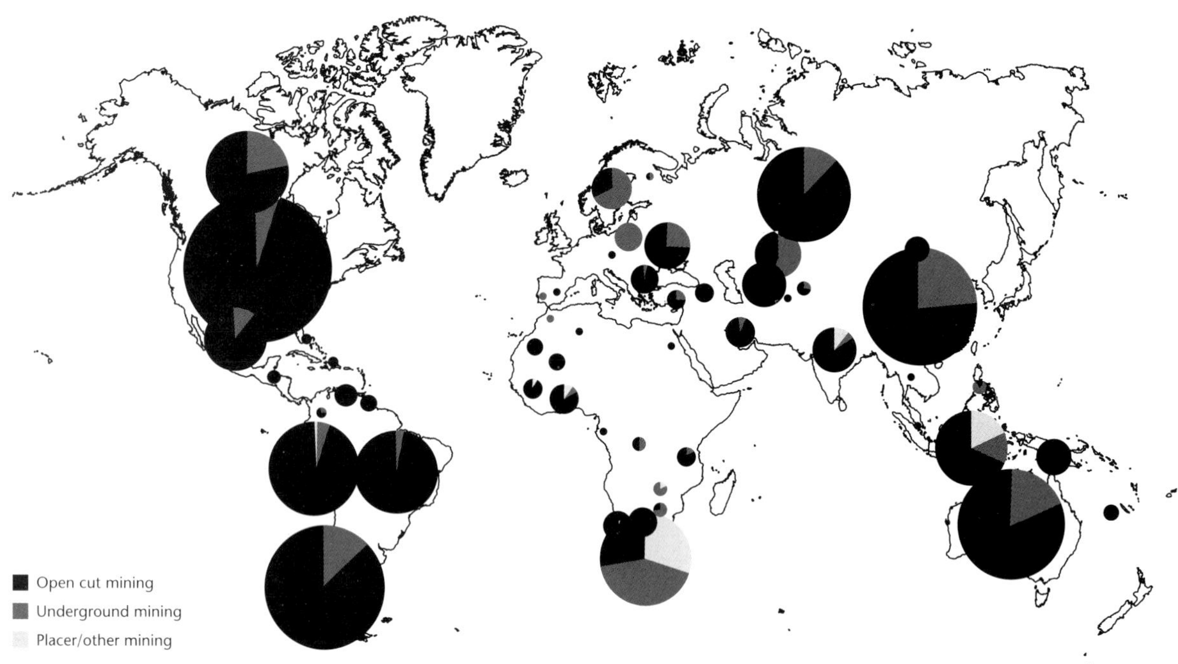

Figure 6.7: The present global distribution of mineral production (after Brett, 2006).

malfunctioning equipment. Most deaths and injuries occur in developing countries, especially in China, and most occur in the coal mining industry (over 3000 per year). Safety is a major concern of minerals companies in developed countries, particularly since the passage, in many jurisdictions, of legislation making company directors and senior managers legally liable for safety. Death and injury rates have fallen dramatically in recent decades but are still high relative to other industry sectors. In the United States, which has the largest mining industry in the developed world, the number of mining fatalities in 2006 and 2007 averaged 69 and the number of injuries was 11 800 per year (USDL, 2009).

Mining can have a huge environmental impact, both at the site where rock is extracted and on its surroundings as a result of the need to store or dispose of the large quantities of solid, liquid and gaseous wastes produced during mining and beneficiation, some of which are toxic. This aspect is examined in detail in Chapters 11 and 12.

6.5.3 Size and structure

If only significant operations are considered, worldwide there are probably a few thousand metallic ore mines and a similar number of coal mines. There may be around 25 000 mines producing various industrial minerals and perhaps 100 000 quarries producing aggregate materials. These figures are estimates; it is difficult to obtain accurate information. Because of the wide range in scale of mining operations, probably a few thousand mines account for the bulk of the world's production of coal and metals. In Africa, Asia and South America there are millions of artisanal and small-scale miners who make a subsistence living by mining, mainly for gold and precious stones (Starke, 2002). The industrial minerals and rocks sector covers scales of operation from artisanal mining, through family-owned and family-operated quarries producing aggregate and stone, to large mining and processing operations, though most are small- to medium-sized (Jeffrey, 2006). It often plays a major role in the local economy. Leaving aside artisanal and small-scale miners, the corporate minerals industry ranges in scale from small single-site companies to huge multinational companies with operations around the world, and from single-commodity producers to multiple-commodity producers. Some companies are vertically integrated and undertake all steps in the value chain – exploration, mining, beneficiation, smelting/refining, fabricating – while others restrict their activities to one or two steps in the value chain.

In 2001, the combined market capitalisation of the top 150 international minerals companies was only US$224 billion, smaller than both General Electric and ExxonMobil (Starke, 2002). The minerals industry, though big, was not large in the global context. It grew rapidly in value between 2002 and 2008, however, as a result of the accelerating demand for commodities, particularly from China. This was also a period of increasing consolidation and globalisation of companies through acquisitions and mergers. Table 6.6 lists the 19 largest minerals and metals companies in 2007, and several other large corporations for comparison. The market capitalisation of these 19 companies alone was US$1200 billion (about US$1000 billion in 2002 dollars). BHP Billiton, now the largest minerals company, had a market capitalisation equal to 70% that of the top 150 minerals and companies in 2001. BHP Billiton, Rio Tinto, Vale and ArcelorMittal are now among the world's largest companies.

For most of the 20th century a large portion of the minerals industry was government-owned, not only in the centrally planned economies of the former USSR, eastern Europe and China but also in India and some Middle Eastern, South American and Central African countries. Many were privatised during the 1980s and 1990s due to the collapse of the USSR and the spreading influence of neo-liberal economic theory. The minerals industry in China is still largely state-owned, the Chilean government owns the copper-producing company Codelco and there are state-owned minerals companies in the Middle East, India and Russia. In most other countries the industry is largely within the private sector, either privately owned or publicly listed on the stock market. In some developing countries the state takes a minority share in mining companies. As result, the minerals industry is now a global industry. Companies operating in the industry do so within the legal, regulatory, political and cultural framework of the countries where they are located.

Small-scale and artisanal miners exploit marginal and small deposits using labour-intensive, low-capital methods. Small-scale and artisanal mining plays an important role in the mineral economy and is estimated to contribute over one-sixth of global non-fuel mineral output (World Bank, 2010). Safety standards are low and the operations often have an adverse impact on the environment. Nevertheless, artisanal and small-scale mining makes an important economic contribution in rural areas of many developing countries. Globally, small-scale and artisanal mining employs many millions of people – gold mining employs 15 million people alone – and provides a livelihood for over 100 million people, almost all in developing countries (World Bank, 2010). Small-scale and artisanal mining involves both formal and informal activities, the latter operating outside the legal framework of the host country.

6.5.4 Minerals companies

There are many types of minerals companies and it is difficult to classify them. It is possible to distinguish several broad types, though there is considerable overlap: junior, middle-sized and multinational minerals companies, metal-producing companies, and recycling companies.

Junior mining companies (Juniors) are small companies which are mainly involved in mineral exploration. They raise capital privately or through equity issued on stock markets. They may explore for mineral deposits with a view to negotiating with larger companies to develop any ore bodies, or develop a small ore body themselves with their own resources so the mine becomes a source of income to support further exploration or expansion. The latter approach is common with gold deposits, which are often relatively cheap to bring into production. There are several thousand Juniors worldwide at any one time but they come and go relatively quickly through mergers, acquisition or failure. The failure rate is high due to the high financial risk associated with exploration.

Medium-sized mining companies operate several mines, sometimes in more than one country. They acquire existing mines or discover new deposits by their own exploration efforts and develop the mine, often with technical assistance from specialised service and engineering companies. Some companies focus almost solely on gold. Those that operate base metal mines usually sell metal concentrates to a trader

Table 6.6: Major minerals and metal commodity companies

Company	Head office	Commodities/products	Market value (US$bn)	Sales (US$bn)	Profits (US$bn)	Assets (US$bn)
BHP Billiton	Australia	Iron ore, aluminium, base metals, coal, manganese, petroleum, diamonds	190.62	39.50	13.42	53.36
Rio Tinto	United Kingdom	Iron ore, aluminium, coal, diamonds, copper, gold and silver, molybdenum, titania, industrial minerals	165.48	29.70	7.31	100.81
Vale	Brazil	Iron ore, nickel, coal, copper, manganese	161.39	33.23	10.26	74.70
ArcelorMittal	Luxemburg	Iron ore, carbon steel, stainless steel	108.82	105.22	10.37	133.65
Anglo American	United Kingdom	Platinum, diamonds, base metals, iron ore, steel, ferromanganese, coal, industrial minerals	85.05	25.47	5.29	44.29
Xstrata	Switzerland	Copper, coal, ferrochromium, nickel, vanadium, zinc	77.01	28.21	5.50	52.24
MMC Norilsk Nickel	Russia	Nickel, platinum, palladium, copper, speciality by-products	51.45	11.93	6.19	16.28
Posco	South Korea	Steel, speciality steels, steel products	49.21	27.91	3.58	33.35
Barrick Gold	Canada	Gold, copper	45.51	6.78	1.20	21.95
Baosteel	China	Steel, speciality steels, steel products	42.16	20.15	1.67	19.26
Freeport-McMoRan	United States	Gold, copper, molybdenum	38.52	16.94	2.98	40.04
Nippon Steel	Japan	Steel, speciality steels, steel products	33.85	36.61	2.99	45.24
Alcoa	United States	Alumina, primary aluminium, fabricated aluminium	30.25	30.75	2.56	38.80
JFE Holdings	Japan	Steel and steel products; also engineering and ship-building	26.24	27.75	2.55	32.84
Angang New Steel	China	Steel, speciality steels, steel products	25.39	6.94	0.91	7.53
Teck	Canada	Copper, coal, zinc, gold, molybdenum and speciality metals, oil sands	17.72	6.42	1.58	13.61
Tata Steel	India	Steel, speciality steels, steel products	14.63	5.83	0.97	11.48
Vedanta Resources	United Kingdom	Aluminium, copper, zinc, lead, iron ore	12.58	6.70	0.96	8.04
Sumitomo Metal Mining	Japan	Copper, iron ore, nickel	12.55	8.23	1.07	7.89
ExxonMobil	United States	Oil and gas	465.51	358.6	40.61	242.08
General Electric	United States	A conglomerate	330.93	172.74	22.21	795.34
Microsoft	United States	Software and IT services	253.15	57.90	16.96	67.34
Royal Dutch Shell	Netherlands	Oil and gas	221.09	355.78	31.33	266.22

Source: Financial data from Forbes Global 2000; 2 April 2008: http://www.forbes.com/lists/2008/18/biz_2000global08_The-Global-2000_Rank.html.
Commodity information from company annual reports.

or custom smelter. Many industrial minerals mines are operated by medium-sized companies.

The large multinational companies have historically been companies that explore, mine, smelt/leach, refine and sell metals and mineral products on world markets, though since the 1990s there has been a trend for these companies to move away from vertical integration. Many have sold their metal production facilities, such as smelters, rolling mills and other downstream operations, and rely to a great extent on Juniors to discover new deposits. Most multinationals are long established, though through mergers and takeovers the number of companies has been decreasing and the company size increasing. There are 30 to 40 companies that fit this category; some of the major ones are listed in Table 6.6. Though new names of companies frequently appear there are few genuinely new entrants into the ranks of the large multinationals, due to the technical complexity of exploration, mining and processing, and the marketing strengths of the existing companies. Name changes, which have been frequent over the past few decades, more often reflect mergers and acquisitions. Multinational iron ore and alumina companies trade on world markets. Iron ore producers are generally not steel producers (e.g. BHP Billiton, Rio Tinto, Vale) whereas alumina producers often also have smelters to produce aluminium metal (e.g. Alcoa, Rio Tinto).

Metal-producing companies purchase ores or concentrates and smelt or leach them to produce metals and alloys. They are usually located in industrialised areas and where there are large consumer markets, mainly in Europe, North America, Japan, South Korea, China and India. The availability of cheap and secure electrical power is a major consideration in the location of aluminium smelters since electricity is the major cost in transforming alumina to aluminium metal. Steel producers usually focus on steel production and often purchase iron ore and coking coal from other companies. They may extend into downstream value-adding through the production of a range of specialty steels and steel products. Some also have heavy engineering, ship-building or transport businesses, which use large quantities of steel. Some smelters, originally established to process locally mined base metal ores and concentrates or iron ore, now rely on imported raw materials since the mines have been worked out and closed. The decrease in the relative cost of transporting ores using bulk shipping has been a major factor in enabling this. Resource-poor countries such as Japan and South Korea established smelting and refining facilities as part of their industrialisation strategy though they lack large mineral resources. The products of metal-producing companies are traded on world metal markets. They are sometimes used in-house to produce semi-fabricated metal products and even consumer products.

Increasingly, primary metal-producing companies also produce secondary metals by processing scrap and other used materials in combination with fresh concentrate or in separate operations. Recycling companies and scrap merchants collect and sort used metal for recycling. Scrap collection varies from small family businesses, often an important part of the economy of developing countries, to large companies with technically sophisticated collection and recovery centres. These may operate internationally. Secondary smelters specialise in treating recycled metals, but much recycled metal is also recovered from scrap by primary smelters and refiners and steel companies.

The minerals industry is supported by a large number of private consultants, contractors, service companies and engineering companies. Service companies range from small specialist companies to large integrated organisations which perform activities such as evaluating new resources, designing mines and processing plants and undertaking environmental evaluation and monitoring, energy audits, testing of raw materials, analysis of production samples and so on. Some professional services companies are highly innovative in the area of systems thinking and planning. They employ staff from a range of professions, such as ecologists, environmental scientists and social scientists, in addition to traditional mining and metallurgical professionals. In recent years, a large professional services industry has developed around issues of sustainability in relation to minerals. Engineering companies undertake construction projects (mines and processing plants) and upgrades. Some mining companies outsource many operational aspects of mining and processing to individual contractors or companies, who contract to perform certain functions. These can include blasting and mining, maintenance, transport and analytical services. The number of people employed in service and support areas far exceeds the number of people directly employed by all the above types of companies. The services sector supporting the minerals industry is an important part of the economy of some countries, such as Canada, Australia and Finland.

6.5.5 Industry associations

Industry associations are organisations founded and financed by companies. They are usually non-profit. Their role is to provide a forum for companies to discuss issues of mutual interest and concern and to develop coordinated strategies in response. They attempt to influence public policy and the activities of regulatory bodies in directions believed to be favourable to the association's members. The types of activities undertaken by industry associations include public relations and advertising, educational programs, and lobbying. The form of lobbying varies widely according to the culture and legal framework of different countries. Advertising is usually undertaken to improve the industry's image or to shape public opinion on specific issues in order to influence public policy.

There are two broad types of mineral industry associations, regionally based and commodity based. Regionally based associations represent the interest of companies in a particular region irrespective of the commodities; for example, the Minerals Council of Australia, National Mining Association in the United States and Chamber of Mines of South Africa. Commodity-based associations represent the interests of companies producing particular commodities. They may be regional, national or international in scope; for example, the World Steel Association, International Council on Mining and Metals, Association of Tin Producing Countries, International Copper Association and International Aluminium Institute. Many international associations have national and/or regional affiliated associations. Most companies belong to more than one industry association.

6.5.6 Industry culture

The culture of companies varies greatly across the minerals sector according to the type of company and varies, to a lesser extent, between companies within the one type of industry. Companies involved in large-scale mining operate in a business environment which is very capital-intensive and has long lead times in developing new mines. Decisions to make major changes to existing technologies for mining and processing are taken rarely, only after long periods of assessment and when the potential financial benefits are high. Radically new technologies are infrequent; R&D and technical development activities are largely focused on incremental changes to existing processes to reduce costs, increase throughput, increase recovery, or improve safety and environmental performance. Historically, the long-term profit margin on most commodities has been small. Any production risk in a large-scale mining operation needs to be minimised, since profitability depends on small margins on large tonnages of material. Changes to one part of the production chain, if unsuccessful, frequently flow on to other parts of the chain, resulting in shut-downs or reduced levels of production. Management tends to focus on the short term. In good times, when prices of commodities are relatively high, management emphasis changes to increasing production with existing equipment and facilities. In bad times, the focus is on maintaining production levels while reducing costs in order to maintain the company's competitiveness. Decisions to expand facilities or to open and develop new mines are taken slowly and in a staged manner because a large investment must be made, often over many years, before production can commence and a return on the investment be realised. These companies adopt a hierarchical command and control type of management and are often perceived by the general public as risk-averse and conservative.

While slow to implement new technologies and open new mines, large-scale mining companies react quickly to market downturns to reduce costs. Historically, in these situations mining companies have frequently acted non-strategically. Expenditure on research and development, and on exploration, is cut back or stopped entirely and the number of employees, particularly in mid-level corporate roles, is rapidly reduced to lower costs even though in the longer term the companies invariably suffer from loss of expertise.

Hence employment in the industry is cyclical, demonstrated a number of times during the 20th century and in the downturn caused by the global financial crisis. The Australian mining industry shed 27 000 jobs in the first half of 2009. Had all industry sectors behaved the same way (and assuming no fall in the rate of participation in the labour force) the Australian unemployment rate would have increased from 4.4% to 19%, rather than to the actual value of 5.8% (Gittens, 2010). The human resource issue is compounded when the economy rises and companies seek to recruit staff. Many people made redundant in the downturn have obtained employment in other industry sectors or commenced early retirement. Fresh graduates will be in short supply – enrolments in tertiary courses that prepare students for the industry respond out of cycle with employment demand because of the time from intake to graduation. Though these patterns are well established and well known, and the industry make claims to have learned from experience, they are invariably repeated during each economic downturn.

Exploration companies and Juniors tend to be entrepreneurial and are sometimes based on the leadership and drive of one key person. They often have little interest in R&D because they have little or no cash flow, but are expert adopters and adapters of the latest technologies and knowledge. Smelting and refining companies and recyclers are more like medium-sized manufacturing companies in culture than like large-scale mining companies. Technology is important because their processes are complex and often unique, they purchase feed materials from others and need to be able to respond to changing grade and impurity levels, and their profit margins are often low. Smelting and refining companies are often more actively involved in research and development.

Different cultures are difficult to integrate within one company and the major culture tends to dominate. Thus, large multinationals which focus on mining have had difficulty maintaining successful exploration arms and R&D and technology groups because these tend to be managed in the same way as the production arms of the company with a focus on short-term results. It is probably partly for this reason that most of the large multinational companies are no longer vertically integrated; they now focus on mining, and process ore only to the extent necessary to produce a tradable commodity. Another reason is that chemical processing, through smelting or leaching, to make high value-added products such as refined metals requires the highest level of technical expertise and adaptability.

6.5.7 Trends shaping the industry

Companies in the minerals industry operate within the broader economy and within the local, national and global social and political environments. They are affected by and respond to trends within these environments. They are also strongly influenced by technological developments that occur outside their industry, particularly those that are transformational or enabling since these affect the uses and demand for their products and the technologies needed to produce them. There have been a number of major trends that have influenced the nature of the industry and that will continue to shape it for the foreseeable future.

Population growth. The global population will continue to grow at least until the middle of this century and this, with the increasing urbanisation of developing countries, will result in growing demand for mineral-based commodities. It will also change the relative demand for particular commodities. China and India, as rapidly developing countries with very large populations, will be the main sources of increased consumption of mineral commodities. Brazil and Russia, with large but significantly smaller populations, are also developing rapidly. Besides being growing consumers of mineral commodities, their large mineral resources mean that they will also become more important producers of mineral commodities.

Economic policy. The influence of neo-liberal economic ideology, which has dominated financial and economic theory for the past three decades, appears to be coming to an end as the social and environmental consequences of deregulation and an unrestrained free market become more apparent. Governments will most likely take a more pro-active role in regulating business and setting goals. This trend has been accelerated by the global financial crisis.

Globalisation. The globalisation of industries and markets, which has been occurring since the 1970s, will continue to affect minerals companies and their business and marketing strategies.

Information technology. Developments in information technology will continue to be an important enabler of technologies in the minerals industry, particularly in areas such as automation (e.g. of mining and processing operations), control of complex systems and handling of large quantities of complex data.

Climate change. The introduction of carbon taxes or emissions trading schemes in response to global warming would have major impacts on the minerals industry, both in the ways in which mineral commodities are produced (lower energy and water) and on the demand for particular commodities (less energy-intensive products). Attempts by minerals industry associations to delay implementation of these and/or seek exemptions are likely to prove counterproductive to the industry's long-term interests and to the economies of countries in which they operate.

Environmental concerns. Awareness of the fragile state of the natural environment, and of the crucial role it plays in sustaining life on Earth, will continue to grow and influence public policy, particularly on mining in environmentally sensitive areas, emissions, wastes and recycling. Major environmental catastrophes accelerate this growing awareness.

Occupational health and safety. Public pressure and government regulation are likely to continue the increased emphasis on health and safety in the workplace.

Companies that respond positively to these trends and seize the opportunities they create are likely to be the successful companies in the decades ahead.

6.6 THE ECONOMIC AND SOCIAL IMPACTS OF MINING

There are two strong views on the economic and social impacts of mining. One holds that mining can be a route to national development for resource-rich countries; Chile and Botswana are given as modern examples. The other view holds that countries or regions that are rich in mineral resources do not benefit from their exploitation to the extent expected, and that the benefits of mining largely go to developed nations while the risks are largely borne by developing nations. It is possible to point to many examples of mineral-rich and mineral-reliant countries (Table 6.5) that are poor, controlled by corrupt and dictatorial governments, politically unstable or remain underdeveloped. The latter view has had a long history. Over 450 years ago, Georgius Agricola (1556), summarising the views of detractors of mining, wrote, 'the inhabitants of these [mining] regions, on account of the devastation of their fields, woods, groves, brooks and rivers, find great difficulty in procuring the necessaries of life ... Thus it is said, it is clear to all that there is greater detriment from mining than the value of the metals which the mining produces'.

6.6.1 Mining as a route to development

Proponents of the view that mining is a route to development use the historical analogy that mining played a major role in the development, and subsequent prosperity, of Australia, Canada and the United States and can therefore play a similar role in underdeveloped countries today. While it is true that mining was significant at various times in the history of those countries, this does not imply that mining was the primary driver of development. Also, Australia, Canada and the United States differed from many of today's resource-rich countries in at least two significant ways (Pegg, 2006). First, each had stable political, legal and financial institutions before their mining industries became significant. Second, the countries developed in a different economic era, when transport costs and trade barriers were much higher. The cheap and rapid transport of ores and semi-finished manufactured goods around the world, and the removal of most trade barriers on manufactured goods during the 1980s and 1990s, has destroyed the historic link between mining and other sectors of local economies that often supported extensive economic development (Power, 2002). While exploitation of rich mineral resources may help a country develop and become prosperous, possessing mineral resources is not a sufficient condition for prosperity. Nor is it a necessary condition, since it is possible to point to many countries which have few mineral resources but which have strong economies and stable government; for example Japan, South Korea, Taiwan, Singapore and many western European countries.

6.6.2 The resources curse

During the 1980s, the idea that countries with natural resources might actually be at a disadvantage relative to countries with fewer natural resources began to gain currency. The term Resources Curse was coined by Auty (1993) to describe this. Numerous studies claimed to find supporting evidence and various explanations have been proposed for the apparent paradox. These may be broadly summarised as follows.

Revenue volatility. The prices of most natural resources are subject to wide fluctuation. When a government's revenues are dominated by royalties from natural resources, long-term planning becomes difficult because the short-term fluctuations in government revenue make it difficult to commit to long-term projects. As a result, income tends to be spent on short-term, poorly-considered projects.

Dutch disease. The revenue from the export of natural resources can cause a country's exchange rate and real wages to increase. This leads to other tradable sectors of the economy becoming less competitive, particularly agriculture and manufacturing. As a consequence, these decline and the economy becomes increasingly reliant on the export of mineral commodities. This occurred in the Netherlands during the 1960s, when it began producing and exporting natural gas from the North Sea fields. It also occurred in the United Kingdom in the 1970s, when it started producing and exporting petroleum from the oilfields off the coast of Scotland.

Government policy. Government policy which aims to protect industry from Dutch disease, for example by high tariffs or industry subsidies, compounds the problem by enabling those industry sectors to maintain inefficient practices. Also, governments with substantial income from natural resource rents may engage in excessive consumption which is not sustainable in the long term. They may misuse revenue by spending it on grandiose or non-productive projects rather than on, for example, education, health and infrastructure. A strategy increasingly adopted by resource-rich countries (e.g. the United Arab Emirates and Norway) is to create sovereign wealth funds, which keep commodity revenues off-shore by investment in international bonds. The objective is to build a revenue base for the future when resources income declines. These have had mixed success.

Corruption. Governments that are heavily reliant on income from natural resources feel less obligation to the people than do governments reliant on income and business taxes for their revenue. As a result, governments become less responsive to the needs and wishes of the people. There is less incentive for government to help raise the incomes of citizens and to develop businesses, since most of its revenue is obtained from royalties from natural resources. This leads progressively to corruption of the political processes and bureaucracy, to loss of liberties by citizens, to authoritarian government and to internal conflict.

More recently, there have been criticisms of the resources curse hypothesis which claim that when data are correctly examined the negative correlation between natural resource endowment and level of economic development either does not exist or is much weaker than claimed. Brunnschweiler and Bulte (2008) pointed out that a correlation between resource dependence, slow growth and conflict does not imply that slow growth and conflict are necessarily caused by resource dependence. Instead, they argue, the prior existence of weak institutions and conflict result in resource extraction becoming the default sector of the economy. Resource dependence is the final outcome and therefore is a symptom rather than a cause of underdevelopment. This seems plausible. Resource-rich countries with robust and relatively corruption-free institutions (e.g. Norway, Canada and Australia) can more easily capture the benefits of natural resource revenues. However, even in such countries it has long been claimed that development has been slowed by their natural resource endowment (see, e.g. Horne, 1964). Brunnschweiler and Bulte (2008) further claimed that the effect of mineral resource wealth (the quantity of mineral resources in the ground rather than the mineral resource revenue) on income growth is positive and significant.

Hartwick's rule

Hartwick (1977) identified theoretical conditions under which use of non-renewable resources is consistent with the indefinite maintenance of a given level of consumption or productive capacity, and hence living standard. His results are based on a mathematical proof, called Hartwick's rule, which built on the work of Hotelling (1931). Hartwick's rule states that if the scarcity rents derived from the extraction of non-renewable resources are saved and completely invested in reproducible (manufactured) capital then, under certain circumstances, the levels of output and consumption will remain constant over time. Hartwick's rule implies that a type of sustainability is theoretically achievable despite the existence and use of non-renewable resources. It is a version of weak sustainability. However, it provides no insight on how to do it globally, nor at the level of individual countries or regions. The key issue with Hartwick's rule is the extent to which other forms of productive assets can substitute for declining stocks of non-renewable resources (discussed in Section 4.3.2). However, it points to the need for underdeveloped resource-rich countries to invest scarcity rents wisely, for example, in infrastructure, education, health and the development of industries to supplant the mining industry as mining declines.

6.6.3 Artisanal and small-scale mining

It is not within the scope of this book to deal extensively with major social issues in relation to the minerals industry, important as they are. Those require examination from a sociological perspective. The focus of this book is on technology and technological approaches to environmental sustainability. Issues to do with artisanal and small-scale mining, however, are particularly important since these forms of mining have an important economic role in many developing countries. Also, the impact of mining on indigenous peoples is an important issue in many regions, including Africa, North and South America and Oceania.

Small-scale and artisanal mining has a role in alleviating poverty, increasing community capital and diversifying the local economy in many rural regions of the developing world, primarily because it is viable in areas with minimal infrastructure where other industries could not function. The relatively high income in small-scale and artisanal mining, compared with agriculture and construction, acts as a lure for financial and social independence in many communities (World Bank, 2010). Small-scale mining often provides employment for workers retrenched from large-scale mines. It is estimated that women account for as much as a third of the sector in some parts of the world. The interactions of small-scale and artisanal miners with large-scale corporate mining companies take a variety of forms. These range from violent confrontation to cooperative support which targets social development and poverty reduction (World Bank, 2010). Social and economic aspects of artisanal and small-scale mining were examined in a joint report by the International Institute for Environment and Development and WBCSD (Hentschel *et al.*, 2003), and the interaction of indigenous peoples with the minerals industry was surveyed in a second joint report (WBCSD, 2003). The reports consider ways in which the minerals industry can work constructively with both groups for mutual benefit.

6.7 THE MINERALS INDUSTRY AND SUSTAINABLE DEVELOPMENT

6.7.1 Industry developments and formation of the ICMM

In response to increasing concerns about the social and environmental impact of the minerals industry, and in line with the principles of corporate social responsibility, nine large minerals companies established the Global Mining Initiative (GMI) in 1998 and commissioned a three-year international project to explore the issues of sustainability in relation to the industry. This resulted in the major report 'Breaking new ground: mining, minerals and sustainable development' (Starke, 2002), often referred to as the MMSD report. Another outcome of the GMI was the formation in 2001 of the International Council on Mining and Metals (ICMM), a CEO-led industry association with headquarters in London, established to provide 'a platform for industry and other key stakeholders to share challenges and develop solutions based on sound science and the principles of sustainable development' (http://www.icmm.com).

The Mining, Minerals and Sustainable Development project examined the role of the minerals sector in contributing to sustainable development, and how that contribution could be increased (Starke, 2002).

Through extensive engagement with stakeholders in workshops and forums organised in several countries, nine key challenges facing the industry were identified. The following list is an extract from the project report (Starke, 2002).

Challenge 1. The minerals industry cannot contribute to sustainable development if companies cannot survive and succeed. This requires a safe, healthy, educated and committed workforce; access to capital; a social licence to operate; the ability to attract and maintain good managerial talent; and the opportunity for a return on investment.

Challenge 2. Mineral development is one of a number of often competing land uses. There is frequently a lack of planning or other frameworks to balance and manage possible uses. As a result, there are often problems and disagreement on issues such as compensation, resettlement, land claims of indigenous peoples, and protected areas.

Challenge 3. Minerals have the potential to contribute to poverty alleviation and broader economic development at the national level. Countries have realised this with mixed success. For this to be achieved, appropriate frameworks for the creation and management of mineral wealth must be in place. Additional challenges include corruption and determining the balance between local and national benefits.

Challenge 4. Minerals development can bring benefits at the local level. Recent trends towards smaller workforces and outsourcing, however, affect communities adversely. The social upheaval and inequitable distribution of benefits and costs within communities can also create social tension. Ensuring that improved health and education or economic activity will endure after mines close requires a level of planning that, too often, is not achieved.

Challenge 5. Minerals activities have a significant environmental impact. Managing these impacts more effectively requires dealing with unresolved issues of handling immense quantities of waste, developing ways of internalising the costs of acid drainage, improving both impact assessment and environmental management systems, and doing effective planning for mine closure.

Challenge 6. The use of minerals is essential for modern living, yet current patterns of use face a growing number of challenges, ranging from concerns about efficiency and waste minimisation to the risks associated with the use of certain minerals. Companies at different stages in the minerals chain can benefit from learning to work together to explore further recycling, reuse and remanufacture of products. They can develop integrated programs of product stewardship and supply chain assurance.

Challenge 7. Access to information is key to building greater trust and cooperation. The quality of information and its use, production, flow, accessibility and credibility affect the interaction of all actors in the sector. Effective public participation in decision-making requires information to be publicly available in an accessible form.

Challenge 8. Many millions of people make their living through artisanal and small-scale mining. It often provides an important, and sometimes the only, source of income. This part of the sector is characterised by low incomes, unsafe working conditions, serious environmental impacts, exposure to hazardous materials such as mercury vapours, and conflict with larger companies and governments.

Challenge 9. Sustainable development requires new integrated systems of governance. Most countries still lack the framework for turning minerals investment into sustainable development; these need to be developed. Voluntary codes and guidelines, stakeholder processes and other systems for promoting better practice in areas where government is unable to exercise an effective role as regulator are gaining favour as an expedient to address these problems. Lenders and other financial institutions can play a pivotal role in driving better practice.

In 2003, the ICMM Council committed corporate members to implement and measure their performance against 10 principles based on the issues identified in the MMSD report and reflected in the Johannesburg Declaration of 2002. The principles are listed in Table 6.7.

6.7.2 Sustainability reporting and sustainability indicators

Many companies produce annual sustainability or environmental reports in addition to the annual report required for public companies. Environmental and

Table 6.7: ICMM principles for member companies

Principle 1	Implement and maintain ethical business practices and sound systems of corporate governance.
Principle 2	Integrate sustainable development considerations within the corporate decision-making process.
Principle 3	Uphold fundamental human rights and respect cultures, customs and values in dealings with employees and others who are affected by our activities.
Principle 4	Implement risk management strategies based on valid data and sound science.
Principle 5	Seek continual improvement of our health and safety performance.
Principle 6	Seek continual improvement of our environmental performance.
Principle 7	Contribute to conservation of biodiversity and integrated approaches to land use planning.
Principle 8	Facilitate and encourage responsible product design, use, reuse, recycling and disposal of our products.
Principle 9	Contribute to the social, economic and institutional development of the communities in which we operate.
Principle 10	Implement effective and transparent engagement, communication and independently verified reporting arrangements with our stakeholders.

Source: http://www.icmm.com/our-work/sustainable-development-framework/10-principles.

social reporting is voluntary in most countries, and there are no regulations concerning form and content. The quality of reporting has been a major issue since its inception. Early reports were unreliable, many companies were selective about what they included in reports, and data were not comparable between reports of different years or between reports from different companies (Hopkinson and Whitaker, 1999).

The Global Reporting Initiative

Sustainability reports are now often prepared using the framework of the Global Reporting Initiative, or GRI (http://www.globalreporting.org). The GRI is a non-profit organisation based in Amsterdam, formed in 1997 with the support of the United Nations Environment Programme with the aim of raising the standard and consistency of reporting. About 1500 organisations from 60 countries used the GRI reporting framework and GRI sustainability reporting guidelines (GRI, 2006) in 2009 to produce sustainability reports. The reporting framework provides detailed guidance on preparing the report and is designed to be applicable to organisations of widely differing size, constituency and location. The sustainability report is structured around a CEO statement, key environmental, social and economic indicators, a profile of the reporting entity, descriptions of relevant policies and management systems, stakeholder relationships, management performance, operational performance, product performance and a sustainability overview. The guidelines specify the information relating to the triple bottom line performance of an organisation that should be reported. The sustainability indicators against which companies are required to report are listed in Appendix III.

Mineral industry reporting

In May 2008, the ICMM Council committed member companies to publicly report their sustainable development performance on an annual basis using the GRI reporting framework. This aimed to strengthen earlier commitments to public reporting and to demonstrate how members are addressing ICMM Principle 10. A Mining and Metals Sector Supplement (GRI, 2010) was developed collaboratively by the GRI and ICMM over a number of years and was released in March 2010. The Mining and Metals Sector Supplement complements the GRI guidelines and includes guidance for reporting on aspects of sustainable development relevant to the mining and metals sector. The supplement is intended to cover all the main activities of the mining and metals sector that may be of interest to stakeholders (mining and primary metal processing, including metal fabrication and recycling)

Table 6.8: Issues and challenges posing business risks for mining and metals companies

Issue	Challenge
Corruption	Given its exposure to corruption, how can industry leaders design broad guiding principles more specific to the industry and ensure that best practices are adopted?
Sustainable development	The mining and metals industry is committed to protecting the environment and public health. Nevertheless, a balance must be struck so the industry remains competitive and innovative.
Social inequality	The mining and metals sector operates in areas with social inequalities, thus the necessity to communicate the global implications of their local decisions to governments. How should local governance complement responsible mining investment to maximise local economic and social benefits?
Supply and demand	How will supply and demand for mineral products develop in light of the failed Doha Round,[a] volatile trade activities and hedge funds? What major risks exist?
Emerging economies	Growth in emerging countries is fuelling rises in commodity prices. Loosening restrictions allow access to foreign investors, while domestic companies from emerging economies are investing abroad to ensure supply and capture new markets.
Talent	It is increasingly difficult for the mining and metals industry to tap into the global talent pool to ensure its growth. How have the options for talent to pursue an attractive career changed in past years?
National champions and resource nationalism	In what ways do tensions between global business aspirations and local political interests manifest?
Role of governments in managing resources	What policy tools can governments employ to ensure the long-term management of resources? How can transitional tensions be avoided? What role should companies play in helping governments build the necessary local capacity?
Reputation	What must the mining and metals industry do to communicate its improved actual performance on social and environmental responsibility and gain a better reputation?
Indigenous peoples	The industry often explores in remote areas inhabited by indigenous peoples. How can the fair treatment of all involved parties be guaranteed?
Supply chain	Not only are prices for key components and services rising, but there is a serious shortage of equipment, shipping capacity, ports and water in several regions. What can be done to secure the products and services required to quench the world's thirst for natural resources?

a) The Doha Round or Doha Development Agenda is the current trade negotiation round of the World Trade Organization, which commenced in November 2001. Its objective is to lower trade barriers around the world.
Source: WEF (2010).

and the entire mining and metals project life cycle (from exploration, through project development, operation, to closure and post-closure).

A review of the social and environmental reports for 2003 of the then top 10 mining companies revealed a patchy performance (Jenkins and Yakovleva, 2006). Some companies trailed well behind others in the quality of reporting and it was not possible to measure the overall performance of the global industry. A review of the 2007 sustainability reports of 25 major mining companies revealed there was still wide variation in the quality of reporting (Mudd, 2009). For example, in a only few reports was energy usage broken down into usage at plant and process level, information which is of most use. In most reports, energy usage was given only for an entire mine site or for the company as a whole. Sometimes only the direct energy consumption was reported, thereby making it impossible to compare the embodied energy in the products produced. Rarely was the quantity of waste rock produced during mining reported, even though waste rock is by far the largest quantity of waste produced at a mine. The trends in the values of indicators over a period of years (e.g. three to five years) were rarely provided, unlike in financial reports, thereby making it difficult to monitor performance over time. It is not apparent why companies choose to present results in ways that cause obfuscation rather than in ways that make comparisons and trends quite clear. By doing this, companies undermine their own claims of supporting and fostering sustainability and corporate social responsibility. The situation may improve when companies start reporting using the Mining and Metals Sector Supplement.

6.7.3 Status of the industry

Medveçka and Bangerter (2007) claimed that, with respect to sustainability, minerals and metals companies are essentially Compliers. In Medveçka and Bangerter's classification (*Compliers*, *Marketeers*, *Innovators* and *Localisers* – see Table 4.6), Compliers are organisations committed to risk reduction. They believe in the concept of weak sustainability. For them the business case for sustainable development is clear – it is a necessity to mitigate risks, both environmental and social. Managing reputation and ensuring optimum conditions to secure licence-to-operate, access to land, finance and markets are prime concerns. This view is reflected in a recent statement from the Future of Mining and Metals Global Agenda Council (a council of the World Economic Forum which consists of senior industry executives and government and academic representatives), that 'The mining and metals industry is driven by shareholder value and wealth creation and guided by supply and demand' and that 'tensions between the industry's global commercial aspirations and the political and social interests of national and local players are considered to be one of the most pressing risks facing the industry' (WEF, 2010). The specific issues and challenges identified by the Council are listed in Table 6.8. The commentary on these issues and challenges reflect a view that the commercial future of mining and metals companies depends on a largely business-as-usual approach, combined with strategies to minimise the associated business risks. In summary, mining companies, while having adopted the principles of corporate social responsibility to varying degrees, are still at an early and tentative stage of making a real commitment to the principles of sustainability and embedding them in company planning, decision-making and operations.

6.8 REFERENCES

Agricola G (1556) *De Re Metallica*. English translation by HH Hoover and LH Hoover. Dover Publications: New York, 1950.

Auty RM (1993) *Sustaining Development in Mineral Economies: The Resource Curse Thesis*. Routledge: London.

Brett D (2006) Trends and players in the mining industry. Presented at *United Nations Conference on Trade and Development, Expert Meeting on FDI in Natural Resources, 20–22 November, 2006*, Geneva. <http://www.unctad.org>.

Broadhurst JL (2006) Generalised strategy for predicting environmental characteristics of solid mineral waste – a focus on copper. PhD thesis. University of Cape Town, South Africa.

Brunnschweiler CN and Bulte EH (2008) Linking natural resources to slow growth and more conflict. *Science* **320**(5876): 616–617.

Delgado L, Catarino AS, Eder P, Litten D, Luo Z and Villanueva A (2009) *End-of-Waste Criteria*. Joint Research Centre, Institute for Prospective Technological Studies, Office for Official Publications of the European Communities: Luxembourg. <http://ipts.jrc.ec.europa.eu/publications/pub.cfm?id=2619>.

Forbes Global 2000. <http://www.forbes.com/lists/2008/18/biz_2000global08_The-Global-2000_Rank.html>.

Gittens R (2010) Miners dig up a load of bulldust. *The Age*, 29 May 2010, Opinion & Analysis, p. 2.

GRI (2006) *Sustainability Reporting Guidelines*. Global Reporting Initiative: Amsterdam, The Netherlands. <http://www.globalreporting.org/ReportingFramework/ReportingFrameworkDownloads>.

GRI (2010) *Sustainability Reporting Guidelines and Mining and Metals Sector Supplement*, version 3.0, March 2010. Global Reporting Initiative: Amsterdam, The Netherlands. <http://www.icmm.com/library>.

Hartwick JM (1977) Intergenerational equity and the investment of rents from exhaustible resources. *American Economic Review* **67**: 972–974.

Hentschel T, Hruschka F and Priester M (2003) *Artisanal and Small-Scale Mining – Challenges and Opportunities.* International Institute for Environment and Development and WBCSD: London. <http://www.iied.org>.

Hooke M (2008) Re-establishing Australia as a global supplier of minerals. Keynote Address, *First International Future Mining Conference and Exhibition*, 19–21 November 2008, University of New South Wales, Sydney. <http://www.minerals.org.au/_data/assets/pdf_file/0018/32760/MHH_Intl_Future_Mining_Conference_191108.pdf>.

Hopkinson P and Whitaker M (1999) The relationship between company environmental reports and their environmental performance: a study of the UK water industry. In *Sustainable Measures: Evaluation and Reporting of Environmental and Social Performance.* (Eds M Bennett and P James) pp. 392–411. Greenleaf Publishing: Sheffield, UK.

Horne D (1964) *The Lucky Country: Australia in the Sixties.* Penguin Books: Melbourne.

Hotelling H (1931) The economics of exhaustible resources. *Journal of Political Economics* **39**: 137–175.

IEA (2009) *Key World Energy Statistics.* International Energy Agency: Paris, France. <http://www.iea.org/publications/free_all.asp>.

Jeffrey K (2006) Characteristics of the industrial minerals sector. In *Industrial Minerals and Rocks.* 7th edn. (Eds JE Kogel, NC Trivedi, JM Barker and ST Krukowski) pp. 3–6. Society for Mining, Metallurgy and Exploration: Littleton, CO.

Jenkins H and Yakovleva N (2006) Corporate social responsibility in the mining industry: exploring trends in social and environmental disclosure. *Journal of Cleaner Production* **14**: 271–284.

Kelly TD and Matos GR (2009) Historical statistics for mineral and material commodities in the United States. *United States Geological Survey Data Series 140.* United States Geological Survey: Reston, VA. <minerals.usgs.gov/ds/2005/140#data>.

Livernois J (2009) The empirical significance of the Hotelling rule. *Review of Environmental Economics and Policy* **3**(1): 22–41.

Mäkinen T and Taskinen P (2008) State of the art in nickel smelting: direct Outokumpu nickel technology. *Mineral Processing and Extractive Metallurgy* **117**(2): 86–94.

Medveçka J and Bangerter P (2007) Engineering sustainable development into industry: unlocking institutional barriers. In *Cu2007. Vol. 6 – Sustainable Development, HS&E and Recycling.* (Eds D Rodier and W Adams) pp. 13–26. Canadian Institute of Mining, Metallurgy and Petroleum: Toronto, Canada.

Mudd GM (2009) Sustainability reporting and mining – an assessment of the state of play for environmental indicators. In *Sustainable Development Indicators in the Minerals Industry.* pp. 77–391. Australasian Institute of Mining and Metallurgy: Melbourne.

Pegg S (2006) Mining and poverty reduction: transforming rhetoric into reality. *Journal of Cleaner Production* **14**: 376–387.

Planning for Integrated Mine Closure: Toolkit (2008) International Council on Mining and Metals: London. <http://www.icmm.com>.

Power TM (2002) *Digging to Development? A Historical Look at Mining and Economic Development.* Oxfam America: Washington DC. <http://www.oxfamamerica.org/files/OA-Digging_to_Development.pdf>.

Ricardo D (1817) On the principles of political economy and taxation. In *The Works and Correspondence of David Ricardo.* 11 vols, 1951–1973. (Ed. P Sraffa). Cambridge University Press: Cambridge. <http://www.econlib.org/library/Ricardo/ricP.html>.

Schmidt RH (1988) Hotelling's rule repealed? An examination of exhaustible resource pricing. *Federal Reserve Bank of San Francisco Economic Review* **4**: 41–54.

Sinclair RJ (2005) *The Extractive Metallurgy of Zinc.* Australasian Institute of Mining and Metallurgy, Spectrum Series Vol. 13. Australasian Institute of Mining and Metallurgy: Melbourne.

Sinclair RJ (2009) *The Extractive Metallurgy of Lead, Melbourne.* Australasian Institute of Mining and Metallurgy, Spectrum Series Vol. 15. Australasian Institute of Mining and Metallurgy: Melbourne.

Slade ME (1982) Trends in natural-resource commodity prices: an analysis of the time domain.

Journal of Environmental Economics and Management **9**: 122–137.

Starke L (Ed.) (2002) *Breaking New Ground: Mining, Minerals and Sustainable Development.* Earthscan Publications: London. <http://www.iied.org/sustainable-markets/key-issues/business-and-sustainable-development/mmsd-final-report>.

Suslick SB and Machado IF (2003) Non-renewable resources. In *Encyclopedia of Life Support Systems (EOLSS).* EOLSS Publishers: Oxford, UK. <http://www.eolss.net>.

TZ Minerals (2008) *Mineral Sands Annual Review.* TZ Minerals International: Victoria Park, WA.

USDL (2009) *Mine Safety and Health at a Glance: Fact Sheet.* Mine Safety and Health Administration, United States Department of Labor: Washington DC. <http://www.msha.gov/MSHAINFO/FactSheets/MSHAFCT10.HTM>.

USGS (2007) *Minerals Yearbook, vol. 1. 2007.* United States Geological Survey: Reston, VA. <http://minerals.usgs.gov/minerals/pubs/myb.html>.

WBCSD (2003) *Finding Common Ground – Indigenous Peoples and Their Association with the Mining Sector.* International Institute for Environment and Development and World Business Council for Sustainable Development: London. <http://www.iied.org>.

WEF (2010) *Shaping the Future of the Mining & Metals Industry.* World Economic Forum. <http://www.weforum.org/en/ip/MiningandMetals/index.htm> (accessed 28/11/10).

World Bank (2010) *Working Together: How Large-scale Mining Can Engage with Artisanal and Small-scale Miners.* International Finance Corporation, World Bank, 2010. <http://www.icmm.com/document/789>.

6.9 USEFUL SOURCES OF INFORMATION

Bleischwitz R, Welfens PJJ and Zhang Z (2009) *Sustainable Growth and Resource Productivity.* Greenleaf Publishing: Sheffield, UK.

Harben PW and Bates R (1990) *Industrial Minerals Geology and World Deposits.* Industrial Minerals Division, Metals Bulletin: London.

Kogel JE, Trivedi NC, Barker JM and Krukowski ST (Eds) (2006) *Industrial Minerals and Rocks.* 7th edn. Society for Mining, Metallurgy and Exploration: Littleton, CO.

Maxwell P and Guj P (Eds) (2006) *Australian Mineral Economics – A Survey of Important Issues.* Monograph no. 24. Australasian Institute of Mining and Metallurgy: Melbourne.

Richards J (Ed.) (2010) *Mining, Society, and a Sustainable World.* Springer-Verlag: Berlin.

Starke L (Ed.) (2002) *Breaking New Ground: Mining, Minerals and Sustainable Development.* Earthscan Publications: London. <http://www.iied.org/sustainable-markets/key-issues/business-and-sutainable-development/mmsd-final-report>.

The InfoMine and Mining.com websites (<http://www.infomine.com> and <http://www.mining.com>) have a wealth of information on minerals companies, mineral technologies, and recent events.

TZ Minerals (2008) *Mineral Sands Annual Review.* TZ Minerals International: Victoria Park, WA.

United States Geological Survey, Minerals Yearbooks and Mineral Commodities Surveys. <http://minerals.usgs.gov/minerals/pubs>.

World Bank (2010) *Working Together: How Large-scale Mining Can Engage with Artisanal and Small-scale Miners.* International Finance Corporation, World Bank. <http://www.icmm.com/document/789>.

7 Producing ores and concentrates

The production stage of a mine during which ore is extracted is part of the life cycle of a mine, as discussed in Section 6.4. It is also part of the materials cycle, as illustrated in Figure 2.2. Activities of the materials cycle other than mining are also performed at the site of a mine. Usually some processing of mined material is undertaken to transform it into a saleable product. This could be very simple physical operations, such as in a quarry where the materials might be crushed and separated into different size fractions or washed to remove fine material. If the products are building stones, they may be cut and shaped to specification. If it is an industrial mineral or metallic ore mine, the mined material may be beneficiated to produce a concentrate of the valuable mineral(s) and the waste material stockpiled or returned to the mine. Further processing may occur on site, such as leaching or smelting to produce metals, or the concentrate may be sold and transported elsewhere for processing. The large-scale operations most frequently undertaken at a mine site are extracting rock from the crust (mining), beneficiating the mined material, and treating, storing and disposing of the solid and liquid wastes produced. These activities are all potentially hazardous to humans and the environment.

This chapter examines how rocks are extracted from the crust and how mined material is beneficiated. The treatment, storage and disposal of the solid, liquid and gaseous wastes is discussed in Chapter 12. The emphasis is on the technologies used, the available alternatives, and the main criteria for choosing between them. Only a brief description is given, sufficient to provide an overview and a simple understanding of the scientific basis of the technologies. More detailed descriptions are readily available in the standard texts for mining engineering and mineral processing listed at the end of the chapter. Some specific aspects are considered as necessary in later chapters.

7.1 EXTRACTING ROCK FROM THE CRUST

The process for extracting material from the Earth is called *mining*. Mining is a large-scale operation involving large equipment and the extraction and movement of huge quantities of material – tens of thousands to hundreds of thousands of tonnes per day at many mining operations. There are three broad methods of mining: surface mining, underground mining and solution or *in situ* mining. The method selected depends on the characteristics of each deposit and the restrictions imposed by safety, technology, environmental concerns and economics. Surface mining is usually used to extract massive deposits that are relatively near the surface. The term massive, as used in this context, has nothing to do with size; it refers to a mineral deposit that is relatively homogeneous and conforms to the structure of the host rock. Quarrying is a form of small-scale surface mining, usually for building stones and aggregate materials. Underground mining is usually used for deep mineral deposits and deposits that are more disseminated. Solution mining involves dissolving the minerals *in situ* by passing a solution through the mineral deposit. The solution is then treated in a separate operation to recover the valuable elements.

Surface mining and underground mining involve two basic tasks: breaking the ore and transporting it to the beneficiation plant. The technique used for breaking rock depends on whether the rock is hard (usually igneous and metamorphic rocks) or soft (usually sedimentary rocks). In hard rock mining, holes are drilled in a pattern in the rock and packed with explosives. These are detonated to break the rock. Drilling is done using rotary or percussion drills driven by compressed air. Drilling muds (see Table 5.5) are used to lubricate and cool the drill bits. The most common type of explosive is a mixture of ammonium nitrate and fuel oil (ANFO). In soft rock mining, the rock is broken up by pneumatically driven hammers or cut by mechanical shears. Transport involves loading the rock and hauling it (horizontally) using trucks, shuttle cars, trains or belt conveyers and hoisting it (vertically) in skips or buckets.

7.1.1 Surface mining

About 80% of all ore and rock extracted from the crust is obtained from surface mines. In surface mining, material lying over the mineral deposit is referred to as *overburden*. This material and, in the case of open pit mining, *waste rock* from the sides of the pit, must be removed in order to gain access to the mineral deposit. This can be a large amount relative to the amount of ore, the ratio being called the *stripping ratio*:

$$\text{Stripping ratio} = \frac{\text{Mass of waste removed}}{\text{Mass of ore extracted}}$$

The stripping ratio is usually greater than 1, and is typically in the range 1–4 for metallic mineral mines (but can be up to 50 or 60) depending on the value of the mineral being mined. The development of a surface mine involves first removing the topsoil then stripping the overburden. To do the latter, blast holes are drilled through the overburden, loaded with explosives, and discharged to shatter the rock which is then removed and stockpiled for backfilling into the mine after ore has been removed. There are three basic approaches to surface mining: open pit mining, strip mining and placer mining.

Open pit mining

Figure 7.1 shows an idealised cross-section of an open pit mine. Most open pit mines are developed in the form of a bowl in which benches (terraces) in the walls help to stabilise the sloping walls. They also enable a number of areas to be worked at the same time. The terraces are often arranged concentrically. The operations in open pit mining are drilling, blasting, loading and hauling. Ore is either blasted off benches, from where it is removed, or faces are fractured *in situ* and the broken rock removed. Loading of ore for haulage is done using mechanical shovels, draglines or front-end loaders according to the scale of the operation. Haulage of ore out of the mine is usually done by trucks following a spiral road around the pit, although railways and belt conveyers are also used. The cost of open pit mining usually increases with depth because of the increasing distance the ore has to be moved to the surface and the increasing quantity of overburden and side wall material that must be removed to expose the ore and maintain stable walls. Deposits generally decline in grade outwards, so the cut-off grade and therefore the final size of the pit vary greatly with the prevailing economic conditions. Sometimes, as the pit becomes deeper, it becomes more economical to convert to underground mining to extract the remaining ore. Remediation of open pit mines is undertaken when mining operations cease. Overburden and other waste materials are then returned to the pit.

Strip mining

Strip mining is generally carried out on a very large scale and is relatively low-cost. It is particularly applicable to shallow, flat deposits of coal, oil-shale, clay, sand and gravel and some placer deposits, such as

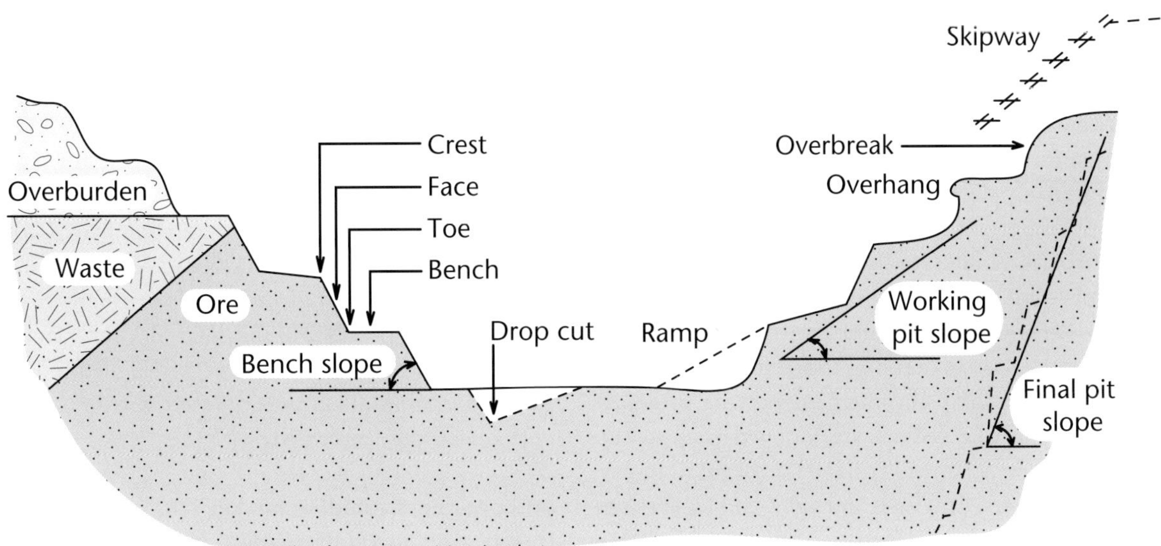

Figure 7.1: Schematic representation of a typical open pit mining operation (after Peters, 1978).

bauxite. As in open pit mining, it involves removing the overburden and extracting the ore but the mining operation progressively moves forward along a front that is perpendicular to the direction of advance. As the overburden is removed from a new area it is backfilled into the trench left by the previous removal (Figure 7.2). Storage of solid mining waste is therefore less of a problem than in open pit mining. Remedia-

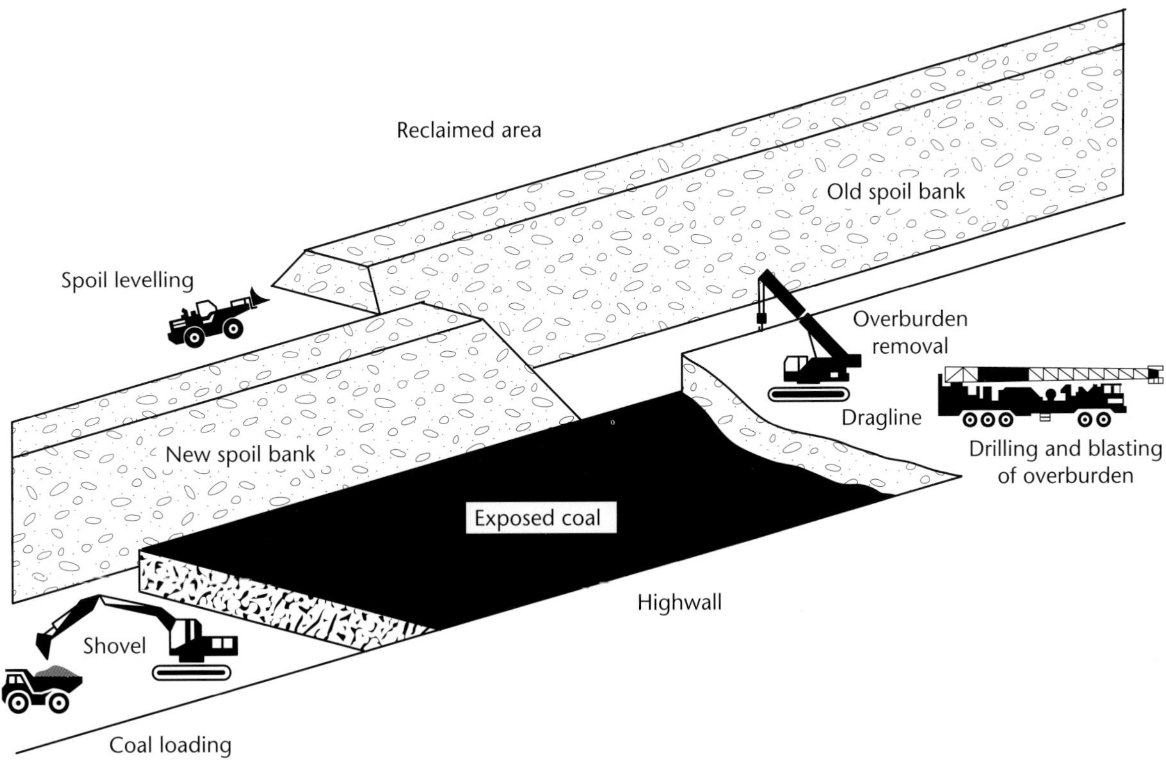

Figure 7.2: Schematic representation of a typical strip mining operation (after Hartman, 1992).

tion is also easier, and can be done to leave little evidence of the mining operation.

Placer mining

Placer mining is a method for recovering valuable minerals from placer deposits at the surface. It is particularly important for recovery of many heavy minerals such as cassiterite, ilmenite, rutile and zircon, and for diamonds and other gemstones. Several methods are used. In hydraulic mining, high-pressure water jets are used to break up the deposit and carry it to the beneficiation plant where the valuable minerals are separated. The tailings are pumped back to the area of the deposit already mined for disposal. In dry areas, bulldozers, draglines and front-end loaders are used to remove the overburden and push the placer material to the beneficiation plant. Dredging is a high-volume mining method used in locations where water is plentiful. Mechanical buckets or draglines are used to extract material from shallow water (which may be natural or artificial ponds) and transport it to the beneficiation plant, which is often mounted on pontoons and located in the lagoon being worked.

7.1.2 Underground mining

The depth of an underground mine depends on the depth of the ore body, and increases as the ore is mined. Mines may be relatively close to the surface or several thousand metres deep. For example, the copper mine at Mt Isa in Australia, a long-established mine, is now around 1800 m deep. The TauTona gold mine in South Africa is 3900 m deep and is the deepest mine in the world. Figure 7.3 shows two idealised sections of underground mines and explains the common terminology. In underground mining, passages are dug from the surface to gain access to the ore body and to bring ore to the surface. Vertical or slightly inclined passages are called *shafts*; horizontal or nearly horizontal passages (e.g. passages driven into a side of a hill or the side of an open pit mine) are called *adits*. Other openings and excavations are made to provide drainage and ventilation, and underground rooms are usually made for workshops, storage, meal areas and so on. Underground passageways used for working the mine are called *drifts* if they run parallel to the geological structure and *cross-cuts* if they cut across it. The workings on one level are joined to those on other levels by passageways called *raises*, if they are excavated from below, and *winzes*, if they are excavated from above. These give access to the *stopes*, which are the zones of ore to be mined between levels. Underground mines are worked in blocks of levels and stopes. Because very much less waste rock has to be removed to gain access to the ore compared with surface mining, the waste rock:ore ratio is very much smaller for underground mining. It is usually much less than 1, typically around 0.05–0.3 for metallic mineral mines. There are three broad methods of underground mining: unsupported methods, supported methods, and caving.

Unsupported methods of mining

Unsupported methods are used for hard rock, tabular-shaped mineral deposits. No artificial supports are used to prevent collapse of worked areas, but rock bolts in the roofs may be used to strengthen them. The most common unsupported method is *room and pillar mining*, in which support for the roof is provided by natural pillars of rock left standing in a systematic pattern. It is a relatively wasteful method, as removal of the pillars to recover more ore is dangerous. Therefore, room and pillar mining is usually used for extracting low-value material. *Stope and pillar mining* is similar and is used for thicker and more irregular mineral deposits. The pillars are spaced irregularly and located in lower-grade regions so that the higher-grade ore can be extracted. *Shrinkage stoping* and *sublevel stoping* are variations used for deposits which are steeply sloping. In shrinkage stoping, the ore is mined in an upwards direction by blasting layers of ore from the exposed overhead surface of the stope. This is allowed to accumulate on the floor, providing a platform for the miners, and is periodically 'shrunk' by drawing just sufficient ore through raises to give miners access to the unbroken ore above. When the last of the stope has been blasted, the remaining broken ore is withdrawn through chutes at the base. In sublevel stoping, the stope is mined horizontally by blasting slices from the face and allowing the rock to fall into sublevels, from where it is removed.

Supported methods of mining

Supported mining methods are usually used for deposits with weak rock structure. *Cut and fill stoping*

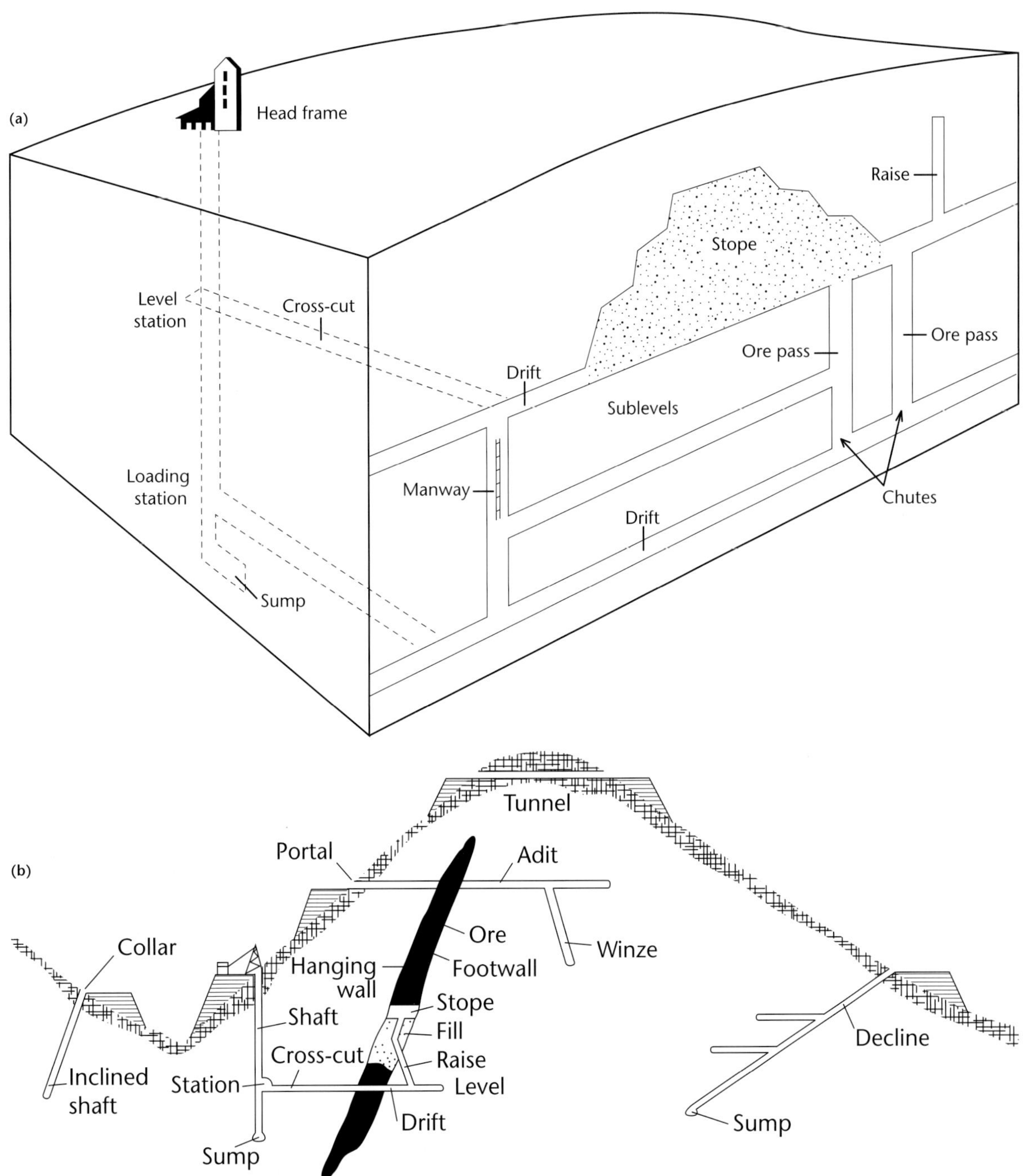

Figure 7.3: Underground mining concepts and terminology. (a) After Peters (1978). (b) After Hartman (1992).

is the most common method. It is used for steeply sloping and irregular deposits and/or where one or both walls lack the strength to stand up during mining. It is similar to shrinkage stoping except that as each slice of ore is removed an equivalent quantity of waste material, such as waste rock or tailings, often with the addition of cement for strength, is placed in the worked-out area below or behind the advancing front. *Square set stoping* is a method in which a small block of ore is removed and immediately replaced by a set or framework of timber which interlocks with other sets. These are filled with rock waste or tailings after the ore is extracted. This method is generally used only for the extraction of high-grade ore bodies where rock walls are not strong enough to support an opening. It is slow and expensive.

Caving methods of mining

Caving methods involve caving the ore and/or overlying rock. This involves allowing the ore or rock to collapse into an open space by drilling and blasting. Subsidence of the surface above the mine usually occurs as a result. *Longwall mining* is a soft rock caving method used mainly in the extraction of coal and other flat-lying deposits which occur in thin seams. Large rectangular blocks, called panels, of coal or ore (typically several hundred metres wide, 1000–3500 m long and several metres thick) are defined, then mined in a single continuous operation. Parallel roadways are driven along each side of a panel and joined to form the starting face, perpendicular to the length of the panel. A shearing machine (with rotating drums with teeth) or plow (a form of mechanical chisel) travels back and forth across the face, cutting a slice on each pass. A conveyor running along the length of the face carries this material away. The area immediately in front of the face is supported by hydraulic roof supports, which are progressively moved forward as the face advances. The roof behind is allowed to collapse into the void. The *block caving* method is used for large, relatively low-grade ore bodies. Adits are driven under the ore to be mined; along each, a series of raises is driven upwards, in turn connected to several smaller finger raises. The block of ore above is blasted; the ore collapses and progressively moves through the finger raises into the adits, from where it is hauled away. As the broken ore is removed, the overlying rock collapses into the space left behind. *Sublevel caving* is a method in which thin blocks of ore are caved by successively undermining small panels. The sublevels are spaced vertically at regular intervals throughout the deposit and mining is commenced from the top level. The broken rock is removed after each blast and the overlying waste rock caves on top of the broken ore. This method is inexpensive and is usually used in massive, steeply dipping deposits. Since recoveries are low, sublevel caving is used to mine low-grade, low-value deposits.

7.1.3 Solution mining

Solution mining, or *in situ* mining, involves the selective leaching of minerals from an ore body without removing the ore from the deposit. It is an alternative to mechanical mining in some applications because of its relatively low capital and operating costs and its short planning and construction times. Solution mining is illustrated in Figure 7.4. It has a small environmental impact at the surface, but groundwater contamination is always a serious potential problem and the method is applicable to only a limited number of commodities and types of deposits. Minerals that are soluble in water can, in principle, be extracted *in situ* (potash, common salt). Minerals that are soluble in aqueous solvents (acids, alkalis, cyanide or others) such as many copper, nickel, uranium and gold minerals can be extracted *in situ* if they are within deposits that trap the solution effectively and prevent it contaminating the groundwater. To date, uranium and copper are the only metals extracted *in situ* on a commercial scale. Sulfur is extracted from salt domes using the Frasch process, the oldest *in situ* mining process: hot water is injected into the deposit to heat and melt the sulfur to enable it to be pumped to the surface. The mineral deposit must be sufficiently permeable for the solution to contact the desired mineral, and react. If a deposit is not naturally permeable this may be induced by various methods, including hydrofracturing (using pressurised water), use of explosives, and undercutting and caving. The pattern and spacing of the addition of the solvent and of sumps for collecting the solution are important design parameters and vary with each deposit. *In situ* leaching is discussed further in Section 8.2.4.

7.2 BENEFICIATING MINED MATERIAL

If the physical state of mined material is changed in any way to increase its value and/or make it easier to process chemically (e.g. to extract a metal or make a pure compound), the material is said to be beneficiated. The process is called *beneficiation* or *mineral processing*. This is usually done at the mine site. It may

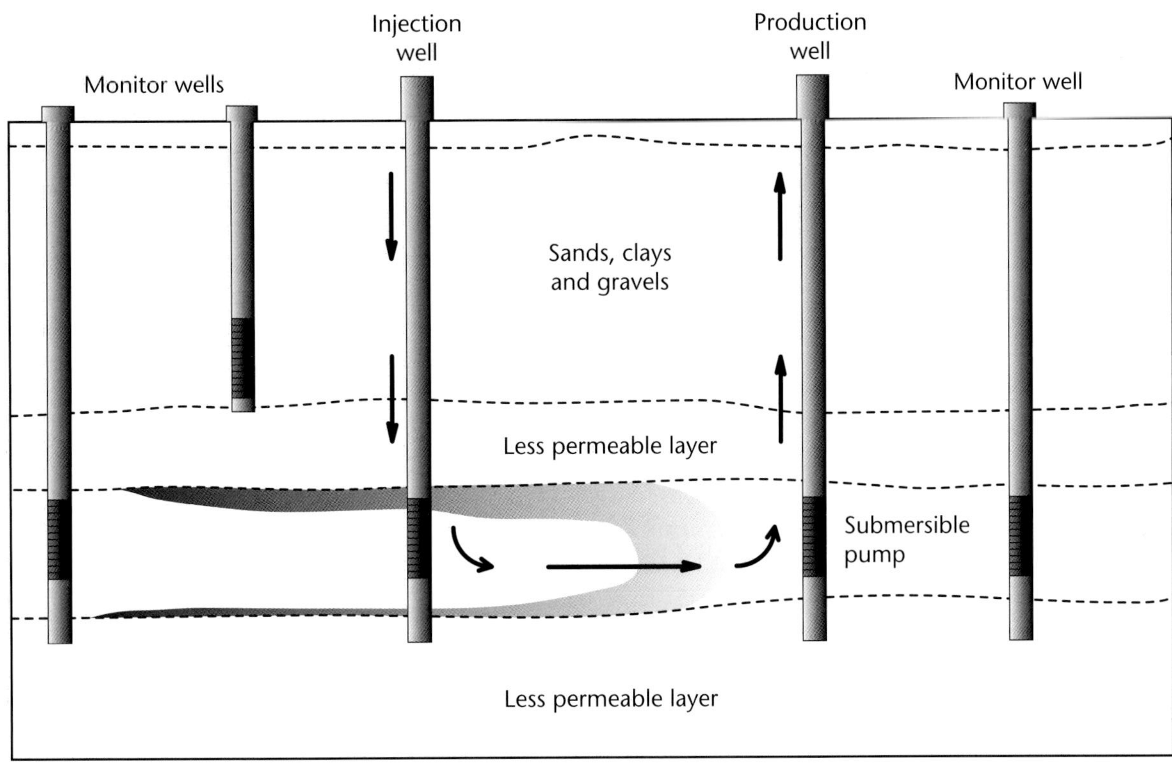

Figure 7.4: Schematic representation of a solution mining, or *in situ* leaching, operation.

involve only crushing and separating the material into various size fractions, as for aggregate and other construction products. For more valuable materials it will involve liberating the valuable mineral(s) from the unwanted minerals then separating the liberated minerals into a concentrate, containing the wanted minerals, and tailings containing the unwanted minerals. The unwanted minerals are called *gangue* (pronounced 'gang') minerals. The term gangue is used to describe any unwanted mineral material from mineral-processing operations. Beneficiation processes are performed on a very large scale and usually involve a number of different operations. The various processing units are connected so material flows continuously from one to the next. Batch operation of processing equipment is rare, and is used only in pilot plant and other developmental-type work and in very small-scale operations. Normally, a mineral-processing operation will run all day every day with material continually flowing into, through and out of pieces of processing equipment linked sequentially to achieve the desired final product. A drawing showing the processing units and the flow of materials through them is called a process *flowsheet*.

7.2.1 Size reduction

The material as mined will vary greatly in size from several metres in diameter down to particles much smaller than 1 mm. Liberation is most commonly achieved by *comminution*, which involves crushing then grinding the mined material. Comminution is generally the most expensive part of mineral beneficiation, because it is energy-intensive. The techniques for crushing and grinding are discussed in the following sections.[16]

Crushing

Various types of machines are used for crushing, the most common being jaw crushers, gyratory crushers

16 Images of various types of crushing and grinding equipment are readily available on the internet, for example at http://www.mine-engineer.com.

and cone crushers. These apply a slow compressive force selectively to the larger rocks in the feed, which is introduced at the top. The larger rocks, under these conditions, tend to break into several smaller pieces of similar size as the material falls though the crusher and is discharged at the bottom. Each of the types of crusher has advantages and disadvantages. They are often used in sequence – jaw crusher, gyratory crusher then cone crusher – to progressively break the rock down to particles less than 10–50 mm in diameter, a size suitable for feeding to grinding mills. The stages of crushing are referred to as primary, secondary and tertiary. Primary crushing is often performed in the pit of an open pit mine or in the lower levels of an underground mine, and is usually performed by jaw or gyratory crushers. This minimises the quantity of waste rock brought to the surface (undersize material of lower grade is left behind) and thus reduces haulage costs. It also reduces the size of rocks to make them easier to handle. Other types of crushers, such as roll crushers and hammer mills, are also used commercially but these are less common.

Grinding

Grinding is performed in tumbling mills, which are essentially large, rotating steel drums where ore is fed in at one end and discharged with a smaller mean particle size at the other. Grinding can be done on wet or dry material but water is usually added – wet grinding is more efficient and doesn't produce dust. Breakage of particles occurs as the material flows from the inlet to the outlet, through a combination of:

- rapid compression of the material, which tends to produce many smaller particles having a wide size range;
- abrasion, which tends to produce one group of particles with size very similar to the original particles and another group of very fine particles.

There are three types of grinding mills in common use: rod mills, ball mills, and autogenous and semi-autogenous mills. With the exception of autogenous mills, the mills contain a dense grinding medium which is added to facilitate the grinding. In rod mills, which are typically 2–6 m in diameter and 3–10 m long, the grinding medium is rods usually made of high carbon steel which is very hard. The rods are 25–150 mm diameter and a few centimetres shorter than the internal length of the mill. They typically occupy 20–30% of the internal volume of the mill. The ore is fed with water into the rotating mill and the rods tumble over each other in parallel alignment. Particles of ore are trapped between the rods and broken. The larger particles of ore are broken preferentially because the force of the rods is applied selectively to them (Figure 7.5) and the rate of breakage of larger particles is greater than for smaller particles. This results in a product with a fairly narrow size range. Rod mills typically reduce the mean particle size to around 0.25–3 mm.

Ball mills have similar geometry to rod mills but are generally shorter. The grinding medium consists of balls, typically 20–100 mm in diameter, made of hard material such as cast iron, high carbon steel, alloy steel or ceramics (the last if contamination by iron may cause a problem in subsequent processing). The balls occupy about 40–50% of the interior volume of the mill. The use of balls rather than rods enables the particles to be ground more finely, since much of the grinding occurs at points of contact of the balls. Ball mills can grind ore particles down to around 20 μm. In autogenous grinding mills (AG mills), larger rocks in the feed, or added separately, act as the grinding medium. Semi-autogenous grinding mills (SAG mills) are much more common than autogenous mills because they are more effective. In these, balls are added to supplement the grinding action of the

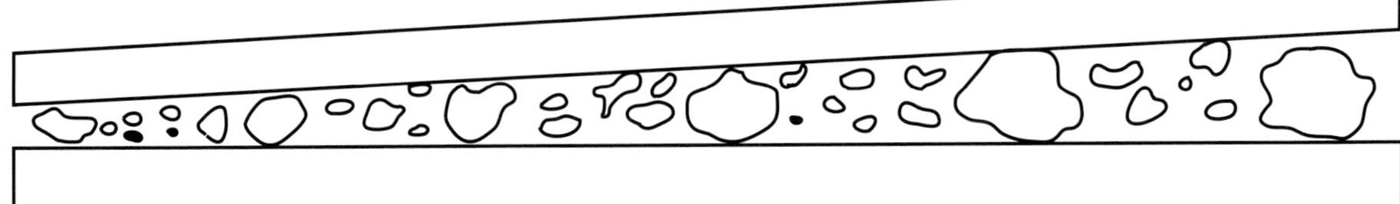

Figure 7.5: The breakage mechanism in rod mills (after Wills, 1992).

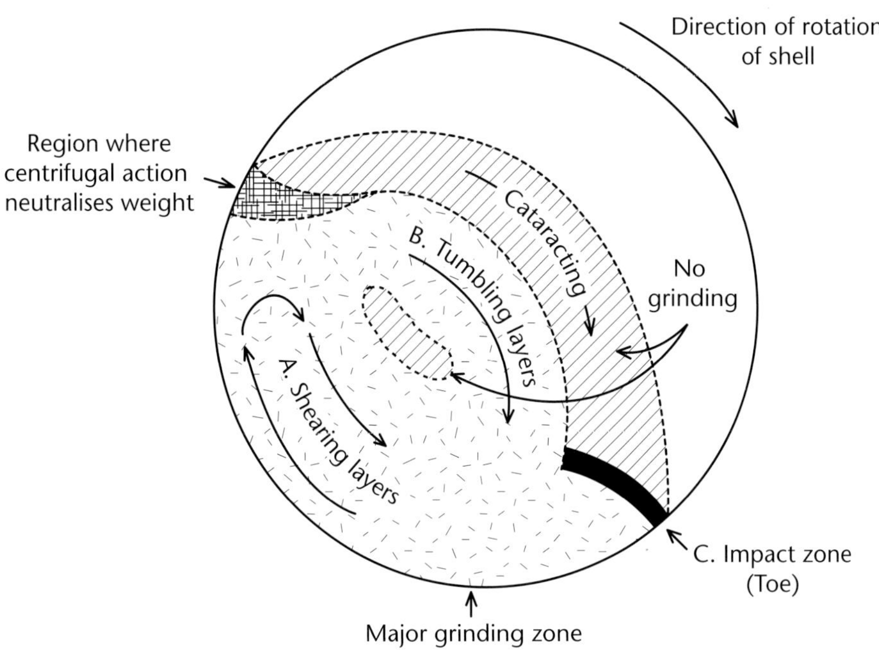

Figure 7.6: Cross-section of a grinding mill showing the movement of charge (after Kelly and Spottiswood, 1982).

larger rocks but the balls occupy only about 5–10% of the volume of the mill. The length:diameter ratio of AG and SAG mills is usually 0.3–0.5, compared with around 1–1.5 for ball mills and 1.5–2.5 for rod mills.

The grinding action in tumbling mills is illustrated in Figure 7.6. The charge in the mill is lifted up the wall in the direction of rotation then falls back along a variety of trajectories under gravity and centrifugal force. In zones A and B, size reduction occurs mainly by abrasion and impact. In zone C, size reduction occurs mainly by impact of particles freely falling from the other side of the mill. The relative contribution to size reduction of the three zones depends on many factors including the geometry of the mill, the size and quantity of grinding media, the speed of the mill and the characteristics of the feed material.

Comminution circuits

Crushing and grinding is usually done in several stages. The equipment is usually configured so that material in the desired size range is removed as it is produced and the oversized material is returned to the crusher or mill for further comminution. This saves energy and minimises the production of very fine particles of valuable mineral. This material, called *slimes*, may end up in the tailings during the separation process. The slimes which end up in tailings represent a loss of valuable mineral and can cause problems in subsequent separation steps, particularly in froth flotation, by coating particles. They also make dewatering more difficult.

Particles are removed from crushing and coarse grinding circuits by *screening*, a simple process involving transporting particles over a surface with perforations of the desired size. Particles less than the opening size fall through, while particles larger than the opening size flow across and off the screen. Screening may be performed wet or dry. Wet screening is more efficient but dry screening has to be used in some situations, particularly when water is scarce or where water reacts undesirably with the ore. In practice, separation is never perfect and undersize particles can remain behind. Industrial-scale screening does not work well for separating particles smaller than about 200 μm, and classification techniques must be used for fine particles. These are discussed in Section 7.2.2. Hydrocyclones are the most commonly used classification device in the minerals industry.

The product from wet grinding circuits is referred to as *pulp*. Pulp consists of the ground ore and water in ratios that optimise the grinding process and enable it to flow and be pumped. The quantity of solids in a pulp is typically in the range 10–60 wt%. Three commonly used configurations of comminution circuits are illustrated in Figure 7.7.

7.2.2 Separating particles

The size to which particles are ground during comminution is determined by the size required to achieve effective liberation, as discussed in Section 5.7. For metallic ores, this is commonly in the range 50–500 μm. The next step is to separate the particles to produce a concentrate which contains the desired minerals and tailings which contain the gangue minerals. This requires a method for selecting particles, either the desired particles or the unwanted particles. In principle, the selection can be based on any useful distinguishing physical or chemical property of a particle. In practice, it must be technically feasible to target the particle and physically remove it in a separate stream. The physical and chemical properties of particles most commonly used to effect separation are size and shape, density, magnetic susceptibility, electrical conductivity and surface chemistry. Given the limitations of technology, it is rarely possible to select particles solely on one property, though each commercial technology uses methods that rely predominantly on one property.

Classification

Classification is a technique for separating particles that exploits the differences in settling velocities exhibited by particles of different size under gravity or an applied centrifugal force. It is thus a technique for separating on the basis of particle size, although particle density also has an effect. The medium of transport of the particles may be a liquid (usually water) or a gas (often air). Classification devices include rake classifiers, which rely on the action of gravity, and gas cyclones and hydrocyclones, which rely on centrifugal force. The principle of the hydrocyclone is illustrated in Figure 7.8. They are the most common classification units and, unlike rake classifiers, have no moving parts. Pulp is injected tangentially at the top; as it swirls around and moves downwards towards the outlet, the larger particles are thrown towards the wall and collect in the conical section at the base, from where they are

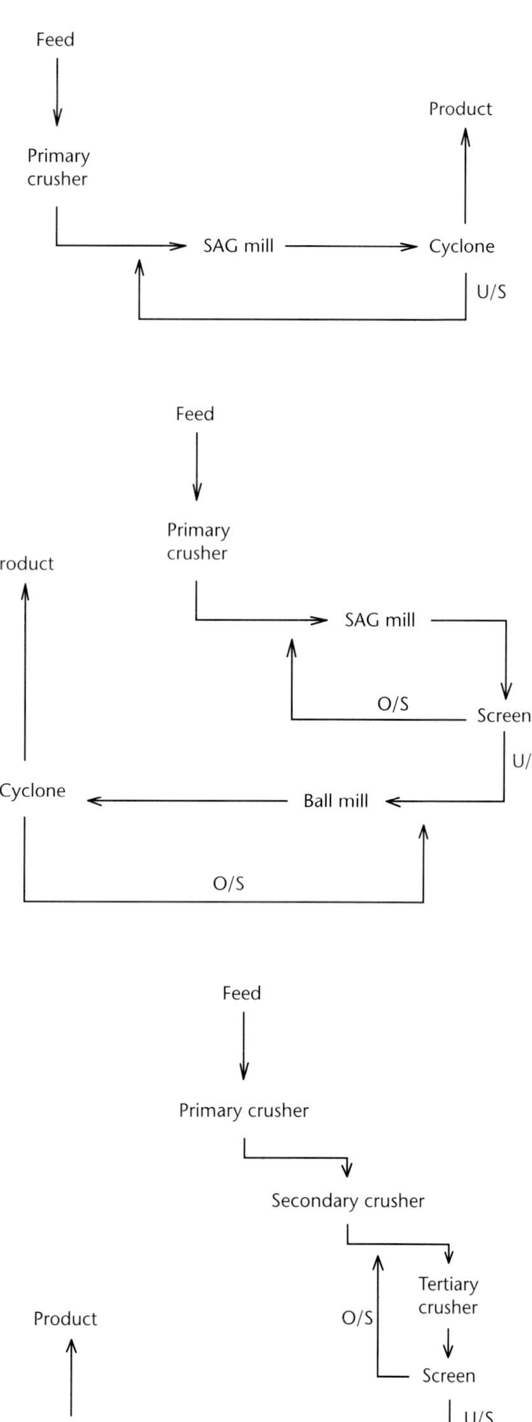

Figure 7.7: Three typical configurations of comminution circuits (after Kelly and Spottiswood, 1982). O/S = oversize material, U/S = undersize material.

discharged. Hydrocyclones are designed so that most water leaves at the top through the vortex finder. Thus there is a flow of water up the centre of the cyclone from the bottom. Fine particles remain in suspension and are carried in the water returning to the top. The inlet of the vortex finder is positioned to optimise the cut-off size of the particles. The centrifugal force within cyclones is more than 10 times the force of gravity, and can be up to several hundred times greater, thus hydrocyclones have much greater productivity per unit size than gravity-operated classifiers. Hydrocyclones are used most commonly in grinding circuits to remove particles in the smaller size ranges, where screening is less efficient. Gas cyclones operate in the same manner as hydrocyclones but the particles are separated without water or other liquid.

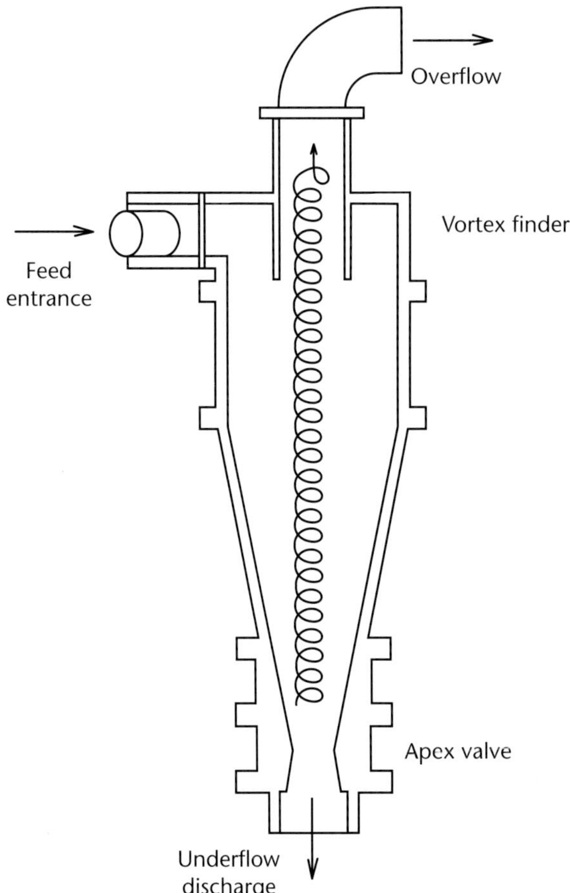

Figure 7.8: Cross-section of a hydrocyclone showing the main components and the flow patterns (after Wills, 1992).

Gravity separation

Gravity separation, like classification, relies on settling rates but the separation is based predominantly on differences in particle density. A number of devices are in common use. Jigs are devices in which a pulse is used to achieve separation. The pulsing motion may be induced mechanically, by lifting and dropping the jig by means of an eccentric drive, or pneumatically by injecting air into the pulp. There are many different designs. Jigs are particularly useful for separating coarse material, around 1–10 mm diameter. Jigs take advantage of the fact that the initial acceleration of particles settling in water is independent of size but is directly proportional to the difference in density between particle and liquid. They work by providing frequent acceleration events for particles, by shaking or pulsing the bed. Denser particles preferentially move to the bottom and the pulp becomes stratified, thereby making it possible to physically separate the top and bottom of the bed. Dense medium separators are frequently used in coal washeries to separate coal from mineral matter. The density of the medium is intermediate between that of the coal and the gangue minerals, so the former floats and the latter sinks. The heavy medium is created by suspending finely ground magnetite or ferrosilicon in water. Pulsing the bed maintains the suspension and facilitates the separation process. The heavy medium is recovered from the underflow and overflow streams, by screening or magnetic separation, and returned to the jig. Pinched sluices and cones are modifications of the simple sluice, a device in which heavy and light minerals become stratified in running water. The heavy fraction is trapped by riffles (raised sections running across the surface over which the pulp flows) or separated by splitters (blades which divide the flow into two or more streams). Shaking tables are another device which rely on stratification in running water.

Magnetic separation

All materials are affected to varying extents by magnetic fields. Magnetic separation relies on the differential movement of materials in a magnetic field. Materials can be divided broadly into two classes according to how they respond to magnetic fields.

Diamagnetic materials are repelled by magnetic fields and move towards positions of low magnetic field potential. Paramagnetic materials are attracted by magnetic fields and move towards positions of high magnetic field intensity. The magnitude of a substance's response to a magnetic field depends on its magnetic susceptibility. High-intensity magnetic fields are used to separate weakly paramagnetic materials. Low-intensity magnetic fields are used to separate materials of high magnetic susceptibility from materials of low magnetic susceptibility. Separation can be done dry or wet. Drum separators are the most common device used. The material to be separated is fed in a stream along the top of a rotating magnetic drum; as it falls, the particles are differentially attracted to or repelled from the drum. The stream discharging from the bottom is segregated accordingly and can be separated into two or more fractions by means of baffles or splitters.

Electrostatic separation

Separation based on differential electrical properties of materials most commonly utilises high tension (typically up to 50 kV) direct current. The technique involves creating a very high voltage difference between an earthed plate, usually in the form of a rotating horizontal cylinder (rotor), and a discharging electrode. The air between them becomes ionised by the discharge; solid particles introduced into the space acquire a surface charge of opposite potential to the ions (by the phenomenon of electrostatic induction). The particles are attracted to the charged surface and become attached. In materials that are good conductors, the induced charge dissipates quickly to Earth and the ions no longer remain attracted to the surface. Particles that are less conducting retain the charge longer and remain attached to the rotor. Thus a separation can be effected by means of a splitter placed towards the bottom of the rotor which divides the stream into a concentrate and tailings. High-tension separation can be used only on very dry material. This limits its application, as it is often not economical to dry the material first. Dust is also a problem in the operation of high-tension separators, and the feed to such units needs to be classified to remove dust.

Froth flotation

Froth flotation is by far the most widely used process for separating minerals. It was developed and implemented by C.V. Potter and G.D. Delprat at Broken Hill, Australia, in 1901–03. It relies on the differential surface property of minerals known as wettability, which is modified and controlled in practice by the addition of organic reagents. In flotation, air bubbles are generated in the pulp in a flotation cell. Some particles (usually those containing the desired mineral) become attached to the bubbles and rise to the top of a cell while others (usually particles of gangue minerals) do not become attached to bubbles and sink to the bottom (Figure 7.9). The scientific basis of flotation is shown in Figure 7.10, which shows a drop of water on a mineral surface. How the water spreads over the surface depends on the balance between three forces of surface tension at the line of contact of the gas, liquid and solid:

$$\gamma_{mineral-air} = \gamma_{water-air} \times \cos\theta + \gamma_{mineral-water}$$

If $\gamma_{mineral-air} - \gamma_{mineral-water}$ is similar in magnitude to $\gamma_{mineral-air}$, then $\cos\theta$ will be equal to about 1 and θ will be approximately zero. In this situation, the water will spread evenly over the surface (it will wet the

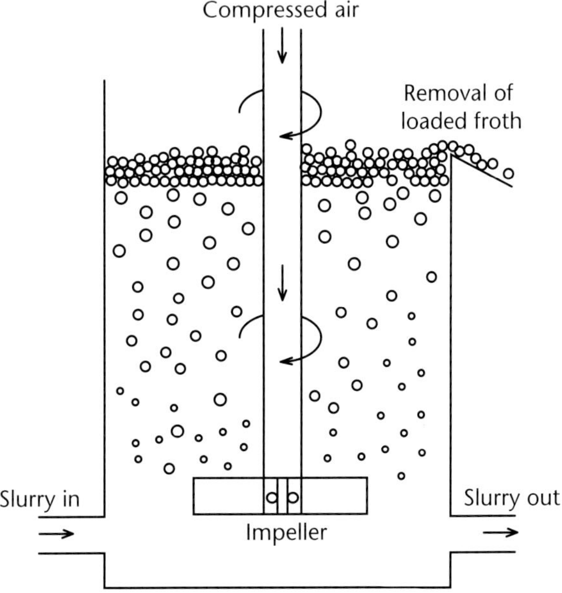

Figure 7.9: Cross-section of a flotation cell, showing the main components.

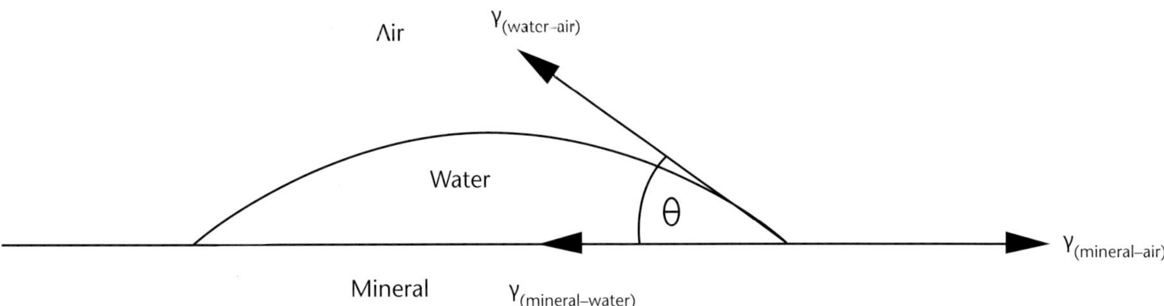

Figure 7.10: Surface tension forces between air, water and a mineral surface.

surface). The surface is said to be *hydrophilic*. On the other hand, if $\gamma_{mineral-air}$ is less than $\gamma_{mineral-water}$, then $\cos\theta$ will have a negative value and θ will be greater than 90°. In this situation, the water will not spread over the surface but remain as a discrete drop. The surface is said to be *hydrophobic*.

The electrical potential of the valence electrons in water molecules is not uniformly distributed; one end of the molecule is slightly positive and the other end is slightly negative (the molecule is said to be polar). As a result of being fractured during comminution, the surfaces of minerals acquire a net charge due to the breaking of chemical bonds. Hence, water molecules are attracted to the surface of most minerals and they are readily wetted. However, if a hydrophobic compound is added to the water, and if it has a polar group in its molecule which is more strongly attracted to the mineral surface than is the water molecule, then these molecules preferentially attach to the surface of the mineral which then becomes hydrophobic. Such compounds are called *collectors*. Xanthates are the most commonly used collectors for sulfide minerals. Xanthates are a group of water-soluble organic compounds which are sodium or potassium salts of xanthic acid. The most commonly used xanthates are potassium ethyl xanthate ($C_2H_5OCS_2K$), sodium ethyl xanthate ($C_2H_5OCS_2Na$), potassium amyl xanthate ($C_5H_{11}OCS_2K$), sodium isopropyl xanthate ($C_3H_7OCS_2Na$) and sodium isobutyl xanthate ($C_4H_9OCS_2Na$).

If air bubbles are introduced into the pulp at the bottom of a flotation cell, hydrophobic particles that encounter a bubble will become attached to it, since they are not wetted, and rise to the surface with the bubble. However, unless the bubbles are stabilised, when they reach the surface the liquid will drain from the film, the bubbles will burst and the minerals, being denser than water, will sink. To prevent this, different organic reagents called *frothers* are added to stabilise the bubbles. Other reagents are also sometimes added. *Modifiers* are added to improve the attachment of a collector to particles. *Depressants* are added to render hydrophilic those mineral particles that would otherwise be coated with collector and float. The use of collectors and modifiers makes it possible to select which minerals will float and which will not, and thus achieve a selective separation.

Flotation is performed in cells arranged in series into banks of cells. Pulp is fed into the first cell and some valuable mineral is removed in the froth. The pulp then flows to the next cell, where more valuable mineral is removed, and so on until the spent pulp (the tailings) discharges from the last bank. The grade of the froth decreases down the bank of cells and the froth from the last cells (the *scavenger cells*) is often returned to the first cell (Figure 7.11a). More often, the froth from the first bank of cells (the *rougher cells*) is diluted then refloated in another bank of cells (the *cleaner cells*) to produce a higher-grade concentrate (Figure 7.11b and c).

7.2.3 Separating solids from water

Most mineral grinding and separation processes involve water and often large quantities of it are used, an aspect that will be examined more fully in Chapter 10. As a result, the products – concentrates and tailings – usually contain large quantities of water. Concentrates are dewatered and dried to produce a product

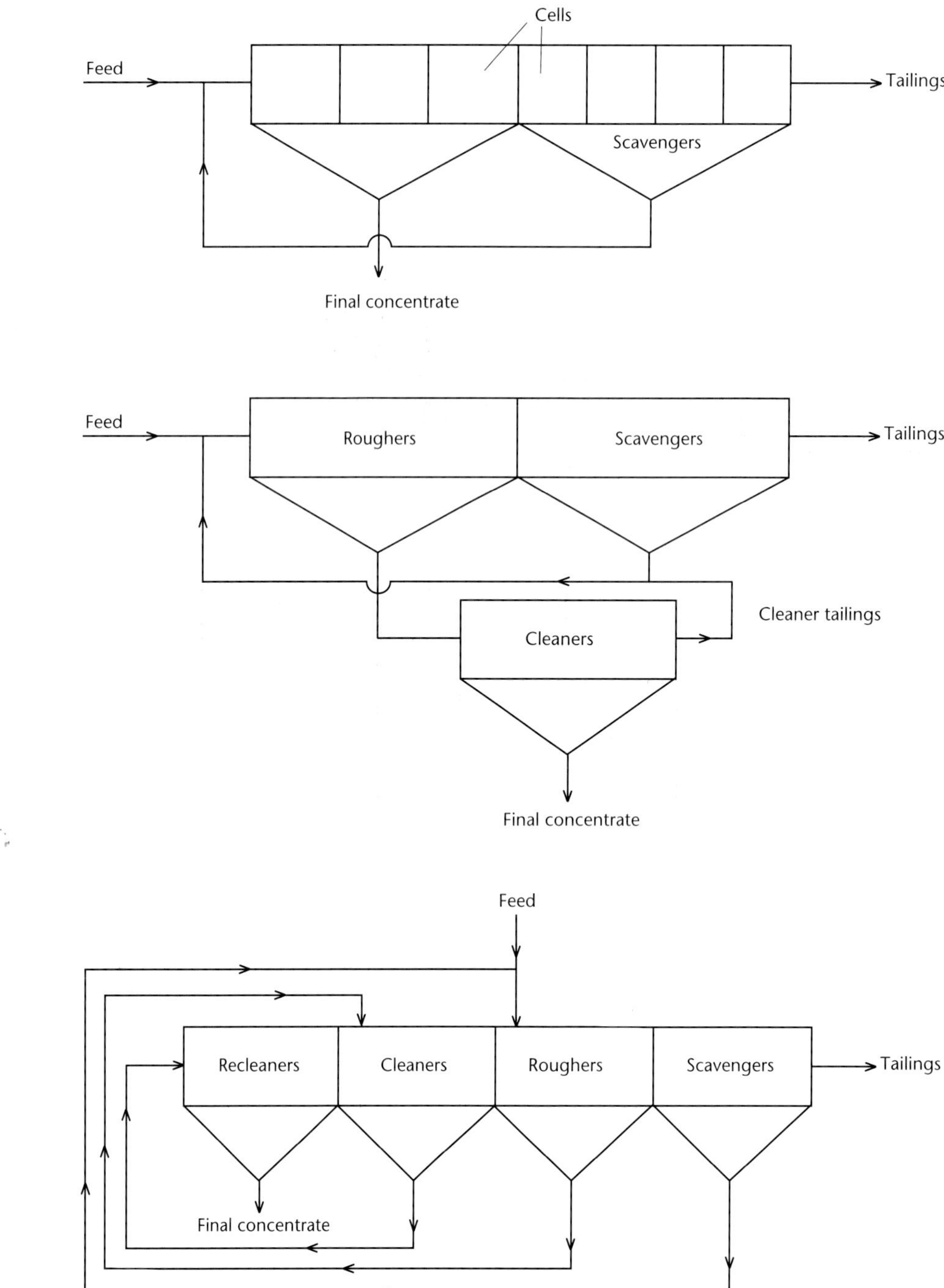

Figure 7.11: Typical configurations of froth flotation circuits (after Wills, 1992).

that can be smelted or otherwise further processed on site, or shipped to a smelter or other processing plant. Tailings are dewatered to recover water for reuse in the plant and to obtain a product that can be more easily disposed of or stored. While dewatering is performed mainly as a final operation on concentrates and tailings, it is also sometimes used between processing stages when water needs to be removed from a stream prior to the next stage. There are three main methods of dewatering: sedimentation (or thickening), filtration and drying. Often, the three methods are used sequentially (Wills, 1992). The bulk of the water is removed by thickening to produce a pulp with perhaps 55–65 wt% solids. Up to 80% of the water can be removed in this stage. Filtration of the pulp produces a filter cake containing 80–90 wt% solids. If less water in the product is required it is then dried, naturally (if climate and weather permit) or artificially.

Thickening

Thickening is the most widely used dewatering technique. It is relatively cheap since it relies on particles settling out of water under the action of gravity to produce a thickened pulp at the bottom and clarified water at the top. However, because settling is slow, thickeners have low productivities and need to be very large to handle the volumes encountered at mining operations. The process is carried out continuously in large cylindrical tanks up to >200 m diameter and 10 m depth (Figure 7.12). Pulp is fed from the surface in the centre through a feedwell. Clarified liquid overflows at the periphery and is collected in launders; thickened pulp is withdrawn from a conical section at the bottom. The solids in the liquid move continuously downwards under the action of gravity and are raked inwards to the discharge point by blades, or rakes, on one or more slowly rotating arms at the base. Reagents called flocculants are added to increase the rate at which particles settle. *Flocculants* are long-chain, high molecular weight, organic polymers which are soluble in water and which bridge a number of particles and hold them together in loose clusters (flocs) which settle more readily than individual particles. The most commonly used flocculants are synthetic compounds called polyacrylamides.

High-rate thickeners are designed to handle more solids per unit area, with a faster settling rate, than conventional thickeners. They use higher flocculant dosages and generally have deeper mud depth, which causes greater compression of the solids and a higher pulp density in the underflow. High-density thickeners have an even deeper mud depth (~2–3 m). This requires a more robust design and increased torque capacity of the rake. The underflow has a correspondingly higher pulp density. Paste thickeners have a still greater depth (>3 m) and produce an even higher pulp density.

Filtration

Filtration is the process of separating solids from water by means of a porous medium which holds back the solids but allows the water to pass. There are many types of commercial filtration equipment, rotary drum filters being the most widely used in the minerals

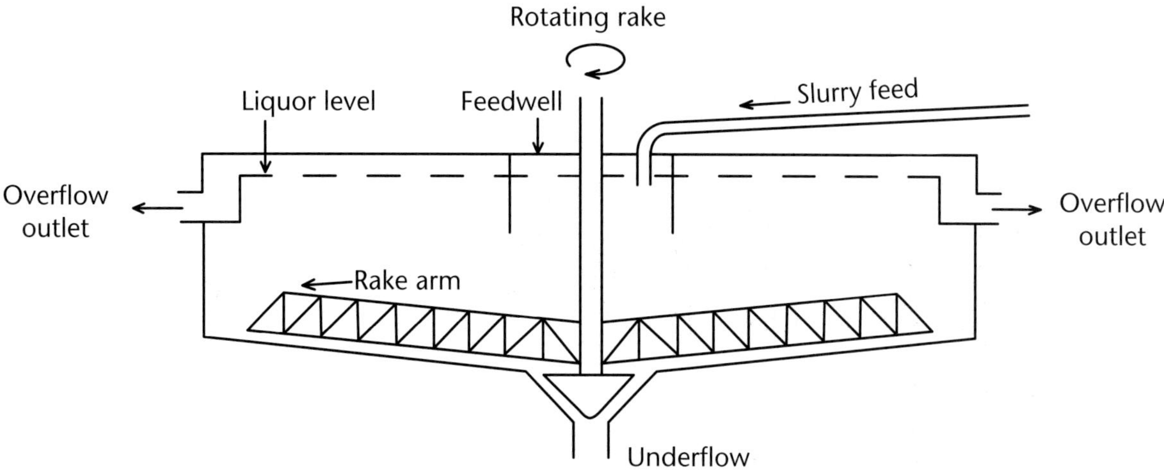

Figure 7.12: Cross-section of a thickener, showing the main components.

industry. A rotary drum filter is a form of continuous vacuum filter. The drum, which is covered with a fabric filter medium (often cotton), is mounted horizontally and rotated partially submerged in a trough into which pulp is fed. The outer part of the drum is divided into compartments, each of which has a number of drain holes in the shell connected internally to pipes. These are in turn connected to a vacuum system through a slip-ring mechanism. As the drum slowly rotates, vacuum is applied to the compartments in the pulp. This forces liquid through the filter medium and a layer of particles builds up on the surface. As this material is lifted out of the trough by the rotating action, water continues to be removed from it. Wash water may be sprayed onto the cake as it moves to the top of the drum. Finally, the compartment is pressurised. This lifts the filter cake off the cloth, allowing it to be removed by means of a scraper or string. That compartment of the drum then returns to the trough and the cycle is repeated. Disc filters, which consist of several large fabric-covered discs on a horizontal drive shaft, are another device for filtration. The principle of operation is similar to that of drum filters.

Drying

Drying is often performed in rotary dryers which consist of a long cylindrical steel shell mounted, with a small incline, on rollers. It is mechanically rotated about its axis at about 25 revolutions per minute. Filter cake is fed in at the upper end and moved progressively to the other end by the rotating motion, where it is discharged. Hot gases, generated by burning a fuel or obtained as waste gas from another nearby process, are passed through the dryer either from the feed end (co-currently) or the outlet end (counter-currently). The rotating action helps expose the solids to the hot gases and helps to transfer heat from the shell to the solids.

7.2.4 Agglomerating particles

Agglomeration is the term used to describe the process of bringing together and bonding a number of particles so that they form a single, larger composite particle. Fine material from beneficiation must be agglomerated to produce material of a larger size if a processing operation requires a coarser feed size. This is usually because a permeable bed of material is needed for reaction with a gas (as in roasting and smelting) or a liquid (as in leaching) which flows through the bed. There are three main techniques for agglomerating mineral particles: pelletising, briquetting and sintering. In pelletising and briquetting the chemical form of the materials is not altered substantially, though solid-state diffusion and limited melting may occur during heating to develop strength in the agglomerates. In the sintering of minerals, both the chemical and physical form of the minerals change substantially. In sintering, fine particles are agglomerated by partial melting of the solids, followed by resolidification. Sintering is discussed in Chapter 8.

An important application of agglomeration is in the smelting of iron ore in the iron blast furnace to produce metallic iron. The iron blast furnace is essentially a tall, counter-current reactor in which iron ore, coke and limestone are continuously fed at the top, air is injected just above the hearth, and molten metal and slag are tapped periodically from the hearth where they collect as a result of the reaction of the air with coke and iron ore (described in Section 8.2.6). Because very large volumes of air are required, the solids in the furnace shaft must be lumpy (around 20–50 mm diameter) so the bed has sufficient connected voids that gas can flow relatively unimpeded and react with each lump. Beneficiated iron ore may be lumpy, in which case it can be fed directly to the blast furnace. Alternatively, it may be fines from a crushing and screening operation (typically less than about 6 mm in diameter), or concentrate from a flotation circuit. In the latter cases, the fine material must be agglomerated before it can be fed to the blast furnace.

Pelletising

Pelletising is used for fine material such as concentrates. The material is mixed with a small quantity of binder, often bentonite or other clay-type mineral, then rolled in a drum or on a disc with some added water. Under appropriate conditions the material forms into balls, the size of which is determined by the residence time in the drum or on the disc. Iron ore pellets are typically 10–20 mm diameter. These are called green pellets, and they harden and gain a certain amount of strength as they dry. They are hardened further, by heating to around 1200–1350°C in a process called *induration*. This gives them sufficient strength to be transported and to maintain their integrity in the blast furnace.

Briquetting

Briquetting involves mixing fine material with a binder and pressing it in a die to form briquettes. This is done in a continuous operation by feeding the material from the top into the gap between two rolls rotating in opposite directions so as to capture the material, compress it and eject it at the bottom. The surfaces of the rolls have matching impressions which fill with the powder, which is then compacted to form briquettes as the rolls rotate. As with pellets, the briquettes can be heat treated, if required, to give them greater strength.

7.3 EXAMPLES OF MINERAL BENEFICIATION FLOWSHEETS

Some flowsheets are presented in the following sections to illustrate the combinations of unit processes used in the production of mineral concentrates. These are purely illustrative – each ore body requires a unique flowsheet though, of course, there are many similarities between flowsheets for concentrates produced from similar types of ores.

7.3.1 Mineral sand concentrates

Typical flowsheets for processing material mined from mineral sands placer deposits are shown in Figures 7.13 and 7.14. There are two major stages. First, the mined material is concentrated into a heavy mineral concentrate by removing silica sand using a wet process involving spirals. The concentrate is dewatered and dried, then sent to a dry processing plant where it is separated into different mineral concentrates (of rutile, ilmenite, zircon and monazite) using electrostatic and magnetic separation.

7.3.2 Production of iron ore fines and lump

A typical flowsheet for treating the low-grade fraction of a banded iron formation deposit of hematite is shown in Figure 7.15. Two products are produced: lump (6–30 mm), which is used directly in iron blast furnaces, and fines (<6 mm), which is agglomerated by sintering before being used.

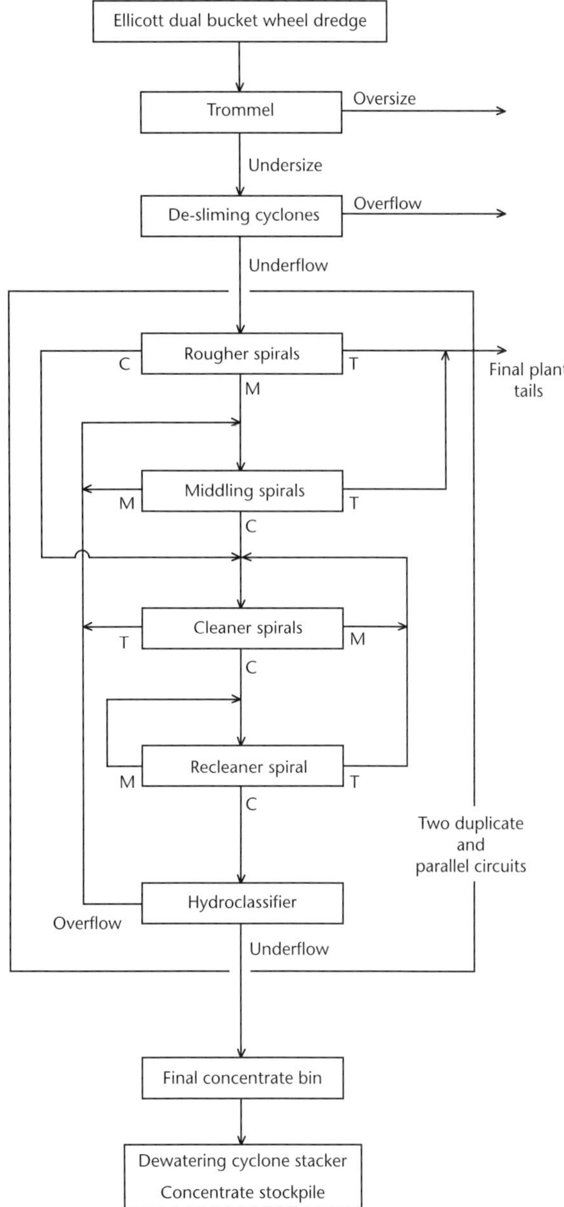

Figure 7.13: Flowsheet of the wet mineral sands concentrating circuit at Tiwest Joint Venture, Cooljarloo, Western Australia (Modified from Hinner, 1993). C = concentrate, T = tailings, M = middlings. Reprinted by permission, Australasian Institute of Mining and Metallurgy.

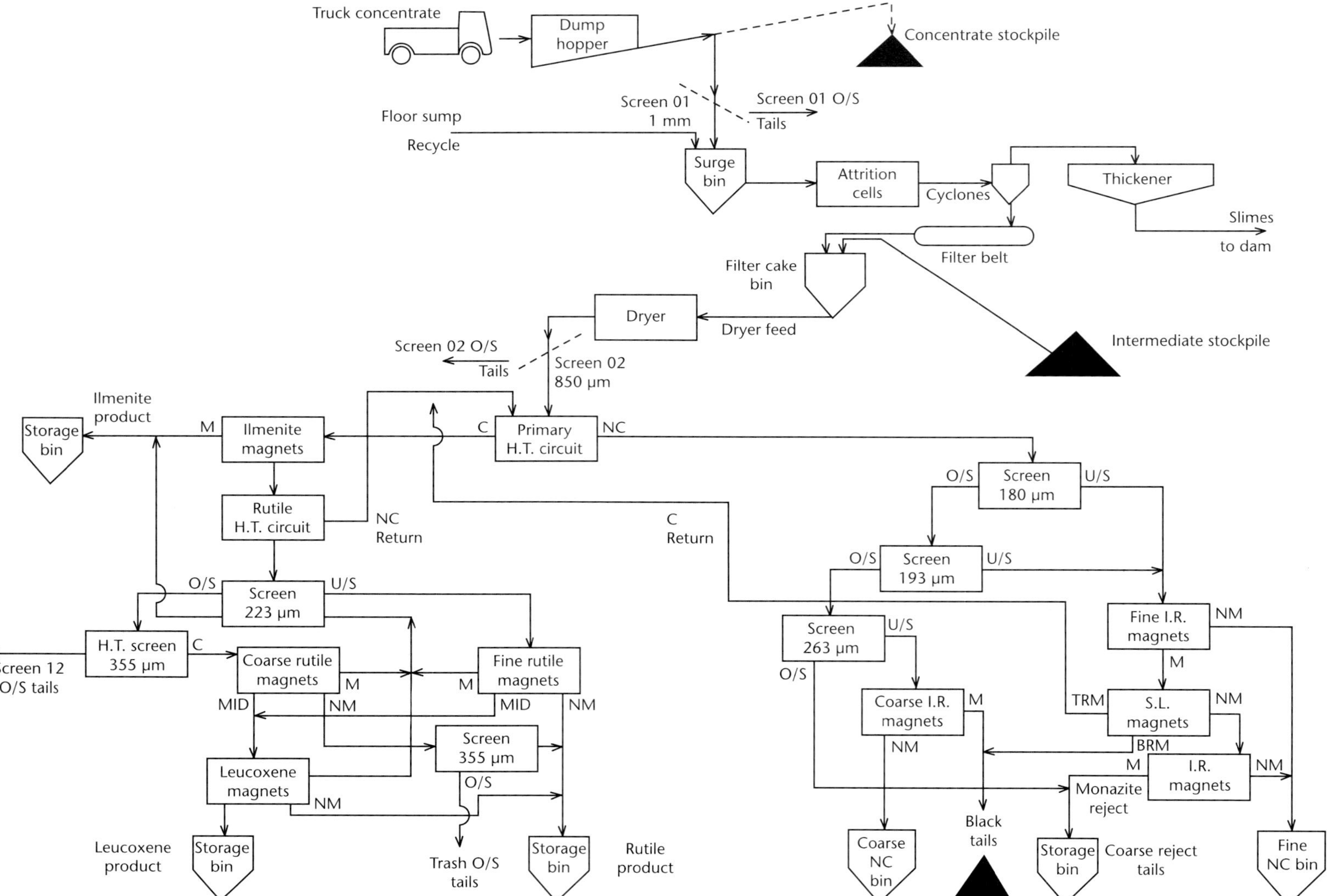

Figure 7.14: Flowsheet of dry mineral separation plant of Tiwest Joint Venture, Chandala, Western Australia (von Horn, 1993). O/S = oversize, U/S = undersize, M = magnetic, C = conductor, NC = non-conductor, HT = high tension, IR = induced roll, SL = semi-lift. Reprinted by permission, Australasian Institute of Mining and Metallurgy.

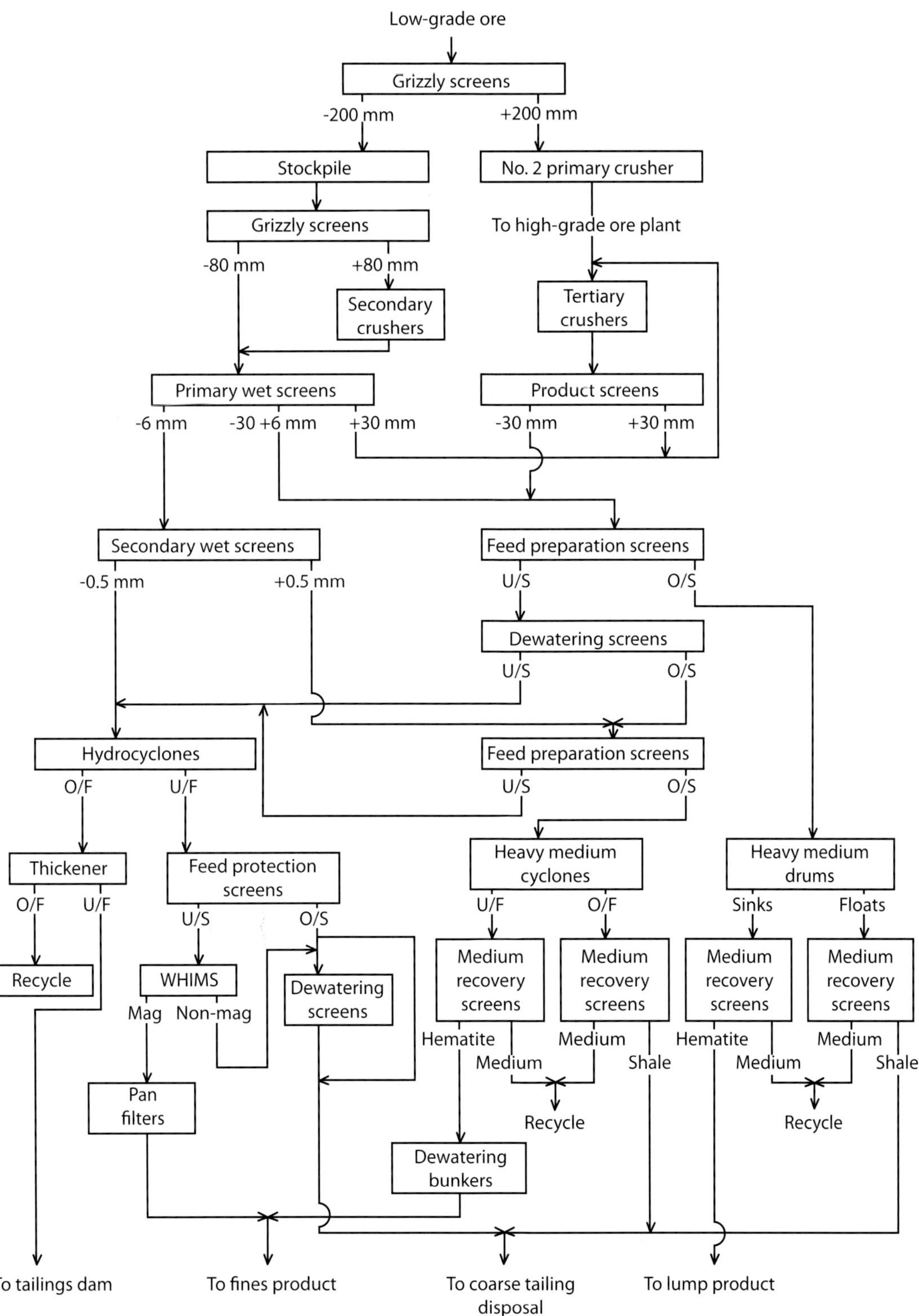

Figure 7.15: Schematic flowsheet of the iron ore concentrator at Mount Tom Price, Western Australia (Petty, 1993). O/S = oversize, U/S = undersize, O/F = overflow, U/F = underflow, WHIMS = wet high intensity magnetic separator. Reprinted by permission, Australasian Institute of Mining and Metallurgy.

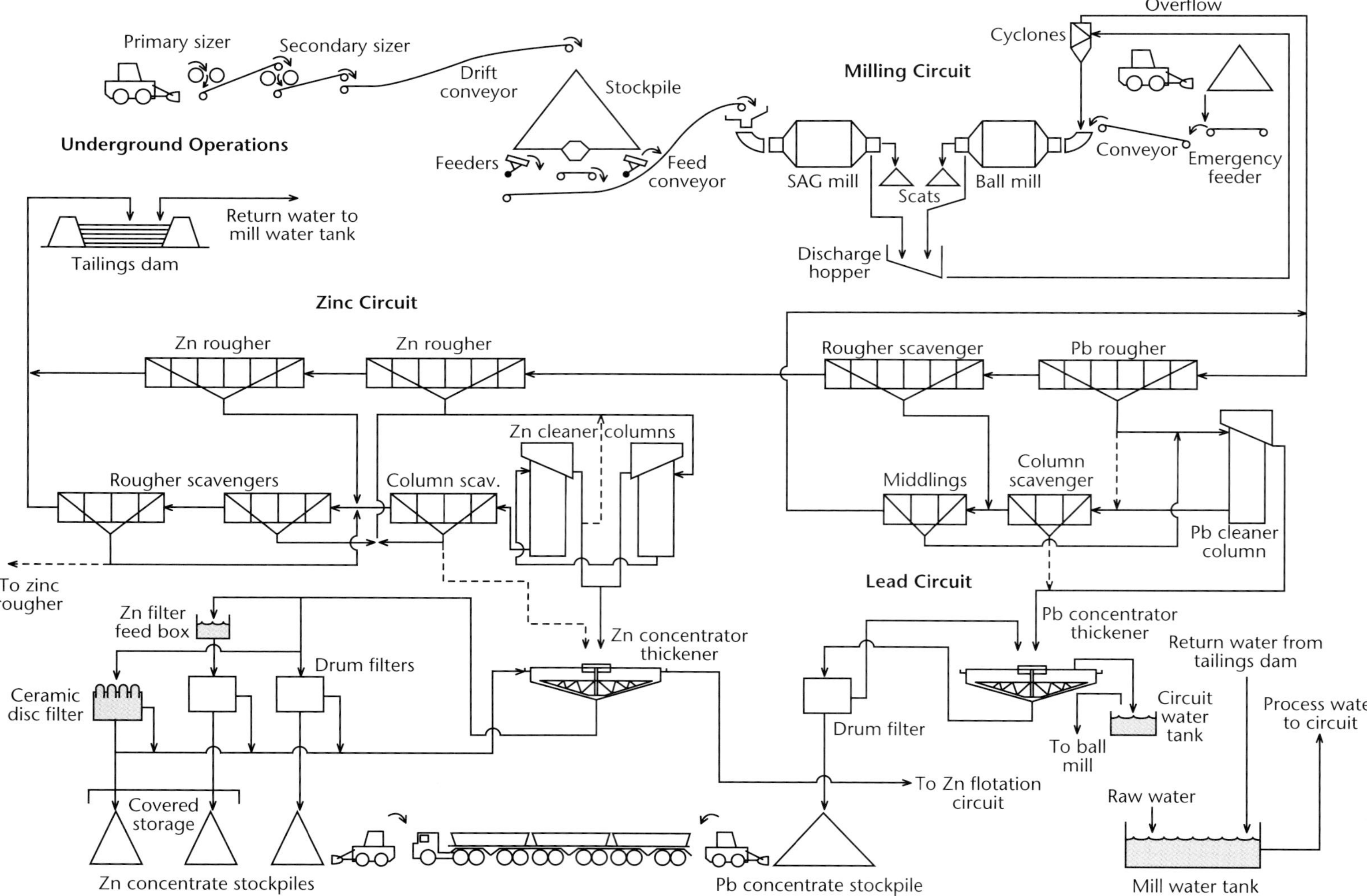

Figure 7.16: Schematic flowsheet of the lead-zinc concentrator at the Cadjebut mine, Cadjebut, Western Australia (Nash, 1993). Reprinted by permission, Australasian Institute of Mining and Metallurgy.

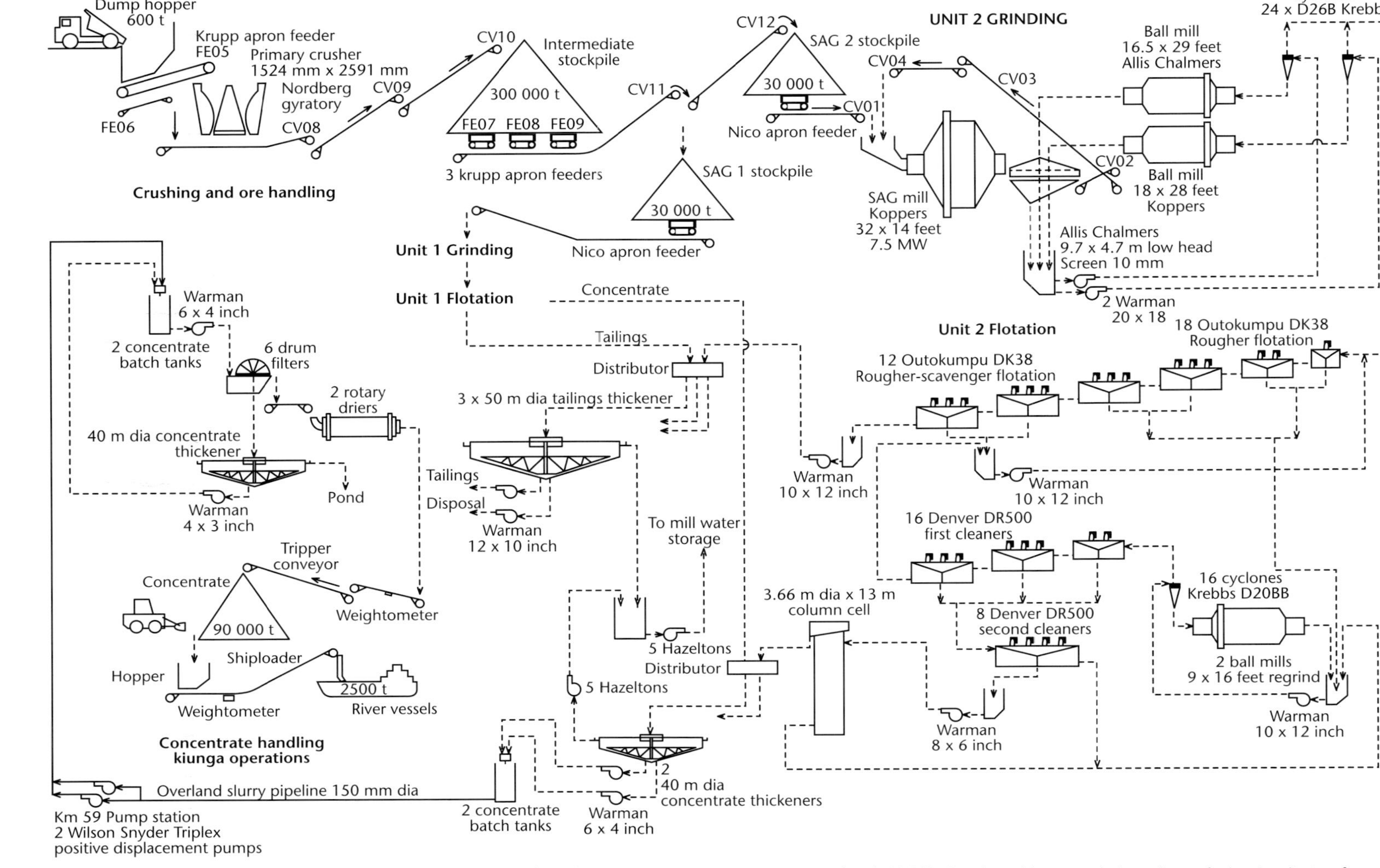

Figure 7.17: Flowsheet of the copper concentrator at the Ok Tedi mine, Papua New Guinea (England, 1993). Reprinted by permission, Australasian Institute of Mining and Metallurgy.

7.3.3 Base metal sulfide concentrates

Figures 7.16 and 7.17 show typical flowsheets for producing zinc and copper concentrates from sulfide zinc and copper ores, respectively.

7.4 REFERENCES

England JK (1993) Copper concentrator practice at Ok Tedi Mining Limited, Ok Tedi, Papua New Guinea. In *Australasian Mining and Metallurgy: The Sir Maurice Mawby Memorial Volume*. Vol. 1 (Monograph no. 19) pp. 661–665. Australasian Institute of Mining and Metallurgy: Melbourne.

Hartman HL (1992) Introduction to mining. In *SME Mining Engineering Handbook*. 2nd edn. Vol. 1. (Ed. HL Hartman) pp. 3–42. Society for Mining, Metallurgy and Exploration: Littleton, CO.

Hinner P (1993) Mineral sands dredging and wet plant of Tiwest Joint Venture, Cooljarloo, WA. In *Australasian Mining and Metallurgy: The Sir Maurice Mawby Memorial Volume*. Vol. 2 (Monograph no. 19) pp. 1269–1273. Australasian Institute of Mining and Metallurgy: Melbourne.

Kelly EG and Spottiswood DJ (1982) *Introduction to Mineral Processing*. John Wiley and Sons: New York.

Nash B (1993) Zinc-lead ore treatment at the Cadjebut mine, Cadjebut, WA. In *Australasian Mining and Metallurgy: The Sir Maurice Mawby Memorial Volume*. Vol. 1 (Monograph no. 19) pp. 535–541. Australasian Institute of Mining and Metallurgy: Melbourne.

Peters WC (1978) *Exploration and Mining Geology*. John Wiley and Sons: New York.

Petty JE (1993) Iron ore beneficiation by Hamersley Iron Pty Limited at Mount Tom Price, WA. In *Australasian Mining and Metallurgy: The Sir Maurice Mawby Memorial Volume*. Vol. 1 (Monograph no. 19) pp. 280–286. Australasian Institute of Mining and Metallurgy: Melbourne.

Von Horn D (1993) Dry mineral separation plant of Tiwest Joint Venture, Chandala, WA. In *Australasian Mining and Metallurgy: The Sir Maurice Mawby Memorial Volume*. Vol. 2 (Monograph no. 19) pp. 1294–1298. Australasian Institute of Mining and Metallurgy: Melbourne.

Wills BA (1992) *Mineral Processing Technology*. 5th edn. Pergamon Press: Oxford, UK.

7.5 USEFUL SOURCES OF INFORMATION

Bartlett RW (1998) *Solution Mining: Leaching and Fluid Recovery of Materials*. 2nd edn. Gordon and Breach Science Publishers: Amsterdam.

Halberthal J. Engineering aspects of solid-liquid separation – filters, thickeners, clarifiers. <http://www.solidliquid-separation.com>.

Hartman HL and Mutmansky JM (2002) *Introductory Mining Engineering*. 2nd edn. John Wiley and Sons: New York.

Hutchison IPG and Ellison RD (Eds) (1992) *Mine Waste Management*. Lewis Publishers: Boca Raton, FL.

Kelly EG and Spottiswood DJ (1982) *Introduction to Mineral Processing*. John Wiley and Sons: New York.

Manual of Acid in situ Leach Uranium Mining Technology (2001) International Atomic Energy Agency: Vienna.

Mine-engineer. This is a general mining and mineral-processing website. The best sections are those on crushing, grinding and dewatering. <http://www.mine-engineer.com>.

Mular AL, Halbe DN and Barratt DJ (Eds) (2002) *Mineral Processing Plant Design, Practice, and Control*. Vols 1 and 2. Society for Mining, Metallurgy, and Exploration: Littleton, CO.

Peters WC (1978) *Exploration and Mining Geology*. John Wiley and Sons: New York.

Williams RE (1975) *Waste Production and Disposal in Mining, Milling and Metallurgical Industries*. Miller Freeman Publications: San Francisco.

Wills BA (1992) *Mineral Processing Technology*. 5th edn. Pergamon Press: Oxford, UK.

Woodcock JT and Hamilton JK (1993) *Australasian Mining and Metallurgy: The Sir Maurice Mawby Memorial Volume*. Vols 1 and 2 (Monograph no. 19). Australasian Institute of Mining and Metallurgy: Melbourne.

Woolacott LC and Eric RH (1994) *Mineral and Metal Extraction: An Overview*. South African Institution of Mining and Metallurgy: Johannesburg, South Africa.

8 Producing metals and manufactured mineral products

The major bulk manufactured mineral commodities are metals (and metalloids), cement, glass, mineral fertilisers and ceramic materials. Production of these requires the chemical transformation of raw materials. The chemistry and technology of the production of metals vary widely, whereas the chemistry and technology of the production of cement, glass, fertilisers and ceramics are more standard. Metals and glass can be relatively easily recycled from secondary sources; the production of these from secondary sources is discussed in Chapter 13. Fertilisers cannot be recycled since they are dissipated during use, and cement cannot be reconstituted and recycled although concrete can be reused in aggregate products. This chapter examines the technology of the production of manufactured mineral commodities from primary raw materials. Only a brief description is given, sufficient to provide an overview and a basic understanding of the scientific bases of the technologies. More detailed descriptions are available in the standard texts listed at the end of the chapter.

8.1 THEORETICAL CONSIDERATIONS

The production of manufactured mineral commodities depends on chemical reactions between the reactants. For example, iron is produced by reacting hematite with carbon at high temperatures:

$$0.5Fe_2O_3 + 1.5C = Fe + 1.5CO$$

This reaction occurs spontaneously when the temperature is greater than about 650°C. The selection of feasible reactions and appropriate conditions are important considerations in the production of chemically transformed mineral products. Any reaction can be written in the form of a chemical equation, but whether the reaction will actually occur is determined by the magnitude of a thermodynamic parameter called the Gibbs energy of the reaction.

The second law of thermodynamics requires that a spontaneous process always be accompanied by an increase in the total entropy of the system and its surroundings (Appendix II). If a reaction occurs at constant temperature (T) and pressure, the surroundings will receive a quantity of heat equal to the enthalpy of the reaction, ΔH. The entropy change of the surroundings will be $-\Delta H/T$ (Equation II.3; Appendix II)).[17] If ΔS is the entropy change of the system, then the total entropy change of the system and its surroundings will be the sum of ΔS and $-\Delta H/T$; namely

17 By convention, the negative sign indicates flow of thermal energy from a system to the surroundings.

$\Delta S - \Delta H/T$. For a chemical reaction to occur spontaneously, $\Delta S - \Delta H/T$ must be greater than zero. Since T is always positive, then for a spontaneous reaction $\Delta H - T\Delta S$ must be negative. The term $\Delta H - T\Delta S$ is equal to the Gibbs energy of the reaction, defined as follows. For the general reaction:

$$kA + lB = mC + nD \qquad 8.1$$

where k, l, m and n are the number of moles of reactants A and B and of products C and D, respectively, the *Gibbs energy of reaction* is defined as:

$$\Delta G = mG_C + nG_D - kG_A - lG_B$$

where G_A, G_B, G_C and G_D are the Gibbs energies of substances A, B, C and D, respectively. When all the reactants and products of a reaction are in their pure natural state at the temperature of interest (called their standard state), we write:

$$\Delta G° = \Delta H° - T\Delta S° \qquad 8.2$$

The values of $\Delta H°$, $\Delta S°$ and $\Delta G°$ for the formation of most common inorganic substances have been determined, usually experimentally, and their values over a wide range of temperatures are available in thermodynamic compilations and databases. For the reduction of Fe_2O_3 by carbon at 25°C (298 K), for example, $\Delta H° = 245.7$ kJ, $\Delta S° = 0.272$ kJ K^{-1}, therefore $\Delta G° = 245.7 - 298 \times 0.272 = 164.7$ kJ. Since $\Delta G°$ is positive, the reaction will not occur spontaneously. However, at 700°C (973 K), $\Delta H° = 235.5$ kJ, $\Delta S° = 0.256$ kJ K^{-1}, therefore $\Delta G°$ is –13.7 kJ. The reaction will occur spontaneously at 700°C; i.e. if a mixture of Fe_2O_3 and carbon is heated to 700°C, the Fe_2O_3 will react with the carbon to form metallic iron and gaseous CO.

The Gibbs energy change is the maximum quantity of work that can be obtained from a spontaneous reaction. Conversely, it is the minimum quantity of work required to drive a reaction in the reverse direction. Thus, a reaction can be forced to proceed in the desired direction even if its Gibbs energy is positive as long as energy is provided to balance the Gibbs energy. For example, aluminium is produced at around 1000°C by the reaction:

$$0.5Al_2O_3 + 1.5C = Al + 1.5CO \qquad 8.3$$

for which $\Delta H^{\circ}_{1273} = 675$ kJ, $\Delta S^{\circ}_{1273} = 0.295$ kJ K^{-1} and $\Delta G^{\circ}_{1273} = 299$ kJ, even though ΔG°_{1273} is positive. This is done by electrolysis (discussed subsequently), by consuming electrical energy of 299 kJ per mol (or about 11.1 MJ or 3.1 kW h per kilogram) of aluminium produced.

Whether a reaction occurs spontaneously or is driven in the desired direction electrochemically, the enthalpy requirement must be satisfied if a constant temperature is to be maintained (as is usually required). Thermal energy (heat) must be supplied or removed according to whether the reaction is endothermic or exothermic. This can be done by burning fuel, by electric resistance heating or by any other appropriate means. In the case of hematite reduction by carbon at 700°C, 235 kJ of heat must be supplied per mole of iron. In the case of electrolytic reduction of alumina in the presence of carbon (Equation 8.3) a total of 675 kJ per mol of aluminium, or 25.0 MJ per kilogram, must be provided to drive the reaction and maintain the temperature. Of this amount, 299 kJ must be supplied electrochemically, the balance being supplied as thermal energy. If all this energy were supplied electrically (through a combination of electrochemical reduction and electrical resistance heating), the theoretical minimum power consumption required for producing aluminium would be about 6.9 kW h per kilogram.

The electrical potential required to cause a reaction to proceed or, in the case of a spontaneous reaction, the electrical potential generated by the reaction (as in a battery or fuel cell), is related to the Gibbs energy change of the reaction by the relation:

$$\Delta G° = -nFE° \qquad 8.4$$

where $E°$ (V) is the cell potential, n is the number of electrons involved in the half-cell reaction forming the substance and F is the Faraday constant (96 485 C mol^{-1}).[18] For example, the half-cell reaction at the cathode in the electrochemical reduction of alumina by means of Equation 8.3 is:

$$Al^{3+} + 3e^- = Al$$

18 The derivation of Equation 8.4 is not considered here; it is available in any standard physical chemistry text.

Hence, $n = 3$. Since $\Delta G^{\circ}_{1273} = 299$ kJ, the cell potential required to reduce alumina is 1.0 V.

While knowledge of the thermodynamics of a reacting system makes it possible to predict whether a reaction will occur, it provides no information about the rate at which the reaction will occur and, therefore, the time required to reach equilibrium. This has to be determined independently by experiments under different conditions to identify the mechanisms by which the reaction occurs. The slowest of the mechanisms in a reaction sequence determines the overall rate of the reaction. The study of the mechanism and rate of reactions is called reaction kinetics, and a large body of theory has been developed to quantify reaction rates. It is beyond the scope of this chapter to consider this in detail.

8.2 METALS

The study of the production of metals from ores and concentrates is called *extractive metallurgy*. There are two processing regimes used in extractive metallurgical processes. In one, the chemical reactions to produce the metal are performed at relatively low temperatures and largely in aqueous solutions. This branch of extractive metallurgy is called *hydrometallurgy*. In the other regime, reactions are performed at high temperatures, typically 500–2000°C, between gases and solids or between gases and molten materials. This branch of extractive metallurgy is called *pyrometallurgy*. Water may be present in pyrometallurgical reactions but only in the form of water vapour as a constituent of the gas phase. A special class of extractive metallurgical reactions is electrochemical in nature and the study of this class is called *electrometallurgy*. Electrometallurgical reactions can be performed in aqueous solutions or at high temperatures, in molten salt mixtures. Electrometallurgy therefore straddles hydro- and pyrometallurgy.

There is no exclusive process for extracting a particular metal and no two metals have exactly the same sequence of steps in their extraction. There are a number of reasons for this.

- The thermodynamic reactivity of metals, and hence the stability of their compounds, as measured by the Gibbs energy of formation, varies widely.
- Metals occur in a variety of mineral forms – as oxides, sulfides, chlorides, silicates etc. Each mineral behaves differently chemically.
- Because metals have widely differing melting and boiling points, the form in which they are produced (solid, liquid or gas) depends on the temperature at which the required chemical reactions take place. For example, zinc is produced as a gas in pyrometallurgical processes and as a solid in hydrometallurgical processes.
- An ore mineral is never associated with exactly the same minerals in different ore bodies, so each ore body must be treated uniquely. In base metal production it is rarely possible to produce a single metal, since many base metal deposits contain several major and many minor metals (see Table 5.13).

Thus, many different technologies are used in the production of metals, and even for the same metal. The fact that very specialised and metal-specific knowledge is needed for the successful operation of metal-producing processes is often forgotten or not fully appreciated even within minerals and metals companies. There is a fundamental difference between extractive metallurgy and its sister disciplines of mining and mineral processing, and between extractive metallurgy and the related discipline of chemical engineering. This section provides an overview of the wide range of technologies used in the production of metals and explains, at an elementary level, the scientific basis for them.

All metals except the noble metals occur in nature as compounds, which have to be reduced to obtain metals in their elemental form. *Reduction* is the addition of electrons to metal cations to form neutral atoms. The decomposition of any compound to yield a metal is a reduction reaction. *Oxidation* is the reverse of reduction. It involves the removal of electrons from anions to form neutral atoms or molecules. Reduction of a metal is accompanied by oxidation of the anion combined with the metal. Thus:

$$MO = M^{2+} + O^{2-}$$

$$M^{2+} + 2e^- = M \text{ (reduction)} \qquad 8.5$$

$$O^{2-} = 0.5O_2 + 2e^- \text{ (oxidation)} \quad 8.6$$

where M is a divalent metal. Summing gives the overall reaction:

$$MO = M + 0.5O_2$$

Reduction can be performed by a variety of methods, discussed below.

8.2.1 The principles of metal extraction

The stability of compounds

The standard Gibbs energy of formation of compounds from their constituent elements, for example

$$2M + O_2 = 2MO$$
$$2M + S_2 = 2MS$$

where M represents a divalent metal, is the measure of their stability. Standard Gibbs energy of formation is a function of temperature, as shown for oxides and sulfides in Figures 8.1 and 8.2, respectively. The relationships are approximately linear and thus have the form:

$$\Delta G° = a + bT$$

Since

$$\Delta G° = \Delta H° - T\Delta S°$$

the slope of the lines in the figures is the average value of $-\Delta S°$ of the reaction and the intercept of the lines at 0 K is the average value of $\Delta H°$ of the reaction. The slopes of the lines for metal oxides and metal sulfides are negative and of similar magnitude; i.e. the entropies of formation of most oxides and most sulfides are comparable. The decrease in entropy (negative value of $\Delta S°$) results mainly from the decrease in disorder due to the consumption of 1 mole of gaseous oxygen or sulfur, respectively. Figures 8.1 and 8.2 show the

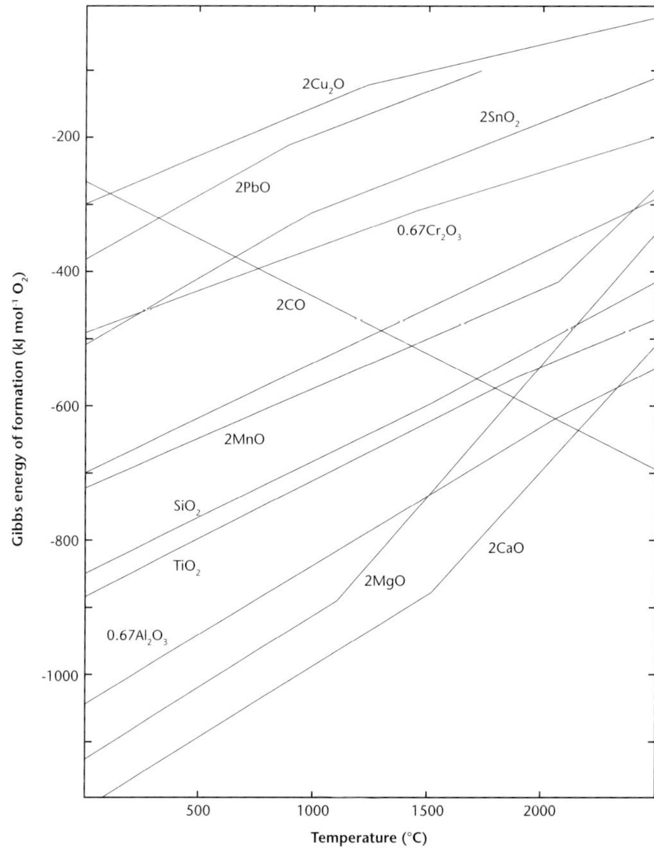

Figure 8.1: The variation with temperature of the standard Gibbs energy of formation of some common oxides. Changes in the slopes of lines indicate change of state of a metal or oxide.

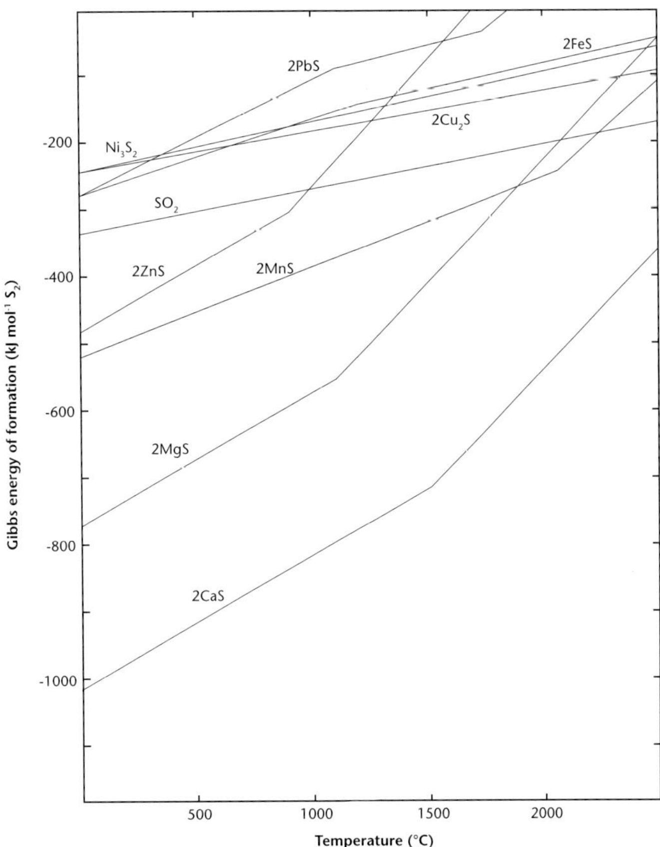

Figure 8.2: The variation with temperature of the standard Gibbs energy of formation of some common sulfides. Changes in the slopes of lines indicate change of state of a metal or sulfide.

relative stability of oxides and sulfides. The lower a line is in these figures, the more negative is the standard Gibb's energy of formation of the oxide or sulfide and the more stable it is; conversely, the more reactive is the metal. Magnesium, for example, is highly reactive and magnesium oxide is highly stable.

Metals can conveniently be divided into three classes: the reactive metals, the noble (inert) metals and those in between (Figure 8.3). The compounds of the reactive metals are the most stable, and the most difficult to reduce. Usually, the production of a metal includes refining steps, in addition to reduction, to remove unwanted impurities in gangue minerals or in the crystal structure of the ore mineral(s). In broad terms, either the concentrate or ore is treated chemically to produce a pure compound which is then reduced to a pure metal, or the concentrate is treated chemically to produce an impure metal which, if necessary, is then refined to the required purity. The two broad approaches are illustrated in Figure 8.4. The refine–reduce sequence is most commonly used for the production of highly reactive metals and the reduce–refine sequence is mainly used for less reactive metals. The reason for this is that an impure form of a reactive metal is difficult to refine using relatively simple oxidation reactions since the host metal, being more reactive than most of the impurity elements, oxidises preferentially. Hence, it is usually preferable to remove most impurities before the metal is reduced.

Heterogeneous reactions

Reduction and refining involve chemical reactions between substances (or species) in two or more phases. These are called *heterogeneous reactions*. In chemistry and physics, a *phase* is any part of a system which is physically homogeneous and bounded by a surface so it can be physically separated from the system. A phase may be composed of a single

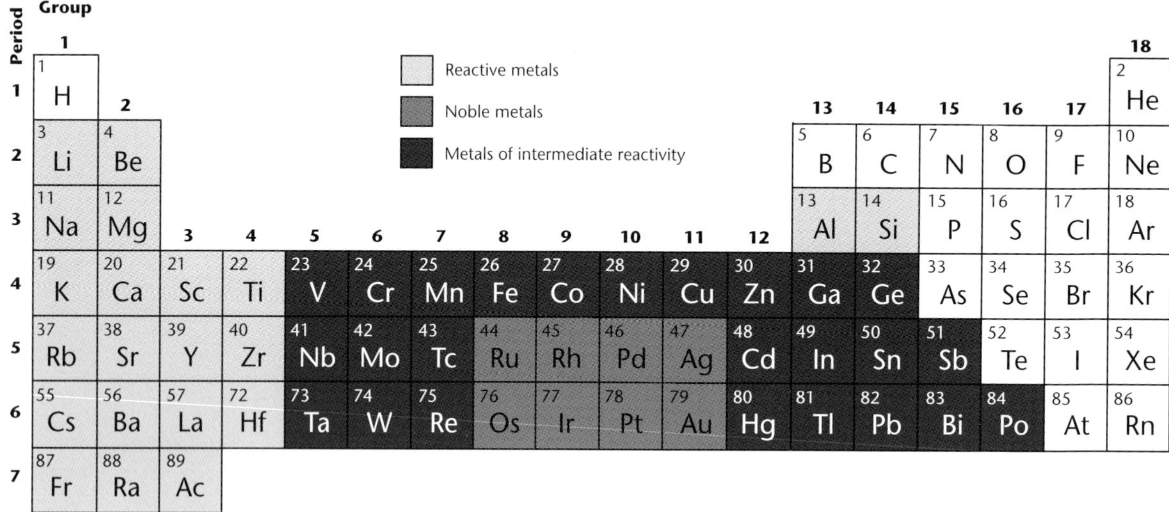

Figure 8.3: Periodic table divided into thermodynamically reactive metals, noble metals and metals of intermediate reactivity.

substance or may be a solution of several substances. Co-existing phases, by definition, are mutually immiscible. For example, in the non-reacting system consisting of ice and water in a glass jar, the ice cubes are a phase, the water is another phase, the air above the water in the glass is a third phase, and the glass of the jar is a fourth phase. The object of chemical reactions in reduction and refining is to change the chemical associations of the desired metal and impurities so that they report to different phases and report preferentially to one phase. It is then possible to physically separate the phases to recover the metal and/or remove the impurities. Common heterogeneous systems encountered in extractive metallurgical processes include aqueous phase/mineral/gas, aqueous phase/organic phase, aqueous phase/metal/gas, slag/matte/gas, and slag/metal/gas. The steps in a heterogeneous reaction are:

- the reactants diffuse from the bulk of the phases to the interface of the phases (mass transfer);
- the reactants react chemically at the interface (kinetics);
- the reaction products diffuse away from the interface back into the bulk phase (mass transfer).

The rate of the slowest of these steps determines the overall rate at which a heterogeneous reaction proceeds. The engineering of systems to perform heterogeneous reactions takes account of this and aims to maximise the rate of the slowest step.

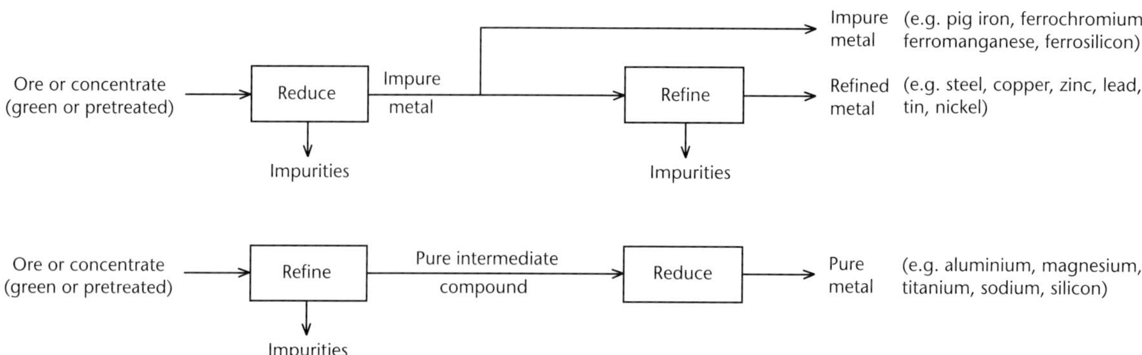

Figure 8.4: The reduce–refine and refine–reduce sequences for the production of metals.

Often in heterogeneous systems the reactant and product species are not in their standard states; they occur in diluted form in solutions or gaseous mixtures. In this case, the Gibbs energy of the reaction is given by:

$$\Delta G = \Delta G^0 + RT\ln\frac{a_C^m a_D^n}{a_A^k a_B^l} \qquad 8.7$$

where $\underline{a}$ represents the activities of the species in solution or, in a gaseous phase, the partial pressure exerted by the particular gas species. This is an equation of fundamental importance in chemical thermodynamics. It is known as the *van't Hoff isotherm.*[19] The *activity* of a species can be thought of as the effective concentration of the species in the solution.[20] In the case of a reaction not initially at equilibrium, ΔG will be non-zero (either positive or negative). If $\Delta G < 0$, the reaction will proceed to the right by consuming reactants. If $\Delta G > 0$, the reaction will proceed to the left by consuming products. In both cases, ΔG moves towards a value of zero. When $\Delta G = 0$, no further reaction will occur and the reaction is at equilibrium. It follows that, at equilibrium:

$$\Delta G^0 = -RT\ln\frac{a_C^m a_D^n}{a_A^k a_B^l} \qquad 8.8$$

where a is the equilibrium activity of the species in the relevant solutions. The quotient:

$$K = \frac{a_C^m a_D^n}{a_A^k a_B^l} \qquad 8.9$$

is called the *equilibrium coefficient* of the reaction. Hence:

$$\Delta G^\circ = -RT\ln K \qquad 8.10$$

It follows, from Equation 8.8, that when substances are present in solutions or gaseous mixtures it is not possible to completely consume any particular substance. There will always be an equilibrium quantity of it in one of the phases involved in the reaction. This is important because it means it is not possible to completely remove impurities or completely recover a valuable metal when they are present in solutions. Data to estimate the activity of species in various types of solutions (aqueous, molten salt, molten silicate etc.) are available in published compilations, as are mathematical models to predict their value in complex, multi-component solutions.

19 The derivation of the van't Hoff isotherm is not considered here; it is available in any standard physical chemistry text.

20 There are several scales used for measuring activity. The most common is the Raoultian activity scale, on which the activity of a pure substance in its standard state is 1. The activity of a substance present in a solution or as a component of a gas mixtures is less than 1; the lower its concentration, the lower its activity. Again, the derivation and formal definition of activity are not considered here. They are available in any standard physical chemistry text.

Phase separation

After desired reactions have occurred, it is necessary to separate the phases. Four types of phase separation are commonly used in extractive metallurgical processes.

Separation of liquids from solids. For relatively coarse solids, liquid separation occurs naturally by drainage under the action of gravity. The separation of finer solids from liquids was discussed in Section 7.2.3 for aqueous systems; techniques include filtration and thickening.

Separation of liquids and solids from gases. In separations involving gases, all that is usually required is to remove gas continuously from the reactor, leaving the solid or liquid behind. This is done using a fan or pump, or simply by leaving the system open to the atmosphere for the gas to escape. Fine particles are often carried out of reactors in the gas stream and usually have to be removed before the gas is released to the atmosphere. Processes for cleaning gases are discussed in some detail in Section 12.3.2.

Separation of two liquid phases. The use of two liquid phases is common in extractive metallurgical processes; for example slag–matte and slag–metal systems in smelting operations and aqueous–organic solutions in hydrometallurgical processes (solvent extraction). If two immiscible liquids are mixed vigorously, droplets of one phase will form within the other continuous phase. Shaking oil and water together in a bottle demonstrates this effect. If allowed to stand, the droplets will coalesce to form a continuous phase as a layer above or below the other phase, according to their relative densities. These two layers can be separated simply by providing drain holes at appropriate heights in the vessel that contains the

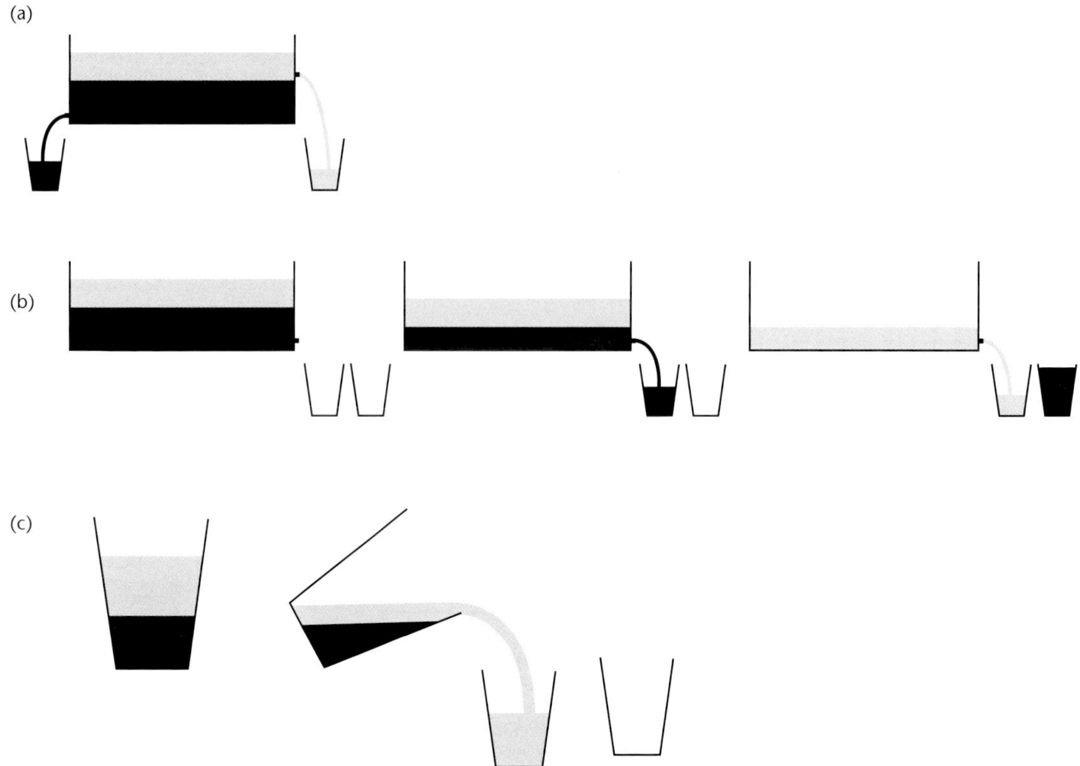

Figure 8.5: Techniques for separating liquid phases. (a) Separating phases from a continuous process. (b) and (c) Separating phases from batch processes.

liquids, or by tilting the vessel to pour off each layer separately (Figure 8.5).

Reduction and refining reactions

The most important reduction reactions are the formation of metals from oxides, sulfides and halides (particularly chlorides). The direct reduction of oxides is performed on an industrial scale at high temperatures using carbon, hydrogen, hydrocarbon gases and reactive metals as the reducing agent. The general reaction can be written as:

$$MO + R = M + RO \qquad 8.11$$

where M represents a metal and R a reducing agent. Some sulfide minerals can be reduced to metal by means of a reaction involving oxygen:

$$MS + O_2 = M + SO_2 \qquad 8.12$$

This is often referred to as an oxidation reaction, but the metal is actually reduced and the sulfur is oxidised. This reaction is important in the pyrometallurgical production of copper. Metal cations are also reduced from aqueous solutions on an industrial scale by the action of a more reactive metal, by hydrogen, or by electrochemical reduction. The ores or concentrates are leached to extract the desired metal into a solution, which is then purified and concentrated. Some metals are also reduced electrochemically from molten salt mixtures.

Impure metals are refined pyrometallurgically (fire-refining) and by electrolysis in aqueous solutions (electro-refining). Fire-refining reactions are frequently oxidation reactions:

$$I + Ox = IOx$$

where I is an impurity element and Ox a substance that reacts preferentially with I. Electro-refining involves transferring metal from the anode (formed from the impure metal) to the cathode, leaving the impurities to remain in solution or form precipitates, by means of an oxidation process at the anode and a reduction process at the cathode:

$$M_{impure} = M^{2+} + 2e^- \text{ (at anode)}$$
$$M^{2+} + 2e^- = M_{pure} \text{ (at cathode)}$$

where *M* is a divalent metal. Summing gives the overall reaction:

$$M_{impure} = M_{pure}$$

8.2.2 Metallurgical reactors

Large-scale chemical reactions are performed in *reactors*. These are specifically designed to hold the reacting substances, facilitate reactions and allow reactants to be added and products removed. Many types of reactors are used in extractive metallurgical processes, including furnaces of many types, fluidised beds, packed beds, pressure vessels and stirred tanks. Some reactors, particularly those used in pyrometallurgical processing, consist of several basic units connected in series or parallel to form a single reactor. The blast furnace for making iron and the mixer-settler used in hydrometallurgical processes for purifying leach solutions are examples (discussed later). An important point of difference between pyrometallurgical processing and mineral beneficiation and hydrometallurgical processing is that the latter two use a relatively small number of basic reactor types to perform operations on a wide range of mineral materials, whereas in pyrometallurgical processsing the reactors are often unique to the particular metal.

The basic types of reactors used in metallurgical and similar types of processing operations are summarised in Table 8.1. Some types of reactors are used exclusively for hydrometallurgical processing and others for pyrometallurgical processing, but many can be used for both. An *in situ* leaching operation and a heap of ore on a pad for leaching are reactors (in this case, fixed bed reactors). However, generally reactors are closed or open vessels. They must be constructed of, or lined with, materials that are resistant to attack by the reacting materials and products. Furnaces are lined with refractories which are resistant to attack by slags, and leaching vessels are often lined with plastic or rubber to reduce corrosion by acids or alkalis. Reactors can be operated continuously or in batch mode. In continuous operation, the rate at which products are removed matches the rate at which reactants are added so the quantity of material inside the reactor remains constant over time. Some of the most common types of metallurgical reactors are illustrated in Figure 8.6. These, and their applications, are discussed in subsequent sections.

8.2.3 Smelting

Smelting of ore or concentrate is a basic operation in the pyrometallurgical extraction of many metals. Melting is the physical process of converting a solid to the liquid state by adding heat. *Smelting* is melting accompanied by chemical change in the substance(s) being melted. The change may be dramatic, yielding an entirely new substance, or there may be a relatively small change in the composition. A *smelter* is a reactor in which a smelting operation is performed. The term applies to reactors as diverse as the blast furnace, flash smelter, top submerged lance (TSL) smelting furnace, and molten salt electrochemical cell. Examples of these are discussed later.

A common product of smelting is *slag*, which is a molten mixture formed mainly from oxides. It is usual to consider slags as molten solutions of oxides (e.g. SiO_2-FeO-CaO) although these compounds do not exist in the molten state. Molten slags are highly ionised. A large proportion of the oxygen atoms are present as O^{2-} anions and most metals are present as cations (e.g. Fe^{2+}, Ca^{2+}). Slags are an essential material in many smelting and refining operations. They are immiscible with metals and their main function is to collect unwanted gangue components and impurities. The composition of slags is chosen so that they have as low a melting point as feasible. Figure 8.7 shows the approximate composition range of a number of common slag types. Another molten product encountered in the smelting of sulfide concentrates is *matte*, which is formed from the sulfide minerals present in the concentrate. Mattes may be thought of as molten solutions consisting of metal sulfides (e.g. Cu_2S-FeS). The major non-metal component of slags is oxygen and of mattes it is sulfur. Metals are produced as molten liquids in most smelting processes. This has the advantage that, being denser than slags and immiscible with them, metals form a separate layer (phase)

Table 8.1: A simple classification of metallurgical reactors and examples of application

Type of reactor	Function	Variants	Applications	Examples
Packed bed reactors	To perform a reaction, or to exchange heat, between coarse solid particles and a fluid (a gas or a liquid)	Fixed bed	Used in both hydrometallurgical and pyrometallurgical applications	Grate sintering of iron ore and of lead concentrate Heap, dump and vat leaching Retort processes[a]
		Moving bed		The shaft section of a blast furnace Retort processes[a]
Fluidised bed reactors	To perform a reaction, or to exchange heat, between fine solid particles and a fluid (a gas or a liquid)		Used in both hydrometallurgical and pyrometallurgical applications	Calcining alumina in the Bayer process Roasting zinc concentrate prior to leaching Removing impurity gases from gas streams
Rotary kilns	To react solid particles with a gas or liquid			Production of cement clinker Treating of dross from aluminium remelting Calcining alumina in the Bayer process
Hearth furnaces	To heat solid material on a hearth	Gas-solid	Used for drying, calcining or roasting solid particles	Multiple-hearth roaster
		Gas-solid-liquid	Used for melting material and, in some cases, carrying out reactions on the molten material	Hearth of a reverberatory furnace Glass melting furnace Aluminium remelting furnace
Suspension and entrained flow reactors	To perform a reaction between fine solid particles and a fluid (a gas or a liquid)			Shaft of a flash smelting furnace Suspension roaster
Settlers	To enable two immiscible liquids to separate into layers so they can be physically removed			Settler zone of a solvent extraction mixer-settler Settler section of a flash smelting furnace Hearth section of a blast furnace
Electric furnaces[b]	To melt solid materials or to melt materials and perform reactions	Open arc[c]	Used where very high temperatures are required and/or where the presence of combustion gases may cause processing problems	Electric arc furnace recycling of scrap steel
		Submerged arc[d]		Carbothermic production of silicon, ferrosilicon, ferromanganese and ferrochromium

Type of reactor	Function	Variants	Applications	Examples
Stirred tank reactors	To increase the reaction rate between solid particles and a liquid or between two liquids.	Mechanically stirred		Mixing zone of a froth flotation cell Agitation leaching Mixer zone of a solvent extraction mixer-settler
		Pneumatically stirred (can be top, bottom or side injection)	Used in both hydrometallurgical and pyrometallurgical applications	BOF steel-making Submerged combustion smelting (e.g. ISASMELT™) Peirce-Smith converter for producing copper
Electrolytic cells	To carry out electrochemical reactions	Aqueous electrolyte	Used for electro-winning and electro-refining of metals	Copper and nickel electro-winning from leach solutions Electro-refining of copper anodes
		Molten salt electrolyte		Electrolytic production of aluminium

a) In retort processes, heat required for the reaction is generated externally and transferred to the bed material through the walls of the retort.
b) An arc furnace consists of a moving packed bed reactor with a settler at the base to allow molten slag and metal to separate.
c) The arc is created in the space above the charge and heat is transferred by radiation to the charge.
d) The arc is created within the bed of solid material and heat is transferred to it directly.

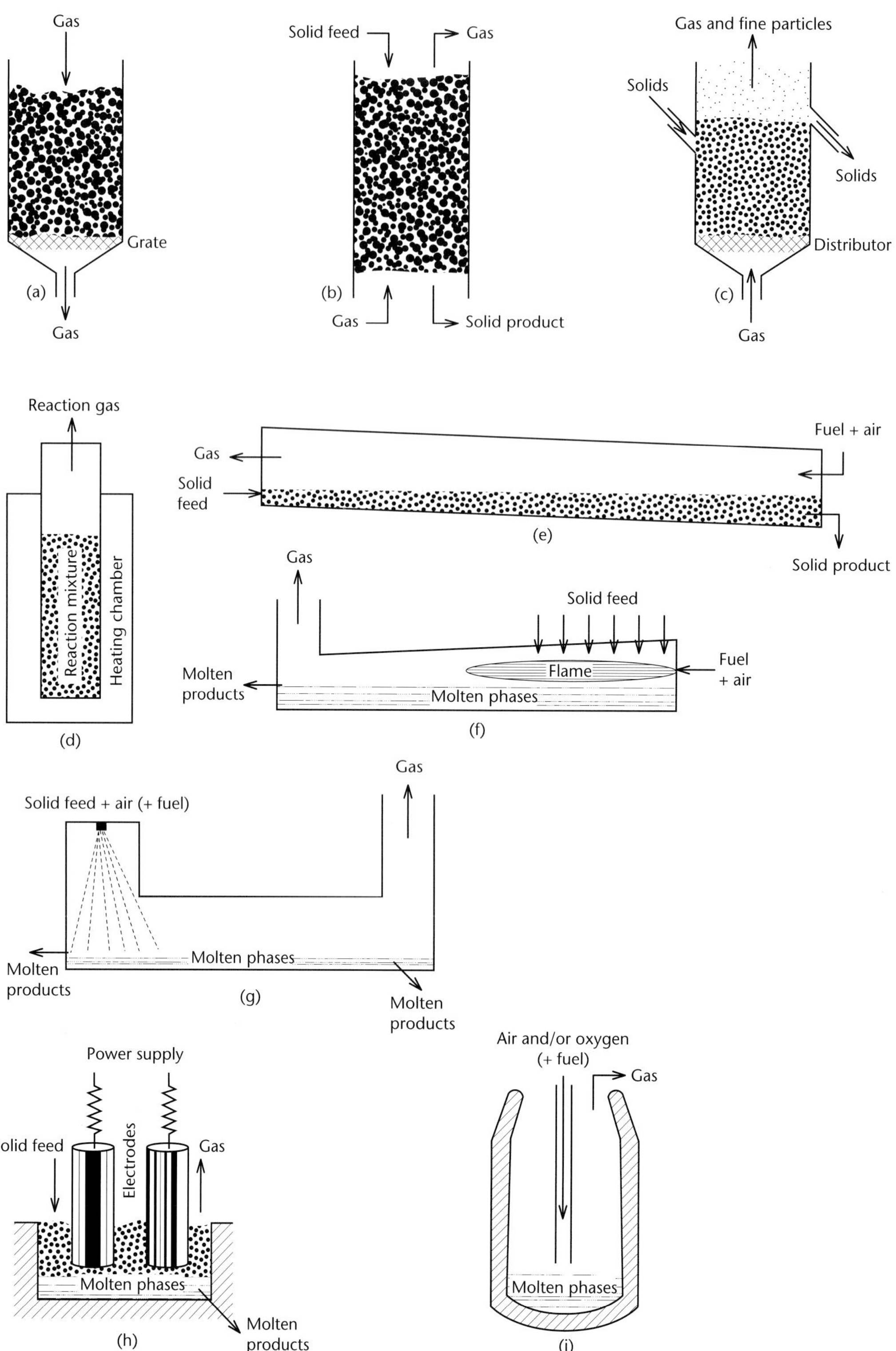

Figure 8.6: Common types of metallurgical reactors. (a) Fixed bed. (b) Moving bed. (c) Fluidised bed. (d) Retort. (e) Rotary kiln. (f) Reverberatory furnace. (g) Flash furnace. (h) Submerged arc furnace. (i) Pneumatic furnace.

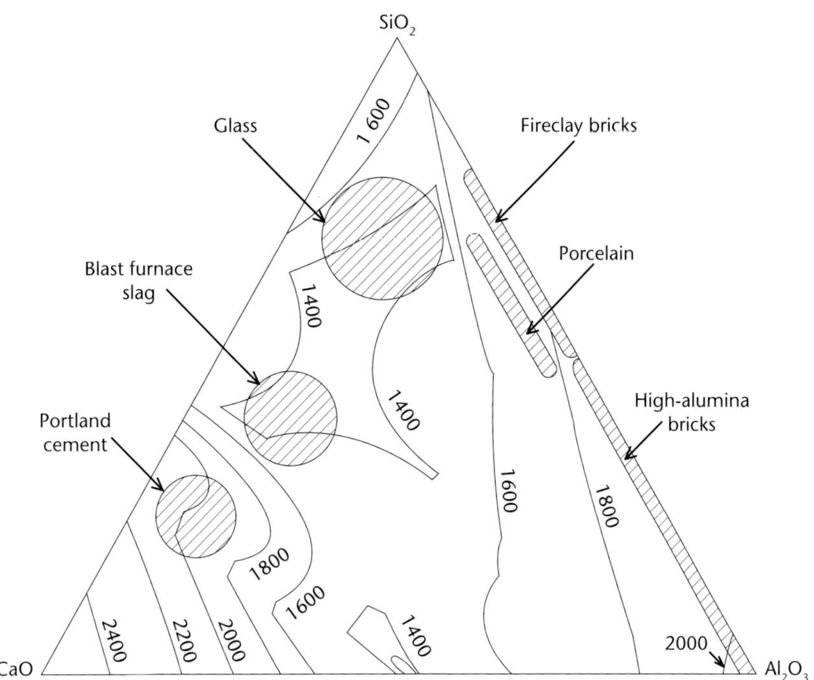

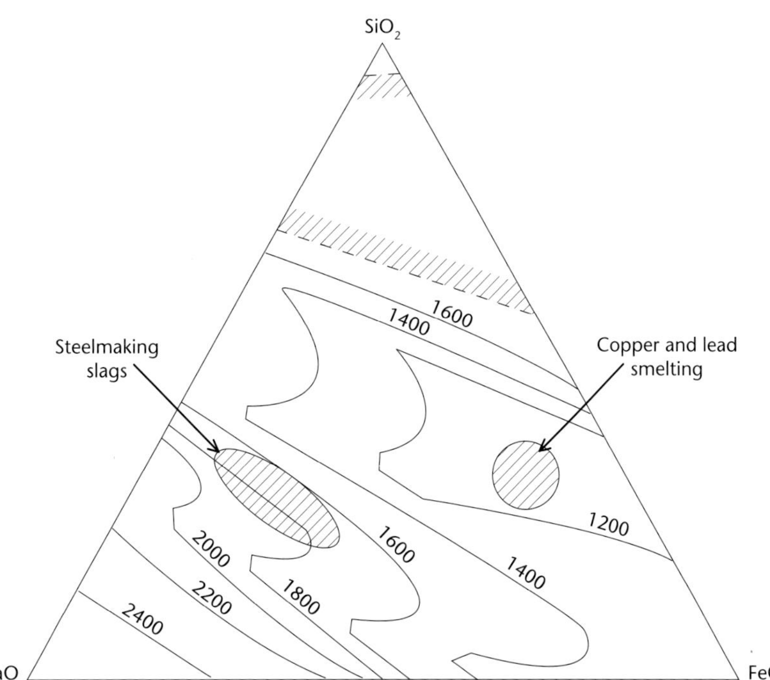

Figure 8.7: Ternary phase diagrams of the CaO-SiO_2-Al_2O_3 and CaO-FeO-SiO_2 systems showing typical composition ranges of common metallurgical slags and manufactured mineral commodities (after Rosenqvist, 1983; with modifications). Composition is expressed in wt%, ranging from 0 wt% along each side of the triangle opposite a species to 100 wt% at the apex. The lines within the figures are lines of constant melting point (isotherms). The area enclosed by any line is molten at the temperature indicated for that line. Temperatures are expressed in degrees Celsius.

in a reactor, below the slag, which makes physical separation from the slag relatively simple.

Smelting reactions

Several types of reactions are used in smelting. Some of the more common are discussed below.

Metallothermic reduction. If the line for a metal in Figure 8.1 lies below that for another metal, the oxide of the first metal will be more stable than the oxide of the second metal. This is also the case for sulfides (Figure 8.2). For example, Al_2O_3 is more stable than Cr_2O_3. It follows that the standard Gibbs energy of the reaction:

$$Cr_2O_3 + 2Al = 2Cr + Al_2O_3$$

is less than zero and that this reaction, therefore, will occur spontaneously. In this case, aluminium is the reducing agent. Reactions of this type are called metallothermic reduction. Metallothermic reactions are highly exothermic.

Carbothermic reduction. Figure 8.1 includes the line for the formation of CO from C and oxygen:

$$2[C] + \{O_2\} = 2\{CO\} \qquad 8.13$$

The conventions adopted in Equation 8.13 (and subsequent chemical equations) to indicate the nature of the phases involved are summarised in Table 8.2. In Equation 8.13, there is an increase in volume of gases (1 mole to 2 moles). Therefore, entropy increases when CO is produced from carbon and oxygen and the line for the reaction has an opposite slope to that for metal oxides. The fact that this line intersects the lines for metal oxides has important consequences for the extraction of metals. Any oxide can be reduced by carbon at any temperature greater than the temperature at which its line intersects with the line for CO, since the standard Gibbs energy change of the reaction:

$$[MO] + [C] = \underline{M} + \{CO\} \qquad 8.14$$

will be less than zero. The minimum temperature for reduction of an oxide by carbon is given at the intersection point of the carbon and metal lines. Approximate values for some common oxides are 600°C, 675°C, 1225°C, 1415°C and 1615°C for SnO, FeO, Cr_2O_3, MnO and SiO_2, respectively. These reactions are referred to as carbothermic reduction. Tin, iron, chromium, manganese and silicon are produced commercially by carbothermic reduction.

Direct smelting. Figure 8.2 includes the line for the formation of SO_2:

$$\{S_2\} + 2\{O_2\} = 2\{SO_2\} \qquad 8.15$$

In this reaction there is a decrease in volume of gases (from 3 moles to 2 moles). This is the same volume change as in the formation of metal sulfides. The entropy change, and therefore the slope of the line for SO_2, is similar to that for metal sulfides. Any metal that lies above the line for SO_2, in principle, can be produced by means of Equation 8.12 since the Gibbs energy change will be less than zero. This is an important reaction in the smelting of copper concentrates.

Electro-winning. Metals present in cationic form in molten salts can be reduced electrochemically (Section 8.1). A compound of the metal is added to an appropriate mixture of molten salts. This technique is used for producing some reactive metals from compounds that are difficult to reduce by other means, or that would be contaminated by the reducing agent (e.g. carbon or a more reactive metal). It is used for producing aluminium from alumina (Al_2O_3) prepared hydrometallurgically from bauxite, and can be used for a number other metals including magnesium, beryllium, calcium, titanium, sodium and potassium.

Table 8.2: Phase type conventions adopted for heterogeneous reactions

Symbol	Phase indicated
[ ]	The substance is present as a solid phase; its activity is 1
( ) and _	The substance is present as a component of a liquid solution; its activity is less than 1 When two immiscible liquids are present: ( ) indicates either an aqueous solution or a molten slag or salt (depending on the context) _ indicates an organic solution, a molten metal or molten matte (depending on the context)
{ }	The substance is present as a component of a gas mixture; its activity is less than 1

Because high temperatures are used and reactions often involve gaseous and liquid phases with large areas of contact, pyrometallurgical reactions are usually fast and smelting processes frequently operate near the chemical equilibrium of the major metal-producing or refining reactions. This means that the rate at which a process performs its function is frequently governed by the rate at which reactants can be supplied to the reaction interface, the rate at which products can be removed, or the rate at which heat can be supplied to sustain the temperature of the process. In other words, the production rate of smelting processes is frequently limited by the rate of mass transfer or heat transfer. This does not mean that chemical equilibrium considerations are unimportant in pyrometallurgical processes. They are vitally important – small changes in the processing conditions can cause major changes in the extent to which the desired reactions occur. Because many competing reactions are possible the selection of process conditions is critical to ensure, first, that the desired reactions occur and, second, that they occur to the maximum extent possible to ensure minimum loss of the valuable metal in the waste products and minimum contamination of the metals with impurities.

Smelting technologies

Blast furnaces, BOF furnaces, reverberatory furnaces, submerged-combustion furnaces, flash furnaces and Peirce-Smith converters are common reactors used for smelting and fire-refining processes, often in very specific applications. These are discussed in Section 8.2.6 in relation to iron and steel production and copper smelting.

8.2.4 Leaching

Leaching is the removal of a substance from a solid using a liquid extraction medium, referred to as the *lixiviant*. Leaching is the primary extraction operation in hydrometallurgical processing. Ideally, the lixiviant is chosen to be as selective as possible so that most of the unwanted components of an ore remain unreacted in the solid. Common lixiviants are sulfuric acid, sodium cyanide, sodium hydroxide and ammonia. Often, commonly used and relatively inexpensive lixiviants are not highly selective and a considerable quantity of unwanted material is also dissolved. If there is no lixiviant which both dissolves a specific metal from its naturally occurring mineral (i.e. ΔG of the reaction is negative) and is acceptably selective, then that metal cannot be extracted by hydrometallurgical means.

The transfer of a metal from an ore or concentrate to a solution is a transfer from a solid to a liquid – a heterogeneous reaction. The solution produced is called *pregnant solution* or *pregnant liquor*. The liquor is separated from the solids in a solid–liquid separation process then is treated in subsequent processes to remove impurities and concentrate the metal. The metal is then recovered by reduction of the cations or by precipitation as a metal compound.

Leaching is a slow process compared with smelting. The extent to which the targeted mineral is leached is determined mainly by mass transfer factors and often equilibrium is not attained within the time available. The unreacted components of the mineral and any non-mobile reaction products progressively form a layer on the surface of the mineral. As reaction proceeds, lixiviant must diffuse through the increasingly thick layer and this often becomes the rate-limiting step. Sometimes, equilibrium cannot be obtained because a mineral particle is surrounded by other unreactive minerals which prevent it being exposed to the lixiviant. Rarely is 100% extraction of the desired metal achieved, and often the recovery is relatively low.

Leaching reactions

Many types of reactions are used in the leaching of metals. Some of the more common are discussed below.

Dissolution of a metal salt. Some metal salts are soluble in water but few of these occur naturally because soluble salts tend to be quickly dissolved and carried into bodies of water, particularly the oceans. However, some minerals can be pretreated to convert them into a water-soluble form. For example, copper sulfide minerals can be converted to soluble copper sulfate by first roasting in air:

$$[CuFeS_2] + 4\{O_2\} = [CuSO_4] + [FeSO_4] \quad 8.16$$

Dissolution by acid. Many metal oxides are soluble in dilute acids (e.g. Cu, Zn, U). The leaching reaction for a divalent metal is given by the reaction:

$$[MO] + 2(H^+) = (M^{2+}) + H_2O \qquad 8.17$$

For metals that occur as sulfide, the oxide can be produced by roasting in air:

$$2[CuFeS_2] + 6.5\{O_2\} = 2[CuO] + [Fe_2O_3] + 4\{SO_2\} \qquad 8.18$$

$$[ZnS] + 1.5\{O_2\} = [ZnO] + \{SO_2\} \qquad 8.19$$

These reactions are called *dead-roasting* since the sulfur is completely oxidised to SO_2.

Dissolution by alkali. Some metals form soluble complex anions in alkaline solutions. The most important example of this is the dissolution of alumina from bauxite in sodium hydroxide solution:

$$[Al_2O_3] + 2(OH^-) = 2(AlO_2^-) + H_2O \qquad 8.20$$

Dissolution by formation of complex ions. Formation of a complex ion may increase the solubility of a weakly soluble metal salt or may permit an oxide to dissolve in pH ranges in which it otherwise has very low equilibrium concentration. The reaction of gold with sodium cyanide is an example of the former. Gold which is too finely disseminated in an ore to be liberated and separated by physical means is extracted by contacting the ore or concentrate with cyanide solution:

$$2[Au] + 4(CN^-) + H_2O + 0.5\{O_2\} = 2(Au(CN)_2^-) + 2(OH^-) \qquad 8.21$$

The gold is then recovered from solution in a separate process. Other metals that form complexes with cyanide include silver, platinum, palladium, copper, nickel, cobalt and other transition metals. Copper, nickel, cobalt and some other transition metals also form complexes with ammonium ions in solution:

$$[CuO] + 2(NH_4^+) + 2(OH^-) = (Cu(NH_4)^{2+}) + 3H_2O \qquad 8.22$$

Dissolution by oxidation. An important reaction for dissolving metals from minerals is oxidation by ferric ions in solution:

$$[Cu_2S] + 2(Fe^{3+}) + 2\{O_2\} = 2(Cu^{2+}) + 2(Fe^{2+}) + (SO_4^{2-}) \qquad 8.23$$

For this reaction to occur, there must be a mechanism for reoxidising ferrous ions to ferric ions:

$$2(Fe^{2+}) + 2(H^+) + 0.5\{O_2\} = 2(Fe^{3+}) + H_2O \qquad 8.24$$

Combining Equations 8.23 and 8.24 gives the overall reaction:

$$[Cu_2S] + 2(H^+) + 2.5\{O_2\} = 2(Cu^{2+}) + (SO_4^{2-}) + H_2O \qquad 8.25$$

Reoxidation of the ferrous ions can be performed by adding an oxidising agent to the solution but it is commonly carried out using bacteria called *Thiobacillus ferrooxidans*, which occur naturally in sulfide ore deposits. These are small, rod-shaped bacteria, about 0.25–1 μm in size, that thrive in the presence of oxygen at ambient temperatures in acidic solutions (pH ~2.5). This is an important mechanism in the heap leaching of copper ores.

Leaching technologies

In leaching on a large scale, a practical way is needed to bring the lixiviant and ore or concentrate into contact, for enough time for reaction to occur, and to recover the pregnant liquor. There are several engineering approaches, summarised in Table 8.3. *In situ*, dump, heap and vat (percolation) leaching are systems that do not use agitation to improve the extraction rate.

In situ leaching. This is also known as solution mining and was discussed in Section 7.1.3. Holes are drilled into the ore deposit, then explosives or hydraulic fracturing are used to create fractures within the deposit and expose the ore to lixiviant. The lixiviant is pumped into the fractured ore body, where it reacts with the ore, is collected in sumps and recirculated. When the concentration of metal reaches a sufficiently high level, part of the collected solution is diverted to the metal recovery process and additional, or make-up, lixiviant is added. This is continued until most of the accessible metal has been extracted. *In situ* leaching is most commonly used for extracting uranium.

Heap and dump leaching. Heap leaching is illustrated in Figure 8.8. Crushed, and sometimes agglomerated, ore is piled into a heap constructed on an impervious bed (pad). Leach solution is sprayed over

Table 8.3: Types of large-scale leaching systems and their characteristics

Type of leaching system	Maximum particle size	Leaching period	Typical copper ore/cons grade (wt% Cu)	Typical pregnant liquor concentration for acid leaching of a copper ore ($g\ mL^{-1}$)	
				Sulfuric acid	Copper
In situ leaching	Fractured ore body	≤20 years	0.5–1	≤2	≤5
Dump leaching	1 m	≤20 years	0.2–1	≤2	≤5
Heap leaching	0.3 m	≤6 months	0.5–1	≤5	≤10
Vat leaching	150 mm	≤10 days	1–2	≤40	≤100
Agitation leaching	1 mm	2–48 hours	20–30	≤50	≤100
Pressure leaching	1 mm	≤5 hours	–	–	–

Source: data from Woollacott and Eric (1994) and Biswas and Davenport (1976).

the top of the heap and allowed to percolate downwards. Sumps or gullies around the base of the heap collect the leach solution, which is recirculated until the concentration of metals is sufficient for further processing, as described for *in situ leaching*. Dump leaching combines characteristics of heap leaching and *in situ* leaching. Ore is dumped into a pit or naturally formed valley and treated in a similar way to heap leaching. The pit or valley contains the lixiviant and enables it to be recovered. Heap and dump leaching are widely used for leaching copper, uranium and gold ores.

Vat leaching. This is also known as percolation leaching. The ore is crushed to the required size and packed to form a permeable bed in a vat or large tank. Lixiviant is introduced from the top or bottom and the solution is recycled to increase the concentration of dissolved metal. After sufficient time, the vats are drained, the solid residue is discarded and the pregnant liquor is sent to the next stage of processing.

Agitation leaching. Finely ground solids are kept in suspension in the lixiviant in a tank by stirring, either mechanically (using a rotating impeller) or by injecting air or other gas into the base of the tank. The larger surface area of exposed mineral (due to the smaller particle size) and the agitation increase the rate of reaction and, usually, the degree of extraction (or recovery) of the metal. Increasing the temperature often increases the reaction rate. This can be done at atmospheric pressure in tanks similar to those used in agitation leaching, by electrical heating or injecting steam. If high pressure is used to increase the rate, a pressure vessel must be used to hold the contents. These vessels are called *autoclaves*. Agitated leaching

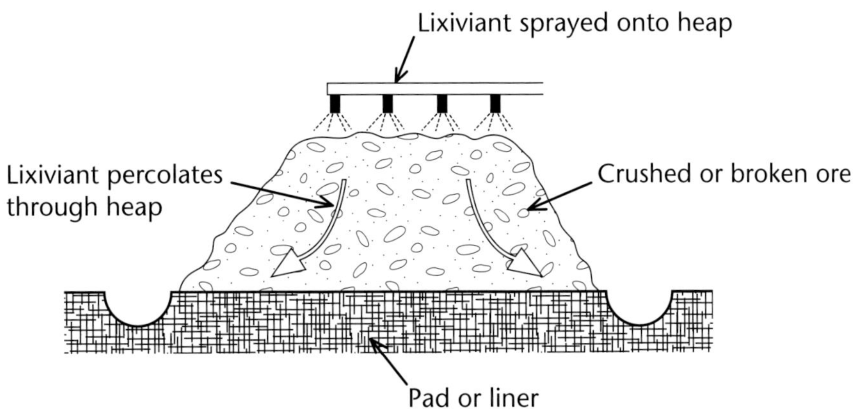

Figure 8.8: Schematic showing the principles of heap leaching.

systems can be operated as batch processes but are usually operated continuously.

8.2.5 The stages in the extraction of a metal

It is rarely possible to produce a metal in one stage of reaction followed by separation of the phases. A series of steps is usually necessary to achieve the desired purity of the product and a sufficiently high recovery. In some cases, it may be possible to achieve separation of phases in the same reactor in which reaction occurs, but in other cases a separate step may be necessary. Figure 8.9 is a generic flowsheet for the production of metals (or pure metal compounds). It includes the beneficiation as well as the chemical stages, which are the subject of interest in this chapter. In a particular operation, the production of a metal follows a specific path from ore to product. For some metals, more than one path can be used. The stages in the extraction of a metal can be grouped into four categories:

- pretreatment;
- extraction;
- refining;
- metal-recovery.

Pretreatment

Some metals are extracted from their ores or concentrates without any chemical pretreatment, although in many cases it is necessary. The objective is to change the physical or chemical nature of the material to facilitate extraction of the metal. Pretreatment processes are almost always pyrometallurgical, but they are usually performed at temperatures below which molten phases form. They are predominantly solid–state reactions or gas–solid reactions. The most common pretreatment is roasting or calcining. Usually, the aim of roasting is to change the structure or composition to make the metal easier to leach, either by converting the metal to a more leachable

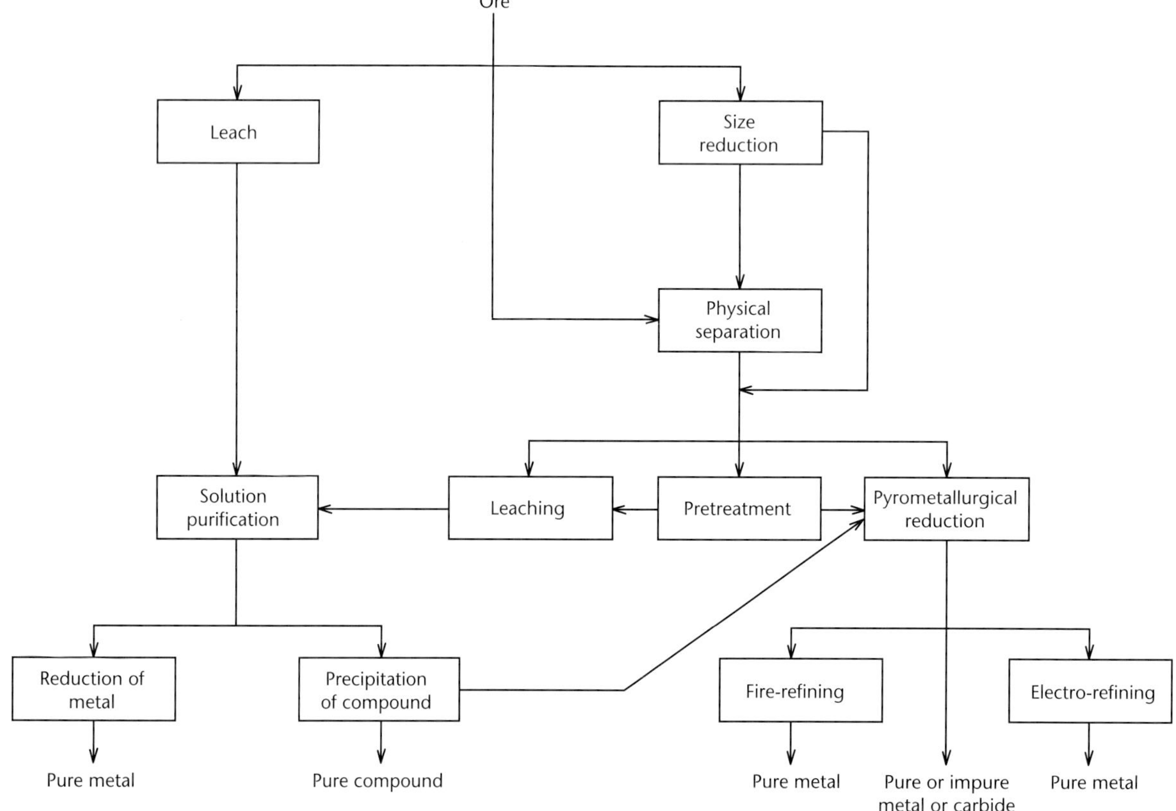

Figure 8.9: Generic flowsheet showing the possible routes for the production of metals from ores.

form, as in roasting of zinc concentrates (Equation 8.18), or by modifying the matrix to make the metal more accessible to the lixiviant, or both. The concentrates of refractory gold ores, in which some of the gold occurs within pyrite grains, are often roasted to make the gold more accessible to the sodium cyanide. This can also be achieved by hydrometallurgical pretreatment using bio-oxidation (involving the use of bacteria) or pressure oxidation in autoclaves, in both cases to promote oxidation of the pyrite. Roasting reactions are usually slow compared to smelting reactions because, in the absence of liquid phases, diffusion of ions through solids is usually the rate-limiting step.

Fluidised bed reactors and rotary kiln reactors (Figure 8.6) are used for performing chemical reactions involving particles and fluids. They are frequently used for calcining and roasting solid materials. A fluidised bed is formed when a bed of small solid particles in a vessel is subjected to conditions that cause the material to behave as a fluid. This is achieved by passing fluid through the bed from the bottom at a rate fast enough to fluidise the particles but not great enough to carry them out of the vessel. Fluidised beds have many properties and characteristics of normal liquids, such as the ability to flow under gravity and provide buoyancy for larger solids. Fluidised bed reactors are usually operated with solid feed continuously added to the bed and product continuously removed as an overflow from the bed and/or as dust in the off-gas. Fluidised bed reactors are also used for transferring heat between gases and solid particles; for example, for recovering heat from waste gases.

Rotary kilns are long (up to 250 m) cylindrical reactors (up to 6 m in diameter). The shell is made of steel plate and is lined with refractory to protect the shell from the high temperatures. The kilns are mounted at a slight inclination to the horizontal and, during operation, are rotated slowly around their axis. Feed is introduced continuously at the elevated end, and product is discharged at the other. Heat is supplied by combustion of a fuel–air mixture introduced through a burner at one end, usually the discharge end. Heat is transferred to the charge and lining from the flame and hot gases. The rotating action serves several purposes: it mixes the charge and exposes it to the hot gases; it moves the charge slowly along the kiln towards the discharge end; and it transfers heat from the lining to the charge, thereby preventing the lining from overheating and improving the overall heat transfer efficiency.

Sometimes the physical state of a concentrate or ore must be altered before it can be processed. This may involve agglomerating the fine particles to increase the particle size so that beds of the material are permeable to gases or liquids to facilitate reaction or heat transfer. Agglomeration can be performed by pelletising and briquetting (Section 7.2.4). Sintering is another technology, discussed in Section 8.2.6.

Extraction

The object of the extraction stages is to remove and discard the large majority of the gangue components while losing as little as possible of the valuable metal(s). In pyrometallurgical extraction, smelting is the most common approach. It results in the production of metal, either of the desired purity or in impure form that requires subsequent refining (e.g. iron, copper, nickel, tin), and slag containing the gangue. In hydrometallurgical extraction, leaching is the process used to extract the metal. The extracted metal is in cationic form in an aqueous solution (e.g. zinc, copper, nickel, gold, aluminium). The solution is separated from the solid wastes and, after purification (refining), the metal is recovered in a separate stage.

Refining

The impure metal from smelting and the metal-containing solution from leaching normally contain impurities extracted along with the metal of interest. The aim of refining is to reduce the concentration of these to levels that meet the specifications for traded metal commodities. The focus in refining is on impurity elements rather than on the valuable metal. Leach solutions can be purified using a number of possible processes, as discussed below.

Chemical precipitation. Precipitation involves the addition of a reagent to form an insoluble compound which precipitates as a solid and which can be removed by a solid–liquid separation process. For example, iron can be removed from solution by adding an alkali to raise the pH and precipitate ferric hydroxide:

$$(Fe^{3+}) + 3(OH^-) = [Fe(OH)_3] \quad 8.26$$

However, precipitation is not a common method for purifying leach solutions.

Solvent extraction. Solvent extraction is a technique for separating substances based on their different solubilities in two immiscible liquids. It is an important technology for purifying and concentrating a range of metals in solutions, particularly uranium, copper, nickel, cobalt and rare earths. An organic solution (usually kerosene) containing a reagent (an *extractant*) is used to transport selected metals from one aqueous solution to another, so that metals are separated and purified. On mixing the organic solution with a pregnant solution containing, for example, copper and iron and other impurities, the organic solution selectively extracts copper. It leaves iron and the other impurities in the aqueous solution (the *raffinate*). This step is called *extraction* (or loading). In the next step, called *stripping*, the copper in the organic solution is stripped by an acidic solution (usually spent electrolyte from the electro-winning circuit in the metal recovery stage) to form a loaded strip liquor (*loaded electrolyte*). Usually, the volume of the strip solution is much smaller than that of the pregnant solution, so the target metal is concentrated as well as purified. The process is illustrated in Figure 8.10. Solvent extraction is carried out as a continuous process in large devices called *mixer-settlers* (Figure 8.11). Mixer-settlers have two zones, one for producing an emulsion of the organic phase in water to create a large interfacial area to ensure rapid reaction and attainment of equilibrium, and a settling zone to allow the emulsion to separate into two layers to enable separation. A solvent extraction circuit will have at least two mixer-settlers, one for extracting and one for stripping.

Ion exchange. Ion exchange serves the same purpose as solvent extraction and relies on similar principles. However, the non-aqueous phase is a solid rather than an organic liquid. Many natural substances have ion exchange properties, such as zeolites, montmorillonite and clay, but synthetic, cross-linked organic polymer resins are usually used in metals extraction. Resins are usually in the form of porous granules, to provide a large area of contact with the solution, and are packed in vertical columns through which the pregnant solution is passed. When the resin is saturated with the metal, as indicated by an increase in the concentration of metal in the solution flowing out of the column, the pregnant liquor is diverted to another column and the saturated column is *eluted* with a strong solution of the displaced anion. This extracts the metal into a pure concentrated solution for metal recovery and regenerates the bed of resin. The cycle is similar to that illustrated in Figure 8.10 for solvent extraction. Ion exchange materials may be cation exchangers (that exchange cations) or anion exchangers (that exchange anions). They may be unselective or highly selective towards particular ions or classes of ions. Ion exchange is usually more expensive than solvent extraction so it is used in applications where solvent extraction is less effective, particularly to separate metals which have very similar chemical properties (e.g. for separating uranium from plutonium and other actinides; and lanthanum, neodymium, ytterbium, samarium and lutetium from each other and the other lanthanides).

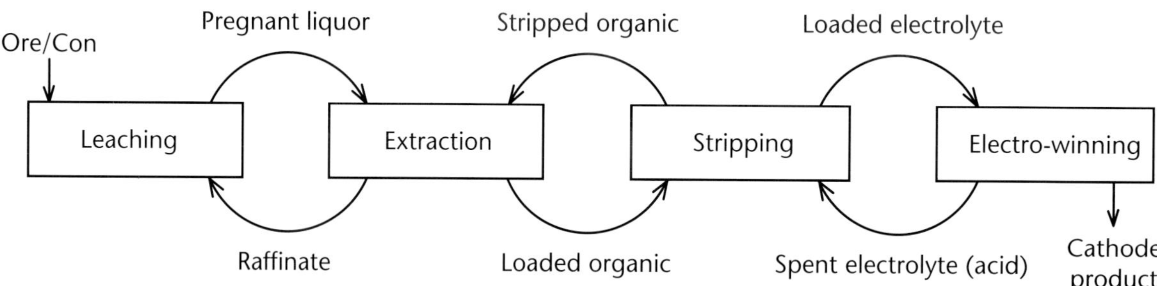

Figure 8.10: The principle of loading and stripping of a metal using solvent extraction, and its connection with the leaching and metal recovery stages.

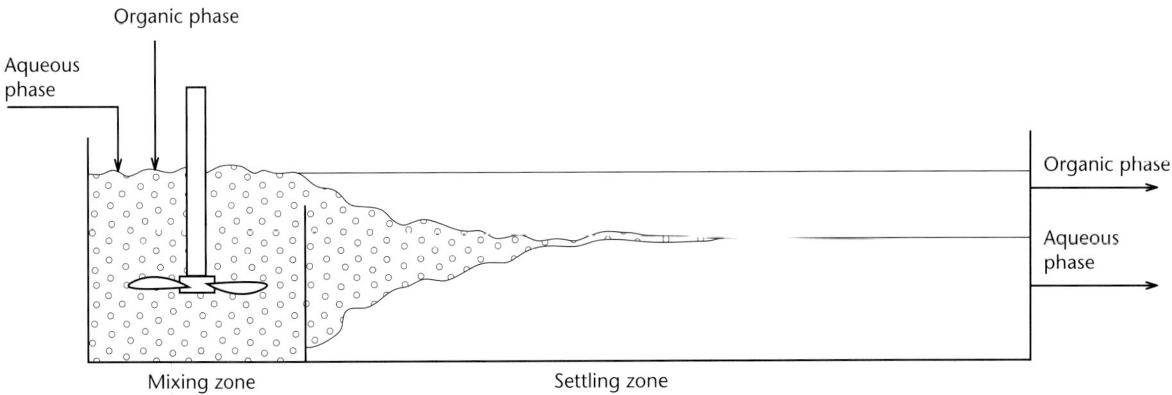

Figure 8.11: Schematic representation of a mixer-settler, used for industrial-scale, continuous solvent extraction (after Evans and De Jonghe, 2002).

Impure metals from smelting operations are refined in two main ways: fire-refining while the metal is molten, and casting the impure metal into anodes which are then refined electrochemically to produce pure metal cathodes. These are discussed below. Metals produced hydrometallurgically can be further refined electrochemically if necessary. To produce very high purity metals, additional refining steps may be needed.

Fire-refining. Selective oxidation of impurities using air, oxygen or another oxidising agent is the preferred method for refining molten metals because it is the cheapest. It is also very efficient for some metals, particularly steel (discussed in Section 8.2.6). When it is not thermodynamically possible, other reactants such as sulfur, chlorine, carbon and lime may be used to react selectively with impurities. The aim is to form a reaction product that will separate as a liquid (or dissolve in the slag) or a gas to enable physical separation from the purified metal. All metals produced in the molten state are purified to some extent by fire-refining, even if only to remove dissolved gases or inclusions of foreign material from the metal.

Electro-refining. Electro-refining involves transferring a metal from the anode to the cathode of an electrochemical cell. It is most commonly used in aqueous systems for refining base metals, particularly copper. It is different from electro-winning (discussed below) in that it does not involve reduction of a cation to metal; it therefore requires much less energy than electro-winning. Electro-refining can also be performed at high temperatures using molten salts. Impure aluminium can be refined by this method.

Metal recovery

For most metals produced pyrometallurgically, the reduction step to form the metal occurs during the extraction stage. The main exceptions are the reactive metals (aluminium, magnesium and titanium). In these cases, a pure compound of the metal is produced in the extraction stage and reduced in the recovery stage. For example, aluminium is reduced electrochemically from alumina and titanium is produced by metallothermic reduction of $TiCl_4$ by magnesium. For metals produced hydrometallurgically, the metal cations extracted into solution during leaching (the extraction stage) are reduced in a separate, metal recovery, stage. Thus, for the less reactive metals produced pyrometallurgically, the recovery step usually consists simply of casting the metal into a form suitable for selling to users of primary metals. For the reactive metals and those produced hydrometallurgically, the recovery stage involves a reduction process. This may be followed by melting and/or casting to produce a saleable form of the metal, though in some cases powder or other forms of the metal are marketed. Sometimes, the metal is recovered as a compound for which there is a market (e.g. MnO_2 for use in batteries and $CuSO_4$).

Several processes are used for recovering metals from aqueous solutions. These are discussed below.

Cementation. This is the oldest method but it has limited application today. It involves using a more reactive metal to displace a metal from solution. The best known example is the use of scrap iron to recover copper from copper sulfate leach solutions:

$$[Fe] + (Cu^{2+}) = [Cu] + (Fe^{2+}) \qquad 8.27$$

though this is rarely done today. Gold is sometimes recovered from cyanide leach solutions using zinc dust.

Hydrogen reduction. The reduction of metal from solution using hydrogen is chemically similar to cementation but is technically more difficult. It is mainly used to recover nickel and cobalt from solution. For nickel, the reaction is:

$$(Ni^{2+}) + \{H_2\} = [Ni] + 2(H^+) \qquad 8.28$$

The reaction is favoured thermodynamically and kinetically by high pressure and high temperature. The reduction of nickel is carried out in autoclaves at about 200 atm pressure and 160°C.

Electro-winning. Electro-winning is the most common method of recovering metals from aqueous solutions, particularly for copper, nickel and gold. Metal cations that can be reduced electrochemically at a potential less than that required to decompose water into hydrogen and oxygen (–1.23 V) can be recovered from aqueous solution by electro-winning. For copper ions in solution, the cathode reaction is:

$$Cu^{2+} + 2e^- = Cu;\ E^\circ = 0.337\ V$$

The anode reaction is:

$$H_2O = 2H^+ + 0.5O_2 + 2e^-;\ E^\circ = -1.229\ V$$

giving the overall cell reaction:

$$H_2O + Cu^{2+} = Cu + 2H^+ + 0.5O_2;$$
$$E^\circ = -1.226 + 0.337 = -0.892\ V$$

Since the potential is less (in absolute terms) than that required to produce hydrogen, it is possible to reduce copper from solution in preference to hydrogen. Some metals, notably zinc, have reduction potentials greater than –1.23 V but can still be reduced from aqueous solutions. This is due to a phenomenon called hydrogen overpotential, which causes the potential required to form hydrogen to increase due to surface effects related to the specific metal. A list of metals that can be produced electrochemically from aqueous solutions is given in Table 8.4.

Table 8.4: Common metals that can be produced electrochemically from aqueous solutions

Metals electropositive with respect to hydrogen	Au, Pt, Ag, Hg, Cu, Bi, Sb, W
Metals electronegative with respect to hydrogen but which can be reduced because of hydrogen overpotential effects	Pb, Sn, Mo Ni, Co, Cd, Fe, Cr, Zn, Mn, Na

Adsorption onto activated carbon. Activated carbon is a highly porous material which acts as a collector for gold and silver cyanide complexes and other complex ions. The carbon is manufactured from coconut shells, which produce a very hard but porous form of carbon when pyrolysed (heated in the absence of air) at 600–900°C. The carbon is activated by exposing it to an oxidising atmosphere at 600–1200°C. In gold recovery, the cyanide complex is adsorbed onto activated carbon particles of 0.5–3.0 mm diameter by contacting the carbon with the agitated pulp. There are two ways in which this is done – either while the gold is being leached, as in the CIL (carbon-in-leach) process, or after leaching, as in the CIP (carbon-in-pulp) process. Because the carbon particles are much larger than the ore particles, the carbon can be separated from the slurry by coarse screens. The carbon is washed then the gold is desorbed (eluted) by contacting the carbon with hot sodium hydroxide solution. The carbon is reactivated then returned to the adsorption circuit and the gold is recovered from the eluate in a separate process, usually electro-winning.

8.2.6 The production of some important metals

This section briefly describes the technologies used in the production of the three metals produced in largest quantities – iron, aluminium and copper. The focus of these descriptions is on production of the metal. The various wastes are mentioned only briefly here, but are discussed in greater detail in Chapter 11. Additionally, simplified flowsheets are given for the main processing options for the common base metals

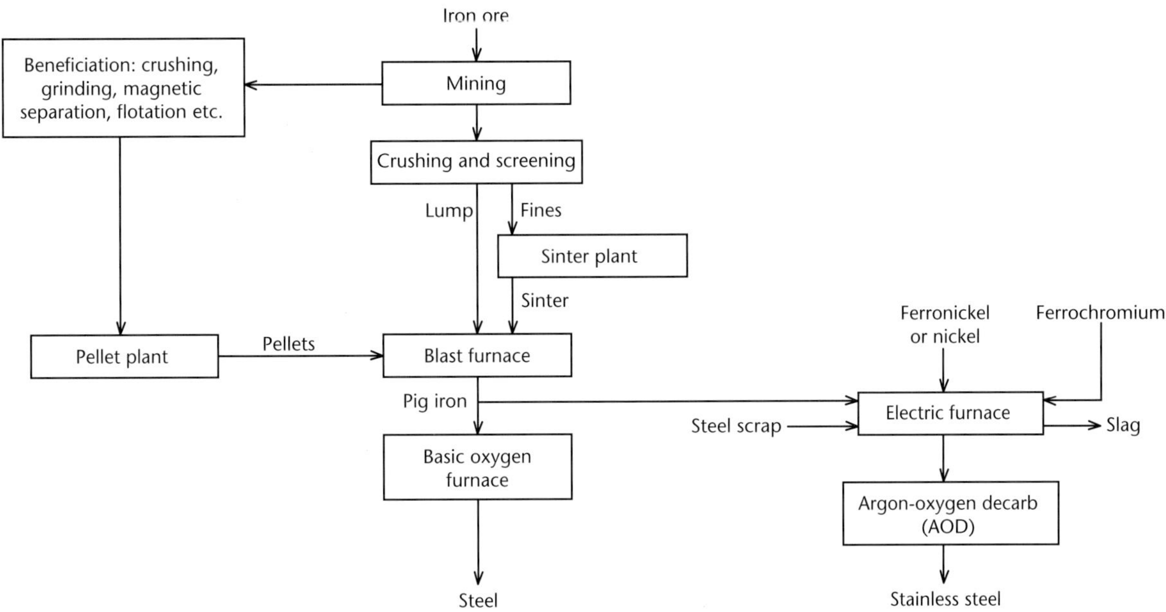

Figure 8.12: The main processing routes for producing steel and stainless steel (after Norgate *et al.*, 2007; with modifications).

nickel, lead and zinc. A summary of the processing routes used in the production of most of the major metals is given in Appendix IV.

Steel and stainless steel

Steel is produced from iron ore in a two-stage process (Figure 8.12). First, the iron oxides in the ore are reduced to impure iron then the iron is converted to steel by refining. The most common route is the blast furnace–basic oxygen furnace (BF–BOF) route, which is responsible for more than 90% of primary production of steel. Plants that produce steel by the BF–BOF route are called integrated steel plants. There are a number of alternative reduction processes, including Finex, Corex and HIsmelt, which produce molten iron. A number of direct reduction processes produce a solid form of iron called direct-reduced iron (DRI). A brief description of the Finex process is given in Section 16.5.1. The blast furnace produces an impure form of molten iron, called *hot metal*, which is conveyed to a BOF for steel-making. The blast furnace is a counter-current shaft furnace which is charged continuously and tapped periodically to remove hot metal and slag. The BOF is a stirred tank, pneumatic reactor which is operated in batch mode.

An iron blast furnace is illustrated in Figure 8.13. It consists of a circular shaft, which is typically about 30 m high with an internal diameter of about 9 m. It is constructed of steel and lined with refractory bricks. Raw materials are charged at the top through a double bell arrangement which prevents gas escaping into the environment. Preheated (900–1300°C), oxygen-enriched air (up to 40 vol% oxygen) is introduced through tuyeres located in the Bosh zone. The raw materials charged at the top are iron ore, coke and fluxes for making slag (particularly limestone). The gas travels upwards from the tuyeres and the solids travel downward; molten slag and hot metal are removed through tap holes in the hearth. Lumpy solids, typically 10–30 mm diameter, are needed to ensure the bed in the furnace is permeable enough for the gas to pass through it and contact the solids. The major reactions, and the regions in which they occur, are shown in Figure 8.13. Coke that survives the descent through the stack is burned in the tuyere zone to produce carbon monoxide and heat. As the hot gases ascend they heat the descending solids and the CO reduces iron oxide. Because coking coals and the coke-making process are expensive, it is common practice to replace some of the coke with pulverised coal of lower quality. This is

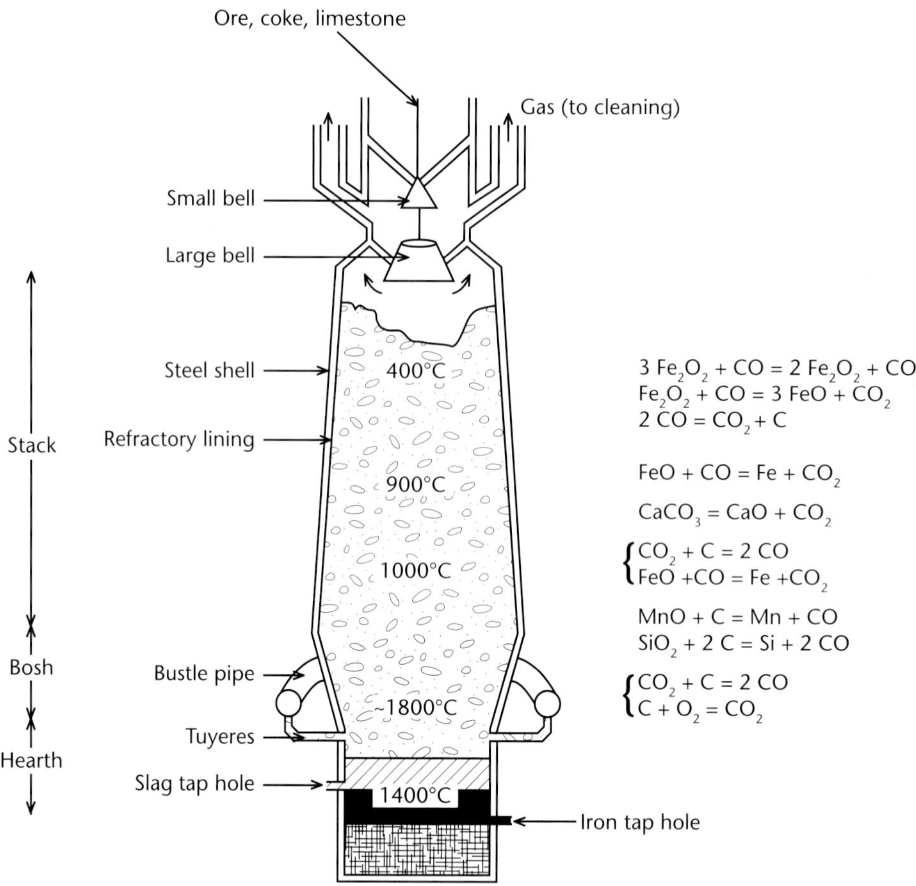

Figure 8.13: An iron blast furnace and the major reactions in iron-making.

injected into the Bosh zone in the air blast through the tuyeres. Alternatively, natural gas or oil can be injected. The metal and the oxide gangue materials begin to melt in the Bosh zone and drip through the coke bed into the hearth, where they separate into a slag and metal layer. Some of the CO_2 produced by reduction of iron is converted back to CO as it ascends, by reacting with carbon. Depending on their stability, some other oxides present in the ore are partially reduced, particularly silica and manganese oxide (Figure 8.1). Carbon is soluble in molten iron and the hot metal formed is a solution consisting typically of 92–93 wt% Fe and 3.5–4.5 wt% C, with the balance mainly Si, Mn, P and S. The gas leaves the top of the furnace at 100–150°C and contains about 25 vol% CO. This gas is combusted in stoves to preheat the air injected through the tuyeres.

The coke for iron-making is prepared from coal in a batch process in which bituminous coal is heated in the absence of air to 1000–1200°C to drive off the volatile matter and leave behind a residue consisting principally of carbon and the mineral matter associated with the coal (the ash). Modern coke ovens can be up to 6.5 m high, 15 m long and 0.5 m wide. The coking time, between charging and discharging, is about 15 hours. At the completion of coking, the coke is discharged from the ovens, quenched in water, then crushed and sized. Undersize material is blended with fresh coal and recycled to the ovens; a small quantity is used in iron ore sintering (discussed later). The ovens are arranged in batteries that contain up to 100 ovens each. Iron ore may need to be agglomerated if it is too fine. The main processes for this are pelletising (Section 7.2.4) and sintering. The first stage of the sintering process is the preparation of the raw (green) sinter feed from iron ores fines, fluxes, in-plant dust and fumes, coke fines and return sinter fines. These are mixed in one or more stages. Water is added and the larger particles become coated with fine particles, in a process called granulation. This ensures the raw mix forms a permeable bed in the next stage. The green mix is fed onto the continuously moving steel grate of a sinter machine, typically 3–5 m wide and up to 100 m long, to a depth of 0.5–0.7 m. A small amount of coke fines is added first to form a layer on which the sinter mix is laid. This protects the grate bars from the high temperatures attained in the sinter mix. Air boxes below the grate draw air down through the bed, which passes under a burner at the feed end to ignite the coke at the surface. As the feed moves horizontally with the grate, the coke combustion zone moves progressively downwards due to the down-draught of air through the bed. This causes partial melting of the finer material which, on resolidifying, cements the larger particles to form a porous mass. The sinter product is dumped at the end of the machine as the grate passes over the return rolls. The sinter passes through a breaker to a cooling bin, then is crushed and sized. Undersize sinter is returned and mixed with fresh raw materials.

In the second stage of steel-making, hot metal is transferred in ladles to a basic oxygen furnace (BOF), which is a refractory lined, vertical, cylindrical steel vessel mounted so it can be tilted during charging and to pour off molten slag and steel. It is operated as a batch process. The molten hot metal is often desulfurised in a separate operation on the way to the BOF by injection of sodium carbonate, lime, calcium carbide or other reagent that reacts preferentially with sulfur. The hot metal is charged, together with scrap steel, mill scale and fluxes such as limestone, dolomite and fluorspar, into the BOF, which typically has a capacity of 250–300 tonnes. Scrap steel, which makes up 25–35 wt% of the charged metal, is an essential component of the charge, not only because it provides a way to recycle steel but because, as it heats and melts, it acts as a coolant to control the temperature rise caused by exothermic steel-making reactions. Oxygen is injected into the furnace through a lance positioned a short distance above the charge. The major refining reactions are the oxidation of silicon, carbon and phosphorus:

$$\underline{Si} + \{O_2\} = (SiO_2)$$
$$\underline{C} + 0.5\{O_2\} = \{CO\}$$
$$2\underline{P} + 2.5\{O_2\} = (P_2O_5)$$

The SiO_2 and P_2O_5 dissolve in the slag which forms a separate phase and the CO escapes as gas from the furnace. Some iron is also oxidised; the iron oxides become constituents of the slag. At the completion of the oxidation reactions, the lance is withdrawn and the slag is poured off. The steel (at ~1600°C) is poured into a ladle and deoxidised, to remove excess oxygen picked up during the oxidation reactions. This is achieved by adding small quantities of reactive metals such as Al, Si and Mn. The molten steel is then poured into moulds to form ingots, which are rolled to form billets, blooms or slabs. Alternatively, and most commonly, the metal is continuously cast to form slabs. This reduces the amount of hot rolling required and reduces energy consumption.

Stainless steel is produced by melting steel scrap, pig iron, ferrochromium and other alloying metals (e.g. nickel or ferronickel) in an electric furnace to produce an alloy of the desired composition. The alloy is tapped into a ladle and refined in a separate operation to lower the carbon content to the required level. Chromium, being relatively reactive, oxidises readily. To minimise this, oxidation of the carbon is done using a dilute argon-oxygen mixture (argon-oxygen decarburisation, or AOD, process) or by oxygen at a reduced pressure (vacuum-oxygen decarburisation, or VOD, process). These lower the pressure of the CO formed during decarburisation well below atmospheric pressure; this, according to Equation 8.7, will favour the oxidation of carbon:

$$\underline{C} + 0.5\{O_2\} = \{CO\}$$

over the oxidation of chromium:

$$2\underline{Cr} + 1.5\{O_2\} = (Cr_2O_3)$$

Aluminium

Aluminium is produced in two stages (Figure 8.14). First, alumina (Al_2O_3) is produced from bauxite using the Bayer process then it is reduced electrochemically using the Hall-Héroult process. The main steps in the Bayer process are the digestion of alumina from bauxite using sodium hydroxide solution by means of Equation 8.20. The reaction is performed in autoclaves at ~145°C for gibbsite, $Al(OH)_3$, and 200–250°C for boehmite, AlO(OH), ores. The bauxite residue is separated from the pregnant (green) liquor in several solid–liquid separation stages. The coarse sand fraction is separated using thickeners and the overflow is filtered. The residue, called red mud because of its colour due to high ferric oxide content, is stored in tailings impoundments or disposed of in the ocean (discussed in Section 12.1.2). In the precipitation step, the liquor is cooled to supersaturate the solution and gibbsite 'seed' particles are added to promote crystallisation of aluminium hydroxide, $Al(OH)_3$. The conditions are carefully controlled to ensure a high recovery of aluminium from the solution and to control the size and shape of the crystals produced. The slurry leaving the precipitators is separated into a coarse and a fine fraction, usually via hydrocyclones. The coarse fraction is sent for calcination and the fine fraction is recirculated as seed. Calcination is done in rotary kilns or fluidised bed calciners at around 1100°C to decompose the hydroxide and produce alumina powder.

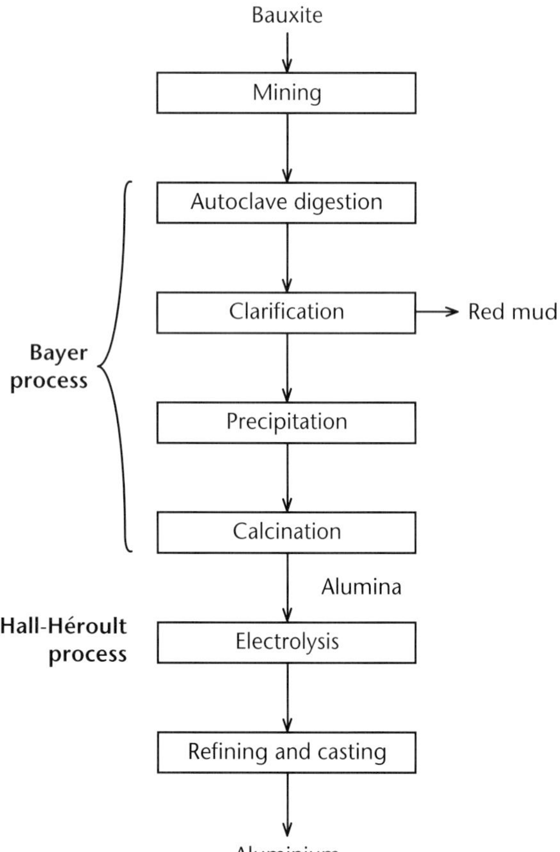

Figure 8.14: The production of aluminium using the Bayer and Hall-Héroult processes.

Aluminium metal is produced from alumina by electro-winning, by means of Equation 8.3. Alumina is dissolved in a bath of molten cryolite, Na_3AlF_6, in a cell. Aluminium metal is deposited on the carbon cathode, and oxygen is deposited on and reacts with the carbon anode. Pure cryolite melts at 1012°C but addition of alumina and small quantities of CaF_2, AlF_3 and other fluorides lowers the melting point and allows the cell to operate at around 960°C, well above the melting point of aluminium (660°C). A modern aluminium reduction cell is illustrated in Figure 8.15. It consists of a rectangular steel shell, typically 9–16 m long, 3–4 m wide and 1–1.3 m deep, lined with refractory insulation and an inner lining of baked carbon (Sanders, 2004). Current enters the cell through the carbon anodes and leaves through the steel collector bars in the carbon cathode at the bottom. The cells are arranged in lines, containing up to several hundred cells, called potlines. An aluminium smelter may have one potline or a number of potlines. Most modern cells use prebaked anodes made in a separate operation on site from petroleum coke and coal tar binder. The mixture is formed into blocks of about 0.7 m × 1.4 m × 0.5 m and baked in the absence of air at 1000–1200°C for several days to form a hard, dense, unreactive block of carbon. Alumina is fed into the top of the cells by breaking the crust and molten aluminium is removed by siphoning into a ladle, usually once a day. The metal produced is 99.6–99.9 wt% Al. Most molten aluminium is refined by bubbling argon and chlorine through the metal to remove small quantities of dissolved impurities (e.g. Fe, Si, Ti, V and Mn) as chlorides. The metal is cast into ingots or directly into billets by means of continuous casting.

Copper

Copper is produced by two main routes (Figure 8.16). Approximately 80% of primary copper is produced pyrometallurgically, the balance being produced hydrometallurgically. Usually, chalcopyrite concentrates are smelted whereas oxide and oxidised copper sulfide minerals (particularly chalcocite, Cu_2S) are more amenable to leaching in sulfuric acid.

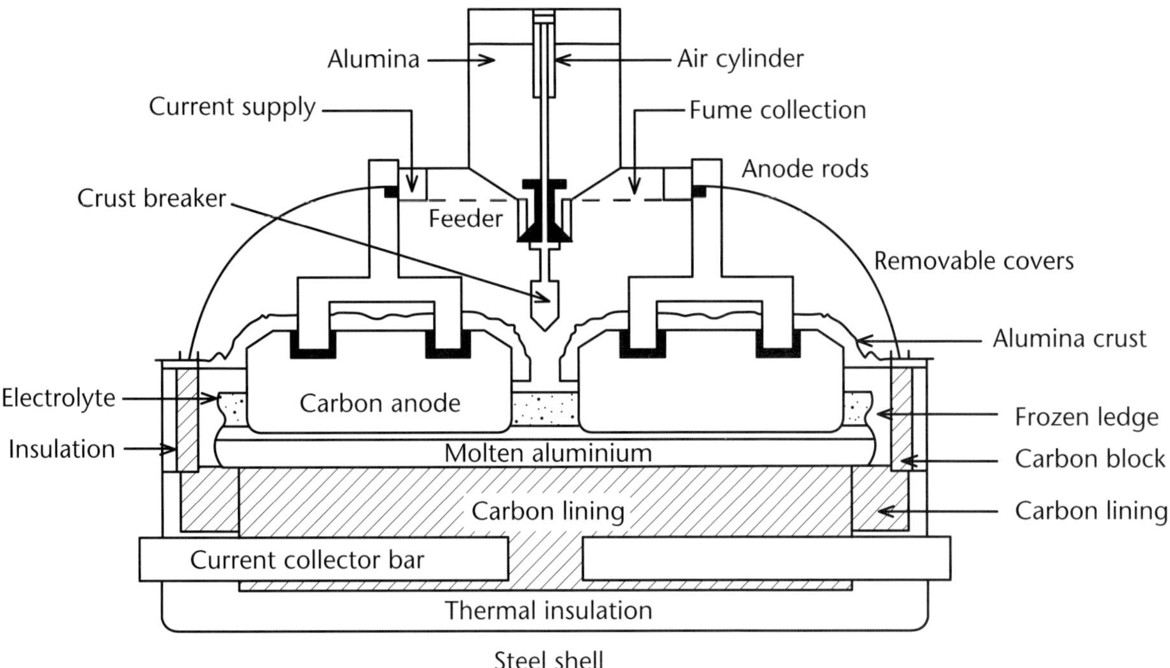

Figure 8.15: A modern aluminium reduction cell (after Sanders, 2004).

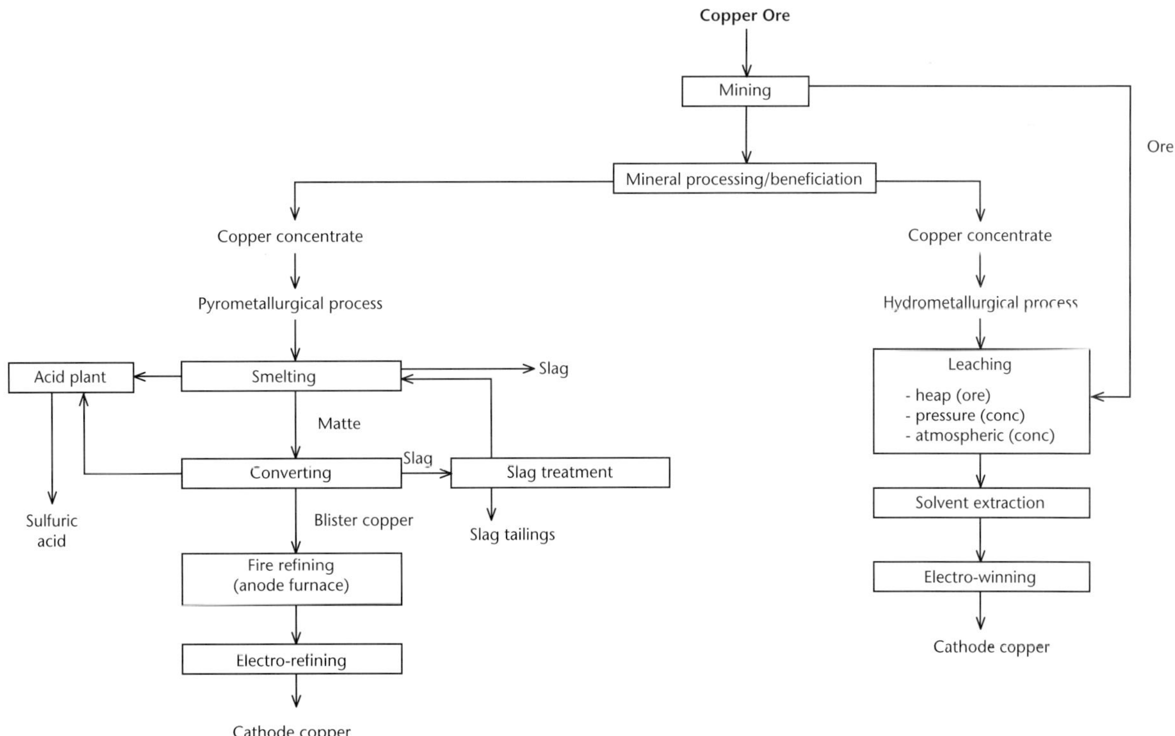

Figure 8.16: The main processing routes for producing copper (after Norgate *et al.*, 2007; with modifications).

Smelting route. Copper concentrate is a mixture of copper and iron sulfides with small quantities of gangue minerals (Table 6.1). Smelting of concentrate is usually performed in two stages: matte smelting, followed by converting. In matte smelting, some of the iron sulfides are oxidised to form gaseous SO_2 and iron oxides, which are fluxed with silica and lime to form a slag (Figure 8.7). The SO_2 is usually captured and converted to sulfuric acid. Other gangue components of the concentrate also enter the slag. Most, if not all, of the thermal energy required for the process is met by the exothermic oxidation reactions; i.e. the process is autogenous, or nearly so. The remaining iron sulfide forms a molten matte phase with the copper sulfides. Matte smelting can be performed in a variety of furnaces. Traditionally, it was carried out in reverberatory furnaces but these are inefficient. Flash furnaces and top submerged lance (TSL) smelting furnaces, such as the ISASMELT™ and Ausmelt processes, are used in modern operations. Matte smelting furnaces operate as continuous reactors.

Reverberatory furnaces (Figure 8.6) are rectangular box-shaped reactors made of steel and lined with refractory bricks, 35–40 m long, 7–10 m high and 3.5–4.5 m high. The roof is usually arched. The material to be heated is charged from hoppers through chutes in the roof along the side walls and forms a layer on the hearth of the furnace. Fuel (oil, natural gas or pulverised coal) is combusted in burners along one end wall in the space above the charge, and heat is radiated from the flame to the charge material and the refractories. The process is thermally inefficient. The combustion gases, and gases formed by reactions of the charge materials, leave the furnace at the opposite end through a flue.

Flash furnaces are similar in construction to reverberatory furnaces, and consist of a suspension smelting section coupled to a settler section. A common configuration is shown in Figure 8.6. Typically, a flash furnace is 25–30 m long. Copper concentrate and fluxes (to form slag) are injected through one or more concentrate burners at the top of the combustion tower and mixed with preheated, oxygen-enriched air. A small amount of fuel may be added if the exothermic reaction of the concentrate does not provide sufficient heat. The particles of concentrate are heated in the shaft to ignition temperature, and partially oxidise. The molten products (and the fluxes) fall to the hearth, where they continue to react and settle out to form a matte and slag layer. The matte and slag are usually tapped from opposite ends of the hearth, to improve the efficiency of copper recovery. The hot gases produced flow through the settler zone in the space above the slag and matte, continuing to heat the products, and leave the furnace at the opposite end through the gas off-take.

TSL smelting technology involves the use of a vertical lance to inject oxygen-enriched air into a molten bath contained in a cylindrical, refractory-lined steel vessel (Figure 8.17). The steel lance is designed so that the gas flowing through it cools it sufficiently to form a frozen layer of slag on the outside, which protects the steel from the high temperature and corrosive environment of the molten bath. Pelletised concentrate and fluxes are continuously fed through a hopper at the top of the vessel and molten product is periodically removed through a tap hole at the base. The molten product flows into a holding furnace, where it separates into slag and matte layers.

In the converting stage, the matte is further oxidised by injecting air. The iron sulfide oxidises preferentially since it is more reactive than copper sulfide (Figure 8.2). Silica is added to flux the iron oxide and form a slag. After removing the slag, the remaining molten copper sulfide, called *white metal*, is converted to copper by means of Equation 8.12 via further injection of air. The most common type of reactor used is the Peirce-Smith converter. This is a refractory-lined cylindrical vessel with a large opening along one side. It is mounted horizontally so it can be rotated about its longitudinal axis. It is operated as a batch process. During charging and tapping, the converter is rolled so its mouth is clear of the gas ducting system. During blowing, the mouth is positioned so that the gases are collected and removed. Matte is charged into the furnace from open ladles, air is blown into the charge through tuyeres located along the side, and silica is added to flux iron in the matte. At completion of the first stage, the slag is poured into ladles for further processing to recover lost copper. Air is then injected into the white metal until most of the sulfur has been oxidised. The product is blister copper containing >98% Cu plus some sulfur, oxygen and other impurities. The blister copper is poured into ladles. Continu-

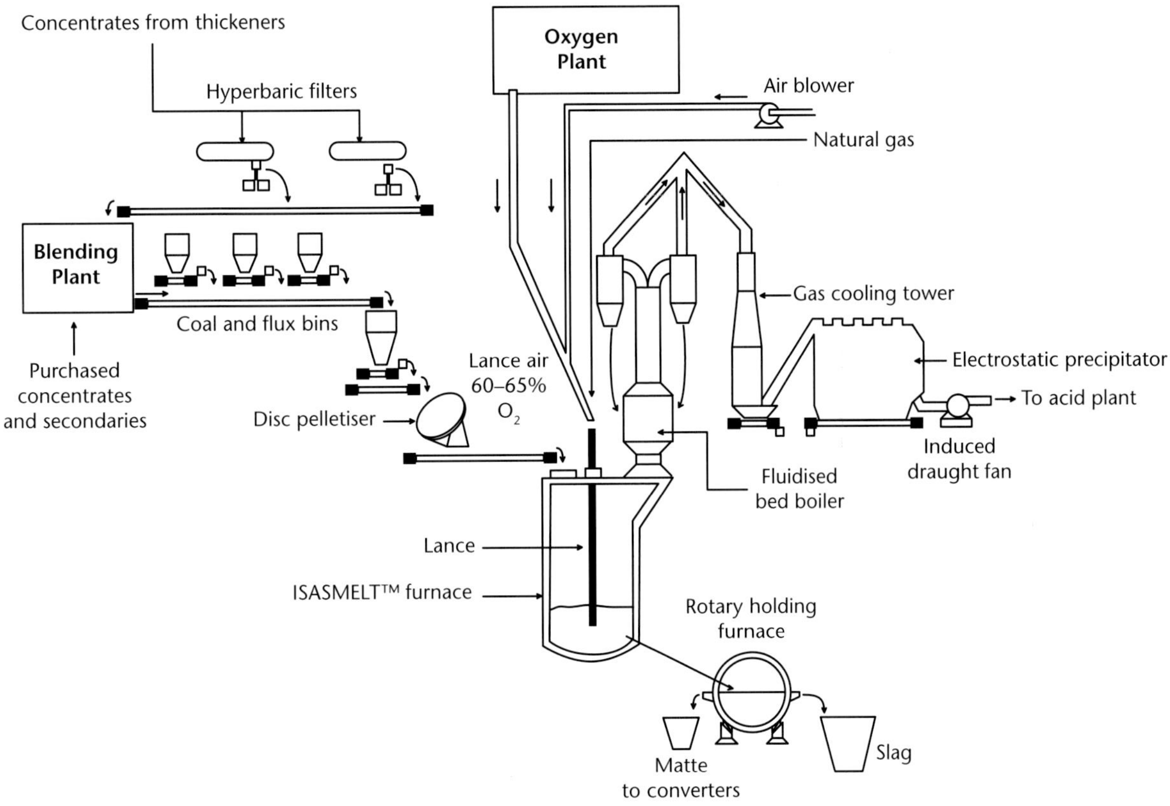

Figure 8.17: The copper ISASMELT™ process as carried out at Mount Isa, Queensland, Australia (Arthur, 2006). Reproduced by permission, Minerals, Metals and Material Society.

ous converter processes, such as flash converting and TSL converting, are used in new smelters.

The blister copper is transferred to an anode furnace for fire-refining. Air is injected into the metal to reduce the sulfur content to around 0.001–0.003 wt% S, then natural gas or propane is injected to reduce the oxygen content to less than 0.2 wt%. The copper is cast into anodes for electro-refining. In electro-refining, the anodes are dissolved electrolytically in acidified copper sulfate solution and the copper is electroplated onto copper or stainless steel starting sheets to produce high-purity copper cathodes. Electro-refining is carried out in large tanks, and the refineries are called tank houses – a large refinery has over 1000 cells in the tank house. The impurities in the blister copper form a solid precipitate, called *anode slimes*, which is treated to recover valuable metals, including Cu, Ni, Se, Te, Au, Ag and Pt.

Hydrometallurgical route. Hydrometallurgical extraction of copper has become more common since sulfuric acid became available in large quantities, due to the increasing requirement to capture SO_2 from smelter gases, and the wide-scale adoption of solvent extraction technology. An increasing portion of copper is being obtained from leaching waste rock from mining and tailings. Leaching of oxidised ores is also becoming common. Leaching of copper concentrates is not common; the preference is to smelt concentrates. The most common hydrometallurgical route is heap leaching of ore or waste rock (often with tailings added), followed by solvent extraction and electro-winning (SX/EW). A typical flowsheet is illustrated in Figure 8.18. The individual operations have been described previously.

Nickel, lead and zinc

The main processing options for nickel, lead and zinc ores are illustrated in Figures 8.19 and 8.20. A simplified flowsheet for pyrometallurgical production of nickel from laterite ores is shown in Figure 8.21.

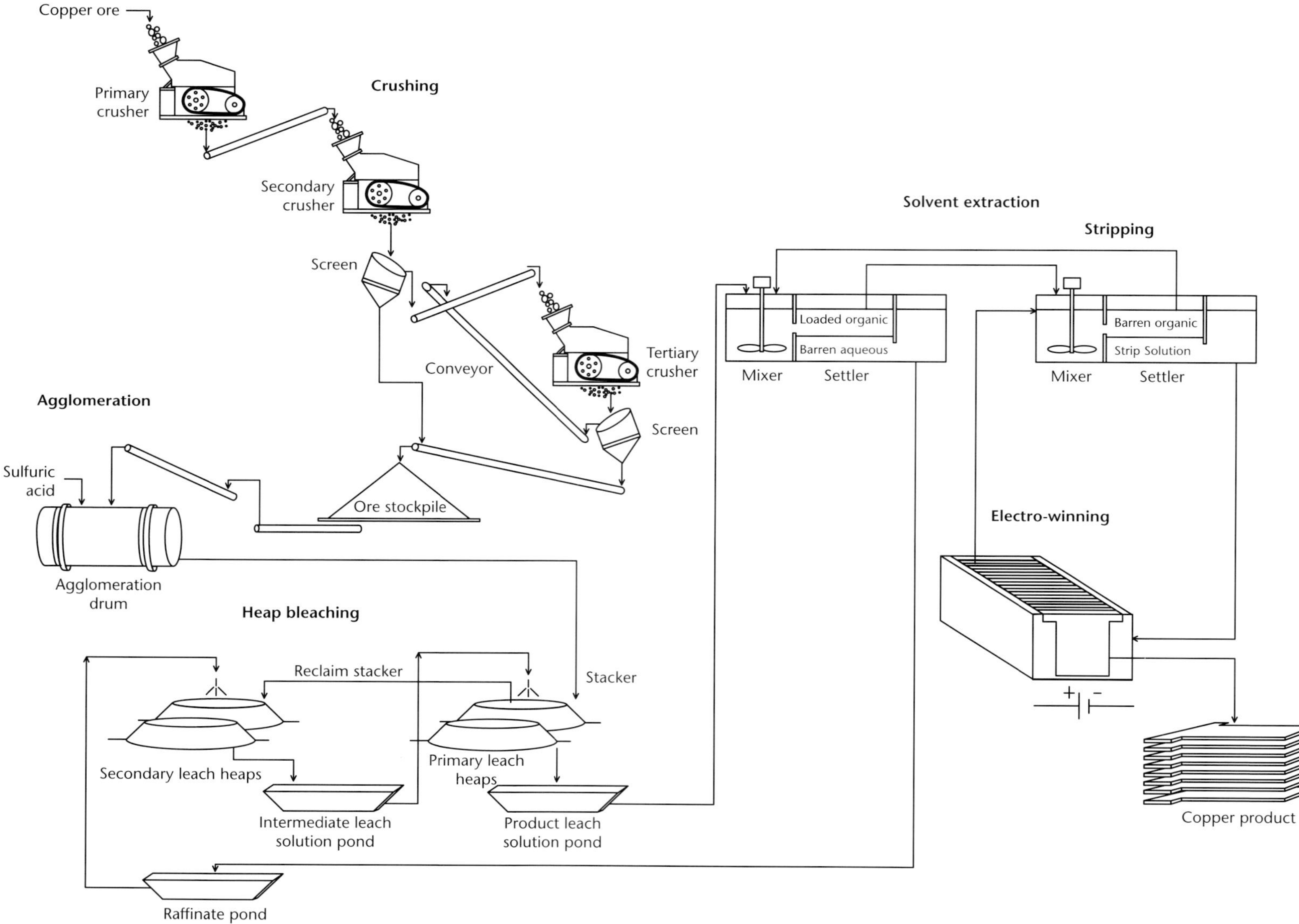

Figure 8.18: A typical copper heap leaching, SX/EW operation. Courtesy P. Bangerter, Hatch, Brisbane.

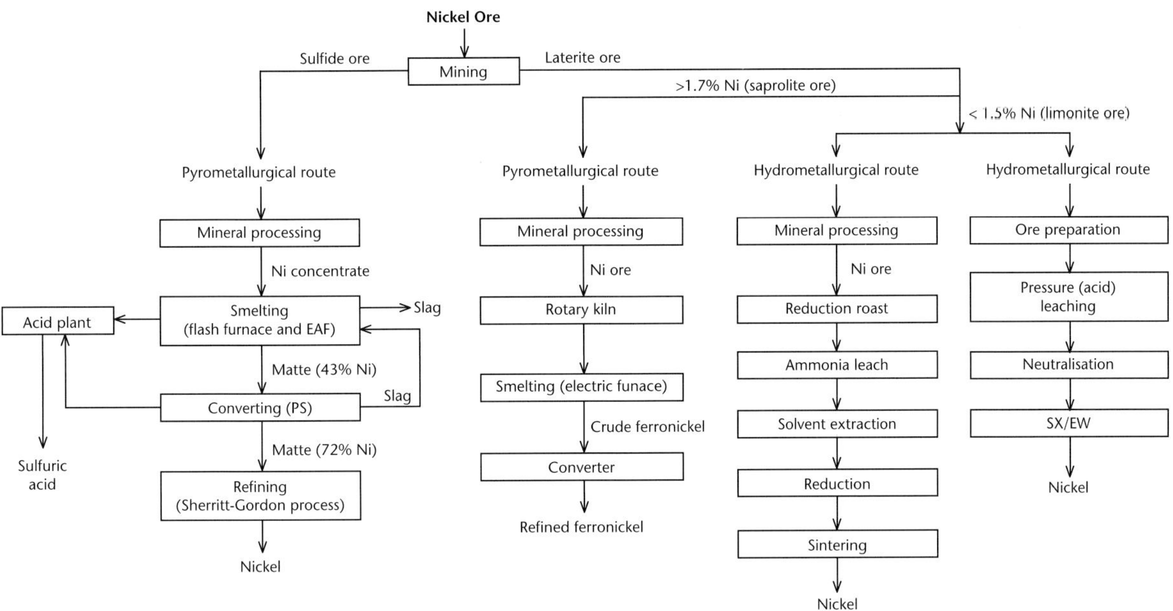

Figure 8.19: The main processing routes for producing nickel (after Norgate *et al.*, 2007).

8.3 CEMENT AND CONCRETE

There are various types of cement. The cement most commonly used for construction purposes is ordinary Portland cement (OPC) or, more simply, Portland cement. It is the basic ingredient of concrete, mortar, stucco and many grouting materials. OPC is a fine grey powder produced by grinding Portland cement clinker (>90 wt%) with small quantities of calcium sulfate (to control the setting time) and up to 5% of other constituents. Portland cement clinker is produced by heating raw materials to around 1450°C. It is a hydraulic (*pozzolanic*) material; i.e. when mixed with water, it reacts and sets within a few hours to a rigid product which progressively hardens. Cement

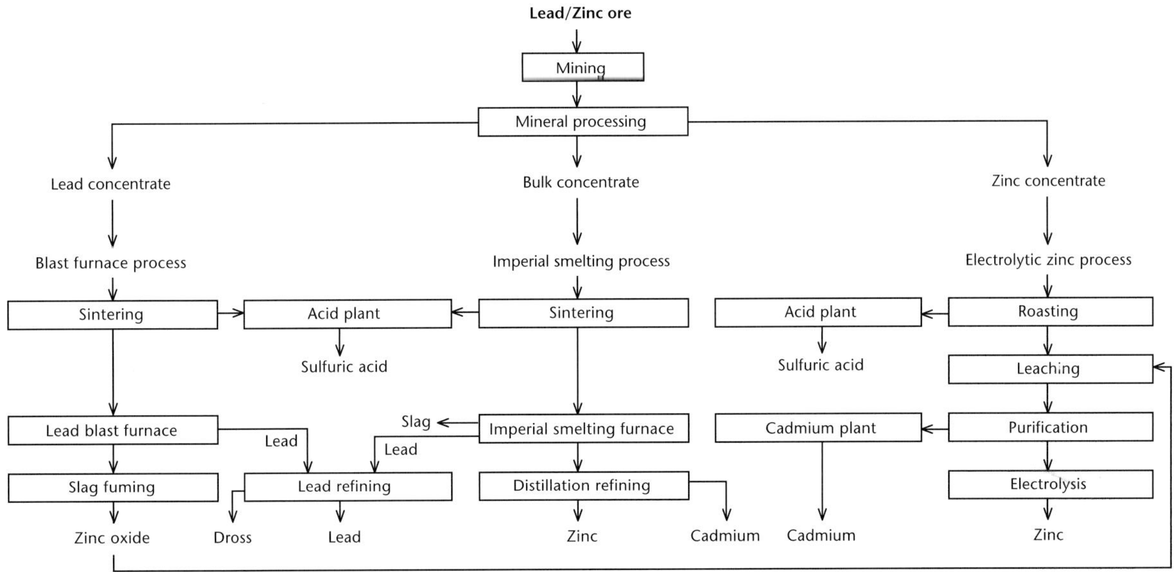

Figure 8.20: The main processing routes for producing lead and zinc (after Norgate *et al.*, 2007; with modifications).

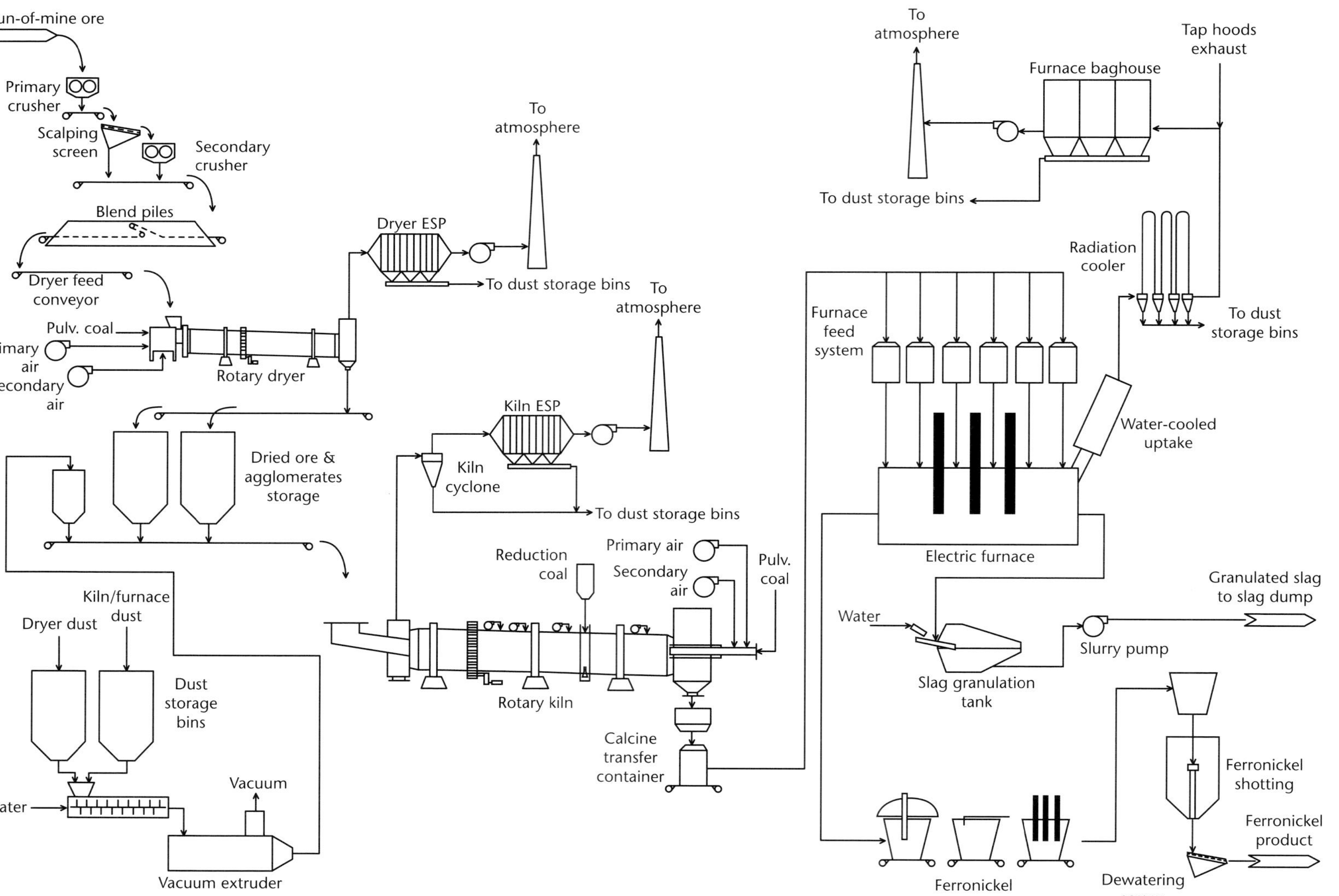

Figure 8.21: A typical nickel laterite smelting operation. ESP = electrostatic precipitator. Courtesy P. Bangerter, Hatch, Brisbane.

clinker consists of at least two-thirds by mass calcium silicates (e.g. $3CaO.SiO_2$ and $2CaO.SiO_2$) with the remainder being calcium- and calcium-iron aluminates (e.g. $3CaO.Al_2O_3$, $4CaO.Al_2O_3.Fe_2O_3$). The chemistry of the setting of cement is very complex; the following is a simplified description. When water is added, calcium silicates react to form calcium silicate hydrate and calcium hydroxide, for example:

$$2(3CaO.SiO_2) + 11H_2O =$$
$$3CaO.2SiO_2.8H_2O + 3Ca(OH)_2$$

Similarly, calcium aluminates react to form calcium aluminate hydrate. The calcium hydroxide formed helps bind the hydrates together. Slowly, over a period of months, the silicate anions (SiO_4^{4-}) polymerise to form chains, their length increasing for up to 15 years.

The raw materials for Portland cement clinker are limestone (~80 wt%) and clay, with small quantities of other materials such as bauxite, iron ore, shale and sand to adjust the composition. The total MgO is kept below 3 wt%. The composition is controlled so that the mass ratio of $CaO:SiO_2$ in the clinker will be 1.5–3.0, depending on the particular specification. The raw materials are quarried and transported to primary and secondary crushers. The crushed material, with a typical size of around 100 mm, is finely ground in a mill to produce *raw meal*. The raw meal is preheated in a series of cyclones, arranged vertically, through which it passes downwards and comes into contact with hot exhaust gases from the clinker kiln, that are moving in the opposite direction. In some operations, the preheated charge leaving the cyclones is precalcined in a small furnace, located above the clinker kiln, to partially decompose the limestone:

$$CaCO_3 = CaO + CO_2$$

The meal then enters an inclined rotary kiln which is heated by burning fuel inside the kiln. The kiln rotates at three to five revolutions per minute; the material slides and tumbles along its length, becoming progressively hotter as it moves towards the burner end. The high temperatures (up to 1450°C) cause a number of chemical reactions and physical changes that create a partially fused material, called *clinker*. The clinker discharges from the hot end of the kiln onto a grate where it is cooled by air entering the burner; this process simultaneously preheats the combustion air, thereby increasing the energy efficiency. The clinker is stored in bins then fed, along with calcium sulfate and cementitious materials such as fly ash and blast furnace slag, to grinding mills. The resulting fine OPC cement powder is stored in bulk or bagged ready for sale.

Concrete is a construction material composed of cement and aggregate, usually a mixture of coarse aggregate (made from gravel or crushed rocks) and fine aggregate (e.g. sand). Often, other pozzolanic materials such as fly ash and ground blast furnace slag are added. Concrete is produced at or near the site where it is to be used, and is prepared as required because there is a relatively short working period before it starts to set. The cement reacts with water to form a continuous matrix which binds the aggregate. This composite nature gives concrete the properties that make it such a versatile construction product.

8.4 GLASS

Technically, a *glass* is any inorganic substance that has been cooled below its melting point without crystallising and thus retains the structure of a liquid. The most common glasses are those based on silica. A large number of silica-based glasses are produced commercially for a variety of applications, but most glass is used for making containers (container glass) and for windows (flat glass). These have similar but slightly different compositions, as shown in Table 8.5. These compositions produce a material that is molten at

Table 8.5: Typical glass compositions (wt%)

	SiO_2	Na_2O	K_2O	CaO	MgO	Al_2O_3	Other
Container glass	72.7	13.8	0.5	11.0	0.1	1.6	0.3
Flat glass	72.8	12.7	0.8	8.1	3.8	1.4	0.4

Source: Boyd *et al.* (2004).

temperatures between around 1400°C and 1600°C (Figure 8.7). Glass does not have a distinct melting point; it softens progressively when heated until its viscosity falls sufficiently that it will flow. The main raw materials for these glasses are highly pure forms of silica sand, soda ash (Na_2CO_3) and limestone ($CaCO_3$). If Al_2O_3 is needed, it is usually added in the form of feldspar. To make coloured glasses, small quantities of metal oxides are added; for example, chromium and iron (green), cobalt (blue) and selenium (red).

The production of glass is a relatively simple melting process. Since many of the raw materials have high melting points, ground recycled glass (*cullet*) is added with the raw material. This melts first and provides a liquid phase into which the raw materials can dissolve. Glass melting furnaces are closed and rectangular, typically 1–1.5 m deep, 10–14 m wide and 50 m long, and lined with refractories. They are usually heated by burning gas in the space above the charge. The furnaces operate continuously – the raw materials are charged at one end, the charge makes its way slowly along the length and is discharged in molten form at the opposite end. Four overlapping processes occur in the furnace. Melting occurs in the first section. Refining occurs in a quiescent zone where gas bubbles, formed by CO_2 release during decomposition of the soda ash and limestone, escape from the melt. The refining zone is separated from the melting zone by a vertical wall under which the melt flows. This prevents solids being carried into the refining zone. Homogenisation occurs throughout the furnace and ensures that the glass leaving the furnace has a uniform composition. Thermal conditioning commences after the glass attains its highest temperature. As the glass moves closer to the outlet end, it begins to cool to the temperature required for forming it.

Molten glass can be moulded, drawn, rolled, cast, blown, pressed or spun into fibres (to form fibreglass). Most flat glass is produced by the float glass process. In this, molten glass flows from the forehearth of the melting furnace onto a pool of molten tin in another furnace. The furnace is designed so that the tin nearest where the glass enters is hotter than the glass. Thus, the glass flows evenly over the surface of the tin and forms a uniformly thick, flat layer with smooth upper and lower surfaces. The other end of the furnace is cooler, so that as the glass moves towards that end it cools and becomes rigid prior to leaving the furnace.

Glass containers are formed by transferring molten glass into moulds by a technique called gob feeding. Two processes are commonly used: the blow and blow method and the press and blow method. In both, a stream of molten glass, at a temperature at which it is plastic rather than liquid (~1050–1200°C), is cut with a shearing blade to form a cylinder of glass (a *gob*). The gob falls through troughs and chutes into a mould. In the blow and blow process, the glass is blown from below, with compressed air, into the empty mould to form a precontainer (parison). The parison is then flipped over into another mould and a second blow forces the glass into the mould to make the final container shape. In the press and blow process, the parison is formed with a metal plunger. The process then continues as for the blow and blow process.

8.5 MINERAL FERTILISERS

Manufactured fertilisers are of two broad types. Mineral fertilisers consist mainly of naturally occurring or manufactured minerals; for example, potassium nitrate (saltpetre), potassium carbonate (potash) and apatite, $Ca_5(PO_4)_3F$, obtained from phosphate rock. Chemical fertilisers are chemical commodities produced as products or by-products of the chemical industry (e.g. ammonia, ammonium sulfate and urea). These materials can be used straight but more commonly they are mixed to produce a more balanced product.

Nitrogen and potassium fertilisers. The main industrial minerals used as nitrogen-rich fertilisers are the naturally occurring alkali-metal nitrates, such as KNO_3 and $NaNO_3$, which occur in natural brines and guano deposits. Manufactured forms are more common; for example, liquid ammonia (produced in the Haber process), ammonium sulfate and ammonium nitrate. Potassium chloride is the major source of potassium for fertilisers. It occurs naturally as the mineral sylvinite, which is a mixture of sylvite (KCl) and halite (NaCl), and in brines and ocean water. It is recovered from sylvinite ores by flotation or by leaching in water followed by fractional crystallisation.

Phosphorus fertilisers. The major source of phosphorus for fertilisers is phosphate rock, a sedimentary rock in which phosphorus occurs principally as the mineral apatite. Other minerals present are quartz, silicates and carbonates, such as calcite and dolomite. These are removed by beneficiation before the apatite is processed. The beneficiation process involves crushing and grinding the rock then separating the apatite by washing, screening or flotation. The apatite concentrate is then reacted with strong acid to form more soluble phosphate compounds. Sulfuric and phosphoric acids are used for the manufacture of superphosphates and nitric acid for nitric phosphates. Ordinary superphosphate is produced by treating apatite with concentrated sulfuric acid to produce a mixture of calcium hydrogen phosphate and calcium sulfate:

$$3Ca_3(PO_4)_2 + 6H_2SO_4 = 6CaSO_4 + 3Ca(H_2PO_4)_2$$

Water, CO_2 (from carbonates in the rock) and fluorine, as fluosilicate (H_2SiF_6) and HF, are produced as gases during the reaction. The rock/sulfuric acid mixture is fed to a reactor called a den. Continuous dens have a moving floor which carries the mixture through to the exit end, over a period of around 30 minutes to allow time for partial reaction. A rotating cutter wheel removes the partially matured product. Batch dens are sometimes used. The product contains around 21 wt% P_2O_5 (dry basis). Superphosphate was the only product made from phosphate rock until the 1950s, when triple superphosphate was introduced. Triple superphosphate is made by reacting apatite with phosphoric acid:

$$Ca_3(PO_4)_2 + 4H_3PO_4 = 3Ca(H_2PO_4)_2$$

The resulting product contains around 46 wt% P_2O_5, of which about 90% is water-soluble. The phosphoric acid required is made in a separate process by reaction of sulfuric acid with phosphate rock, as in ordinary superphosphate manufacture:

$$Ca_3(PO_4)_2 + 3H_2SO_4 = 2H_3PO_4 + 3CaSO_4$$

The calcium sulfate is removed by filtration and the phosphoric acid solution is concentrated by evaporation.

Both ordinary and triple superphosphate are granulated after exiting the den to produce the desired handling characteristics. The material is pulverised then fed to an inclined rotating drum, where it is agglomerated to form granules of <6 mm. The product matures in storage for several weeks before it is shipped to customers.

Despite the relatively low phosphate content of ordinary superphosphate, which increases transportation and handling costs, it has some advantages over triple superphosphate. Its production requires less energy and can utilise by-product sulfuric acid, it avoids the need to dispose of calcium sulfate waste, and superphosphate retains some impurities in insoluble form. This last consideration is likely to become more important as sources of higher-grade phosphate rock become depleted.

8.6 COMMODITY CERAMICS

Bulk commodity ceramics include bricks, roof tiles and pottery. These are made from clay-based raw materials. The production of bricks illustrates the relatively simple technology of commodity ceramic manufacture. The raw materials are blended to produce a consistent feedstock, then milled and screened to remove coarse materials and impurities. Water is added and the clay and water mixture is kneaded in a pug mill, which consists of a mixing chamber with one or more revolving shafts with blade extensions, to form a plastic mass. This is then formed into brick shapes. The mixture is often subjected first to a reduced pressure to remove air pockets, to increase the strength of the brick and the workability of the clay–water mixture. Several methods are used for forming bricks. In the stiff-mud process, which is the most common, the mixture is extruded into a long ribbon of the desired cross-section. The ribbon is cut to individual brick lengths by rotating wire knives. In the soft-mud process, more water is added to make a thinner paste which is packed into moulds. The green bricks are removed from the moulds when they have set enough to be handled. In the dry-press process, used to make good-quality face brick, enough water is added to make the clay damp. The damp clay is then put into moulds under pressure and the green bricks removed when they have set sufficiently. In all three processes, the green bricks are stacked and dried in chambers through which hot air (up to 200°C) is circulated. The bricks must be dried slowly to minimise shrinkage and cracking; drying times are typically 24–48 hours. Next, the bricks are put in a kiln and slowly heated, over 10–40 hours, to 900–1350°C depending on the temperature needed to cause partial melting of the clay. Two types of kilns are used, the tunnel kiln and the periodic kiln. The tunnel kiln has the highest production rate and uses conveyors or cars on rails to carry the dried bricks through the firing process. The periodic kiln is a batch kiln that is loaded, put through the cycle, then emptied. During firing, the iron oxide impurity in the clay is oxidised to ferric oxide, which gives bricks and tiles their characteristic red colour.

8.7 REFERENCES

Arthur PS (2006) ISASMELT™ – 6,000,000 tpa and rising. In *Sohn International Symposium on Advanced Processing of Metals and Materials.* (Ed. F Kongoli) pp. 275–290. Minerals, Metals and Materials Society: Warrendale, PA.

Biswas AK and Davenport WG (1976) *Extractive Metallurgy of Copper.* Pergamon Press: Oxford, UK.

Boyd DC, Danielson PS, Thompson DA, Velez M, Reis ST and Brow RK (2004) Glass. In *Kirk-Othmer Encyclopedia of Chemical Technology.* 4th edn. Vol. 12. (Eds J Kroschwitz and M Howe-Grant) pp. 565–626. John Wiley and Sons: New York.

Evans JW and De Jonghe LC (2002) *The Production and Processing of Inorganic Materials.* Minerals, Metals and Materials Society: Warrendale, PA.

Norgate TE, Jahanshahi S and Rankin WJ (2007) Assessing the environmental impact of metal production processes. *Journal of Cleaner Production* **15**: 838–848.

Rosenqvist T (1983) *Principles of Extractive Metallurgy.* 2nd edn. McGraw-Hill: Auckland, NZ.

Sanders RE (2004) Aluminum and aluminum alloys. In *Kirk-Othmer Encyclopedia of Chemical Technology.* 4th edn. Vol. 2. (Eds J Kroschwitz and M Howe-Grant) pp. 279–343. John Wiley and Sons: New York.

Woollacott LC and Eric RH (1994) *Mineral and Metal Extraction: An Overview.* South African Institute of Mining and Metallurgy: Johannesburg.

8.8 USEFUL SOURCES OF INFORMATION

Bodsworth C (1994) *The Extraction and Refining of Metals.* CRC Press: Boca Raton, FL.

Gilchrist JD (1989) *Extraction Metallurgy.* 3rd edn. Pergamon Press: Oxford, UK.

Rosenqvist T (1983) *Principles of Extractive Metallurgy.* 2nd edn. McGraw-Hill: Auckland, NZ.

Woollacott LC and Eric RH (1994) *Mineral and Metal Extraction: An Overview.* South African Institute of Mining and Metallurgy: Johannesburg.

9 Energy consumption in primary production

Energy and water are essential inputs in the production of mineral and metal commodities and the minerals industry consumes large quantities of both to bring about the transformations of matter required to make tradable commodities. This chapter and the next are concerned with these inputs. Energy is conserved as it undergoes changes but it degrades in quality after each change. While energy can be recovered and reused, the fraction of it available for useful work decreases with each transformation (second law of thermodynamics), usually because the temperature at which it is available progressively decreases. This chapter examines the energy inputs required to produce mineral and metal commodities, their nature, where they are added along the value-adding chain and the factors that affect energy consumption. It concludes with a survey of sustainability indicators for energy used by the minerals industry.

9.1 DIRECT AND INDIRECT ENERGY AND GROSS ENERGY REQUIREMENT

Energy is consumed at all stages of value-adding; for example, diesel fuel consumption during mining, electrical energy consumption during grinding, and fuel combustion during smelting. Figure 9.1 shows the flow of materials and the inputs and outputs of energy along the value-adding primary stages of production; i.e. to the stage at which a material is in a form suitable for use as an input to construction or manufacturing and fabrication. This is an expanded version of the cradle-to-gate part of Figure 2.7. Two stages of chemical processing have been included, as two major chemical processing stages, such as reduction and refining, are sometimes required to produce the final product. Bauxite, for example, is leached in the Bayer process to produce alumina, which is then smelted to produce aluminium metal; hot metal is produced in the iron blast furnace from iron ore or sinter, which is then converted to steel in a separate operation. For any particular commodity, the processing chain may stop at any point after mining, depending on the degree of processing required for the final product (Figure 5.9). Some industrial mineral and construction products are not chemically processed; the processing chain stops after beneficiation.

Since energy is conserved, any energy added during processing must be released into the environment (usually as waste heat) along the value chain and/or be converted into a form of potential energy (usually chemical) in the value-added product. These forms of energy are directly related to the production of a

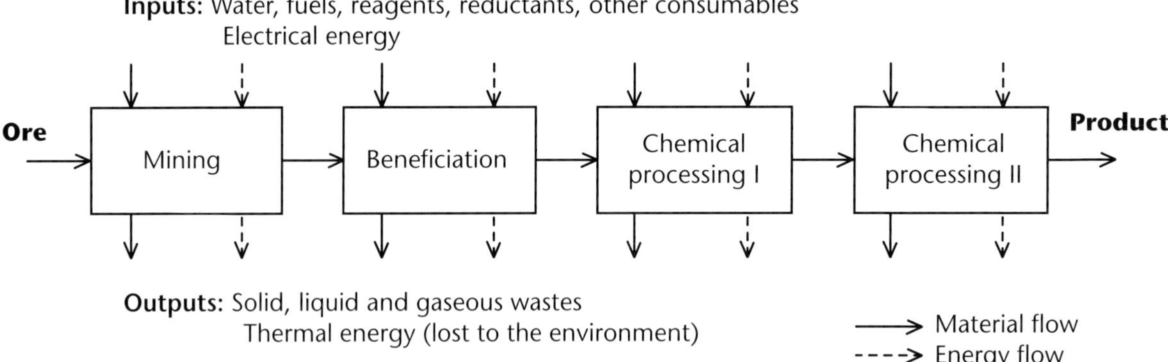

Figure 9.1: Material and energy flows in value-adding stages in the production of a mineral or metal commodity.

commodity. They are of particular interest to minerals companies since they have a direct impact on company operations; for example, costs associated with purchase of energy, and CO_2 emissions as a result of combustion of fossil fuels. These forms are referred to as *direct energy*. The total energy required to produce a commodity also includes energy not directly used in the value-adding process but needed to produce inputs to the process. These forms are referred to as *indirect energy*. The largest is usually the inefficiency associated with the generation of electrical energy using fossil fuels. For example, coal-fired power stations typically convert only about 35% of the heat produced by burning coal into electrical energy. The remainder is dissipated as waste heat in the gases released to the atmosphere. For every kilowatt hour of electrical energy consumed directly in the production of a commodity, an additional 65/35 = 1.86 kW h (equivalent) of energy is lost at the power station. This should be included in calculations of the energy required to produce a commodity. Other sources of indirect energy are the energy required to manufacture reagents used in processing and explosives for mining, the energy for transporting reagents and so on, and the energy required to make the machinery and buildings used for mining and processing.

The sum of the direct and indirect energy required for a particular operation or process (e.g. mining, transportation, crushing, leaching) is called the *gross energy requirement* (GER). Some typical GER values for common mining and beneficiation operations are summarised in Tables 9.1 and 9.2, respectively. These values mostly date from the 1970s and, since technology continually improves, represent the high end of the range that would be encountered today. The rela-

Table 9.1: Typical gross energy requirements for mining operations

	MJ per tonne rock moved
Drilling and blasting	
– Underground drilling	10–200
– Underground blasting	20–100
– Open cut drilling	2–20
– Open cut blasting	3–20
Loading	
– Underground	40–200
– Open cut	5–30
Transport	
– Conveyor (per tonne km^{-1})	2–8
– Truck (per tonne km^{-1})	0.4–2.0
– Train (per tonne km^{-1})	0.2–0.6
– Typical underground mine	60
– Typical open cut mine	40
Miscellaneous	
– Water pumping	5–20
– Ventilation (underground mine)	30–60
Entire mining operation	
– Underground mining	110–400
– Open cut mining	25–50

Source: Chapman and Roberts (1983).

Table 9.2: Typical gross energy requirements for beneficiation operations

	MJ per tonne ore treated
Crushing and grinding	
– Crushing to <15 mm	20–50
– Crushing and grinding to <200 μm	50–100
– Crushing and grinding to <100 μm	150–200
– Autogenous grinding to <100 μm	200–250
– Fine grinding to <50 μm	200–400
– Fine grinding to <25 μm	250–600
Separation and concentration	
– Gravity separation	30–100
– Magnetic separation	30–150
– Electrostatic separation	150–300
– Flotation (sulfide ores)	100–180
– Flotation (oxide ores)	75–200
Dewatering	
– Thickening	5–25
– Filtration	25–100
– Drying (from 5 wt% water)	250–500
– Drying (from 30 wt% water)	~1500
Tailings disposal (per tonne tailings)	10–50
Agglomeration	
– Granulation	25–100
– Pelletisation and cold bonding	~1000
– Pellet induration	1200–2000
– Downdraught sintering (sulfide concentrate)	2000–3000

Source: Chapman and Roberts (1983).

Table 9.3: Typical gross energy requirements for metal production from concentrates

	MJ per kg metal produced
Titanium (Kroll process) – sponge	381
Magnesium (electrolytic)	313
Aluminium (Bayer + Hall-Héroult process)	212 (31 + 181)
Nickel (sulfide smelt, hydromet processing of matte)	93
Zinc (calcine sulfide concentrate, leach, electro-win)	50
Copper (sulfide smelt, electro-refine)	10.9
Copper (leach, SX/EW)	46.3
Steel (blast furnace, basic oxygen furnace)	21.9
Lead (sinter, smelt, fire-refine)	16.2

Source: Norgate and Jahanshahi, CSIRO internal report; private communication, July 2009.

tivity of the values is, however, unlikely to have changed much. Typical GER values for the chemical processing of beneficiated products to produce some common metals are listed in Table 9.3. Note that the GER values in the tables are expressed on different bases: per tonne of rock moved (for mining), per tonne of ore treated (for beneficiation) and per tonne of metal produced (for metal production). Some general observations can be made. Physical operations (e.g. drilling, loading, transporting, crushing, grinding, separation) are much less energy-intensive than operations involving chemical processing (e.g. sintering, smelting, leaching, electro-winning). Underground mining is typically an order of magnitude more energy-intensive than open cut mining. This is largely due to the energy required for mine ventilation, water removal and haulage, and a greater use of explosives. When compared on a per tonne of ore basis (rather than per tonne of rock moved), the difference between the mining methods decreases because the waste rock:ore ratio of open cut mining is much larger than for underground mining. The energy required to make metals from their concentrates varies widely, by more than an order of magnitude. This is due to the relative stabilities of the compounds from which the metals are produced, as discussed in Section 8.2.

9.2 EMBODIED ENERGY

The sum of the GERs of the individual stages along the value chain in the production of a commodity, from cradle-to-gate, is called the *embodied energy* of the commodity. It is the total energy required to produce a commodity in a saleable form from its natural state. Embodied energy is a useful parameter for comparing the energy intensiveness of commodities.

9.2.1 Calculation of embodied energy

Some simple relationships can be developed to calculate the quantity of energy required to produce a mineral or metal commodity and see how it is distributed across the value chain in Figure 9.1. Initially, we will do this on the basis of 1 tonne of ore in the ground (i.e. unmined ore). The equations, and their derivation, are given in Box 9.1. The total energy per tonne of ore (in the ground) is given by Equation 9.9 which is the sum of the energy required at each stage (Equations 9.2, 9.4, 9.6 and 9.8). A more useful basis for comparison purposes is the energy per tonne of 'value' in the final product rather than per tonne of ore in the ground. This is obtained by dividing Equation 9.9 by the mass of 'value' in the final product to yield Equation 9.10.

The factors that determine the energy required to produce a particular quantity of product can be seen from Equation 9.10. The first term is the energy required for mining and beneficiation. The second and third terms are the energies required for the two chemical processing steps. If desired, the first term can be separated into two terms, one for the energy required for mining and one for the energy required for beneficiation. The first term indicates that the lower the grade of the ore body and the greater the waste rock:ore ratio, the greater will be the energy required to produce one tonne of product. The explanation is that more material has to be mined and more ore has to be processed in the beneficiation plant to produce a particular quantity of product. The grade of an ore body is determined by the geology. The waste rock:ore ratio is also determined by the geology of the deposit, particularly its depth below the surface, but it is also dependent on the mining technique. The second and third terms show that, unlike for mining and beneficiation, the energy required for chemical processing is independent of the ore grade. This is because the beneficiated product used as input to the chemical processing step is usually controlled to the same composition irrespective of the grade of the starting ore; for example, a copper concentrate typically contains around 25 wt% Cu irrespective of the grade of ore from which it was produced.

Equation 9.10 also indicates that the lower the recovery at each stage of value-adding, the greater will be the overall energy required. Again, this follows from the fact that more material must be mined and processed to recover a particular quantity of product as more of the product is lost in wastes produced along the value chain. The magnitude of the energies required at each value-adding stage, E_M, E_B, E_{CI} and E_{CII}, clearly affect the total energy. These values, and recovery efficiencies, are dependent on technology; technological developments and improvements can change them to reduce energy consumption. There are, however, limits to which improvements in these can be made due to the nature of the materials treated and the laws of thermodynamics. This is considered further in Section 9.5.

Application of the embodied energy equation to some common commodities

The application of Equation 9.10 can be demonstrated for three common metals using values from Tables 9.1 and 9.2, for mining and beneficiation respectively and from Table 9.3, for chemical processing. These examples should be considered as illustrative only – the situation for each particular mineral deposit and processing operation is unique.

- *Aluminium.* It is assumed that bauxite ore is mined from placer deposits, lightly beneficiated (by screening) then processed by the Bayer process (chemical processing stage 1) to produce high-purity alumina which is then smelted (chemical processing stage 2) in an electrolytic cell to produce aluminium.
- *Copper.* It is assumed that ore is obtained from an open cut mine and is coarsely crushed and screened (the beneficiation stage). The ore is then heap leached, the leach solution is purified and concentrated using solvent extraction, and the copper is recovered by electro-winning to produce cathode grade copper (chemical processing stage).
- *Steel.* It is assumed a hematite iron ore is obtained from an open cut mine. It is crushed and screened and the fines are sintered (the beneficiation stage). The sinter is then smelted in a blast furnace (chemical processing stage 1) and the impure hot metal is converted to steel in a basic oxygen furnace (chemical processing stage 2).

Box 9.1: Calculation of the energy required to produce mineral and metal commodities

Let the energy required for mining be E_M MJ per tonne of rock moved (waste rock plus ore), the waste rock:ore ratio be W, the grade of the ore body (as a fraction) be G_O and the fraction of ore recovered during mining be R_M. The remaining fraction of unmined ore is that part of the ore body that is left unmined due to the mining technique used, or is diluted with waste rock and lost, again due to the mining technique. Then, for 1 tonne of unmined ore:

$$Mass\ of\ material\ moved\ during\ mining = 1 + W \tag{9.1}$$

Therefore:

$$Energy\ required\ for\ mining = E_M(1 + W) \tag{9.2}$$

For each tonne of ore in the ground R_M tonnes are recovered and sent to the beneficiation plant:

$$Mass\ of\ ore\ beneficiated = R_M \tag{9.3}$$

Let the energy required for beneficiation of the ore be E_B MJ per tonne of ore treated. Then

$$Energy\ required\ for\ beneficiation = E_B R_M \tag{9.4}$$

Let the fractional recovery for the beneficiation stage be R_B. Let the energy required for the first chemical processing stage be E_{CI} MJ per tonne of 'value' in the intermediate product and the corresponding fractional recovery be R_{CI}. Then:

$$Mass\ of\ `value'\ in\ intermediate\ product = G_O R_M R_B R_{CI} \tag{9.5}$$

and

$$Energy\ required\ for\ chemical\ processing\ stage\ I = E_{CI} G_O R_M R_B R_{CI} \tag{9.6}$$

Let the energy required for the second chemical processing stage be E_{CII} MJ per tonne of 'value' in the intermediate product and the corresponding fractional recovery be R_{CII}. Then:

$$Mass\ of\ `value'\ in\ product = G_O R_M R_B R_{CI} R_{CII} \tag{9.7}$$

and

$$Energy\ required\ for\ chemical\ processing\ stage\ II = E_{CII} G_O R_M R_B R_{CI} R_{CII} \tag{9.8}$$

The total energy, E_T MJ per tonne of ore (in the ground), is obtained by summing the energy required at each stage (Equations 9.2, 9.4, 9.6 and 9.8):

$$E_T = E_M(1 + W) + E_B R_M + E_{CI} G_O R_m R_B R_{CI} + E_{CII} G_O R_M R_B R_{CI} R_{CII} \tag{9.9}$$

The energy required per tonne of 'value' in the final product is obtained by dividing Equation 9.9 throughout by the mass of 'value' in the final product (G_O R_M R_B R_{CI} R_{CII}). When this is performed and the resulting equation simplified, the following is obtained:

$$E_{Em} = \frac{1}{G_O R_M R_{CI} R_{CII}} \left(\frac{E_M(1 + W)}{R_M} \right) + \frac{E_{CI}}{R_{CII}} + E_{CII} \tag{9.10}$$

where E_{Em} is the energy per tonne of 'value' in the products, or the embodied energy.

Table 9.4: Gross energy requirement (GJ per tonne or MJ per kg) for the production of aluminium, copper and steel calculated using Equation 9.10. Values are indicative only

	Al	Cu	Steel
Assumptions			
Ore grade	35 wt% Al	1.0 wt% Cu	60 wt% Fe
Waste rock:ore ratio	1	3	2
Recovery – mining	85%	90%	90%
Recovery – beneficiation/sintering	85%	95%	99%
Grade of beneficiated product	40 wt%	1.2 wt%	62 wt%
Recovery – 1st chemical stage	80%	85%	98%
Grade of product – 1st chemical stage (water-free basis)	53 wt%	94 wt%	94 wt%
Recovery – 2nd chemical stage	100%	99%	95%
Grade of product – 2nd chemical stage	100 wt%	99.5 wt%	98 wt%
Energy for mining (MJ per tonne of rock moved)	20	25	25
Energy for beneficiation/sintering (MJ per tonne of ore treated)	30	20	250
Energy for 1st chemical stage (MJ per kg of value in product from stage 1)	30	45	20
Energy for 2nd chemical stage (MJ per kg of value in final product)	65	–	0.35
Calculated values			
Gross energy requirement (MJ per kg metal produced)			
– Mining	0.2	13.9	0.15
– Beneficiation/sintering	0.13	2.5	0.45
– 1st chemical stage	30	45.5	21.1
– 2nd chemical stage	186	–	1.05
Total (embodied energy)	*216*	*61.9*	*22.7*

The results are summarised in Table 9.4. Typical values were assumed for ore grade, waste:rock ratio, recoveries along the value chain and product grades, as indicated in the table. It can be seen that the mining and beneficiation stages require the least energy and that the energy required for chemical processing is one to two orders of magnitude greater. Also, the total (or embodied) energies of the three commodities vary by an order of magnitude. This reflects the relative stabilities of alumina, copper sulfides and iron oxide from which the metals are produced.

9.2.2 Values of embodied energy

Table 9.5 lists typical values of embodied energy for a range of common commodities as estimated by a number of authors. Norgate's data is broken down further in Table 9.6 to show individual GER contributions along the value-adding chain. These are also graphed in Figure 9.2. The variation in values in Table 9.5 reported by different authors for the same commodity is mainly due to the different ore grades. The mining method, the technologies used for processing, the recoveries at each stage and the efficiency of generation of the electrical energy consumed in the process are also important variables that affect the final value. Norgate's data are mainly for recent Australian practice, whereas the Battelle-Columbus and Kellogg data are for operations as they were performed in the United States in the 1970s. The data in Table 9.5 lead to some interesting observations. Products

Table 9.5: Embodied energy (GJ per tonne or MJ per kg) of some common mineral commodities

Commodity	Form of product	B-C[a]	Others	Norgate and Rankin[b]
Sand and gravel	Washed and sized	0.065		
Concrete	Blocks		1.5[c]	
	Cast, *in situ*		1.9[c]	
Rock salt		0.20		
Building bricks		4.1	2.5[c]	
Portland cement		8.8	5.6[c]	
Glass	Containers	20	12.7[c]	
Iron	Steel slabs	28		22.7
	Galvanised sheet		38[c]	
Lead	Refined ingots; from smelting sulfide ore	31		20
Zinc	Slabs (electrolytic); from sulfide ore	70	50[d]	48
Copper	Wire bar; from smelting sulfide ore	128	125[e]	33[f]
Nickel	Electrolytic; from smelting sulfide ore	167		114
Tin	Refined ingots; from oxide ore	221		
Aluminium	Ingot; produced from bauxite	284	170[b]	212
Magnesium	Ingot (electrolytic; seawater)	416		
Titanium	Sponge metal; from alluvial rutile	474		
Uranium	Yellowcake (U_3O_8)	902		
Silver	Refined bars	1710		
Gold	Refined bars	68 400		
Non-mineral materials				
Softwood	Kiln dried, sawn		3.4[c]	
Hardwood	Kiln dried, sawn		2.0[c]	
Plywood			10.4[c]	
Plastics	General		90[c]	
	PVC		100[c]	

a) Battelle-Columbus Laboratories (1975a, 1975b,1976).
b) Norgate and Rankin (2000, 2001, 2002a, 2002b).
c) DCCEC (2008).
d) Kellogg (1982).
e) Kellogg (1974); ore grade 0.7 wt% Cu.
f) Ore grade 3.0 wt% Cu.

requiring little beneficiation and no chemical processing have low embodied energy contents (sand and gravel, rock salt). Products that are produced from geochemically abundant resources and are easily processed chemically have moderate values of embodied energy (e.g. bricks, cement, glass, steel). Geochemically scarce and reactive metals have high embodied energy contents. Gold has by far the highest embodied energy of the common metals, primarily because it occurs at such extremely low concentrations.

Table 9.6: The contributions of mining, beneficiation, metal production and refining to the embodied energy content (GJ per tonne or MJ per kg) of some common metals

Commodity	Process	Mining	Beneficiation	Production of metal, matte or chemical intermediate	Smelting and/ or refining	Total
Steel		0.11	0.45	21.21	0.93	22.7
Aluminium		0.18	0.18	30.58[a]	180.57	211.51
Copper	Pyro	7.91	11.69	10.33	3.09	33.02
Copper	Hydro	16.59	1.58	46.29		64.46
Lead		1.85	2.93	13.17	1.62	19.57
Lead	ISP	1.98	3.14	25.77	1.62	32.51
Nickel	Pyro	11.52	19.23	33.33	49.44	113.52
Nickel	Hydro	9.3		184.47		193.77
Zinc	Electrolytic	2.11	3.35	42.98		48.44
	ISF	1.92	3.04	4.96		35.85

a) Bayer process.
Source: T. E. Norgate (unpublished data, private communication, July 2009).

9.3 EMBODIED ENERGY AND GLOBAL WARMING POTENTIAL

Energy consumption and greenhouse gas emissions are closely related since much energy used in primary extraction and processing is fossil fuel based. Reducing energy consumption not only lowers the operating cost and therefore the cost of production of a mineral commodity, but it has environmental and sustainability benefits. Using less energy results in less greenhouse gas production and extends the life of the reserves of fossil fuels.

The quantities of carbon dioxide resulting from the various fossil fuel based energy inputs, direct and indirect, used in the production of the more common

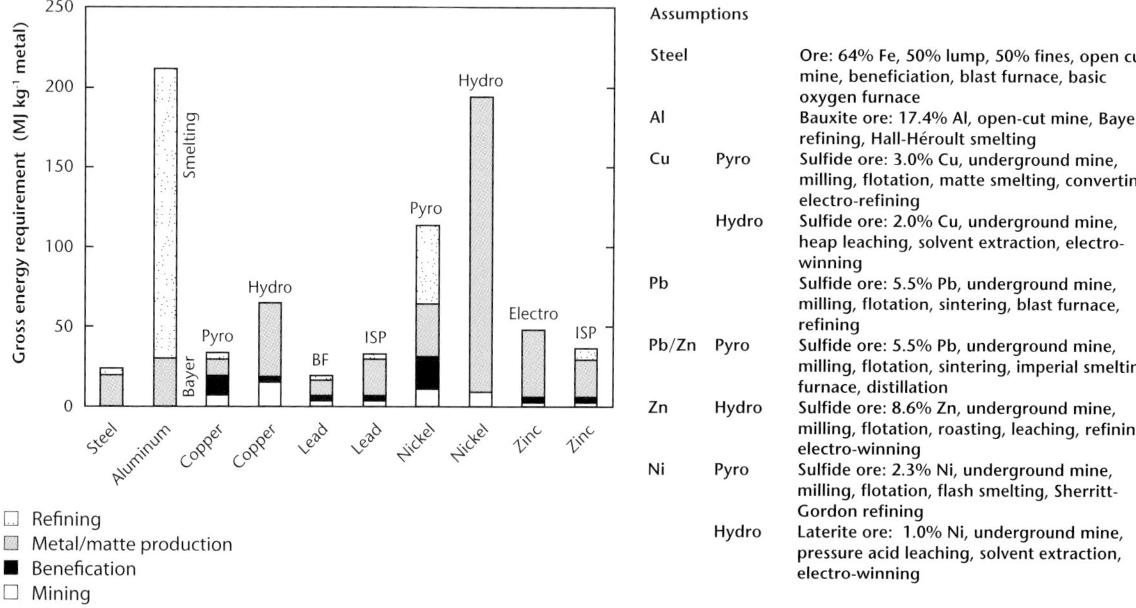

Figure 9.2: Gross energy requirement for cradle-to-gate production of common metals from ores using conventional, state-of-the-art technologies and black coal generated electrical energy (after Norgate and Rankin, 2000, 2001, 2002a, 2002b; Norgate unpublished data, private communication, July 2009).

Table 9.7: The heats of combustion and quantities of CO_2-e produced for combustion of oil, black coal, coke and natural gas

	Heat of combustion (MJ kg^{-1})	Tonnes of CO_2-e per tonne
Oil	41	3.40
Black coal	30	2.86
Coke	31	3.81
Natural gas	53	3.21

metals are summarised in Figure 9.3. These data are for Australian conditions. Greenhouse gas emissions are conventionally expressed in terms of CO_2-equivalents (written as CO_2-e), or global warming potential. This comprises the contribution of the main greenhouse gases (other than water vapour), namely, CO_2, CH_4 and N_2O, aggregated using the appropriate equivalence factors (Table 2.8). The energy data and greenhouse gas emissions were calculated using the combustion values of fossil fuels listed in Table 9.7. It was assumed that the average thermal efficiency of Australian coal-fired power stations is 35%. The values of CO_2-e in Figure 9.3 range from around 2 tonnes of CO_2-e per tonne of metal for lead and steel to around 22 tonnes of CO_2-e per tonne of metal for aluminium. For comparison, the CO_2-e emissions associated with the manufacture of cement, the manufactured mineral commodity produced in greatest quantity, is about 0.8–1.0 tonne of CO_2-e per tonne (WBCSD, 2009). It is apparent that greenhouse gas emissions largely reflect the corresponding energy consumption values (Figure 9.2). As expected, the data show that metal production processes that use energy in a more direct manner (particularly using coal as a fuel) generally have lower energy consumptions and greenhouse gas emissions than processes that utilise the energy indirectly as electricity.

9.3.1 Hydrometallurgy *versus* pyrometallurgy

While copper, nickel and zinc are produced commercially by both hydrometallurgical and pyrometallurgical methods, the data of Figure 9.2 reveal that significantly less energy is required to produce these metals pyrometallurgically. This appears to be counter-intuitive, but there are several reasons why this

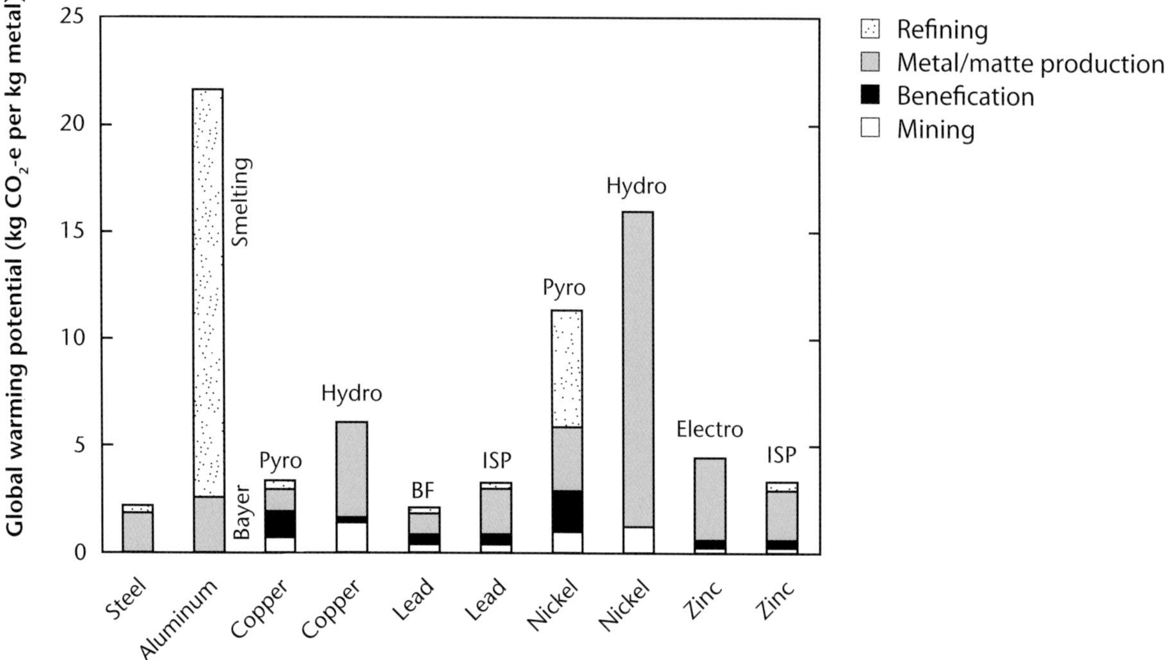

Figure 9.3: The global warming potential for cradle-to-gate production of common metals from ores using conventional, state-of-the-art technologies and black coal generated electrical energy (after Norgate and Rankin, 2000, 2001, 2002a, 2002b; Norgate, unpublished data, private communication, July 2009). Assumptions are the same as in Figure 9.2.

can be the case. When an aqueous solution is heated for leaching (from, say, 25°C to 80°C), heat must be provided not only for the minerals in the ore or concentrate (as in pyrometallurgical processes) but for the water, the mass of which may be up to two orders of magnitude greater than that of the minerals. Furthermore, leaching reactions are slow and residence times in reactors are long, thus heat losses are correspondingly larger compared with pyrometallurgical processes. For metals produced from sulfide ores (e.g. copper, nickel, zinc), much of the heat required for pyrometallurgical processing is provided by oxidation reactions to form SO_2 thereby reducing, or even eliminating, the need to burn fossil fuels for heat. Finally, some metals produced hydrometallurgically are reduced from the leach solution by electro-winning, which is energy-intensive. If the electrical energy required for this is supplied from coal-fired power stations, their low thermal efficiency results in high indirect energy consumption.

9.3.2 Global greenhouse gas production

The annual global quantity of greenhouse gases resulting from the production of individual metals, calculated using the values from Table 9.6 and the metal production values in Table 6.4, are listed in Table 9.8. Corresponding values for ordinary Portland cement (the most energy-intensive of the non-metallic manufactured mineral commodities) are also given. Although steel has the second-lowest energy consumption per tonne for the metals considered, Table 9.8 shows that steel production consumes by far the largest quantity of energy on an annual basis and generates the largest quantity of greenhouse gases. This is because of the large quantity of steel produced. Aluminium, which is energy-intensive to produce and is produced in quantities second only to steel, consumes the second-largest quantity of energy and generates the second-largest quantity of greenhouse gases, though less than half that caused by steel production. These values are compared with the 2007 global annual production of greenhouse gases of 28 962 million tonnes (IEA, 2009) in Table 9.8. Steel and aluminium production account for around 7% and 3% of annual global greenhouse gas production, respectively, and all the metals listed together account for over 10%. Cement production accounts for about 8% of global greenhouse gas production.

9.3.3 Impact of the source of electricity used

The quantity of greenhouse gases produced per kilowatt hour of electricity varies according to the source of electricity. Some typical values are listed in Table 9.9. This affects the quantity of greenhouse gases formed during the production of metals, particularly those produced with a high input of electrical energy. Table 9.10 illustrates this for aluminium production using electricity from three sources. Historically, aluminium smelters have been located where there is plentiful, cheap and reliable electrical power because electricity is the major cost in smelting alumina. Increasingly, consideration is being given to locations where power can be generated using low greenhouse gas emitting technologies, particularly natural gas (in the Middle East), hydroelectricity (in Canada and Scandinavia) and nuclear energy.

9.4 THE EFFECT OF DECLINING ORE GRADE AND LIBERATION SIZE ON ENERGY CONSUMPTION

The overall recovery, R_T, if the material passes through all the stages shown in Figure 9.1, is given by the mass of 'value' in the final product (Equation 9.7) divided by the mass of 'value' in 1 tonne of unmined ore (G_O):

$$R_T = R_M R_B R_{CI} R_{CII} \tag{9.11}$$

For an ore of grade G_O yielding an overall recovery of R_T the quantity of ore in the ground, M_T, required to yield 1 tonne of product is given by:

$$M_T = \frac{1}{G_O R_T} = \frac{1}{G_O R_M R_B R_{CI} R_{CII}} \tag{9.12}$$

As the grade of ore decreases, the energy required to produce a particular quantity of product increases since more material has to be treated to recover the same quantity. Taking copper as an example, and using the assumptions in Table 9.4, the quantity of energy required to produce 1 tonne of copper as the ore grade falls from 3.0 wt%, calculated using Equation 9.10, is shown in Figure 9.4. When the ore grade

Table 9.8: Global annual energy consumption and greenhouse gas emissions for the production of various metals and cement from raw materials

Metal		% of total global metal production[a]	Total global annual production[b] (Mt)	Embodied energy[c] (GJ per tonne)	Global warming potential[d] (tonnes CO_2-e per tonne)	Global annual energy consumption (GJ)	Global annual GHG emissions (tonnes CO_2-e)[e]	% global greenhouse gas production[f]
Copper	Pyro	80	15.6	33.02	3.25	6.13×10^8	6.0×10^7	0.21
	Hydro	20		64.46	6.16			
Nickel	Pyro	60	1.66	113.52	11.45	2.42×10^8	2.2×10^7	0.08
	Hydro	40		193.77	16.08			
Lead	BF	89	3.55	19.57	2.07	7.5×10^7	7.8×10^6	0.03
	ISP	11		32.51	3.18			
Zinc	Electrolytic	90	10.5	48.44	4.61	4.95×10^8	4.7×10^7	0.16
	ISP	10		35.85	3.34			
Aluminium		100	38	211.51	21.81	8.0×10^9	8.3×10^8	2.9
Steel	BF/BOF	70	924	22.7	2.19	2.1×10^{10}	2.0×10^9	7.0
Cement			2600	5.69[g]	0.8–1.0[h]	1.46×10^{10}	2.3×10^9	8.1

a) Estimate by Norgate, internal CSIRO report, November 2006.
b) Data from Table 6.4.
c) Data from Table 9.6 unless otherwise indicated.
d) Data used to construct Figure 9.2 unless otherwise indicated.
e) Assuming coal-based electricity at 35% generation efficiency.
f) Estimated global CO_2 production in 2007 was 28 962 Mt (IEA, 2009).
g) Table 9.5.
h) World Business Council for Sustainable Development, http://www.wbcsd.org.

Table 9.9: Life cycle greenhouse gas emissions caused by electricity generation

Source	g CO_2-e per kW h	Reference
Black coal	960, with scrubbing 1050, without scrubbing	Sovacool (2008) Sovacool (2008)
Diesel	788	Sovacool (2008)
Heavy oil	788	Sovacool (2008)
Natural gas	443	Sovacool (2008)
Hydro	10 for a 3.1 MW station 15–25	Pehnt (2006) Lenzen (2008)
Nuclear	1.4–288 over life time of plant; average 66	Sovacol (2008)
	10–130 over life time of plant; average 65	Lenzen (2008)

falls below about 0.5 wt%, the energy required to produce copper begins to increase very rapidly.

The particle size to which an ore must be crushed and ground to liberate minerals sufficiently for an acceptable recovery is an important factor in determining the energy required to produce a mineral commodity. The Bond equation (Bond, 1952) is widely used to estimate the energy required for grinding:

$$E = \frac{10W_i}{\sqrt{P_{80}}} - \frac{10W_i}{\sqrt{F_{80}}}$$

where E is the energy for grinding (kW h per tonne of ore), F_{80} and P_{80} are the size[21] (mm) of the feed to the grinding mill and product from the mill, respectively, and W_i is the Bond work index. The Bond equation assumes the net energy needed to break a particle is proportional to the length of new cracks required to be formed during breakage. It is empirical to the extent that the Bond work index has to be determined experimentally for each type of material. Energy for grinding is often expressed as kW h per tonne because grinding mills are usually electrically powered, but it can be converted to Joules using the conversion 1 kW h = 3.60 MJ. Consider a sulfide ore (e.g. of copper, nickel, lead or zinc) which is ground to liberate the minerals which are then separated by froth flotation. Assume the size of the particles entering the mill is 80% <5 mm. A typical value of Bond work index for sulfide ores is 15.

21 The size of particles is often expressed as 80 wt% less than a certain size. Thus, a milled product in which 80 wt% of the material has a particle size less than 200 μm has a P_{80} value of 200 μm.

Substituting these values into the Bond equation for a range of product sizes yields the curve in Figure 9.5, which shows how the energy required for grinding increases rapidly as the particle size required for liberation decreases. In summary, as ore grades decrease and ores become more complex, the quantity of ore required to be mined per unit of metal, and the fineness of the grind required to achieve liberation, will increase. As a result, the energy required to produce mineral concentrates will steadily increase over time. This is a concern for the long-term supply of mineral commodities.

9.5 THE LOWER LIMITS OF ENERGY CONSUMPTION

From an energy perspective, the processes used in producing mineral and metal commodities can be grouped into three broad classes: those involved with moving material, those involved with sorting or separating material, and those involved with producing an element from a mineral. The first two classes involve physical

Table 9.10: Greenhouse gas emissions for aluminium production from different sources of electrical energy

	Efficiency	Global warming potential (kg CO_2-e per kg Al)
Black coal	35%	22.4
Natural gas (combined cycle)	54%	13.3
Hydroelectricity	80%	9.9

Source: Norgate and Rankin (2001).

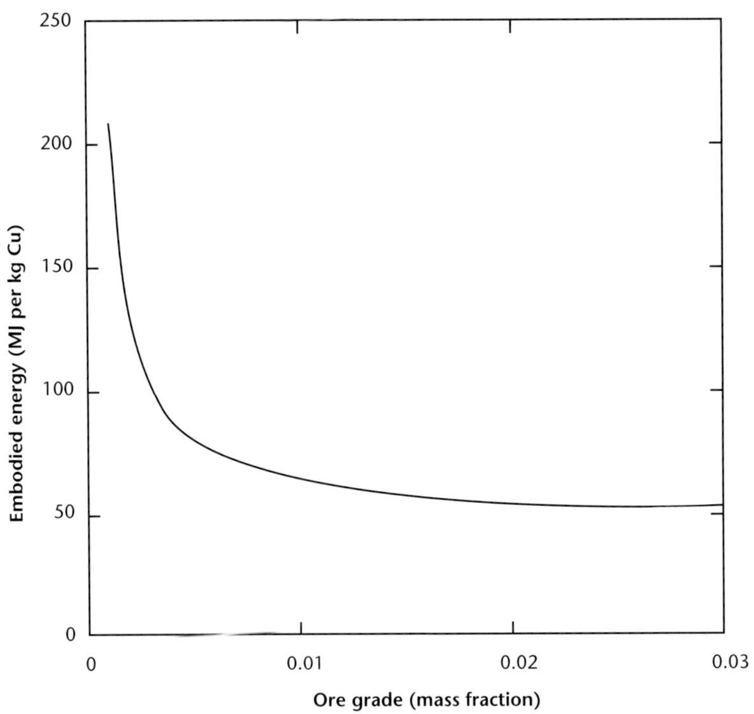

Figure 9.4: Gross energy requirement to produce 1 tonne of copper from ore as a function of ore grade.

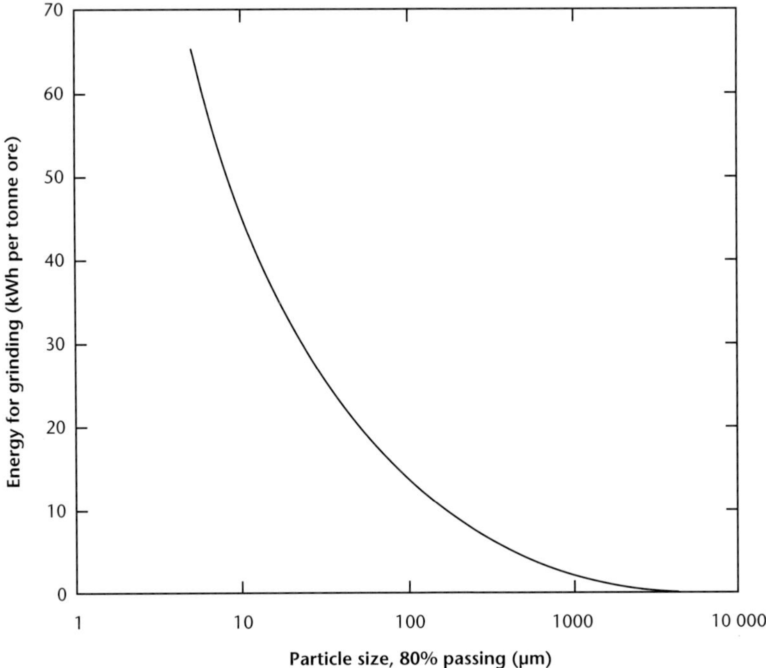

Figure 9.5: Energy required for grinding as a function of grind size calculated using the Bond equation (feed size assumed 80% <5 mm).

processes – the chemical form of the material does not change – while the last involves chemical processes. The processes in each class involve the use of energy in different ways. It is instructive to examine the theoretical minimum values required for each, since these set the limit to technological improvements. These values will never be attained in practice but the exercise serves to demonstrate, first, that there are limits below which it is theoretically impossible to go and, second, that the theoretical limits are so far below present practice that there is great scope for technological developments to improve energy efficiency.

9.5.1 Energy required for moving materials

Large quantities of material are moved in mining and processing operations, the major steps being raising the ore from the mine to the surface, transporting it to the processing plant, and moving the tailings and other processing residues to the disposal or storage site. In the absence of frictional resistance, moving material horizontally requires no expenditure of energy beyond that to produce the initial acceleration. This is a consequence of Newton's first law of motion (formulated in 1686). The theoretical energy required to accelerate 1 tonne of material to a velocity of, for example, 5 m s^{-1}, is equal to the kinetic energy of the body moving at that velocity:

$$E = \frac{1}{2}mv^2 = 0.5 \times 1000 \times 5 = 2.5 \text{ KJ}$$

Because of frictional resistances in the machinery, however, force must be applied to maintain movement, hence more work is done and energy consumed. This is typically in the range 0.2–8 MJ per tonne of rock moved, according to the mode of transport (Table 9.1). Raising and lowering materials vertically requires work against the force of gravity. In the absence of frictional resistances in the equipment, the energy required to raise 1 tonne of rock from a depth of, for example, 200 m is given by:

$$w = fs = mas = 1000 \times 9.81 \times 200 = 1.96 \text{ MJ}$$

The actual energy required is typically around 40 MJ per tonne of rock in an open cut mine (Table 9.1), due to frictional resistances.

The total energy used in transport is determined by the need to accelerate the material to a certain speed, the friction which must be overcome, and the efficiency of energy conversion. Nothing can be done about the first, but friction and the efficiency of conversion are amenable to technological improvements. Truck engines, for example, are only about 30% efficient. Friction can be reduced by decreasing the number of moving parts in contact throughout the transport system. Reducing the quantity of material to be moved, however, probably has the greatest potential for reducing energy consumption for transport. For example, if some or all of the beneficiation were performed in the mine and waste were disposed directly into the mine workings, only beneficiated material would need to be brought to the surface and less wastes would have to be returned from the surface to the workings.

9.5.2 Energy required for sorting and separating material

The second class of processes is concerned with sorting or separating; for example, separating valuable minerals from gangue minerals. Consider an idealised copper ore containing 1 wt% copper present as chalcopyrite with quartz as the only gangue mineral. The Gibbs energy of mixing these particles is given by the relation:[22]

$$\Delta G = x_C \mathrm{R}T\ln x_C + x_Q \mathrm{R}T\ln x_Q \qquad 9.14$$

where x_C and x_Q are the mole fractions of chalcopyrite and quartz in the mixture, respectively, T is the temperature (K) and R is the universal gas constant (8.314 J K^{-1} mol^{-1}). The mole fractions of chalcopyrite and quartz are 0.0096 and 0.9904, respectively. If it is assumed that the temperature is 25°C (298 K), then ΔG is –123.3 J per mole of mixture, or approximately –2 MJ per tonne of ore. The value is negative since random mixing of particles is a natural process. Gibbs energy is the quantity of work which can be obtained when a process is carried out spontaneously, or the minimum quantity of work required to drive a process that is not spontaneous. Separation of the mixture into chalcopyrite and quartz fractions, therefore, will require the input of at least 2 MJ per tonne of ore. This sets the approximate theoretical limit for separation processes after the minerals have been liberated. The

22 The derivation of this equation is not considered here; it is available in any standard physical chemistry text.

energy used by separation processes in practice is typically one to two orders of magnitude greater (Table 9.2). This is due to the need to overcome accelerational and frictional forces of the particles and in the equipment used to perform the separation.

An even greater hurdle is that, usually, the minerals have to be liberated by crushing and grinding before they can be separated. On a larger scale, rock has to be broken up in the mine so it can be removed. The latter is achieved by blasting with explosives and the former is undertaken mechanically in crushers and mills. In both cases, the fundamental process is the creation of new surfaces. The energy required to create new surfaces in perfectly brittle solids can be calculated from the Griffith equation (Griffith, 1921), a fundamental equation based on calculating, from thermodynamic principles, the increase in surface energy due to the propagation of a crack. The theory explained the observation that the energy required to fracture solids is much less than the energy required to break the atomic bonds holding the material together. Real materials have flaws, such as pores, cracks and grain boundaries, which provide sites for crack propagation. However, the Griffith equation underestimates, by two or more orders of magnitude, the energy required to fracture most materials because energy is absorbed by other processes occurring at the same time as creation of new surfaces, particularly plastic deformation around the tip of the crack. Griffith assumed solids are perfectly elastic and that any energy used in deformation during fracture is recovered as the material returns to its original shape. Very few, if any, substances are perfectly elastic.

Energy is also consumed in grinding processes due to friction between particles as they move during and after breakage. Friction between the particles during pulverising of coal may absorb 60% of the total energy supplied, and plastic deformation and other phenomena within the particles may account for a further 30% (Brown, 1966). The remainder is used to create new surfaces. Typically, the energy used to create new surface area during crushing and grinding of rocks is 4–10% of the actual energy supplied (Brown, 1966; Harris, 1966). In practice, crushing of rock requires around 20–50 MJ per tonne, and grinding to <100 μm requires 150–250 MJ per tonne. Energy consumed in breakage other than for creation of new surface area ultimately appears as thermal energy; this accounts for the temperature rise in pulps during grinding. Rocks contain an abundance of flaws which provide breakage sites. However, as particles are ground smaller, the flaws are progressively eliminated as breakage along them occurs. A size will ultimately be reached at which all flaws have been removed. The energy to create new surface area will rise towards the theoretical energy required to break the chemical bonds in the minerals of the rock. This situation probably begins to occur in the ultrafine size range (<15 μm). As shown in Figure 9.6, at sizes below around 15 μm the energy required for grinding begins to increase rapidly as the particle size decreases.

There have been considerable advances in fine grinding technology, and a number of mill types that were previously used only on a small scale for specialty materials have been scaled up for application in the commodity minerals industry. The most widely used of these are stirred media mills. Figure 9.6 indicates that, at coarser grinds, stirred mills require about 30% less energy than ball mills. In the ultrafine range, this advantage increases to more than 50%. In contrast to tumbling mills, with rotating outer shells, stirred media mills impart motion to the charge via an internal stirrer. There are two broad types. Tower and low-speed vertical stirred mills have a double-start helical rotating screw inside a stationary vertical, cylindrical grinding chamber filled with balls or pebbles. The pulp is fed at the base. As the particles work their way upwards they are ground by attrition and abrasion, by the action of the grinding medium. They discharge at the top. High-speed stirred mills are used for ultrafine grinding. A series of discs (often with pins) mounted on a shaft are rotated at speeds up to 2000 rpm within the cylindrical mill, which contains fine sand or ceramic material as the grinding medium.

9.5.3 Energy required for chemical processing

The third class of processes, the production of an element from a mineral, is fundamentally different from the others in that it involves chemical reactions. The theoretical minimum energy required to produce a metal, or any other substance, is the enthalpy change of the reaction. For comparative purposes, the chemical form of the metal is taken to be that most

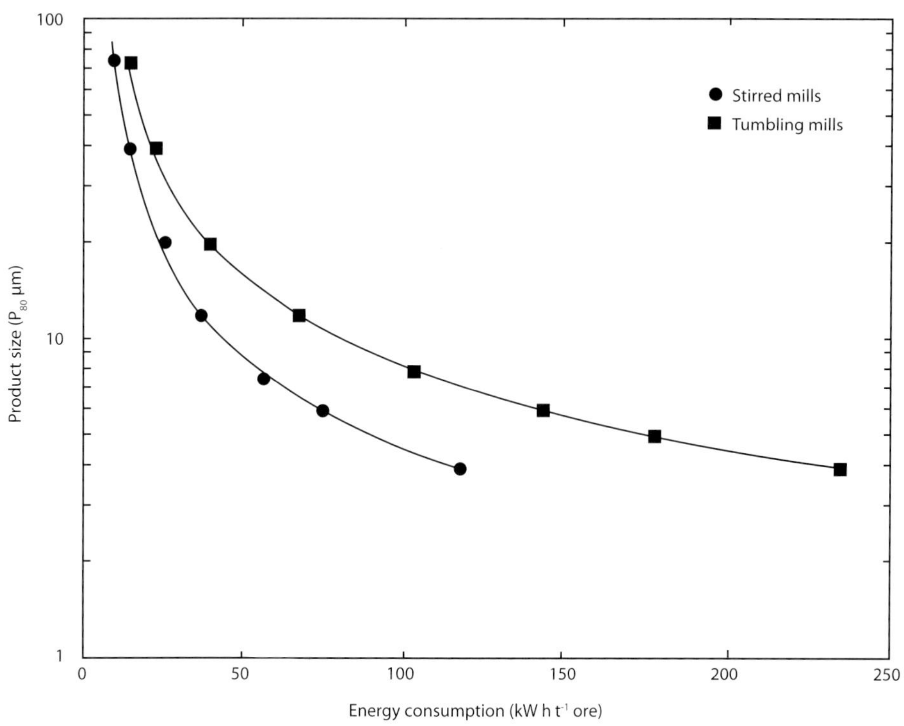

Figure 9.6: A comparison of the energy required for grinding in tumbling mills and stirred mills (after Lichter and Davey, 2002).

commonly found in nature (e.g. Fe_2O_3 for iron, Al_2O_3 for aluminium, PbS for lead) and all the reactants and products are assumed to be in their standard state; i.e. pure substances at 1 atmosphere pressure and 25°C (or 298 K). Since:

$$\Delta G^{\circ}_{298} = \Delta H^{\circ}_{298} - T\Delta S^{\circ}_{298}$$

it is apparent that the enthalpy change of a reaction is made up of two terms:

$$\Delta H^{\circ}_{298} = \Delta G^{\circ}_{298} + T\Delta S^{\circ}_{298}$$

The magnitude of ΔH°_{298} may be positive, in which case the reaction is endothermic and thermal energy must be supplied to maintain a constant temperature. If it is negative, the reaction is exothermic and thermal energy must be removed to maintain a constant temperature. ΔG°_{298} is the maximum quantity of work that can be obtained from a spontaneous reaction, or the minimum quantity of work required to drive a reaction. When ΔG°_{298} is negative, a reaction will occur spontaneously but when ΔG°_{298} is positive energy must be supplied to drive the reaction. This energy must be supplied chemically or electrochemically. Often, the temperature of the products is greater than 25°C, particularly in smelting processes, and this must be allowed for in the calculation of total energy required. Thus, the minimum theoretical energy required to produce a metal from a compound is the sum of three energy terms.

1 The ΔG°_{298} component of the enthalpy of the reaction. When ΔG°_{298} is positive, electrical energy equivalent to the magnitude of ΔG°_{298} must be applied to drive the reaction through an electrowinning type process, or a reactant must be added to cause an alternative reaction with a negative value of ΔG°_{298}. When ΔG°_{298} is negative, the reaction occurs spontaneously and ΔG°_{298} contributes to the thermal energy required or produced by the reaction.
2 The $T\Delta S^{\circ}_{298}$ component of the enthalpy of the reaction. This energy must be supplied thermally, or removed thermally if it is a negative quantity.

3 The thermal energy required to heat the products of the reaction to the temperature at which they are discharged from the reaction vessel.

The temperature of the reactor vessel may be raised for any of several reasons:

- so that ΔG of the metal-forming reaction becomes negative and the reaction can proceed without the need for electrical energy to drive it (as for carbothermic reduction of iron oxides at temperatures greater than about 700°C);
- to increase the rate of the reaction;
- to form molten phases so that physical separations can be made.

Some of this heat is often recovered, particularly from gas streams in smelting processes, but in most processes much is lost to the environment.

Table 9.11 lists the above terms for a number of common metals. The theoretical energies have been expressed in both MJ and kW h per kilogram of metal produced for ease of comparison. The most striking observation is the wide range of energy required to make metals from their naturally occurring compounds by dissociation into their constituent elements. This simply reflects the relative stability of these compounds. Another observation is that, by utilising commonly available reactants, the energy required for production of metals can be decreased considerably. For example, aluminium and iron oxides are reacted with carbon and copper sulfides are reacted with oxygen in practice. The latter actually produces a net amount of heat.

Typical values of the actual energy used in the production of these metals from their concentrates (from Table 9.3) are listed in the last column in Table 9.11 for comparison. In all cases, the actual values are very much greater than the minimum theoretical values. This is expected since the values listed do not include the many complicating factors encountered when real concentrates are processed (discussed in Chapter 8). Because of impurities in the concentrates, many additional reactions other than the principal, primary metal-forming reaction occur. These impurities require additional processing steps for removal to acceptable levels. Materials also have to be handled, agitated and moved between stages, and separated, and this consumes energy. Raw materials often have to be pretreated or prepared in some way to make them suitable for the process; for example, coal is converted into coke and iron ore fines are sintered for iron making, and carbon anodes have to be made for aluminium smelting. This requires additional energy. As an example, consider Hall-Héroult cells for producing aluminium. Modern cells consume around 13.5 kW h per kilogram of aluminium produced. The minimum theoretical energy is around 6.4 kW h per kilogram of Al, just under half of present best practice, indicating there is still much room for improvement. Great gains have been made; Figure 9.7 shows the trend in energy consumption since aluminium was first produced commercially in the 1880s. However, Hall-Héroult cells use less than one-third of the total energy consumed in smelting (~50 MJ kg^{-1} compared to ~180 MJ kg^{-1}), indicating that there are potentially greater gains to be made in other areas of the smelter, particularly anode production.

9.6 ENERGY SUSTAINABILITY INDICATORS AND REPORTING

The Global Reporting Initiative guidelines have five indicators for energy consumption, against which companies are required to report. These are listed in Table 9.12. Two of these are mandatory (core) and three are optional (voluntary). The quantities of greenhouse gases emitted, both direct and indirect (EN16 and EN17), are also relevant indicators. These are discussed in Section 12.4. According to the survey by Mudd (2009), among 25 major mining companies which produced sustainability reports in 2007 there was wide variation in how companies reported data. Some simply provided total energy consumption, others provided subtotals for direct and indirect energy sources, and others provided detailed energy inputs by fuels or sources. In a only few reports was energy usage broken down into usage at plant and process level, information which is of most use. In most it was reported only for an entire mine site or for the company as a whole. Sometimes only the direct energy consumption was reported, making it impossible to compare the embodied energy in the products.

Table 9.11: The theoretical minimum energy required to produce some common metals from their naturally occurring mineral

Metal-forming reaction	ΔG^{o}_{298} (KJ)	ΔH^{o}_{298} (KJ)	$T\Delta S^{o}_{298}$ (KJ)		Theoretical			Actual[a]
					KJ mol^{-1}	MJ kg^{-1}	kW h kg^{-1}	MJ kg^{-1}
$0.5Fe_2O_3 = Fe + 0.5O_2$	370.5	411.5	41.0	*Products discharged at 25°C*				
				ΔG^{o}_{298}	370.5	6.63	1.84	
				$T\Delta S^{o}_{298}$	41.0	0.73	0.20	
				Total	411.5	7.37	2.05	
$0.5Fe_2O_3 + 1.5C = Fe + 1.5CO_2$	74.7	116.4	41.6	*Products discharged at 25°C*				
				ΔG^{o}_{298}	74.7	1.34	0.37	
				$T\Delta S^{o}_{298}$	41.6	0.74	0.21	
				Total	116.4	2.08	0.58	
				Products discharged at 700°C				
				ΔG^{o}_{298}	74.7	1.34	0.37	
				$T\Delta S^{o}_{298}$	41.6	0.74	0.21	
				Sensible heat in Fe	23.3	0.42	0.12	
				Sensible heat in CO_2	47.9	0.86	0.24	
				Total	187.6	3.36	0.93	
				Products discharged at 1600°C				
				ΔG^{o}_{298}	74.7	1.34	0.37	
				$T\Delta S^{o}_{298}$	41.6	0.74	0.21	
				Sensible heat in Fe	75.4	1.35	0.37	
				Sensible heat in CO_2	125.7	2.25	0.63	
				Total	317.5	5.67	1.58	22
$0.5Al_2O_2 = Al + 0.75O_2$	791.1	837.8	46.7	*Products discharged at 25°C*				
				ΔG^{o}_{298}	791.1	29.32	8.15	
				$T\Delta S^{o}_{298}$	46.7	1.73	0.48	
				Total	837.8	31.05	8.63	

Metal-forming reaction	ΔG°_{298} (KJ)	ΔH°_{298} (KJ)	$T\Delta S^{\circ}_{298}$ (KJ)		Theoretical			Actual[a]
					KJ mol^{-1}	MJ kg^{-1}	kW h kg^{-1}	MJ kg^{-1}
				Products discharged at 1000°C				
				ΔG°_{298}	791.1	29.32	8.15	
				$T\Delta S^{\circ}_{298}$	46.7	1.73	0.48	
				Sensible heat in Al	39.6	1.47	0.41	
				Sensible heat in O_2	48.6	1.80	0.50	
				Total	926.1	34.32	9.54	
$0.5Al_2O_3 + 0.75C = Al + 0.75CO_2$	495.4	542.7	47.4	*Products discharged at 25°C*				
				ΔG°_{298}	495.4	18.36	5.10	
				$T\Delta S^{\circ}_{298}$	47.4	1.75	0.49	
				Total	542.7	20.12	5.59	
				Products discharged at 1000°C				
				ΔG°_{298}	495.4	18.36	5.10	
				$T\Delta S^{\circ}_{298}$	47.4	1.76	0.49	
				Sensible heat in Al	39.6	1.47	0.41	
				Sensible heat in CO_2	36.5	1.35	0.38	
				Total	618.8	22.94	6.37	181
$0.5Cu_2S = Cu + S_2$	63.0	71.9	8.9	*Products discharged at 25°C*				
				ΔG°_{298}	63.0	0.99	0.28	
				$T\Delta S^{\circ}_{298}$	8.9	0.14	0.04	
				Total	71.9	1.13	0.31	
$0.5Cu_2S + O_2 = Cu + 0.5SO_2$	–106.8	–108.9	–2.1	*Products discharged at 25°C*				
				ΔG°_{298}	–106.8	–1.68	–0.47	
				$T\Delta S^{\circ}_{298}$	–2.1	–0.03	–0.01	
				Total	–108.9	–1.71	–0.48	

Metal-forming reaction	ΔG°_{298} (KJ)	ΔH°_{298} (KJ)	$T\Delta S^{\circ}_{298}$ (KJ)		Theoretical			Actual[a]
					KJ mol^{-1}	MJ kg^{-1}	kW h kg^{-1}	MJ kg^{-1}
				Products discharged at 1200°C				
				ΔG°_{298}	–106.8	–1.68	–0.47	
				$T\Delta S^{\circ}_{298}$	–2.1	–0.03	–0.01	
				Sensible heat in Cu	46.5	0.73	0.20	
				Sensible heat in SO_2	30.9			
				Total	–31.4	–0.49	–0.14	11
$NiO = Ni + 0.5O_2$	211.6	239.7	28.1	*Products discharged at 25°C*				
				ΔG°_{298}	211.6	3.60	1.00	
				$T\Delta S^{\circ}_{298}$	28.1	0.48	0.13	
				Total	239.7	4.08	1.13	185
$PbS = Pb + 0.5S_2$	137.6	163.8	26.1	*Products discharged at 25°C*				
				ΔG°_{298}	137.6	0.66	0.19	
				$T\Delta S^{\circ}_{298}$	26.1	0.13	0.03	
				Total	163.8	0.79	0.22	
$PbS + O_2 = Pb + SO_2$	–202.3	–197.3	5.0	*Products discharged at 1200°C*				
				ΔG°_{298}	–202.3	–0.98	–0.27	
				$T\Delta S^{\circ}_{298}$	5.0	0.02	0.01	
				Total	–197.3	–0.95	–0.26	
				Products discharged at 1200°C				
				ΔG°_{298}	–202.3	–0.98	–0.27	
				$T\Delta S^{\circ}_{298}$	5.0	0.02	0.01	
				Sensible heat in Pb	39.0	0.19	0.05	
				Sensible heat in SO_2	61.8	0.30	0.08	
				Total	–96.5	–0.47	–0.13	16

Metal-forming reaction	ΔG°_{298} (KJ)	ΔH°_{298} (KJ)	$T\Delta S^{\circ}_{298}$ (KJ)		Theoretical			Actual[a]
					KJ mol^{-1}	MJ kg^{-1}	kW h kg^{-1}	MJ kg^{-1}
				Products discharged at 25°C				
$SnO_2 = Sn + O_2$	515.8	577.6	61.8	ΔG°_{298}	515.8	4.35	1.21	
				$T\Delta S^{\circ}_{298}$	61.8	0.52	0.14	
				Total	577.6	4.87	1.35	
$TiO_2 = Ti + O_2$	889.4	944.7	55.3	*Products discharged at 25°C*				
				ΔG°_{298}				
				$T\Delta S^{\circ}_{298}$	55.3	1.16	0.32	
				Total	944.7	19.74	5.48	381
$MgO = Mg + 0.5O_2$	569.4	601.6	32.2	*Products discharged at 25°C*				
				ΔG°_{298}	569.4	23.42	6.51	
				$T\Delta S^{\circ}_{298}$	32.2	1.33	0.37	
				Total	601.6	24.75	6.88	313

a) From Tables 9.3 and 9.6.
Source: Thermodynamic values are from HSC (2002).

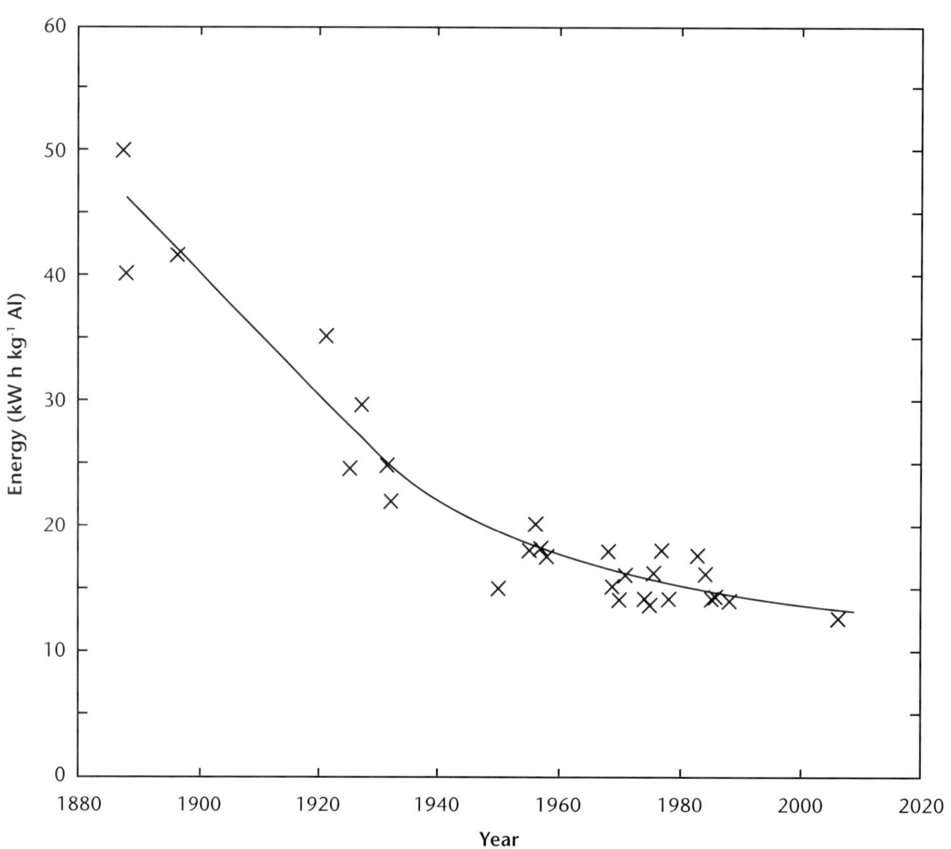

Figure 9.7: The historical trend in the energy required to produce aluminium from alumina (adapted from Van Leeuwen, 2000). Value for 2006 is 'best practice' (Keniry, 2008).

Reporting against the voluntary energy indicators was relatively uncommon. Mudd noted that 'the overall variability in reporting detail is problematic since it limits the ability to compare various companies and mine sites in an accurate manner, especially to assess the effects of ore grade, mine type, project scale, type of energy resources consumed (especially electricity source) and so on'.

9.7 REFERENCES

Battelle-Columbus Laboratories (1975a) *Energy Use Patterns in Metallurgical and Non-metallic Mineral Processing. High Priority Commodities.* PB 245 759. United States National Technical Information Service: Springfield, VA.

Battelle-Columbus Laboratories (1975b) *Energy Use Patterns in Metallurgical and Non-metallic Mineral*

Table 9.12: GRI sustainability indicators for energy consumption

EN3	Direct energy consumption by primary energy source	Core
EN4	Indirect energy consumption by primary energy source	Core
EN5	Energy saved due to conservation and efficiency improvements	Voluntary
EN6	Initiatives to provide energy-efficient or renewable energy based products and services, and reductions in energy requirements as a result of these initiatives	Voluntary
EN7	Initiatives to reduce indirect energy consumption and reductions achieved	Voluntary

Source: GRI (2006).

Processing. Intermediate Priority Commodities. PB 246 357. United States National Technical Information Service: Springfield, VA.

Battelle-Columbus Laboratories (1976) *Energy Use Patterns in Metallurgical and Non-metallic Mineral Processing. Low Priority Commodities*. PB 261 150. United States National Technical Information Service: Springfield, VA.

Bond FC (1952) The third theory of comminution. *Transactions of the American Institute of Mining Engineers* **193**: 484–494.

Brown RW (1966) Energy distribution in pulverizing. *Transactions of the Institution of Mining and Metallurgy* **C75**: 173–180.

Chapman PF and Roberts F (1983) *Metal Resources and Energy*. Butterworth: London.

DCCEC (2008) *Your Home Technical Manual*. 4th edn. Department of Climate Change and Energy Efficiency, Commonwealth of Australia: Canberra. <www.yourhome.gov.au>.

GRI (2006) *Sustainability Reporting Guidelines*. 3rd edn. Global Reporting Initiative. <http://www.globalreporting.org>.

Griffith AA (1921) The phenomena of rupture and flow in solids. *Philosophical Transactions of the Royal Society of London* **A221**: 163–198.

Harris CC (1966) On the role of energy in comminution: a review of physical and mathematical principles. *Transactions of the Institution of Mining and Metallurgy* **C75**: 37–56.

HSC (2002) *HSC Chemistry for Windows v5.0*. Outokumpu Research Oy: Pori, Finland.

IEA (2009) *Key World Energy Statistics*. International Energy Agency: Paris, France. <http://www.iea.org/publications/free_all.asp>.

Kellogg HH (1974) Energy efficiency in the age of scarcity. *Journal of Metals* **26**(6): 25–29.

Kellogg HH (1982) The state of nonferrous extractive metallurgy. *Journal of Metals* **34**(10): 35–42.

Keniry J (2008) Aluminium smelting greenhouse footprint and sustainability. In *Light Metals 2008*. (Ed. DH De Young) pp. 369–373. Minerals, Metals and Materials Society: Warrendale, PA.

Lenzen M (2008) Life cycle energy and greenhouse gas emissions for nuclear energy: a review. *Energy Conversion and Management* **49**: 2178–2199.

Lichter JKH and Davey G (2002) Selection and sizing of ultrafine and stirred grinding mills. In *Mineral Processing Plant Design, Practice and Control*. Vol. 1. (Eds AL Mular, DN Halbe and DJ Barrett) pp. 783–800. Society for Mining, Metallurgy and Exploration: Littleton, CO.

Mudd GM (2009) Sustainability reporting and mining – an assessment of the state of play for environmental indicators. In *Sustainable Development Indicators in the Minerals Industry*. pp. 377–391. Australasian Institute of Mining and Metallurgy: Melbourne.

Norgate TE and Rankin WJ (2000) Life cycle assessment of copper and nickel production. *MINPREX 2000 International Congress on Mineral Processing and Extractive Metallurgy*, Melbourne, 11–13 September. pp. 133–138. Australasian Institute of Mining and Metallurgy: Melbourne.

Norgate TE and Rankin WJ (2001) Greenhouse gas emissions from aluminium production – a life cycle approach. *Proceedings of the International Symposium on Greenhouse Gases in the Metallurgical Industries: Policies, Abatement and Treatment*, 26–29 August 2001, Toronto, Ontario, Canada. pp. 275–290. Metallurgical Society, Canadian Institute of Mining and Metallurgy and Petroleum.

Norgate TE and Rankin WJ (2002a) The role of metals in sustainable development. In *Green Processing 2002*. pp. 49–55. Australasian Institute of Mining and Metallurgy: Melbourne.

Norgate TE and Rankin WJ (2002b) An environmental assessment of lead and zinc production. In *Green Processing 2002*. pp. 177–184. Australasian Institute of Mining and Metallurgy: Melbourne.

Pehnt M (2006) Dynamic life cycle assessment (LCA) of renewable energy technologies. *Renewable Energy* **31**: 55–71.

Sovacool BK (2008) Valuing the greenhouse gas emissions from nuclear power: a critical survey. *Energy Policy* **36**: 2950–2963.

Van Leeuwen T (2000) 'An aluminum revolution, equity'. Research report. Credit Suisse First Boston Corporation: Boston, MA.

WBCSD (2009) World Business Council for Sustainable Development. <http://www.wbcsd.org>.

10 The role of water in primary production

Large quantities of water are used in the minerals industry to bring about the transformations of matter required to make tradable mineral commodities. Water used in the industry may be chemically transformed, and consumed, if it takes part in chemical reactions. However, much of the water is used in physical processes. In such applications, its form does not change though its quality may degrade due to contamination. Water often becomes dispersed into different streams, thereby making its collection and recycling difficult. Nevertheless, the quality, and in large part the quantity, of water can be restored in full, though (as required by the second law of thermodynamics) this requires the expenditure of energy. A large quantity of water can be recovered in minerals operations by relatively simple physical processes such as thickening and filtration. This chapter examines the availability and usage patterns of water, the sources of water used by the minerals industry and the quantities of water required to produce metals. It concludes with a survey of sustainability indicators for water. The potential environmental impacts of contaminated water and approaches to minimise these at mine sites are discussed in Chapters 11 and 12.

10.1 GLOBAL WATER RESOURCES

Water is essential to all life and for many human activities, including the production of mineral and metal commodities. However, equitable access to water is far from real in many parts of the world. The United Nations report 'Water in a changing world' (United Nations, 2009) states:

> *Water is essential for achieving sustainable development ... Properly managing water resources is an essential component of growth, social and economic development, poverty reduction and equity, and sustainable environmental services ... Water is linked to the crises of climate change, energy and food supplies and prices, and troubled financial markets. Unless their links with water are addressed and water crises around the world are resolved, these other crises may intensify and local water crises may worsen, converging into a global water crisis and leading to political insecurity and conflict at various levels.*

Though the world's resource of water is approximately 1400 million km^3 (Table 3.3), most of this is saline water present in the oceans. Only about 2.75 vol% is fresh water, a large fraction of it relatively inaccessible ice in the polar ice caps. Of total fresh water, groundwater makes up about 25 vol% (8.2×10^6 km^3) and fresh water in lakes and rivers makes up about 0.33 vol% (130 000 km^3). Lakes and rivers offer the most accessible forms of fresh water for human use but are also vital to water-based ecosystems. A

Table 10.1: Renewable water resources and per capita availability of water by continent

	Area (10^6 km^2)	Population (millions) 1994	Renewable water resources (km^3 per year)			Potential water availability (10^3 m^3 per year)	
			Average	Max.	Min.	Per km^2	Per capita
Europe	10.46	685	2900	3410	2254	277	4.23
North America	24.3	453	7890	8917	6895	324	17.4
Africa	30.1	708	4050	5082	3073	134	5.72
Asia	43.5	3445	13 510	15 008	11 800	311	3.92
South America	17.9	315	12 030	14 350	10 320	672ç	38.2
Australia and Oceania	8.95	28.7	2400	2880	1891	269	83.7
Total	*135*	*5633*	*42 780*	*44 750*	*39 780*	*316*	*7.60*

Source: Shiklomanov (2000).

more useful concept than water resources, when considering water for consumption, is the *renewable water resource* – the quantity of water that is replenished annually. This is the actual water resource that is potentially available for human use. It comprises the water that is continuously recharged in the water cycle and is composed mainly of run-off water which enters rivers, groundwater that flows into rivers, and upper aquifer groundwater that is not drained by river systems. The mean global volume of renewable water resources is estimated to be around 42 780 km^3 per annum (Shiklomanov, 2000). However, there is huge variation from continent to continent (Table 10.1), and within continents and countries. The values include some run-off which occurs during floods and is not available for use, thus they overestimate the potentially usable quantity of water.

It is important to distinguish between the quantity of water withdrawn from the hydrosphere and the quantity actually consumed. *Water consumption* is defined as the irretrievable loss of water after taking into account the water that is returned to the biosphere (Brown, 2003). This water is not consumed in the chemical sense (it still exists!) but it is no longer part of the renewable water resource. It remains part of the total inventory of water, however, and ultimately re-enters the water cycle. The global consumption of renewable water is shown in Table 10.2. This shows that in 1995, the year for which the data apply, around 3790 km^3 of water was withdrawn from the hydrosphere and around 2070 km^3 (61%) was consumed. By far the largest quantity of water is consumed in agriculture. Industrial use, which includes production of mineral and metal commodities, is a small consumer of water in relative terms. It is clear from Tables 10.1 and 10.2 that, globally, less than 10% of the renewable water resource is utilised each year and that therefore there is no global shortage of fresh water. Rather, the problems alluded to in the United Nations report 'Water in a changing world' seem to be due to the mismatch at country and regional levels of the demand for water for agriculture and the availability of water for other purposes.

Water availability is a complex problem which can be exacerbated or eased by international trade. Food grown in one region and transported to another consumes water in the region where it is grown but saves

Table 10.2: Global water withdrawal and consumption (km^3) in 1995

	Water withdrawal (km^3)	Water consumption (km^3)
Agricultural use	2504	1753
Municipal use	344	49.8
Industrial use	752	82.6
Reservoirs[a]	188	188
Total	*3788*	*2073*

a) Loss through evaporation
Source: Shiklomanov (2000).

water in the region where it is sent. Since the 1960s, many of the world's least developed countries have changed from being self-sufficient or net exporters of agricultural commodities, to net importers, even though most are not short of fresh water. Self-sufficiency in food production is not a requirement for individual countries as long as water resources remain adequate at the global level, though many countries, for strategic and nationalistic reasons, seek to be largely self-sufficient. Importation of agricultural products by countries which have scarce water resources is usually necessary but it is economically possible only if those countries have other tradable commodities to offset the cost. An example is the dry but oil-rich Middle East countries.

The *water footprint* of a region or country is the total volume of water used to produce the goods and services consumed by its inhabitants (Hails, 2008). It includes water withdrawn from rivers, lakes and aquifers that is used in agriculture and industry and for domestic purposes, as well as the water from rainfall that is used to grow crops. The water footprint is analogous to the ecological footprint discussed in Section 4.1.2. The latter is the total area of productive space required to produce the goods and services consumed by the population of a country or region, and the water footprint is the volume of water required to produce those goods and services. The total water footprint of a country or region is made up of two components.

1 *The internal water footprint.* This is the volume of water needed to grow and provide the goods and services which are produced and consumed within the country or region. It excludes the water needed to produce the goods and services which are exported.
2 *The external water footprint.* This is the volume of water used for the production of goods in the countries from which goods are imported.

Table 4.1 lists the internal, external and total water footprint for a selection of countries and for the entire world. In 2001, the world average water footprint was 1.24 million litres per person per year. The external water footprint accounted for 16% of this amount, but the actual percentage varies greatly within and between countries. A large number of countries have an external water footprint which makes up more than 50% of their total.

The impact of a water footprint depends on where and when water is extracted. Water used in an area where it is plentiful will probably not be harmful to the environment. In an area experiencing water shortages, the same level of water use could result in aquifers being depleted, rivers drying up, and loss of ecosystems (Hails, 2008). Externalising the water footprint may be an effective strategy for a country experiencing internal water shortages but it may also mean externalising environmental impacts. The trade in virtual (or embodied) water is influenced by global commodity markets and agricultural policies which tend to overlook the environmental, economic and social costs to the exporting countries (Hails, 2008). This points to the need for greater international cooperation on water resource management.

10.2 WATER IN THE MINERALS INDUSTRY

Water is essential to the minerals industry because its properties, together with its general availability, make it ideal for many functions. These include lubricating, cooling and agglomerating; for carrying particles; as a medium enabling particles to be acted on (particularly for grinding and separating); and as a medium or reactant for chemical reactions. The minerals industry consumes a relatively small quantity of water at national and global levels. For example, in Australia, which has a large mining industry sector and a low average rainfall, the minerals industry accounts for 2–3% of the total water consumed.

Overall, the direct consumption of water at a mine site, consisting of a mine and beneficiation plant, is often 0.4–1.0 m^3 per tonne of ore. Water consumption for a major mine may be up to 30 billion litres (or 30 million cubic metres) per year for many decades (Brown, 2003). Where a smelter or leach operation is also present, water consumption will be much higher. Water is an important issue at mine sites and companies maintain plans to manage and optimise their use of water (DITR, 2008).

Although mining operations usually consume a relatively small quantity of water at the national level, they can have a major impact on water use at local and regional levels if their use of water requires a large proportion of the renewable water resource for the area. In many cases, water used by a mine reduces the quantity of water potentially available for other, usually agricultural, purposes. This is particularly a problem in dry and arid regions. In other cases, mines use water that is either not required for agriculture or that cannot be used because it is too saline or otherwise unsuitable. When mining progresses below the upper level of the watertable, the pit (in an open cut mine) or mined section (in underground mines), creates a sink into which water flows. This water must be continually removed to enable mining to continue; this can alter the flow pattern of the groundwater through aquifers and affect water availability from bores elsewhere.

Contaminated water may be discharged from mine sites and enter the groundwater or run into streams. This can remain a problem long after a mine has ceased operation. The activity of mining itself uses little water but larger quantities are used during beneficiation and in chemical processing, particularly hydrometallurgical processing. Table 10.3 summarises some important uses of water in the minerals industry, the purity and relative quantities of water required, mechanisms by which water can be lost, and the potential impact of these waters on the environment. The problem of contaminated water on mine sites is discussed in detail in Chapters 11 and 12.

Water for mining operations can be obtained from various sources, including from a third party (local authority or private supplier), purpose-built dams to collect precipitation and run-off, rivers, natural lakes and groundwater (both sub-artesian and artesian). Water from these sources is referred to as *raw water*. *Potable water* is water of sufficient purity for human consumption. Water is lost from mining operations by seepage into the ground, evaporation and releases to the environment, either controlled or accidental. The flows of water around a typical mine site are shown in Figure 10.1. There is often considerable recycling of water within a mine site (see Section 12.2). The net consumption of water is the difference between the quantity of water entering the site and the quantity of water leaving the site. Therefore, in any given period (e.g. a month, a year):

Net consumption of water = *Precipitation* + *Groundwater flow* + *Run-off* − *Evaporation* − *Seepage* − *Discharge (to the environment)* − *Water in product*

10.3 THE EMBODIED WATER CONTENT OF METALS

The total quantity of water consumed in the production of a material, or its *embodied water content*, is the quantity of water consumed directly (by the types of activities listed in Table 10.3) and indirectly (e.g. in power generation) in the production of the material from resources in their natural state. The embodied water contents of a number of metal commodities are listed in Table 10.4. These estimates are based on published water consumption data for mining, beneficiation, leaching and smelting of particular commodities assuming electricity is generated using conventional coal-fired power plants which consume 1.76 m^3 water per MW h, a typical value for Australian power stations. Table 10.4 shows that gold production requires by far the largest water consumption (252 087 $m^3\,t^{-1}$ Au), followed by nickel produced hydrometallurgically from laterite ores. Gold requires a large quantity of water for both concentrating and leaching; nickel produced from laterite requires a large quantity of water for leaching but not for beneficiation since the ore is leached after minimal beneficiation. Steel requires the smallest quantity of water (2.9 $m^3\,t^{-1}$). Most of the water consumed is in the steel-making step, a large fraction for water-quenching of slag and steel billets. The metals with the highest quantities of indirect water consumption (in absolute and relative terms) are aluminium and titanium; this reflects the high electrical energy consumption in the smelting stages for these metals. Copper, nickel and zinc produced by electro-winning from leach solutions also have a high proportion of indirect water consumption because of the electrical energy consumed by electro-winning.

Table 10.3: Aspects of water usage in the production of mineral and metal commodities

	Mining	Beneficiation	Chemical processing	
			Aqueous processing (hydrometallurgical)	High-temperature processing (pyrometallurgical)
Uses of water	In drilling mud (for lubrication and cooling) For suppressing dust formation (by spraying on dusty areas) For equipment wash-down For human consumption	For transporting fine particles As a medium for grinding and separating particles of ore For human consumption	For transporting fine particles As a medium for chemical reactions For human consumption	For cooling gases, ingots, billets etc. For granulating slag For gas scrubbing to remove impurities For human consumption
Inventory volume	Small	Very large	Large	Small to moderate
Proportion recycled	High	Very high	High	Moderate
Water quality	Raw	Raw to potable	Raw to potable	Raw to potable
Mechanisms of losses	Seepage Evaporation	Entrainment in tailings Evaporation from stored tailings Seepage from stored tailings Spills and releases to the environment	Entrainment in leach residues Evaporation from stored residues Seepage from stored residues Spills and releases to the environment	Evaporation Spills and releases to the environment
Potential environmental impacts	Consumption of renewable water resources Drainage from mine and waste rock piles resulting in acidity and eco- and human toxicity	Consumption of renewable water resources Suspended solids in released water Drainage from tailings resulting in acidity and eco- and human toxicity	Consumption of renewable water resources Eco- and human toxicity due to acidity, heavy metals, cyanide etc. from spills and releases	Consumption of renewable water resources Eco- and human toxicity from spills and releases

Source: After Pulles *et al.* (1995); with modifications.

The consumption of water per tonne of ore for the metals listed in Table 10.4 (column 5) does not vary widely. The average value is 2.1 $m^3\,t^{-1}$ ore. This is two to five times greater than the range given by Brown (2003) for a typical mine with a beneficiation plant (0.4–1.0 $m^3\,t^{-1}$). However, the values in the table include chemical processing steps to produce a metal (which are water-intensive) and water used indirectly (particularly in power generation). The relative constancy of water consumed per tonne of ore irrespective of the nature of the ore and the metal being produced implies that water consumption is largely a function of ore grade; as grade decreases more ore must be treated to extract the same quantity of metal, therefore more water will be consumed. Figure 10.2 shows the data in the table plotted on logarithmic

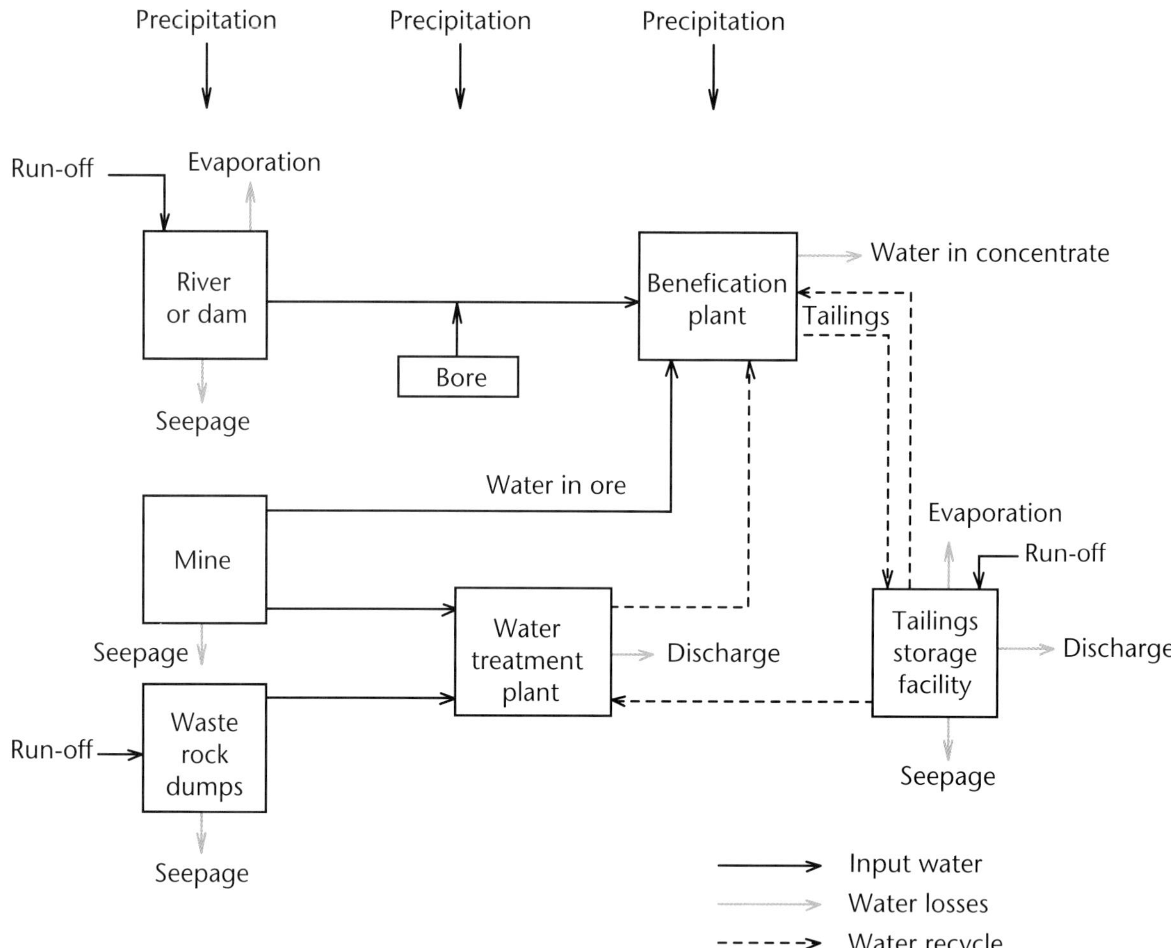

Figure 10.1: A simplified mine site water balance.

scales. It follows that water demand at a mine site will grow with time, even at constant output, as ore grade decreases.

10.4 WATER SUSTAINABILITY INDICATORS AND REPORTING

The Global Reporting Initiative has four indicators for water consumption against which companies are required to report. These are listed in Table 10.5. Two of these are mandatory (core) and two are optional (voluntary). Mudd (2009) found that, among 25 major mining companies which produced sustainability reports in 2007, the reporting of water was even more variable than for energy. Most companies reported only total water volumes, not the volumes by source. The reporting of the extent of water recycling was poor, with most companies not giving any data or other information. Only 16 companies reported water discharges, despite EN21 being a core indicator. The manner in which this was addressed varied – some gave volumes involved while others reported volumes and selected quality indicators (e.g. salinity or particular metals). Some companies reported on the health of water resources and ecosystems downstream from their operations, ranging from qualitative statements to water quality data. Mudd concluded that 'given that impacts on water resources and ecosystems is often widespread and (of) fundamental concern for local communities and interested stakeholder groups, there

Table 10.4: Water consumed (direct plus indirect) in the production of some common metals using conventional, state-of-the-art technologies and black coal generated electrical energy

	Ore	Process	Water consumption (m^3)		
			per tonne metal	per tonne ore	% Indirect
Steel	Oxide, 44 wt% Fe	Mining – beneficiation – blast furnace – basic oxygen furnace	2.9	1.8	17
Aluminium	Bauxite,17.4 wt% Al	Mining – Bayer process – electro-winning (Hall-Héroult)	35.9	6.2	77
Copper	Sulfide, 3 wt% Cu	Mining – comminution – flotation – smelting – converting – electro-refining	25.9	0.7	13
	Sulfide, 2 wt% Cu	Mining – heap leaching – SX – electro-winning	38.0	0.5	21
Nickel	Sulfide, 2,3 wt% Ni	Mining – comminution – flotation – flash smelting – Sherritt-Gordon refining	79.0	1.4	11
	Laterite, 1 wt% Ni	Mining – pressure acid leaching – SX – electro-winning	376.6	3.5	2
Zinc	Sulfide, 8.6 wt% Zn, 5.5 wt% Pb	Mining – comminution – flotation – sintering – blast furnace (ISF)	21.2	1.5	9
		Mining – comminution – flotation – roasting – leaching – electro-winning	26.3	1.8	29
Lead	Sulfide, 8.6 wt% Zn, 5.5 wt% Pb	Mining – comminution – flotation – sintering – lead blast furnace	12.6	0.5	14
		Mining – comminution – flotation – sintering – blast furnace (ISF)	21.7	0.9	9
Titanium	Mineral sands, 9.8 wt% Ti	Dredging – beneficiation – Becher process – chlorination – Kroll process	110	5.4	38
Gold	Gold ore, 3.6 g Au t^{-1}	Mining – comminution – flotation – CIL cyanidation – electro-winning	252 087	0.8	3

Source: Norgate and Lovel (2006); with additional process information.

Table 10.5: GRI sustainability indicators for water usage

EN8	Total water withdrawal by source	Core
EN9	Water sources significantly affected by withdrawal of water	Voluntary
EN10	Percentage and total volume of water recycled and reused	Voluntary
EN21	Total water discharge by quality and destination	Core

Source: GRI (2006).

is clear room for improvement on this area in sustainability reporting across the mining industry'.

10.5 REFERENCES

Brown ET (2003) Water for a sustainable mining industry – a review. In *Water in Mining 2003*. pp. 3–13. Australasian Institute of Mining and Metallurgy: Melbourne.

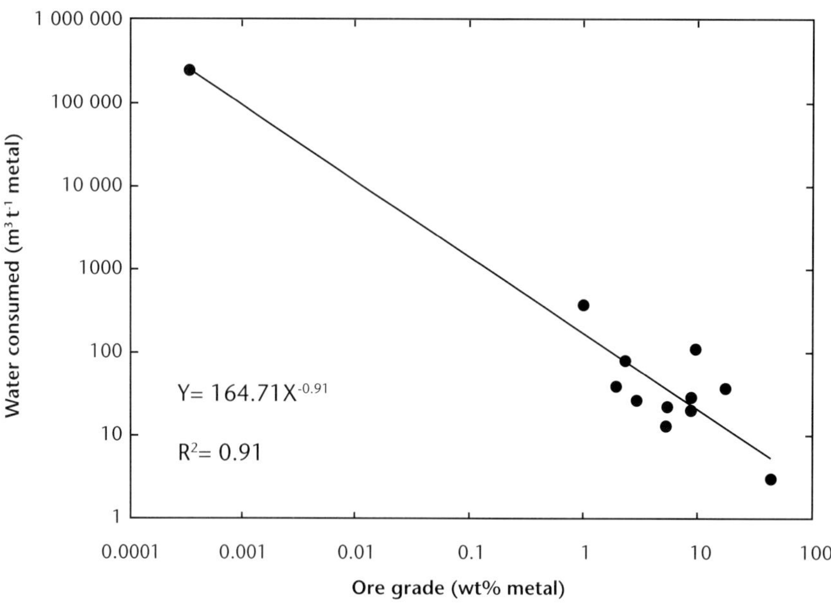

Figure 10.2: Water consumed (direct plus indirect) in the production of some metal commodities as function of ore grade.

DITR (2008) *Water Management.* Leading Practice Sustainable Development Program for the Mining Industry. Department of Industry, Tourism and Resources, Commonwealth of Australia: Canberra. <http://www.ret.gov.au/resources/resources_programs/lpsdp/>.

GRI (2006) *Sustainability Reporting Guidelines.* 3rd edn. Global Reporting Initiative. <http://www.globalreporting.org>.

Hails C (Ed.) (2008) 'Living planet report 2008'. WWF International: Gland, Switzerland. <http://assets.pnda.org/downloads/living_planet_report_2008.pdf>.

Mudd GM (2009) Sustainability reporting and mining – an assessment of the state of play for environmental indicators. In *Sustainable Development Indicators in the Minerals Industry.* pp. 377–391. Australasian Institute of Mining and Metallurgy: Melbourne.

Norgate TE and Lovel RR (2006) Sustainable water use in minerals and metal production. In *Water in Mining 2006.* pp. 331–339. Australasian Institute of Mining and Metallurgy: Melbourne.

Pulles W, Howie D, Otto D and Easton JA (1995) 'Manual on mine water treatment and management practices in South Africa'. Report No. TT80/95, June 1995. Water Research Commission: Pretoria, South Africa.

Shiklomanov IA (2000) Appraisal and assessment of world water resources. *Water International* **25**(1): 11–32.

United Nations (2009) 'World water development report 3: Water in a changing world'. Earthscan: London. <http://www.unesco.org/water/wwap/wwdr/wwdr3/pdf/WWDR3_Water_in_a_Changing_World.pdf>.

10.6 USEFUL SOURCES OF INFORMATION

DITR (2008) *Water Management.* Leading Practice Sustainable Development Program for the Mining Industry. Department of Industry, Tourism and Resources, Commonwealth of Australia. <http://www.ret.gov.au/resources/resources_programs/lpsdp>.

United Nations (2009) 'World water development report 3: Water in a changing world'. Earthscan: London. <http://www.unesco.org/water/wwap/wwdr/wwdr3/pdf/WWDR3_Water_in_a_Changing_World.pdf>.

Water in Mining 2003 (2003) Australasian Institute of Mining and Metallurgy: Melbourne.

Water in Mining 2006 (2006) Australasian Institute of Mining and Metallurgy: Melbourne.

Water in Mining 2009 (2009) Australasian Institute of Mining and Metallurgy: Melbourne.

World Economic Forum (2009) 'Thirsty energy – water and energy in the 21st century'. <http://www.weforum.org/en/ip/energy/Publications/index.htm>.

11 Wastes from primary production

Mining and processing of ores produces huge quantities of wastes of many forms. The handling, storage or disposal of these is a major logistical exercise for mining and processing operations. It is often the area of greatest concern for the local community and regulators because of potential impact of the materials on the environment and human health. The issues associated with mining wastes have long been recognised. In 1556, Georgius Agricola, writing about detractors of mining, stated, 'the strongest argument of the detractors (of mining) is that the fields are devastated by mining operations ... Further, when the ores are washed, the water which has been used poisons the brooks and streams, and either destroys the fish or drives them away'. Mining waste remains one of the major issues facing the minerals industry. This chapter is concerned with the nature of the wastes arising from the primary production of mineral and metal commodities. Of particular interest are the types and quantities produced, where along the value chain they are produced, and the impact they can have on humans and the environment. Chapter 12 examines how these wastes are managed to reduce human and environmental harm. Chapter 16 considers how waste production can be reduced or even eliminated, to prevent many of the potential problems.

11.1 WASTES AND THEIR ORIGIN

Wastes are substances produced as part of the materials cycle (see Figure 2.2) for which there are no present uses. They are usually discarded by discharging to the environment or stored in purpose-built facilities. Figure 2.3 shows the major types of wastes produced in the materials cycle. The largest quantities are produced during the mining, beneficiation and chemical processing stages. Mineral deposits usually occur at depth in the Earth's crust and waste rock must be removed to gain access. Mined material usually has to be beneficiated to remove gangue minerals, to obtain a product which can be used in its own right or further processed. Chemical processing is often necessary to produce a material of required form and composition. The formation of wastes is therefore inherent in the production of mineral and metal commodities using present technologies and under current economic paradigms which place little or no value on most of the material extracted during a mining operation. The concept of waste is dynamic; what was considered a waste in the past may now be considered a product, and what is a waste now may become a raw material in the future. For example, although slags are still usually referred to as wastes, most iron blast furnace slag produced today are used in construction products. Hence the term co-product, rather than waste, is often preferred.

Wastes are produced directly as part of the value-adding processes and indirectly through the use of an input to the value-adding processes. The production of waste rock during mining and of tailings during beneficiation are examples of direct wastes. Wastes produced during the manufacture of explosives and the generation of electricity for use in mining and processing are examples of indirect wastes. Table 11.1 lists some of the major types of direct and indirect wastes produced in the value-adding stages of the materials cycle: mining, beneficiation, chemical transformation, fabrication and recycling. These are generic types of wastes. Not all will be produced during the production of a particu-

Table 11.1: Common types of wastes produced directly and indirectly as a result of mining, beneficiation, processing, manufacturing and recycling

Mining	Direct	*Solids*	Waste rock; dust
		Liquids	Acid mine drainage (from mine and waste rock heaps)
		Gas	CO_2 from fuel consumption
	Indirect	*Solids*	Chemical wastes from explosives manufacture; fly ash from electrical power generation
		Gases	CO_2 from electrical power generation; CO_2 from explosives manufacture
Beneficiation	Direct	*Solids*	Tailings
		Liquids	Acid rock drainage (from tailings)
	Indirect	*Solids*	Chemical wastes from reagent manufacture; fly ash from electrical power generation
		Gases	CO_2 from electrical power generation; CO_2 from reagent manufacture
Chemical transformation	Direct	*Solids*	Leach residues; electro-refining residues; slag; fumes from furnaces; spent furnace refractories
		Liquids	Acid and alkaline solutions from leach residues; cyanide-containing solutions (from gold extraction)
		Gases	CO_2, SO_2, NOx etc. from smelting
	Indirect	*Solids*	Fly ash from electrical power generation
		Gases	CO_2 from electrical power generation
Fabrication and manufacture	Direct	*Solids*	In-plant scrap; dross; mill scale
		Liquids	Pickling liquids, electro-plating liquids
		Gases	CO_2 from furnaces
	Indirect	*Solids*	Fly ash from electrical power generation
		Gases	CO_2 from electrical power generation
Recycling	Direct	*Solids*	Numerous types of wastes produced during recycling of scrap metals, electronic scrap, consumer goods etc. (e.g. electric arc furnace dusts, dross from aluminium recycling)
		Liquids	Waste process solutions containing metals
		Gases	CO_2 NOx etc. from furnaces
	Indirect	*Solids*	Chemical wastes from reagent manufacture; fly ash from electrical power generation
		Gases	CO_2 from electrical power generation; CO_2 from reagent manufacture

lar mineral or metal commodity, or by a particular processing technology. Many other wastes not listed are also produced, often in quite small quantities, in specific processes.

This chapter is concerned only with wastes produced directly in the production of mineral and metal commodities. The total quantities of these are usually much larger than the quantities of valuable products. This is illustrated in Figure 2.4, which shows average global values for some common commodities. Solids make up the largest quantity of primary production waste materials. The quantities can range from several times the mass of valuable element for abundant elements (e.g. iron and aluminium ores) to many thousands, and even millions, of times for some scarce elements. The quantity of directly produced wastes decreases along the value chain from mining to manufacturing to recycling. The largest quantities of solid and liquid wastes are produced during mining and beneficiation; the major quantities of gaseous wastes are produced during high-temperature chemical processing (particularly smelting of metals and cement manufacture). Wastes from mining and beneficiation have the largest potential environmental impact on land and water, and chemical processing wastes have the largest potential impact on the atmosphere. Some valuable final product is lost at each stage of value-adding and forms part of the waste from that stage. These losses are due to technological limitations, such as the need to leave some ore in mines as support material, incomplete liberation and separation during beneficiation, and bound or inaccessible minerals not amenable to leaching.

11.2 SOLID WASTES

The major solid wastes produced during the primary production of common mineral and metal commodities, and representative composition ranges, are listed in Table 11.2. The types of wastes, and their compositions, vary widely from commodity to commodity, from ore body to ore body and from process to process. Solid wastes are predominantly inorganic in nature, reflecting their mineralogical origin, though some contain residual quantities of reagents added during processing steps. This is the case for tailings produced by flotation, which contain residual quantities of collectors and frothers, and tailings that have been thickened, which contain residual quantities of flocculants. Leach residues contain residual quantities of reagents added to dissolve the valuable mineral; for example, red mud contains considerable quantities of sodium hydroxide, gold leach residues contain sodium cyanide, and copper and zinc leach residues contain sulfuric acid. Slags from smelting processes contain large quantities of inorganic fluxes added during processing to modify the slag properties. Many solid wastes have to be treated and stabilised before they are disposed of or stored.

11.2.1 Calculation of the quantities of solid wastes

Some simple relationships can be developed to estimate the quantities of direct wastes produced along the value-adding chain in Figure 9.1. These consist of the unwanted substances (gangue) in the ore, the part of the valuable minerals or substances that is lost, and the solid products of any reagents or other consumables added during processing. Initially, we will do this on the basis of 1 tonne of ore in the ground (i.e. unmined ore). The equations, and their derivation, are given in Box 11.1. The quantity of solid wastes per tonne of mined ore (on a dry basis) is given by Equation 11.14 which is the sum Equations 11.4, 11.7, 11.10 and 11.13. This assumes all direct wastes from chemical processing are solids. A more useful basis for comparison purposes is the quantity of waste per tonne of 'value' in the final product rather than per tonne of ore in the ground. This is obtained by dividing Equation 11.14 by the mass of 'value' in the final product to yield Equation 11.15.

Equations 11.4, 11.7, 11.10 and 11.13 give the mass of solid wastes on a dry basis. In reality, waste rock, tailings and leach residues contain significant quantities of water. Furthermore, in pyrometallurgical processes, some elements, particularly oxygen and sulfur, are removed as gases (usually CO_2 and SO_2). In these cases, Equations 11.10 and 11.13 overestimate the mass of solid waste. Also, during chemical processing, reagents are added and their reaction products can form part of the solid waste. Equations 11.10 and 11.13 do not take this into account.

Box 11.1: Estimation of the quantities of wastes produced during the production of mineral and metal commodities

Mining stage. Let the waste rock:ore ratio be W, the grade of the ore body (as a fraction) be G_O and the fractional recovery be R_M. On the basis of 1 tonne of unmined ore:

$$\textit{Mass of 'value' in unmined ore} = G_O \qquad 11.1$$

$$\textit{Mass of ore recovered} = R_M \qquad 11.2$$

$$\textit{Mass of 'value' in the mined ore} = G_O R_m \qquad 11.3$$

$$\textit{Mass of waste from mining} = W + (1 - R_M) \qquad 11.4$$

Physical beneficiation stage. Let the grade (fractional) of the beneficiated product be G_B and the fractional recovery be R_B. Then:

$$\textit{Mass of 'value' in beneficiated product} = G_O R_M R_B \qquad 11.5$$

$$\textit{Mass of product} = \frac{G_O R_M R_B}{G_B} \qquad 11.6$$

and

$$\textit{Mass of waste from beneficiation} = R_M - \frac{G_O R_M R_B}{G_B} \qquad 11.7$$

Chemical processing, stage I. Let the grade (fractional) of the beneficiated product be G_{CI} and the fractional recovery be R_{CI}. Then:

$$\textit{Mass of 'value' in product} = G_O R_M R_B R_{CI} \qquad 11.8$$

$$\textit{Mass of product} = \frac{G_O R_M R_B R_{CI}}{G_{CI}} \qquad 11.9$$

and

$$\textit{Mass of waste} = \frac{G_O R_M R_B}{G_B} - \frac{G_O R_M R_B R_{CI}}{G_{CI}} \qquad 11.10$$

Chemical processing, stage II. Let the grade (fractional) of the beneficiated product be G_{CII} and the fractional recovery be R_{CII}. Then:

$$\textit{Mass of 'value' in product} = G_O R_M R_B R_{CI} R_{CII} \qquad 11.11$$

$$\textit{Mass of product} = \frac{G_O R_M R_B R_{CI} R_{CII}}{G_{CII}} \qquad 11.12$$

and

$$\textit{Mass of waste} = \frac{G_O R_M R_B R_{CI}}{G_{CI}} - \frac{G_O R_M R_B R_{CI} R_{CII}}{G_{CII}} \qquad 11.13$$

The total mass of solid waste (on a dry basis) per tonne of unmined ore, assuming all wastes from chemical processing are solids, is obtained by adding the mass of wastes from each stage (Equations 11.4, 11.7, 11.10 and 11.13). This results in the equation:

$$Total\ mass\ of\ waste = W + 1 - \frac{G_O R_M R_B R_{CI} R_{CII}}{G_{CII}} \tag{11.14}$$

The total mass of wastes per tonne of final product is obtained by dividing the total mass of solid wastes by the mass of the final product:

$$Total\ mass\ of\ waste\ (per\ tonne\ of\ product) = \frac{(W+1)\,G_{CII}}{G_O R_M R_B R_{CI} R_{CII}} - 1 \tag{11.15}$$

The overall recovery is given by the mass of 'value' in the final product divided by the mass of 'value' in 1 tonne of unmined ore:

$$Overall\ recovery\ efficiency = R_M R_B R_{CI} R_{CII} \tag{11.16}$$

Table 11.2: Examples of solid wastes produced during the production of some common mineral and metal commodities, and representative compositions. With the exception of the zinc leach residues, the species listed express composition only and do not indicate the mineralogical species present in the waste material

Commodity	Typical composition (wt%)
Steel (produced by blast furnace reduction of iron ore followed by BOF steel-making)	
– Blast furnace slag	SiO_2 35, CaO 40, Al_2O_3 11, MgO 8, S 1.2, Fe 0.4, MnO 0.5 (CaO/SiO_2 = 1–1.5)
– BOF slag	SiO_2 12–18, CaO 45–50, Al_2O_3 1–4, MgO 1–4, Fe 14–19, P_2O_5 2.5 (CaO/SiO_2 = 2.8–4.4)
– BOF fumes[a]	Fe 50–80 (as metallic and oxides), Mn 1–3, SiO_2 1–3, CaO + MgO 2–8, P_2O_5 0.3–1.0, Zn 1–4 (particle size: 0.05–100 μm)
Aluminium (Bayer plus Hall-Héroult processes)	
– Red mud[b]	Fe_2O_3 30–60, SiO_2 3–50, Al_2O_3 10–20, Na_2O 2–10, CaO 2–8, TiO_2 ≤10
– Spent pot linings[c]	C 20–60, F 10–15, Al 3–15, Na 3–15, Ca 1–2, Si 1–2, cyanide 0.1–0.3, small quantities of S, Fe and other elements
Copper (produced by smelting sulfide concentrate followed by converting and electro-refining)	
– Waste rock[d]	Cu 0.1–0.4, Fe 3–7, S 1–40, SiO_2 28–35, Al_2O_3 11, MgO 1–2, CaO 3–5, small quantities of heavy and other metals (Cu, Ni, Zn, Pb, Cd, Fe, V, Mn, In, Ge, Se, Te, Mo, As, Sb etc.)
– Flotation tailings[d]	Cu 0.03–2.0, Fe 0.1–74, S 0.3–11, Al_2O_3 1–1.5, MgO 0.5–4.5, CaO 0.3–8.5, plus small quantities of Ni, Zn, Pb, Cd, Fe, V, Mn, In, Ge, Se, Te, Mo, As, Sb etc.
– Final (discard) slag	Cu 0.3–0.7, Fe (as oxide) 37–44, SiO_2 28–33
– Smelter dusts and fumes[d]	Cu 0.5–27, Fe 4–29, S 7–12, As 1–13, Pb 0.3–45, Mo 0.6–1.6, Zn 1.5–12, Sb 0.02–0.4, Bi 0.04–2, Cd 0.2–0.8
Zinc (produced by roasting sulfide concentrate – leaching – electro-winning)	
– Leach residues[e]	Jarosite ($R_2Fe_6(OH)_{12}(SO_4)_4$, where R is any of K^+, NH_4^+, Na^+, Ag^+ or R_2 is Pb^{2+}) with Fe ~29, Zn ~3.5, Pb ~1.9; or goethite (FeO.OH), with Fe ~34, Zn ~13, Pb ~2.2
Titania, TiO_2 (produced from ilmenite from beach sands by the Becher process)	
– Becher process residue[f]	Fe 65, Ti 2.3, Si 0.65, Mn 0.28, Al 0.14, V 0.03, Cr 0.01

a) Geiger *et al.* (1982).
b) Hill and Sehnke (2006).
c) Marvis (1995).
d) Broadhurst (2006).
e) Sinclair (2005).
f) T.E. Norgate, CSIRO internal report, July 2008.

11.2.2 Quantities produced

Three examples will demonstrate the application of the above equations. These should be considered representative only – the situation for each mining and processing operation is unique, depending on the nature of the deposit and the mining and processing technologies used.

- *Gold.* Assume the ore is extracted from a deep underground mine and is crushed and ground then leached with cyanide. The gold is recovered from the leach solution using carbon-in-pulp, and finally electro-won.
- *Copper.* Assume the ore is obtained from an open cut mine. In one case the ore is beneficiated to produce a copper concentrate which is then leached in tanks, in another the ore is crushed and screened then heap leached. Assume in both cases the leach solution is purified and concentrated using solvent extraction, then the copper is recovered by electro-winning to produce cathode-grade copper.
- *Aluminium.* Assume the bauxite ore is mined from placer deposits then treated using the Bayer process to produce alumina, which is then smelted in an electrolytic cell to produce aluminium.

The masses of wastes produced, calculated using the above equations, are summarised in Table 11.3. Typical values were assumed for the grade of ore, waste rock:ore ratio, recovery efficiencies and product grade. The overall recoveries are also given. The values demonstrate the huge quantities of solid wastes produced directly during the production of metals. In the case of gold, which occurs at extremely low concentrations in the crust, about 1 million tonnes of material has to be mined to produce 1 tonne of gold. Even in the case of aluminium, which is a geochemically abundant element, about 6 tonnes of solid waste are created.

Table 11.3 shows that in each case the greatest quantity of waste is created by the mining operation. Beneficiation to produce a concentrate produces the next largest quantity, in cases where a considerable degree of concentration of the 'value' is achieved. In heap leaching of copper ores and leaching of bauxite, only a small quantity of concentration is achieved during beneficiation, which for these involves simply crushing and screening. A relatively small quantity of slimes is produced from residual elements in the electro-winning solution in the case of copper electrowinning, but this is not waste as it usually contains a high concentration of valuable minor elements including silver, gold, tellurium, selenium and others. An interesting observation is that, in copper production, the quantity of waste produced, and therefore the quantity of material mined, is less for the heap leaching option than for the vat leaching option. This is because the overall recovery is higher (72% compared with 64%), which allows less material to be mined to produce the same quantity of copper. The importance of recovery efficiency is illustrated further in Table 11.4, which presents several recovery scenarios for aluminium. Improving the efficiency of recovery along the value-adding chain produces less waste and requires less material to be mined, handled and processed to produce a given quantity of product. This results in lower energy and reagent consumption. The benefits are both environmental and financial.

11.3 LIQUID WASTES

As discussed in Chapter 10, large quantities of water are used in the production of mineral and metal commodities, a large proportion of it used directly. Direct loss of water occurs by evaporation, seepage and releases to the environment, both intentional and unintentional (Figure 10.1). Of interest is contaminated water released to the environment, either intentionally or unintentionally – this constitutes a waste when the water contains impurities beyond limits that are safe for the environment or human consumption. Contaminated waste water is often referred to as *effluent*.

11.3.1 Waste water

Table 11.5 lists the types of water produced on a mine site and their associated quality, and Table 11.6 lists the types of water required at a mine site. Water has to be managed carefully to minimise, and ideally prevent, the release of contaminated water to the environment, and to return excess water to the environment only after it has been treated to local

Table 11.3: Quantities of solid waste produced directly during the production of 1 tonne of gold, copper and aluminium (indicative values only)

	Gold	Copper (vat leaching)	Copper (heap leaching)	Aluminium
Assumptions				
Ore grade	2 ppm Au	1.0 wt% Cu	1.0 wt% Cu	35 wt% Al
Waste rock:ore ratio	0.5	3	3	1.0
Recovery – mining	90%	90%	90%	90%
Recovery – beneficiation	100%	85%	95%	90%
Grade of beneficiated product	2 ppm	25 wt%	1.2 wt%	40 wt%
Recovery – leaching and solution purification	85%	85%	85%	90%
Grade of product – leaching and solution purification (dry basis)	100 wt%	94 wt%	94 wt%	53 wt%
Recovery – electro-winning	100%	99%	99%	100%
Grade of product – electro-winning	100 wt%	99.5 wt%	99.5 wt%	100 wt%
Calculated values				
Solid wastes (tonnes per tonne of metal produced)				
– Rock waste and ore left unmined	0.39×10^6	479	429	4.31
– Beneficiation tailings (dry basis)	–	134	25.9	0.75
– Leaching and solution purification residues (dry basis)	0.59×10^6	3.7	97.5	0.89
– Electro-winning residues/slimes	–	0.1	0.1	–
Total waste (tonnes per tonne product)	0.98×10^6	617	552	5.95
Overall recovery	77%	64%	72%	73%

legislated standards of purity. The major sources of contaminated water released to the environment are seepage from tailings dams, release or spillage from tailings dams, seepage from waste rock heaps and leach residue heaps, and seepage from mines. The relative importance of these sources varies from mine to mine and according to the technology used (e.g. the methods used to prevent seepage from heaps and dams), and varies during the life of a mine. Many sources of release of contaminated water continue, and may become more severe, after a mine has ceased operation. Water from tailings dams contains suspended mineral matter and dissolved matter, either residual chemicals that were used during processing or reaction products, including dissolved metals. Of particular concern are residual flotation reagents from beneficiation plants, sulfuric acid from oxide leaching operations such as nickel laterite, sodium cyanide from gold leaching operations and sodium hydroxide from bauxite leaching. Acid and metalliferous drainage (AMD), also called acid mine drainage and acid rock drainage (ARD), is the most ubiquitous liquid waste produced in mining operations. It is discussed below.

11.3.2 Acid and metalliferous drainage

The term acid and metalliferous drainage is preferred to acid mine drainage and acid rock drainage, though those are in common use, since it recognises that not all the problems associated with AMD are due to its

Table 11.4: The effect of efficiency of recovery on the quantity of waste and mined material in the production of aluminium

	Case 1	Case 2	Case 3
Assumptions			
Ore grade	35 wt%	35 wt%	35 wt%
Waste rock:ore ratio	1.0	1.0	1.0
Grade of beneficiated product	40 wt%	40 wt%	40 wt%
Grade of product – Bayer process	53 wt%	53 wt%	53 wt%
Grade of product – electro-winning	100 wt%	100 wt%	100 wt%
Recovery – mining	85%	90%	95%
Recovery – beneficiation	85%	90%	95%
Recovery – Bayer process	80%	90%	95%
Recovery – electro-winning	100%	100%	100%
Calculated values			
Solid wastes produced (tonnes per tonne Al)			
– Rock waste and ore left unmined	5.68	4.31	3.50
– Beneficiation (tailings)	1.08	0.75	0.53
– Red mud	1.24	0.89	0.74
– Electro-winning	–	–	–
Total waste (tonnes per tonne Al)	8.00	5.95	4.77
Overall recovery	58%	73%	86%

Table 11.5: Types of water produced on a mine site and associated quality

Type of water	Quality
Groundwater	Salinity varies from low to very saline (above seawater salinity); pH is variable; may contain heavy metals of natural origin
Run-off water	Slightly acidic pH; low metal content; usually high suspended solids due to erosion
Mine water	High salinity and likelihood of metal contamination
Acid metalliferous drainage	Very low pH; high metal content; sulfates; clear
Process water	Generally the most contaminated; contains heavy metals; other salts; process chemicals such as flotation reagents, frothers, SX reagents
Rainwater	Slightly acidic
Evaporation	
Sewage	Contains pathogens and elevated contents of nitrogen, phosphorus and organics

Source: Lévy *et al.* (2006); with modifications.

Table 11.6: Types of water used on a mine site and associated quality requirements

Type of water	Quality requirements
Process water	Needs to comply with process operational requirements
Human consumption	Drinking water quality (potable)
Dust suppression	Needs to comply with health and safety requirements
Equipment washing	Needs to comply with health and safety requirements
Discharged water	Needs to comply with environmental discharge regulations
Cooling towers	Limited by scaling potential, low dissolved and suspended solids

Source: Lévy *et al.* (2006); with modifications.

acidic nature (DITR, 2007). AMD forms wherever iron sulfide minerals, particularly pyrite (FeS_2), which is geologically abundant and occurs in many types of deposits, is exposed to air and water, resulting in the formation of sulfuric acid and iron hydroxide:[23]

$$[FeS_2] + 3.75\{O_2\} + 3.5H_2O = [Fe(OH)_3] + 2(H_2SO_4)$$

AMD forms wherever pyrite, oxygen and water co-exist and its formation is aided by the action of naturally occurring bacteria. Reaction is particularly rapid when the exposed mineral is present in small particles rather than in massive form, because there is then a much larger surface area of mineral available for contact and reaction with oxygen and water (Section 5.7.3). AMD can form in waste rock dumps, ore stockpiles, heap leach residue dumps, tailings storage facilities, the pits of open cut mines, and in mined areas of underground mines where air is accessible.

As the acidified water migrates through a mine site, waste rock pile or tailings dam it reacts with other minerals and may dissolve a range of metals and salts. The acid is partly neutralised as a result but this is at the expense of increased toxic element concentrations. The resulting solution can have devastating environmental effects (Section 11.5.1). In some situations, the AMD is neutralised by reaction with carbonate rocks (usually limestone and dolomite) it comes in contact with:

$$(H_2SO_4) + [CaCO_3] = [CaSO_4] + \{CO_2\} + H_2O$$
$$(H_2SO_4) + [MgCO_3] = (MgSO_4) + \{CO_2\} + H_2O$$

$CaSO_4$ is virtually insoluble in water and precipitates as a solid whereas $MgSO_4$ is soluble. Hence, AMD neutralised by calcium carbonates has low sulfate concentrations whereas AMD neutralised by magnesium carbonates has high sulfate concentrations and is more of a potential environmental problem.

AMD forms not only from the mining of sulfide ores but from any deposit that contains pyrite. It is particularly common in coal mining and can occur in mines from which oxide ores are extracted, for example iron ore mines. AMD is acidic, with a pH typically in the range 1.5–4, has a high acidity[24] (50–15 000 mg L^{-1} $CaCO_3$ equivalent), a high sulfate content (typically 500–10 000 mg L^{-1} of SO_4^{2-}), often a high content of dissolved metals (e.g. Fe, Al, Mn, Cd, Cu, Pb, Zn and As) and low content of dissolved oxygen (usually <6 mg L^{-1}). Its presence is often indicated by red-coloured and unnaturally clear water, orange-brown precipitates in drainage lines, the death of fish and other aquatic animals, precipitate formation on mixing with other water at stream junctions, poor growth of revegetated areas such as waste rock dumps, vegetation dieback, and corrosion of concrete and steel structures (DITR, 2007).

23 The conventions adopted in this and subsequent chemical equations to indicate the nature of the phases involved are summarised in Table 8.2.

24 Acidity is a measure of the capacity of a solution to neutralise a strong base, such as $CaCO_3$. It is determined analytically by titration. The acidity of a solution generally increases as its pH decreases but solutions with similar pH values may have very different values of acidity.

11.4 GASEOUS WASTES

Most gaseous waste products are produced during high-temperature operations such as drying, roasting, sintering, smelting and refining. When they are released into the atmosphere, either intentionally or unintentionally, gaseous wastes are often called *emissions*.

11.4.1 The types of gases produced in smelting

Table 11.7 lists the common major and minor gases, and the types of particulate matter often suspended in them, produced during the production of some commonly manufactured mineral and metal commodities. When fuel (oil, natural gas or coal) is burned to produce heat, CO_2 and H_2O are the major gaseous products. When carbon (coal or coke) is used as a reductant to reduce oxides, large quantities of CO_2, and sometimes CO, are produced. When sulfur-containing materials (particularly sulfide concentrates) are roasted or smelted, sulfur dioxide (SO_2) is a major gaseous product. Small quantities of sulfur trioxide (SO_3) may also be produced. Collectively, the oxides of sulfur are referred to as SOx (pronounced 'sox'). In non-oxidising processes, elemental sulfur may be volatilised as gaseous S_2, which condenses to solid sulfur as the gases cool.

The gas from high-temperature reactors is almost always a mixture of several gases, including nitrogen if air has been input. Nitrogen, being relatively inert, passes through the reactor largely unreacted. Nitrogen is not considered a waste product since it is an input which simply flows through and out of the process, returning to the atmosphere. However, it carries heat as it leaves the reactor and imposes a thermal load on the furnace. Almost invariably the gas contains entrained, fine solid or molten particles of matter arising from the solid or molten materials in the reactor, or formed by reactions within the reactor. The gas often contains small quantities of other, potentially harmful, gases in addition to N_2, CO_2, H_2O and sometimes SO_2, which make up the bulk of most gas streams. A small quantity of nitrogen in the air introduced into a reactor may react at high temperatures to form gaseous nitrogen oxides (NO, NO_2 and N_2O), collectively called NOx (pronounced 'nox'), which if released contribute to acid rain and eutrophication (see Section 3.4). Dioxins (polychlorinated dibenzene-para-dioxins, or PCDDs), furans (polychlorinated dibenzofurans, or PCDFs) and polychlorinated biphenyls (PCBs) can form in any oxidation process (e.g. combustion or incineration) where chlorine or fluorine are present. Sintering of iron ores (an oxidising process) is a major source of dioxin production in the minerals industry. Chlorine is often a minor component of the coal used to make coke for the blast furnace (a reducing environment) and the fines formed comprise a constituent of the feed to the sinter plant (an oxidising environment). Polycyclic aromatic hydrocarbons (PAHs) can be formed during pyrolysis of coal, for example, during coke-making for the iron blast furnace and in making anodes for aluminium smelting. PCDDs, PCDFs, PCBs and PAHs are highly toxic substances (see Section 11.5.2). Perfluorocarbons, mainly CF_4 and C_2F_6, form from reactions around the anode during aluminium smelting, particularly as a result of the so-called anode effect during which the electrolyte, which contains high levels of fluorine, becomes temporarily depleted of alumina due to poor feeding practices. Perfluorocarbons are extremely powerful greenhouse gases with very long residence time in the atmosphere.

The gas produced in reduction processes using carbon-based reductants contains combustible compounds, particularly CO, and sometimes H_2. These gases are flared (burned off) to make them safe to release to the atmosphere, are used as fuel in another process on the site, or are burned to generate electricity (co-generation). The off-gas from iron blast furnaces, for example, contains about 25 vol% CO and 1 vol% H_2, in addition to N_2, CO_2 and H_2O, and has value as a fuel. It is combusted on site in 'stoves' to preheat the blast furnace air and to fire the coke ovens. When used in some useful way, these gases are by definition not wastes. Only gas that is finally released to the atmosphere is waste.

11.4.2 The quantities of gas produced in smelting

The quantity of gas produced in smelting operations can be very large, thus very large ducts and other gas handling facilities are required. The gas handling and

Table 11.7: Typical compositions of some gas streams produced during production of steel, aluminium, copper, zinc, titanium dioxide and cement

Commodity	Typical chemical composition: Major gases (vol%)	Minor gases	Particulate matter
Steel (produced by blast furnace reduction of iron ore followed by BOF steel-making)			
Blast furnace gas	11% CO_2, 60% N_2, 27% CO, 1% H_2		Iron ore and sinter dust, coke breeze, limestone dust
Sinter plant gas	CO_2, N_2, H_2O	NOx, dioxins	Iron ore and sinter dust, coke breeze, limestone dust
Coke oven gas	1.5% CO_2, 4% N_2, 6% CO, 55% H_2, 28% CH_4, balance other hydrocarbons		Tars and light oil
BOF gas	CO, CO_2	NOx, SOx	Metallic iron and iron oxide fume, slag droplets, calcined limestone dust
Soaking furnace gas	N_2, CO_2, H_2O		
Aluminium (Bayer plus Hall-Héroult processes)			
Alumina calcination	N_2, CO_2, H_2O		Alumina dust
Anode baking furnaces	N_2, CO, CO_2, H_2O	SOx, polycyclic aromatic hydrocarbons (PAH), tars	Coke dust
Pot room	CO_2, CO	HF, SOx, CS_2, H_2S, perflurocarbons, hydrocarbons, PAH	Carbon dust, alumina dust, cryolite, Al fluoride
Copper (produced by smelting sulfide concentrate followed by converting and electro-refining)			
Matte smelting	SO_2 (3–30%), N_2, CO_2	SO_3	Concentrate fines
Copper converting	SO_2 (3–7%), N_2, CO_2	SO_3	Slag droplets, quartz (sand), calcined limestone dust
Electro-refining			Acid mist
Zinc (produced by roasting sulfide concentrate – leaching – electro-winning)			
Concentrate roaster	SO_2, N_2	SO_3	ZnO, PbO, ZnS, Pb, Zn sulfates
Electro-winning	–	–	Sulfuric acid fume
Titania, TiO_2 (produced by chlorination of synthetic rutile produced from ilmenite from beach sands)			
Synthetic rutile kiln (Becher process)	N_2, CO, CO_2, H_2O		Ilmenite, rutile, quartz
Cement (produced from limestone and quartz)			
Limestone calcination	N_2, CO_2, H_2O		Lime dust
Clinker production	N_2, CO_2, H_2O		Calcium alumino-silicate dust

cleaning facilities usually comprise around 20% of the total capital cost of a smelter. The quantities of gas produced can be illustrated with two examples. One is the smelting of sulfides to produce metals such as copper, nickel, lead and zinc, in which sulfur is eliminated mainly as SO_2. The second is the production of steel by the reduction of iron ore in a blast furnace, in which the oxygen from iron oxide in the ore is eliminated by reaction with coke (carbon), to produce hot metal, which is further refined to make steel, and ultimately released to the atmosphere as CO_2.

Sulfide smelting

Consider a copper concentrate containing 25 wt% copper and 30 wt% sulfur. To make 1 tonne of copper requires that 0.30/0.25 = 1.2 tonnes of sulfur be removed. During processing, virtually all this sulfur is ultimately converted to SO_2. Though the sulfur is present as sulfides, particularly copper and iron sulfides, the overall relation is:

$$S + O_2 = SO_2$$

From stoichiometry, 32 tonnes of sulfur combine with 2 × 16 = 32 tonnes of oxygen to form 32 + (2 × 16) = 64 tonnes of SO_2. Therefore, in making 1 tonne of copper, 1.2 × 64/32 = 2.4 tonnes of SO_2 are produced. The ideal gas law relates the pressure, volume and temperature of gases as follows:

$$PV = nRT \qquad 11.17$$

where P (Pascals) is the pressure, V (m^3) is the volume, T (K) is the temperature, n is the number of moles of gas, and R is the universal gas constant (8.314 J mol^{-1} K^{-1}). At ambient temperature (25°C) and atmospheric pressure (101.3 kPa), the volume of SO_2 produced is:

$$V = \frac{nRT}{P} = \frac{2.4 \times 10^6}{64} \times 8.314 \times \frac{(25 + 273)}{101\,300}$$
$$= 917 \text{ m}^3$$

The gases in copper smelting are produced at around 1200°C. Gases expand with temperature and the actual volume at the smelting temperature can be calculated using the ideal gas law as follows:

$$V = \frac{nRT}{P} = \frac{2.4 \times 10^6}{64} \times 8.314 \times \frac{(1200 + 273)}{101\,300}$$
$$= 4534 \text{ m}^3$$

For a smelter producing 1000 tonnes of copper per day, this is equivalent to 190 000 m^3 of SO_2 per hour. SO_2 makes up about 5 vol% of the off-gases in the reverberatory furnace copper smelting process, the balance being mainly CO_2 and H_2O from combustion of fuel and nitrogen from the air introduced into the furnaces for reaction purposes. The off-gas from converters is typically 5–10 vol% SO_2. Hence the total volume of hot gas to be handled will be of the order of 4 million cubic metres per hour, or around 10 000 m^3 per tonne of copper. More modern processes, such as flash smelting, top submerged lance (TSL) smelting and the Mitsubishi process, use oxygen-enriched air and do not require fuel. The SO_2 content of the gas from these processes is much higher, typically 25–50 vol%, accordingly the volume of gas to be handled will be proportionally smaller. Until the 1970s, SO_2 was routinely discharged to the atmosphere. However, increasing concern about the problem of acid rain (see Section 3.4.5) has resulted in most regulatory authorities requiring SO_2 to be captured. This is discussed in Chapter 12.

Iron and steel making

The iron blast furnace–BOF integrated steel-making process produces a large volume of gas. While the iron blast furnace off-gas contains considerable CO as well as CO_2, the CO is combusted to provide heat in other operations on the steelworks site. Modern blast furnaces typically consume about 400 kg of coke (containing ~90 wt% C) to produce 1 tonne of hot metal (containing ~95 wt% Fe). Therefore, making 1 tonne of steel requires around 400/0.9/0.95 = 470 kg carbon. Assuming hot metal contains 5 wt% C, 420 kg C will be removed in the blast furnace and 50 kg C in the BOF. This carbon ultimately ends up as CO_2 in the gases released from the steelworks:

$$C + O_2 = CO_2$$

From stoichiometry, 12 tonnes of carbon combines with 2 × 16 = 32 tonnes of oxygen to form 12 + (2 × 16) = 44 tonnes of CO_2. Therefore, in making 1 tonne of steel, about 0.470 × 44/12 = 1.723 tonnes of CO_2 will be produced. Using the ideal gas law, the volume of this at ambient temperature and atmospheric pressure is:

$$V = \frac{nRT}{P} = \frac{1.723 \times 10^6}{44} \times 8.314 \times \frac{(25 + 273)}{101300}$$
$$= 958 \text{ m}^3$$

The air blast to an iron blast furnace is enriched with oxygen, typically to a total oxygen content of around 40 vol%. Therefore, the total volume of gas produced (measured at room temperature and atmospheric pressure) will be 958 × 420/470/0.4 = 2140 m^3 per tonne of steel.

11.5 THE IMPACT OF WASTES ON HUMANS AND THE ENVIRONMENT

Mining, beneficiating and chemical processing of mined materials produces a range of solid, liquid and gaseous waste streams. Elements not recovered during value-adding report to these streams. These elements are considered *contaminants* if, due to human activity, they are present at a higher concentration than in the material in which they originally occurred or in a more readily available form. For example, in the case of solids, the smaller particle size due to grinding, and the resulting large surface area available for reaction, means elements are more easily leached into the environment than from the host rock. Contaminants are considered to be *pollutants* if they are at a high enough concentration to cause harm or damage to humans or the environment. Many waste streams contain valuable elements and compounds that could be recovered, as well as elements and compounds of no real value. Some of these elements and compounds are toxic and can cause environmental problems and health-related problems for humans, though many are relatively benign.

Less spectacular than collapses of mines or tailings dams, and other catastrophic failures, the effects of the release of tailings, solutions generated from solid wastes (particularly AMD), liquid effluents from processing operations, and gaseous emissions from smelting and other high-temperature processes are often more insidious and act over long periods. Some of their impacts include:

- degradation and pollution of the soil, water and atmosphere of large areas surrounding a mine or processing plant;
- smothering of topsoil (in the case of solid residues);
- destruction of ecosystems;
- reduction of biodiversity;
- adverse effects on the health and well-being of local communities.

Figure 11.1 illustrates how environmental impact and economic value build along the value-adding

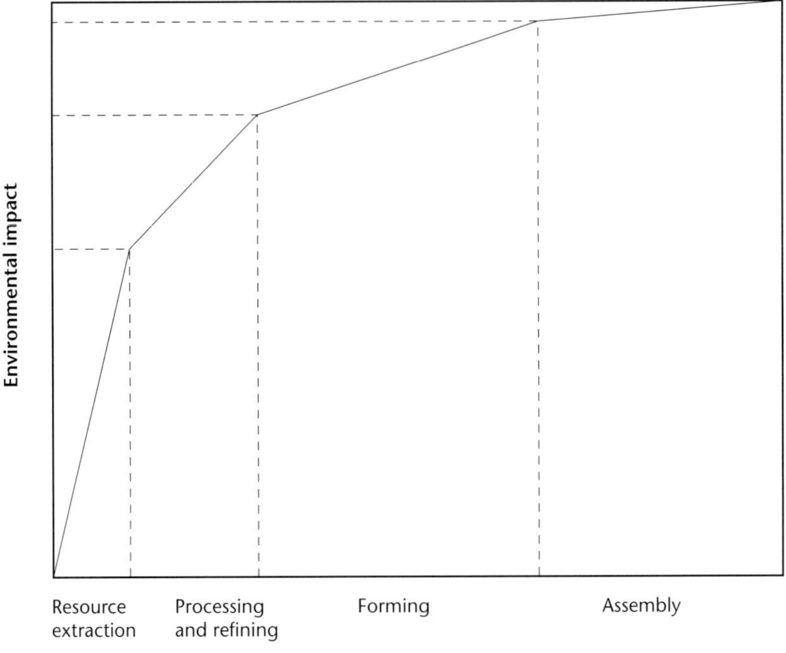

Figure 11.1: The accumulated environmental impact and added value along the value-adding chain (after Clift and Wright, 2000).

Table 11.8: The potential magnitude of impact of wastes and emissions from metal production on the environment

	Land	Water	Atmosphere
Mining and beneficiation	High	High	Low
Chemical processing (smelting, refining, leaching, electro-winning etc.)	Moderate	Moderate	High
Forging and rolling (to make bars, rods, slabs, sheets)	Low	Low	High
Manufacturing	Low	Low	Low

Source: Villas-Bôas (2002); with modifications.

chain from primary resource to finished product. Least value is added during the mining and processing stages, but these have the greatest environmental and human impact. Table 11.8 gives a qualitative ranking of the potential severity of the impact of mining, processing and manufacturing wastes and emissions on land, water and atmosphere along the value-adding chain.

11.5.1 Examples of the impacts of mining wastes

The following examples illustrate some of the types of harm wastes can cause if released into the environment. These are extreme examples and are not intended to be representative of the impacts of all mining and processing wastes. The examples cover solid, liquid and gaseous wastes.

Tailings

Mineral processing tailings are sometimes discharged into nearby surface watercourses, particularly in regions of high rainfall and steep and unstable terrain where the building of containment ponds may be difficult (Engels and Dixon-Hardy, 2009). The Ok Tedi mine in Papua New Guinea and the Grasberg mine in Indonesia are examples of large mining operations where tailings and waste rock are discharged directly into the local watercourse. This type of disposal creates major immediate and long-term issues, including environmental and human health liabilities and costs associated with remediation and reclamation. At the Ok Tedi mine 70 million tonnes of tailings, overburden and mine-induced erosion have been discharged into the Ok Tedi river system each year since operations commenced in 1984. This discharge has caused widespread harm, both environmentally and socially. Around 50 000 people living in more than 100 villages downstream from the mine have been affected. Toxic components in the tailings kill or contaminate fish, consumption of which causes harm to other animal species in the area as well as the indigenous people. Build-up of mine sediment in the lower Ok Tedi and the resulting rise in the river bed have caused flooding and sediment deposition on the floodplain. This has led to the destruction of forest. To date, about 1630 km^2 have been affected by dieback but up to 2400 km^2 of forest may ultimately be affected (Ok Tedi Mining). In 1999, BHP, then the major shareholder and operator of the mine, admitted that the project was the cause of 'major environmental damage' (Burton, 1999). BHP Billiton, the successor to BHP, sold the mine, which is expected to operate until at least 2020 (Ok Tedi Mining).

Acid and metalliferous drainage

Two examples of serious AMD problems are Mt Lyell in Australia and the Berkeley Pit in the United States. The Mt Lyell copper mine is located in mountainous country at Queenstown near the west coast of Tasmania. It commenced operation in 1883 and operated continuously until 1994. There are two sources of AMD: underground workings and the waste rock dumps. A study undertaken at the time of closure (Koehnken, 1997) found that large volumes of acid drainage were leaving the lease site. They contained significant concentrations of metals. The median copper load leaving the site was ~2000 kg per day and of that more than 70% was from underground workings. Nearly all the copper entered the nearby Queen River. Millions of tonnes of waste rock were present on the lease site in waste rock dumps. The acid drainage generated from these accounted for about 20% of the total load leaving the site. The dumps were oxidising at a maximum rate and were likely to continue producing acid drainage for 600 years. The study also found that the Queen and lower King rivers were essentially lifeless, due to acidity and high metal concentrations, and that the biological communities in Macquarie Harbour, into which the King River flows, were impoverished compared with similar coastal environments. The Mt Lyell Remediation Research and Demonstration Program was established jointly by the Tasmanian state government and the Australian government to investigate options for remediation. Two options were identified: long-term neutralisation of acid drainage and copper removal followed by release of clean water to the river; and removing acid drainage from the catchment through collection and discharge to the ocean using a pipeline. The former was the preferred option but, to date, no action has been taken.

The Berkeley Pit (MBMG, 2010), a former open pit copper mine in Butte, Montana, is about 1.6 km long by 0.8 km wide and 600 m deep. The mine opened in 1955 and closed in 1982. At the time of closure, water pumps at a depth of 1200 m in the nearby Kelly shaft were turned off and groundwater from the surrounding aquifers began to slowly fill the pit. The level had risen 950 m by the end of 2006. As a result of oxidation of the surrounding rocks, the water became acidified (pH ~2.5) and metals, including arsenic, cadmium and zinc, were leached from the rocks. When the pit water reaches the level of the natural watertable, estimated to occur around 2020, the pit water will flow into the groundwater and into Silver Bow Creek, which flows into Clark Fork River. The Butte underground mines and the Berkeley Pit are now part of a federal government Superfund Site (USEPA, 2010). A long-term monitoring program has been established to ensure that the pit remains the sink, or terminal pit, for groundwater entering the mine workings. The maximum water level that will be allowed in the Berkeley Pit has been established by the United States Environmental Protection Agency to prevent mine-water discharge into nearby aquifers and surface waters. A treatment plant was constructed in 2003 to treat water diverted from the Berkeley Pit after the water level reaches the critical point.

Lead

Contamination around lead mines and smelters is an example of a long-term health problem caused by mining. High lead levels in the blood of people, particularly children, living near lead mines and smelters is an endemic problem which persists long after the operations have been cleaned up or closed. Cases of lead contamination are regularly reported. For example, Reuters news service reported (3 September 2009) that more than 800 children living near a lead smelter in Shaanxi province in China had high levels of lead, 174 of whom were admitted to hospital, and 1354 children living near the Wugang smelter in Wenping, Hunan province, also tested positive for high levels of lead. Villagers blocked roads, to plead for treatment and compensation. *The Australian* newspaper reported (28 June 2009) a number of lead-related issues in Australia:

- 11% of children aged between one and four years living in the lead mining and smelting town of Mt Isa (north-west Queensland) had lead levels higher than the internationally accepted limit of 10 mg dL^{-1} of blood;
- nearly 40% of children in Port Pirie (a South Australian lead smelting town) were at or above the 10 mg dL^{-1} limit;
- a lead reduction program begun in the New South Wales mining town of Broken Hill in 1991 resulted in the percentage of children whose blood lead level exceeded 10 mg dL^{-1} being reduced from 85% to 25% by 2007.

The problem at Mt Isa appears to have been exacerbated by the reluctance of the mine owner and the Queensland state government environment and health authorities to acknowledge and respond effectively to the fact that the main environmental source of lead is mining and smelting (Munksgaard *et al.*, 2010).

Cyanide

Cyanide is used as a leaching reagent in the extraction of gold, and as a flotation reagent. It is a toxic substance; a fatal dose for humans can be as low as 1.5 mg kg^{-1} body weight. Worldwide, there are frequent small spills of cyanide and dozens are reported every year. Some well publicised very large spills and their cause are listed below.

1985–91	Summitville, USA	Seepage from a leach pad
1995	Omai, Guyana	Collapse of a tailings dam
1998	Kumtor, Kyrgyzstan	A truck carrying cyanide drove off a bridge
2000	Baia Mare, Romania	Collapse of a tailings dam
2000	Tolukuma, Papua New Guinea	A helicopter dropped a crate of cyanide into rainforest

In the collapse of the tailings dam at Baia Mare, about 100 000 m^3 of tailings rich in cyanide, metals and metalloids escaped into the river system (Lottermoser, 2003). The contaminants travelled into local and regional rivers and eventually into the Danube River and the Black Sea.

Cyanide breaks down quite rapidly when exposed to sunlight but the less toxic compounds, cyanates and thiocyanates, may persist for some years. Large-scale releases tend not to be fatal since people can be warned about the danger. However, cyanide spills often have a devastating effect on rivers, killing everything downstream and affecting the livelihoods of people dependent on the river. The entire food chain, from phytoplankton to fish and birds, may collapse. The spill at Baia Mare had a devastating effect on 2000 km of the Danube catchment area. Fortunately, cyanide is quickly flushed from river systems by fresh water and affected areas can soon become repopulated. In the Somes River below Baia Mare, plankton levels returned to 60% of normal within 16 days of the spill (UNEP, 2000).

Sulfur dioxide

Industrial emissions of sulfur dioxide can produce locally elevated, but still relatively low, concentrations in the atmosphere around the source. Sulfur dioxide can cause respiratory problems in humans (e.g. bronchitis), it can irritate the nose, throat and lungs and may cause coughing, wheezing, phlegm and asthma attacks. Low concentrations of sulfur dioxide, which can persist hundreds of kilometres downwind from smelters, can harm plants and trees and reduce crop productivity. Higher levels, and acid rain, adversely affect land and water ecosystems. Sulfur dioxide released to the atmosphere is ultimately absorbed directly by soils and plants or is returned to the ground in the form of acid rain. The latter is a serious environmental problem in Europe and North America, but less serious in dry climate regions such as Australia. Around 60% of the world's copper smelters have gas treatment systems that recover more than 90% of the sulfur in gases. Most of these smelters use high-intensity smelting technologies, such as flash smelting and ISASMELT™. These use highly oxygen-enriched air which makes capture of SO_2 technically easier (Section 12.3.3). In smelters in Japan, 99% of SO_2 is recovered. There is increasing pressure, particularly in the United States and Europe, to achieve similar levels of sulfur recovery.

11.5.2 Toxicity

Toxicity is the property or properties of a material that produces a harmful effect upon a biological system. The terms *ecotoxicity* and *human toxicity* are commonly used in discussing the toxicity of substances. The Basel Convention (Section 11.6.1) defines ecotoxic substances as follows:

> *Eco-toxic substances are substances or wastes which, if released, present or may present immediate or delayed adverse impacts to the environment by means of bioaccumulation and/or toxic effects upon biotic systems (Annex III, H12).*

The Basel Convention does not define human toxicity, but distinguishes between acute poisons and toxic substances:

> *Acute poisons are substances or wastes liable either to cause death or serious injury or to harm human health if swallowed or inhaled or by skin contact (Annex III, H6.1).*
>
> *Toxic substances are substances or wastes which, if they are inhaled or ingested or if they penetrate the skin, may involve delayed or chronic effects, including carcinogenicity (Annex III, H11).*

Table 11.9: The essentiality of elements for living matter

Non-essential	Essential	
	Humans and animals	**Plants**
	Micro-nutrients	*Micro-nutrients*
Al, Sb, Be, Cd, Pb, Bi, Hg, Ag, Li, Br, Rb, Sr, Cs, Sr, Sc, Ti, Zr, Hf, Ta, Nb, Re, PGMs, Au, In, Tl, Br, Te	Cr, Cu, Fe, Mn, Mo, Ni, Se, (Sn), F, I, Co, Zn, (V), (Si), (As), (B), (W), (Ba)	B, Cu, Fe, Mn, Mo, Zn, V, (Ni), Cl
	Macro-nutrients	*Macro-nutrients*
	Na, Ca, S, P, Mg, K, Cl	Mg, K, P, Ca, S

The elements enclosed in brackets are possibly essential but this is unconfirmed. Micro-nutrients are required in relatively small amounts (typically <100 mg per day for humans); macro-nutrients are required in much larger quantities.
Source: Broadhurst (2006)

Potentially hazardous substances associated with solid wastes from the primary production of mineral and metal commodities are predominantly metals and metalloids carried into the wastes from the original rock, and various anions, occurring naturally in the solids or forming from reagents added during processing. Toxic metals are often referred to as heavy metals but, while heavy metals are usually toxic, not all toxic metals are heavy. Hence, the term is misleading. In addition to elevated metal concentrations and high anion contents, solutions associated with solid mineral wastes often have extreme pH values. Acidic solutions typically form from solid wastes produced during mining and beneficiation of sulfide ores (AMD). Alkaline solutions often form in dumps of fly ash from coal combustion for power generation, slags from smelting operations and tailings from Bayer processing of bauxite.

The essentiality of elements

Some elements are essential to life, others are not. An element is considered essential when:

- it is present in living matter;
- it is able to interact with living systems;
- a deficiency of the element results in a reduction of a biological function that is preventable or reversible by appropriate quantities of the element.

These are called *essential elements*. The elements carbon, hydrogen and oxygen make up the major part of all living matter but about 23 other elements, in lesser quantities, are essential. Other elements, for which no biological, nutritional or biochemical function has been identified, are also found in living organisms. These are termed *non-essential elements*. A list of the major essential and non-essential elements is given in Table 11.9.

For every organism, there is a concentration range within which the requirements of the organism for a given essential element are met – a window of essentiality. This is illustrated in Figure 11.2. Within this window, organisms are able to regulate their internal essential element concentration without experiencing excessive stress, by means of homeostatic mechanisms. Below the concentration limit an organism suffers from deficiency of the element and above the limit the element becomes toxic. In contrast, non-essential elements have a negligible effect on organisms at a below-threshold level but become increasingly toxic as the dose increases above this level. This is also illustrated in Figure 11.2.

The toxic effect of elements and compounds

Some elements have a tendency to accumulate in the body; that is, they increase in concentration over time. This is referred to as *bio-accumulation*. The effect of elements is frequently modified by the simultaneous exposure to another element, called interaction. Many such interactions have been identified. For example, potassium and thallium ions have a common cellular receptor. Thus, potassium deficiency augments the toxicity of thallium, which in turn can be reduced by the application of potassium (Wadenbach, 2006). Table 11.10 summarises the effect of some metals on the body after chronic exposure. The table is not complete, nor does it take into account interaction effects and other modifying factors.

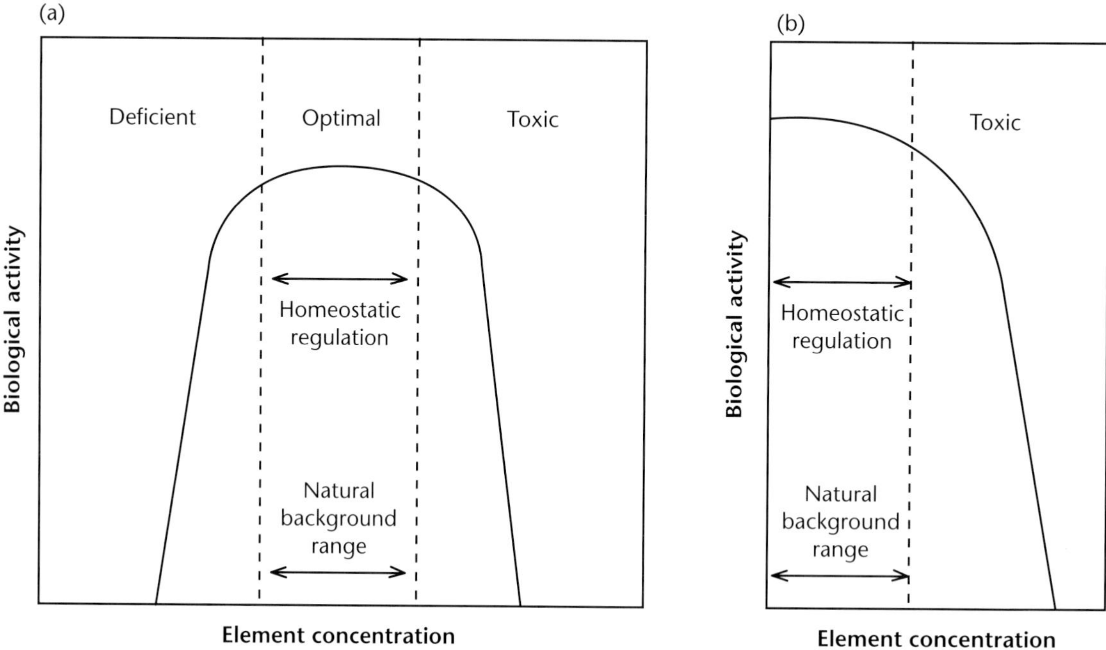

Figure 11.2: The qualitative effect of element concentration in living matter on biological activity. (a) Essential elements. (b) Non-essential elements.

Cyanide, used in the minerals industry for leaching gold and as a flotation reagent, is highly toxic to humans and the environment. Inhalation of 100–300 ppm of hydrogen cyanide or ingestion of 50–200 mg, or 1–3 mg per kilogram of body weight (calculated as hydrogen cyanide), will cause 50% fatality within a human population (International Cyanide Management Code, 2009). Many organic compounds are toxic but most are not relevant in the present context because they are prepared synthetically and do not occur naturally in the crust. However, a few very toxic organic compounds are formed in small quantities during chemical processing at high temperatures, particularly when coal and coke are involved (Mukkinger and Jenkins, 2008). PCDDs, PCDFs, PCBs and PAHs are known to be carcinogenic. When released to the atmosphere these compounds accumulate in the soil through rain; human exposure to them is usually via the food chain. They are stable, do not degrade readily and have a tendency

Table 11.10: Organs affected by chronic exposure to some metals

Organ	As	Be	Pb	Cd	Co	Cr	Cu	Fe	Hg	Mn	Mo	Ni	Se	Tl
Liver	✓							✓			✓			✓
Respiratory tract	✓	✓		✓	✓	✓		✓				✓		✓
Blood	✓		✓				✓						✓	✓
Nerves			✓						✓	✓				✓
Kidney	✓		✓	✓					✓		✓			✓
Skin	✓					✓								
Reproduction			✓											
Heart								✓						✓

Source: Wadenbach (2006).

Table 11.11: Typical solid mineral waste components grouped and ranked on the basis of hazard potential

Components	Toxicity characteristics
Major soluble salts	
Na, K, Mg, Ca, SO_4^{2-}, Cl^-, HCO_3^-, NO_3^-	All are essential macro-nutrients, low to moderate toxicity (only toxic at relatively high concentrations), relatively abundant in natural environments
Trace to minor soluble constituents (metals, metalloids, minor salts)	
Group A: High to severe hazard potential	
Cd, Hg	Toxic to all organisms at low concentrations; non-essential and relatively scarce in natural environments
Sb, Pb, As, Se, Ag, Tl, Te, In, Cr, CN^- (cyanide ion)	Relatively high toxicity; partial to non-essential; scarce in natural environments
Bi, Re, Ge	Toxicity uncertain (generally moderate); non-essential; scarce in natural environments
Group B: Moderate to high hazard potential	
Ni, Cr, Cu, Co, B, U, Zn, Be, Mo	Mainly essential at low concentrations; toxic at higher concentrations; present in trace quantities in natural environments
Mn, Fe, Fe, Al, Ba, V, F	Majority micro-nutrients; toxic or aesthetic effects at elevated concentrations; relatively abundant in natural environment
Sn, Br, Ga, W, Li, Zr, Ta, Hf, Sc, I, rare earth elements	Toxicity uncertain (generally low to moderate); predominantly non-essential; present as trace elements in natural environment
Group C: Low to moderate hazard potential	
Ti, P, Rb	Low toxicity; relatively abundant in natural environment
Acidity	
H^+, OH^-	Negligible direct toxic effects over quite a wide range (pH 5.5–8.0)

Source: Broadhurst and Petrie (2009).

to bio-accumulate in the body fats of humans and animals. Table 11.11 lists major contaminants commonly associated with solid mineral wastes, grouped and ranked according to their hazard potential, taking into account their toxicity and essentiality, aesthetic (taste, colour, odour) and physical (corrosion, fouling) properties and natural background levels. This list is a qualitative guide to the major hazardous components encountered in mineral wastes.

11.5.3 Bioavailability

To pose an environmental risk, contaminants must both have a capacity to cause harm or damage and be available for release to the surrounding environment. In solids, the availability of a contaminant is determined by its potential to react with water and oxygen in its pores or on its surfaces. Reaction results in a deposit of secondary solids, which are stable under the disposal conditions, and a solution containing released contaminants. Subsequent migration of the contaminants within the surrounding environment may have an adverse effect on the quality, and consequently the usability, of ground and surface waters and soils. The chemical reactions involving the major components of mineral wastes, such as the dissolution of silicate minerals, carbonation of the pore waters by CO_2 from the atmosphere, and oxidation of pyrite, are often slow and continue to influence the pore water composition and mobility of trace and minor elements over long periods (Broadhurst and Petrie, 2009). The time for complete oxidation of the pyrite in a waste dump, for example, is typically 200–500 years (Hansen, 2004).

Even if an element is released by chemical reaction and is toxic, it is only considered a hazard if it is in a form that can be taken up by living matter. Many geochemical factors influence the distribution of metal species in water, sediment and soil. The formation of organic and inorganic metal complexes and metal absorption onto solid particles reduces the bioavailability of elements and therefore their toxicity. As a result, the relationship between toxicity and the total or dissolved concentration of elements can be highly variable; neither total nor dissolved aqueous elemental concentrations are good predictors of metal bioavailability and toxicity. However, dissolved, free ionic metal species are much more bioavailable than most complexed or adsorbed metal species.

Theoretical models predict the bioavailability of elements in different environments, with varying degrees of success (ICMM, 2007). However, experimental measurement of bioavailability is still the most reliable method and numerous tests have been developed. The approach is to react a sample of the waste with an acid (in some cases, distilled water) for a specified time, then filter the sample and analyse the filtrate to determine the concentration of elements. The tests may involve one or multiple extractions. The reaction is assumed to go to completion within the time of extraction, therefore these tests aim to measure the maximum concentrations under a given set of conditions. The most aggressive bioavailability test is the Toxic Characteristic Leach Procedure developed by the US Environmental Protection Agency, which uses acetic acid, buffered to maintain a constant pH (USEPA, 1992). Less stringent tests use unbuffered acetic, citric or mineral acids or even distilled water. All are intended to simulate and accelerate the removal of toxic contaminants from wastes in domestic landfill environments and hence are not directly applicable to mining and processing wastes, which are rarely mixed with organic matter. A recent approach to the characterisation of the leaching behaviour of waste materials is a pH-dependence leaching test which involves parallel extractions of the material at a liquid:solids ratio of 10 L kg^{-1} for 48 hours at a number of pH values (Van der Sloot, 2007). Information derived from the test can be used for geochemical speciation. The test also provides information on the sensitivity of leaching under externally imposed changes in pH under specific scenarios.

11.6 THE INTERNATIONAL REGULATION OF WASTES

11.6.1 The Basel Convention

With the tightening of environmental laws and regulations in developed nations during the 1970s, the costs of treating and disposing of hazardous waste materials increased significantly. At the same time, globalisation of shipping made movement of wastes over long distances cheaper and easier. A trade in hazardous wastes rapidly developed, particularly from developed to less developed countries with less stringent regulations for disposing of wastes. The Basel Convention on the Control of Transboundary Movements of Hazardous Wastes and Their Disposal, usually referred to as the Basel Convention, was established in 1989 in response to this trend. It is an international treaty to reduce the movement of hazardous wastes across borders. It is also intended to minimise the quantity and toxicity of wastes generated in the first place, to ensure their sound management as close as possible to their source of generation, and to assist less developed countries in management of their own hazardous wastes. Since 1989, 172 countries have ratified the treaty; the only major country yet to do so is the United States. The text of the Convention is available on the Convention's website (http://www.basel.int).

The types of wastes covered by the Basel Convention are defined in Annexes to the text. A waste is covered by the Convention if it falls in the categories listed in Annexes I and II (where wastes are classified according to their source and whether they contain specific substances) and if it also exhibits one of the hazardous characteristics listed in Annex III. The types of wastes that are excluded are described in Annex IX. The definition of the term 'disposal' is given in Article 2.4 which refers to a list of operations, in Annex IV, which are considered to be forms of disposal. The definition is broad and includes recovery, recycling and reuse. By restricting the international movement of wastes, the Basel Convention encourages countries not only to keep wastes within their borders and close to the source of generation, but to reduce the quantity of waste produced. This is reflected in Article 4.2, which requires each country to ensure that the generation of hazardous and other wastes is reduced to a minimum, taking into account social, technological and economic aspects.

The Basel Convention has important implications for the minerals and metals production industries. Waste rock, tailings and AMD, by their nature and quantities, have to be treated at their source; other wastes could, except for the Basel Convention, be shipped internationally for disposal. Of course, a lot of potentially toxic elements are shipped internationally in the form of traded mineral commodities and manufactured products. The wastes from the further processing of imported mineral commodities and the use, recycle and disposal of manufactured products is of increasing concern in many countries. The European Parliament's response to this is discussed below.

11.6.2 REACH and the European Chemicals Agency

The European Parliament has passed legislation regulating the movement of chemical substances into, out of and between member countries. The relevant regulation (EC 1907/2006), Registration, Evaluation, Authorisation and Restriction of Chemical substances (REACH), came into effect on 1 June 2007 with phased implementation over 10 years. The aim of REACH is to improve the protection of human health and the environment through the better and earlier identification of the intrinsic properties of chemical substances. REACH is wide in scope. It covers all substances manufactured within the European Union, imported, used as intermediates or placed on the market on their own, in preparations or in products.

The REACH Regulation puts responsibility on industry to manage the risks from chemicals and to provide safety information on substances. Manufacturers and importers are required to gather information on the properties of their chemical substances which will allow their safe handling, and to register the information with the European Chemicals Agency. The Regulation also calls for the progressive substitution of the most dangerous chemicals when suitable alternatives have been identified. Waste substances are exempted because they are covered by the Basel Convention. REACH has been criticised by some countries, particularly the United States, India and Brazil, who claim it will restrict global trade. The European Chemicals Agency in Helsinki manages the technical, scientific and administrative aspects of the REACH system and aims to ensure that the legislation can be properly implemented and has credibility with stakeholders (European Chemicals Agency, 2009). The Agency manages the databases necessary to operate the system, coordinates the evaluation of suspicious chemicals and maintains a public domain database on which consumers and professionals can find information on hazardous substances.

For substances on their own or in preparations, such as mineral concentrates, REACH imposes an obligation on manufacturers and importers to submit a registration (to the European Chemicals Agency) for each substance manufactured or imported in quantities of 1 tonne or more per year (European Commission, 2007). Different rules apply for manufactured goods. These bear in mind the need to adopt a proportionate approach for the many products on the market, and the potential of some to harm human health and the environment due to the chemical substances contained in them. In these cases, REACH requires substances that are intended to be released from articles during normal and reasonably foreseeable conditions of use to be registered if those substances are present in the articles above a total of 1 tonne per year. All substances of very high concern present in articles above a concentration of 0.1 wt% and above a total of 1 tonne per year must be notified to the Agency except where exposure to humans and environment can be excluded during normal conditions of use, including disposal.

11.6.3 Implications of the Basel Convention and REACH

The combined effects of the Basel Convention and REACH on the trade of mineral and metal commodities are not yet clear. However, there are two drivers that are certain to influence future trends: the restriction on the export of mineral and other wastes from mineral-producing countries; and the increasing restriction on the importation of hazardous substances into the European Union. The latter includes mineral commodities, particularly base metal concentrates which contain, in addition to the primary metal of interest, most of the elements in the periodic table.

Financial penalties have been routinely imposed by purchasers on mineral commodities which contain impurities which cause processing and/or disposal problems. These are likely to increase as REACH is implemented. Future regulations may make the export of some products, containing particular impurities, impossible. The logical response would be to process mined commodities to the fullest extent at the point of mining and store all the associated wastes at the site. This is counter to the trend of recent decades, during which the major minerals companies have retreated from value-adding, essentially to mining and processing mined ore only enough to produce a saleable commodity (e.g. iron ore, base metal concentrates, alumina). Companies may need to reassess this trend.

11.7 REFERENCES

Agricola G (1556) *De Re Metallica*. English translation by HH Hoover and LH Hoover. Dover Publications: New York, 1950.

Basel Convention. <http://www.basel.int>.

Broadhurst JL (2006) Generalised strategy for predicting environmental characteristics of solid mineral waste – a focus on copper. PhD thesis. University of Cape Town, South Africa.

Broadhurst JL and Petrie JG (2009) Ranking and scoring potential environmental risks from solid mineral wastes. *Sustainability through Resource Conservation and Recycling '09*, Cape Town, South Africa. MEI Conferences.

Burton B (1999) BHP admits Ok Tedi mine is environmental disaster. *Asia Times Online*, 13 August 1999.

Clift R and Wright L (2000) Relationships between environmental impacts and added value along the supply chain. *Technological Forecasting and Social Change* **65**: 281–295.

DITR (2007) *Managing Acid and Metalliferous Drainage*. Leading Practice Sustainable Development Program for the Mining Industry. Department of Industry, Tourism and Resources, Commonwealth of Australia: Canberra. <http://www.ret.gov.au/resources/resources_programs/lpsdp>.

Engels J and Dixon-Hardy D (2009) Tailings.info – the website all about tailings. <http://tailings.info/index.htm>.

European Chemicals Agency (2009) <http://echa.europa.eu>.

European Commission (2007) *REACH in brief*. European Commission, Department Directorate General, October 2007. <http://www.vet.uu.nl/nca/userfiles/other/REACH_in_brief.pdf>.

Geiger GH, Kozakevitch P, Olette M and Roboud PV (1982) Theory of BOF reaction rates. In *BOF Steelmaking*. Vol. 1. (Ed. JM Gaines) pp. 302–431. Iron and Steel Society: Warrendale, PA.

Hansen Y (2004) Environmental impact assessment of solid waste management in the primary industries – a new approach. PhD thesis. University of Sydney: Sydney.

Hill VG and Sehnke ED (2006) Bauxite. In *Industrial Minerals and Rocks – Commodities, Markets and Uses*. 7th edn. (Eds JE Kogel, NC Trivedi, JM Barker and ST Krukowski) pp. 227–261. Society for Mining, Metallurgy and Exploration: Littleton, CO.

ICMM (2007) *MERAG: Metals Environmental Risk Assessment Guidance*. International Council on Mining and Metals: London. <www.icmm.com>.

International Cyanide Management Code (2009) <http://www.cyanidecode.org>.

Koehnken L (1997) 'Final report'. Mount Lyell Remediation Research and Demonstration Program, Supervising Scientist Report 126. Supervising Scientist: Canberra.

Lévy V, Fabre R, Goebel B and Hertle C (2006) Water use in the mining industry – threats and opportunities. In *Water in Mining 2006*. pp. 289–295. Australasian Institute of Mining and Metallurgy: Melbourne.

Lottermoser BG (2003) *Mine Wastes: Characterization, Treatment and Environmental Impacts*. Springer-Verlag: Berlin.

Marvis J (1995) Aluminum industry. In *Pollution Prevention Handbook*. (Ed. TE Higgins) pp. 377–387. Lewis Publishers: Boca Raton, FL.

MBMG (2010) Berkeley Pit facts. Montana Bureau of Mines and Geology. <http://www.mbmg.mtech.edu/env/env-berkeley.asp>.

Mukkinger P and Jenkins B (2008) Emissions and environmental impact. In *Industrial and Process Furnaces: Principles, Design and Operation*. pp. 386–390. Elsevier: Amsterdam.

Munksgaard NC, Taylor MP and Mackay A (2010) Recognising and responding to the obvious: the source of lead pollution at Mount Isa and the likely health impacts. *Medical Journal of Australia* **193**(3): 131–132.

Ok Tedi Mining. <http://www.oktedi.com>.

Sinclair RJ (2005) *The Extractive Metallurgy of Zinc.* Spectrum Series Vol. 13. Australasian Institute of Mining and Metallurgy: Melbourne.

UNEP (2000) *UNEP/OCHA Environment Unit, UN assessment mission – Cyanide Spill at Baia Mare,* March 2000. <http://www.reliefweb.int/ochaunep/edr/index.htm>.

USEPA (1992) *Test Methods for Evaluating Solid Waste, Physical/Chemical Methods (also known as SW-846), Method 1311, Toxicity Characteristic Leaching Procedure.* United States Environmental Protection Agency, July 1992. <http://www.epa.gov/osw/hazard/testmethods/sw846/online/index.htm>.

USEPA (2010) *Cleaning up the Nation's Hazardous Wastes Sites.* United States Environmental Protection Agency. <http://www.epa.gov/superfund/index.htm>.

Van der Sloot HA (2007) <www.leaching.net>.

Villas-Bôas RC (2002) Current issues on sustainable development that impacts the minerals extraction industries. In *Indicators of Sustainable Development for the Mineral Extraction Industry.* (Eds RC Villas-Bôas and C Beinhoff) pp. i–xxvi. CNPq/CYTED: Rio de Janeiro.

Wadenbach P (2006) Toxic effect of metals and metal compounds. In *Sustainable Metal Management.* (Eds A von Gleich, RU Ayres and S Gößling-Reisemann) pp. 393–402. Springer: Dordrecht, The Netherlands.

11.8 USEFUL SOURCES OF INFORMATION

MERAG: Metals Environmental Risk Assessment Guidance (2007) International Council on Mining and Metals: London. <www.icmm.com>.

Williams RE (1975) *Waste Production and Disposal in Mining, Milling and Metallurgical Industries.* Miller Freeman Publications: San Francisco.

12 Management of wastes from primary production

The impacts from the wastes produced during mineral and metal production can have lasting environmental and social consequences and be difficult and costly to address through remedial measures. Wastes from mining and processing therefore need to be managed carefully to ensure the long-term stability of storage and disposal facilities and to prevent or minimise air, water and soil contamination. All legal jurisdictions have regulations concerning the safe management of wastes from the minerals industry, but these are often inadequate or not strictly enforced. Many undesirable practices still persist. The inappropriate or unsafe management of wastes generates opposition from local communities, the general public and non-governmental organisations interested in environmental, social and sustainability issues. More than anything else, this has contributed to the negative public perception of the mining industry (Engels and Dixon-Hardy, 2009).

The usual approach to managing wastes is to contain and collect them at the point of production, treat them to the extent necessary to make them safe, then dispose of them to the land, to water or to the atmosphere. Sometimes they are processed before disposal, to recover valuable materials. This chapter examines the conventional approaches and technologies used to manage the wastes produced during mining and processing of mined materials to the stage of producing tradable mineral or metal commodities. It does not consider whether the quantities of these wastes could be reduced, eliminated or used as raw materials for producing other value-added commodities. Those issues are the subject of Chapter 16.

12.1 MANAGEMENT OF SOLID WASTES

Solid wastes include materials that are removed to gain access to the mineral resource, such as overburden and waste rock, the tailings remaining after valuable minerals have been extracted from the ore, and the residues produced from any further processing, such as leaching residues, water treatment residues and slags. The handling, storing and disposal of these, and recovery of the water associated with them, is a major part of mining and processing operations and one which potentially has great human and environmental impact. Figure 12.1 summarises the range of conventional options available for managing solid mining wastes. Mine wastes can be used on- or off-site, stored in mine waste heaps, or used in leaching operations to recover additional valuable elements. Similarly, tailings may be used on- or off-site, stored

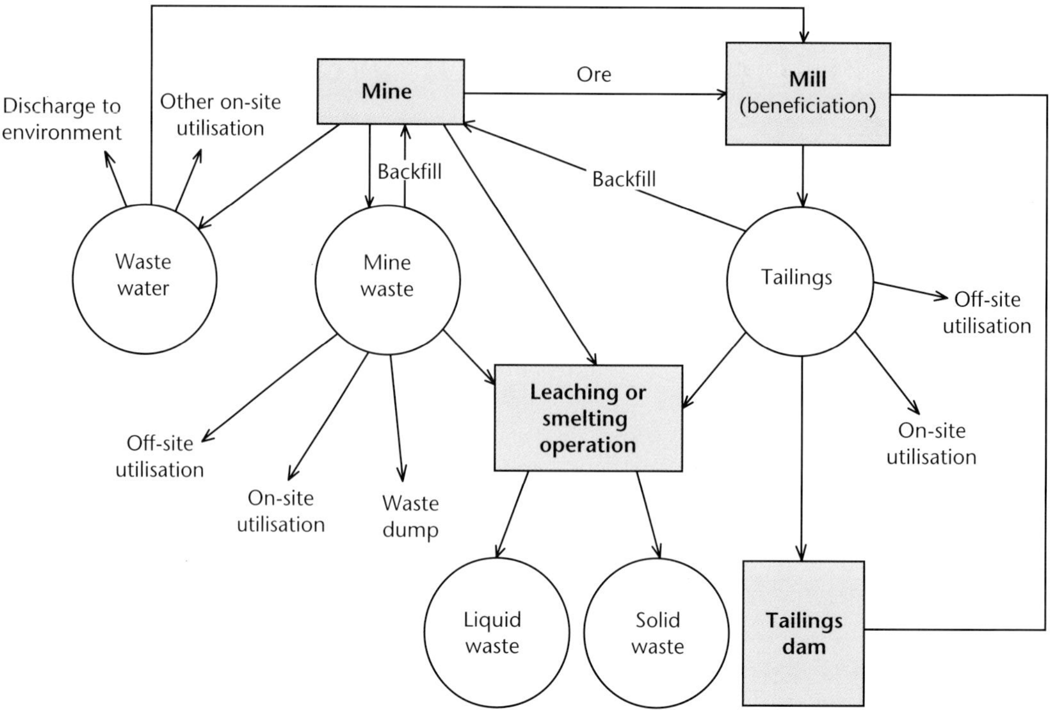

Figure 12.1: Management of solid mining waste (after Wilmoth *et al.*, 1991).

in tailings ponds or used in leaching operations. The presence of waste disposal and storage facilities can result in loss of land for productive purposes or in loss of natural ecosystems; erosion and leaching of material from such facilities by wind and rain can cause environmental and human problems. The collapse of tailings facilities and heaps, which occurs not infrequently, can have catastrophic impacts on the environment and human health and safety.

12.1.1 Waste rock

Overburden and waste rock from mining operations are stored on the surface. Good engineering practice requires that dumps be located where they will not prevent future expansion of the mining operation so that the full value of the deposit can be recovered, even if this may not happen in the foreseeable future. Similarly, waste rock that may in the future become economical to process for its valuable minerals should be dumped in locations which make it as cheap as possible to transport to the beneficiation plant. Unfortunately, these points have often been overlooked for short-term benefit during mine development. Ideally, all waste rock should be returned to the mine when the mine has been exhausted and there is no likelihood of it being reopened.

Waste rock is broken up sufficiently to be moved to the disposal site by truck or conveyor, where it is dumped to form heaps and levelled by machines such as bulldozers. Some is pushed over the edge to form slopes at the material's natural angle of repose, to ensure stability. Important considerations are ensuring stable slopes and controlling the flow of water in and around the waste in order to minimise erosion, protect the structure and prevent infiltration, the last being particularly important since acid drainage is the most pervasive problem associated with many waste dumps. Where rainfall is high, particular care needs to be taken to ensure physical stability of the dumps – if they become waterlogged they can fail, with catastrophic consequences. In dry climates, the surface may need regular wetting to reduce dust generation. This is not a long-term solution. At the time of closure, a permanent method of rehabilitation needs to be implemented if the material is not to be returned to the mine. This may be a cover of vegetation; in arid regions, it may be necessary to form a crust on the surface (Starke, 2002).

Overburden from open pit mines is usually stored in piles near the mine. Ideally, this material should be returned to the pit in a stable form when the mining operation ceases. The overburden from strip mining is returned directly to the mine as the mining operation progresses. The preference in underground mining is to return material immediately as backfill to worked-out stopes to avoid the need to store waste on the surface, either permanently or temporarily, and to minimise environmental damage. However, this is more expensive than surface storage. Waste rock can be returned to underground mines in two main ways (Engels and Dixon-Hardy, 2009).

1 *Dry rock fill.* Dry rock fill consists of a mixture of waste rock, surface sands, gravels or dried tailings which is dropped down a raise or tipped into an open stope.
2 *Cemented fill.* Cemented fill consists of thickened tailings and waste rock. The tailings act as cement, filling the voids between the larger lumps of waste rock and binding them. Alternatively, a slurry of Portland cement can be poured over the waste rock to fill voids and bind the rocks, although this is more expensive.

12.1.2 Tailings

Tailings are usually stored on the surface in purpose-built tailings dams. In-pit, co-disposal, riverine and off-shore disposal methods are also used. In-pit storage involves filling mined-out open pits with tailings. Co-disposal involves mixing tailings with coarse mine waste which is then backfilled into underground or open pit mines, used to cover conventional tailings dams or used to construct elevated waste heaps. Off-shore and riverine tailings disposal involve the discharge of tailings to natural bodies of water. Tailings with high slimes content do not dewater easily and for disposal on land are separated into a slimes and a coarser sand fraction using hydrocyclones. The slimes are usually stored in a surface impoundment and the sand is pumped underground. This is referred to as hydraulic sand. Hydraulic sand fill can be mixed with binders if necessary to obtain the required strength. As the sand settles and consolidates, excess water is expelled and bled off.

Conventional tailings storage

When tailings are stored on the surface in purpose-built impoundments, the embankments are usually strong enough to hold back both the tailings and water so the water can be recovered and reused. However, when thickened, paste or dry stack methods of storage are used, the embankments are built to retain only run-off, bleed water and tailing fines rather than the entire weight of the tailings, since the solids are self-supporting. Two basic configurations are used for surface storage: ring dykes and valley impoundment. The ring dyke configuration (also known as a paddock or cell) consists of an embankment extending completely around the impounded tailings whereas the valley impoundment takes advantage of natural topography to help contain the tailings. In the valley configuration, an embankment is placed across a valley in a similar manner to that for a conventional water storage reservoir. A variation is the hill-side configuration, in which the tailings are contained on three sides by embankments and on the fourth side by the side of a hill. The advantage of the ring dyke is that it can usually be constructed close to the processing plant and the only water contained within it is process water and direct rainfall. This is not the case with valley and hill-side impoundment since, being low-lying, water run-off from surrounding land will collect in the impoundment area. This has to be collected and treated before it is released.

Embankments for tailings can be built in two ways. In a water retention type dam, the embankment is constructed to its full height before tailings are discharged into it. In a raised embankment dam, the embankment is built progressively higher as more storage is required. Water retention dams are similar in structure to conventional water storage dams and are used mainly at operations where large volumes of water must be stored; for example, to enable the plant to operate through the dry season or where flooding can occur. The raised embankment design is more common because, since the embankment is raised at intervals, the initial capital cost is lower. The choice of construction material is also greater, since smaller quantities are needed at any one time.

Raised embankments can be constructed in several ways, as shown in Figure 12.2. An upstream embankment is commenced with a free-draining starter wall, or dyke. The tailings are discharged from the crest of the dyke wall to form a beach that becomes the foundation for subsequent embankment raises (Figure 12.2a). Most failures involve this type of embankment, because the embankments are built on an

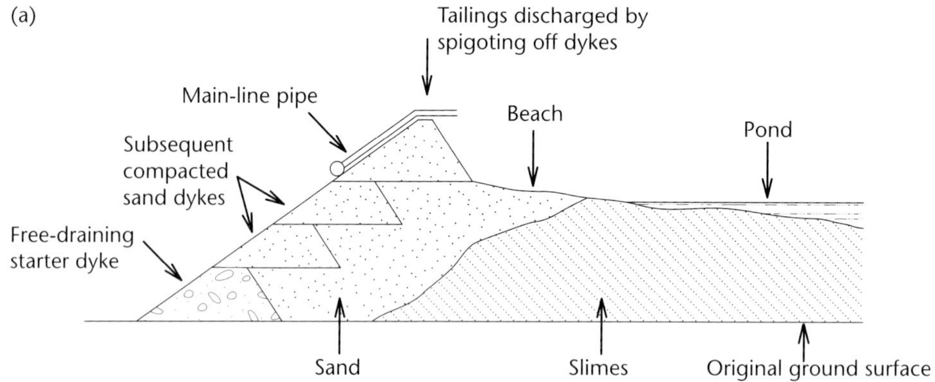

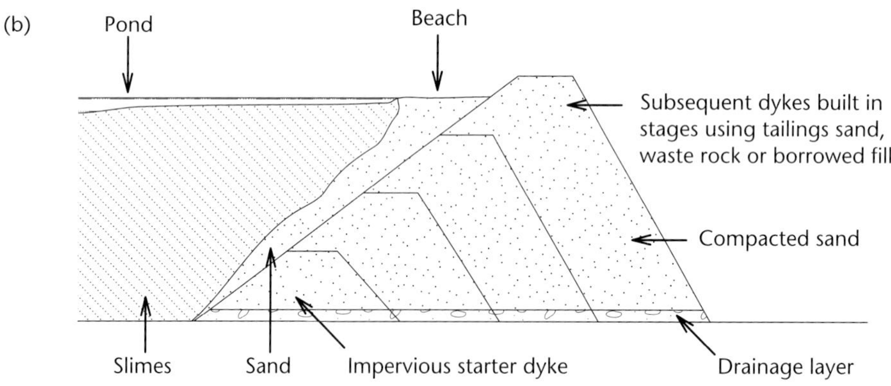

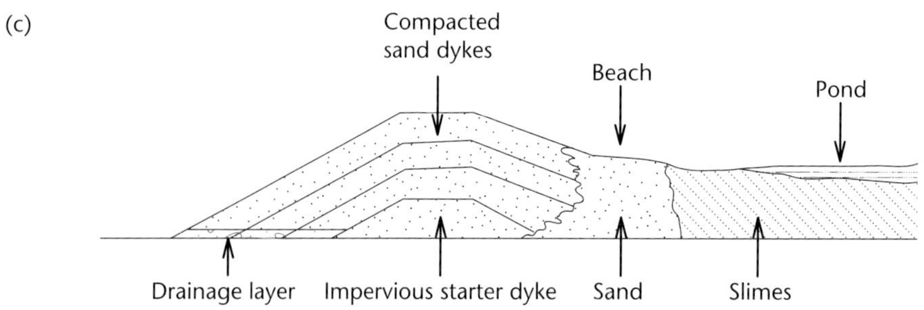

Figure 12.2: Design of raised embankments for tailings storage. (a) Upstream method. (b) Downstream method. (c) Centre-line method (after Wills, 1992).

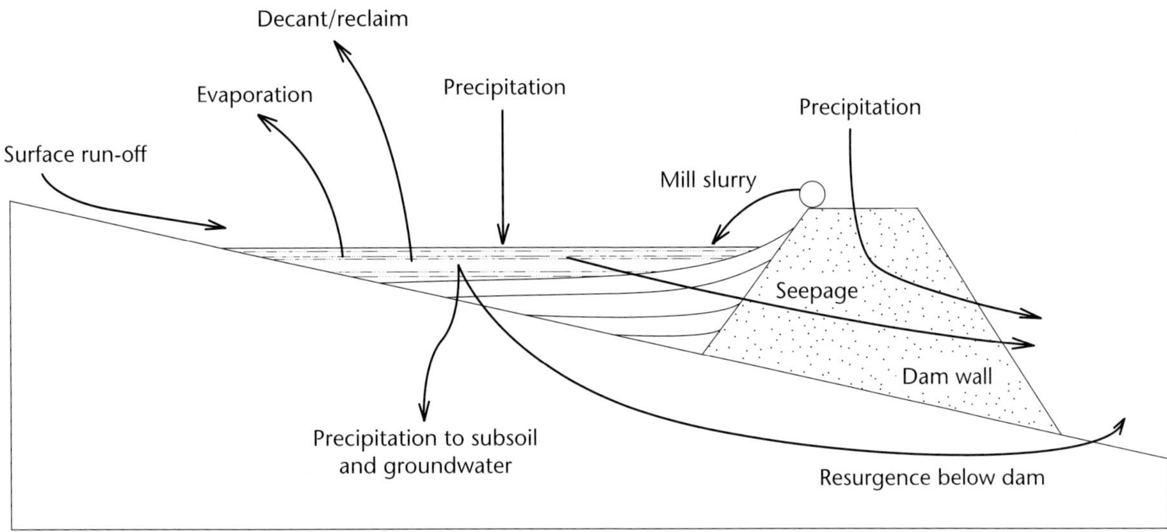

Figure 12.3: Water gain and loss mechanisms in a tailings dam (after Wills, 1992).

unconsolidated foundation of earlier tailings deposits. The downstream embankment was developed to reduce the risk of failure. A downstream embankment commences with an impervious starter dyke; tailings are deposited behind the dyke (Figure 12.2b). When the embankment is raised the new wall is supported on top of the downstream slope of the previous section. Clearly, more construction material is required. The centre-line method is a compromise between the upstream and downstream designs. It is more stable than the upstream design but does not require as much construction material as the downstream design. Like the upstream method, the tailings are discharged at the crest of the embankment to form a beach behind the dam wall. When raising is required, material is placed on both the tailings and the existing embankment (Figure 12.2c). As a result, the embankment crest is raised vertically and does not move in the upstream or downstream direction.

Water management in tailings storage facilities

Water stored in tailings dams is decanted as it separates from the settling solids, and stored separately in ponds, or is sent directly to the processing plant after pH adjustment or purification to the extent required for reuse. Reuse is particularly important in mines located in arid regions where water is scarce. If the plant does not require recycled water, or not all that is potentially available, the stored water will be sent to evaporation ponds or treated and released to the environment. Treating water to the purity required for discharge to the environment can be expensive; internal recycling of water even where rainfall is high may be the most economical option.

Figure 12.3 shows the main mechanisms by which water is gained and lost in a conventional tailings impoundment. The relative magnitudes of each mechanism vary greatly according to climate, geography, design of the impoundment and nature of the tailings. However, all can be controlled except precipitation and evaporation but even these can be moderated, for example by use of surface coverings. Water management involves preventing surface run-off entering the impoundment and preventing seepage of tailings water into the groundwater and through the dam wall into surface water. In the last case, since it is difficult to build completely impervious walls, a better strategy can be allowing water to seep through the dam wall, collecting it in a drainage pond and recycling it to the process or back to the tailings dam.

The most serious problems associated with tailings dams are the release of polluted water into the environment through seepage and overflow and through the catastrophic failure of dams. The presence of too much water in the impoundment is the fundamental cause of both. The liquefaction of the solids, internal and external erosion, seepage and overflowing are the main water-induced failure mechanisms of tailings facilities. The risk of water seepage and physical instability in conventional tailings facilities can be reduced by good drainage and maintaining little, if any, ponded water (Engels and Dixon-Hardy, 2009). Hence, effective management of water in and around tailings impoundments is an important responsibility of a mine operator. These problems can be virtually eliminated by using paste and dry stack facilities, since these contain little or no water.

Surface thickened, paste, and dry stack methods

Surface thickened, paste, and dry stack methods involve dewatering the tailings to varying extents before depositing them on the surface at the mine site (Engels and Dixon-Hardy, 2009). These methods are increasingly used because of their environmental benefits although they are more expensive than traditional methods.

Surfaced thickened tailings are tailings that have been dewatered using thickeners (or a combination of thickeners and filter presses) to the point where they form a homogeneous mass when deposited. When deposited layer on layer, thickened tailings dry to a high degree and become very mechanically stable. The method involves stacking the pulp in self-supporting conical piles which are self-draining and easily reclaimable. Operating costs are high, due to the cost of dewatering. Thickened tailings disposal has significant advantages. Water is conserved and evaporation is minimised in arid climates; a starter dam is not required, which reduces capital cost; reclamation costs are lower than for impoundments that store tailings and water; the risks associated with the retaining embankments of conventional storage are reduced; there is little or no separation of the solids and liquids, which results in less oxygen ingress and therefore less generation of acid from sulfur-bearing tailings (AMD); and the tailings properties can be modified, if necessary, by adding binders to increase stability, reduce erosion and minimise seepage.

Paste tailings are tailings that have been dewatered even further, to the point where they do not segregate when deposited and produce little, if any, bleed water. Their high viscosity requires the use of positive displacement pumps for transportation, which limits the distance over which paste can be moved economically. Paste disposal at the surface evolved from technology used for backfilling in underground mining. It utilises dewatering technology developed to produce tailings with lower water content, particularly high-density and paste thickeners that have large height:diameter ratios (often greater than unity). Paste is deposited in a manner similar to thickened tailings, to form a conical pile. In some situations, paste disposal can be cheaper than conventional disposal methods depending on the additives used, the distance the paste is pumped and the storage space available. It has similar advantages to thickened tailings disposal. However, surface paste disposal is rarely used and is relatively unproven as a technology. It is likely to become more common for environmental reasons and because the daily management requirements are less demanding than those of conventional storage techniques.

Dewatering of tailings to even higher degrees using a combination of filtration systems produces a cake with typically less than 20 wt% moisture. This cannot be transported by pipeline by pumping, due to its low moisture content. It is usually transported by conveyor or truck to the storage site, where it is spread and compacted to form a stable deposit called a dry stack. As with thickened and paste tailings, the mechanical dewatering process increases costs compared to conventional tailings deposition. However, dry stacking has many advantages. It requires no water-retention bunding; it can be used where water conservation is critical; it is suited to areas of seismic activity or where the construction of retention embankments is difficult; it avoids pipe freezes and frosting problems in cold climates; groundwater contamination through seepage is virtually eliminated; the higher recovery of water by filtration allows better recovery of dissolved metals and process chemicals (e.g. gold and cyanide); and dry stack facilities are easier to close and rehabilitate, require a smaller area than other surface tailings storage options, and can be used in difficult environments (e.g. undulating and steep terrain).

Closure of tailings storage facilities

The main mine-related risks to public health and the environment from mining operations are from tailings storage facilities. This is reflected in the high level of community concern about their closure, decommissioning, rehabilitation and aftercare (DITR, 2007a). Contaminants in tailings can be mobilised through a number of mechanisms: airborne transport as dust, dissolution in water and movement of the water, and water-borne transport of suspended solids. The objective of tailings storage facility closure, decommissioning and rehabilitation is to leave the facility safe and stable, with little need for ongoing maintenance. This requires the following to be undertaken (DITR, 2007a):

- containing or encapsulating the tailings to prevent their escape into the environment;
- minimising seepage of water from the tailings into surface and groundwaters;
- providing a stabilised surface cover to prevent erosion of the facility;
- designing the final landform to minimise post-closure maintenance.

Guidelines have been developed for this and other aspects of mine closure (UNEP, 2005).

River and sub-sea disposal of tailings

River and sub-sea disposal methods reduce the engineering requirements of tailings disposal and reduce the tailings storage footprint on-shore. They can also, under the right conditions, stabilise the tailings by minimising or preventing oxidation and mobilisation of contaminants by covering them with water (Norman and Rathorth, 1998). Their major disadvantages are the unpredictable nature of tailings flows as they leave a discharge outlet, the unknown effects on aqueous environments, and the potential migration of the contamination (Engels and Dixon-Hardy, 2009). River discharge, as practised for example at Ok Tedi (Section 11.5.1), creates major environmental liabilities and costs associated with remediation and reclamation and generates strong public opposition. However, in some situations tailings can be discharged off-shore with much less environmental impact than those of conventional land-based methods, and with considerable cost saving. In these cases, tailings should be deposited below the level in a water body where photosynthesis and reproduction of marine plants occur and at a depth at which they are not disturbed by turbulence caused by wind and waves. Even when off-shore disposal is damaging to a marine ecosystem, it may be the best option in a particular situation compared to the risks associated with land disposal. Surface storage has an ongoing environmental impact and maintenance cost; off-shore disposal may have only a temporary environmental impact and no maintenance requirement once mining activities cease. Tailings placed underwater do not have to resist erosion from post-mining activities since they will eventually be covered by land-derived sediments.

12.1.3 Residues from leaching operations and water treatment

Water treatment processes and leaching operations produce solid residues that have to be managed carefully because of their toxic element content. There are many possible disposal and storage strategies, ranging from off-site transport to disposal in mine workings. The choice depends on many complex factors. Where on-site storage of water treatment residues is not permitted, or is not possible for environmental reasons, a high-density sludge process (see Figure 12.4) followed by filtration to further thicken the sludge prior to transport is preferable. Disposal in mine workings is usually the least expensive option when storage on site is permitted and is acceptable environmentally. Other options include co-disposal with tailings in engineered ponds, on waste rock piles, under a water cover, or in natural land depressions. Disposal with tailings can be done at operating mines. Another option is to dispose of sludge over the top of a closed tailings storage facility. This may help reduce oxidation of the tailings (and AMD formation) by providing a wet barrier similar to that of engineered soil covers. Engineered ponds can be designed to both drain water from the bottom and allow for evaporation from the surface. Most sludges will densify to several times their original solids content when disposed of this way. Alternatively, sludge can be allowed to densify in an engineered pond for a year or two then used to cover tailings or deposited in worked-out sections of mines. Sludge stored underwater will not densify as much as surface-disposed sludge, but this option may be viable as the sludge will remain stable.

When surface disposal is used, revegetation is important to minimise dusting and improve aesthetics. It may be possible to revegetate sludge directly, but it can be done more easily with a layer of topsoil.

12.1.4 Slags

Slag by-products from the smelting of metals have historically been considered wastes. However, they are finding increasing use as aggregate materials, as raw materials for cement manufacture and as an additive to concrete, as well as in more specialised applications. Slags are produced at temperatures in the range 1200–1600°C. Their heat content is not recovered although it is a potentially important source of high-grade (high-temperature) heat. Some slags produced are intermediate products and are recycled within the process to further extract valuable elements, the final waste being tailings, leach residue or slag depending on how it is processed. The focus of this section is on slags that are considered as outputs, either waste or as inputs to another industry (e.g. construction or cement manufacture).

Iron blast furnace slag

The largest quantities of slag are produced by the iron and steel industry. In blast furnace production of iron, around 250–300 kg of slag is produced per tonne of hot metal and each tonne contains about 1.8 GJ of sensible heat. This means that 1.8 GJ of thermal energy is removed, and could be recovered, when 1 tonne of slag cools to 25°C. Iron blast furnace slag consists of a molten mixture of silica, lime, magnesia, alumina and small quantities of other oxides (Table 11.2), and a little entrained iron. It is environmentally benign and resistant to leaching. Conventionally, the slag removed through the tap hole in the hearth of the iron blast furnace is collected in pits adjacent to the furnace or in ladles which are then transported to a pit where the molten slag is poured out. The slag is allowed to solidify by natural cooling. Often, water is sprayed over the surface to accelerate cooling and induce cracking. This makes the slag easier to break up when cooled. The solidified slag is then removed and taken to a landfill site, or crushed and screened to produce aggregate products in the size range 5–20 mm, such as concrete aggregate and road base. The fines created during crushing can be used as sand in concrete.

Globally, about half of all iron blast furnace slag is water-quenched rather than air-cooled. Water-quenching cools the slag so rapidly that the molten structure of the slag is largely preserved, resulting in a glassy, rather than crystalline, product. Slag that is greater than about 90% glass is pozzolanic and is suitable as a substitute for Portland cement (see Section 8.3) and is a significantly higher-value product than aggregate. In this case, the slag is poured into inclined launders, or troughs, and high-pressure, high-volume jets of water are played onto the surface as it flows. This has the effect of creating small solid granules of slag about 1–5 mm in diameter. These are carried down the launder into a pond, from which they are recovered and dried. The dried material is then ground in a mill and sold as a product for blending with Portland cement. This process is water- and energy-intensive. Typically, about 10 tonnes of water per tonne of slag is required to achieve the needed quench rate. Of this, about 1.5 tonnes of water is lost through evaporation and 8.5 tonnes are recycled after filtering. Energy, usually electrical, is required for pumping the water through the jets, for solid–liquid separation and water purification. All the sensible heat content of the slag is lost to the environment.

In both air-cooling and water-quenching, sulfurous gases and acid mist are produced when the water reacts with sulfur which is present in small quantities in the slag. These can cause environmental and occupational health problems. Acid mist also causes corrosion of steel. The high water and energy consumption, the loss of the sensible heat from the slag, and sulfur emissions make present conventional technologies for blast furnace slag handling and treatment quite unsatisfactory. While some technologies are available for enclosing the granulation process to minimise sulfur emissions, there is great potential to develop new technologies to recover the heat and produce a valuable by-product without the high water consumption and recycle rate. This is discussed further in Section 16.5.2.

Steel-making slag

Hot metal from a blast furnace is converted to steel by reaction with oxygen to remove impurities. This is most commonly done in basic oxygen furnaces (BOFs). About 100 kg of slag is produced per tonne of steel. BOF slags consist largely of silica, lime and iron oxides (Table 11.2) but, unlike blast furnace slags, contain significant quantities of entrained metal, typically 5–15 wt%. The entrained metal is the most valuable component and is recovered and recycled to the BOF. Slag from the BOF is poured into a ladle then tipped out to cool. The solidified slag is crushed to liberate the entrained metal, which is then removed using magnetic separators. The slag is then usually ground finely and the product sold for aggregate-type applications (e.g. asphalt aggregates, road base and sub-base materials) and as general fill. Some can be recycled within the steelworks, to the sinter plant or blast furnace, but the quantity is limited by the requirement for the slag to be an outlet for undesired elements in steel. BOF slags have high lime contents (~45 wt% CaO) and contain around 1 wt% phosphorus. They could be used as a soil additive to improve productivity. About 6% of BOF slag produced in Germany is used in this way (Janke *et al.*, 2006). The presence of heavy metals in the slag, such as manganese and chromium, may preclude wide application as a soil additive.

Base metal smelting slags

Globally, the quantity of slag produced from base metal smelting is much smaller than from iron and steel-making because of the lower production rates of these metals. Also, the scale of operation of base metal smelters is usually smaller than that of integrated iron and steel producers so the volume produced at any one site is relatively small. There is much greater variation in the types of slags produced, for the reasons discussed in Section 8.2. Both air and water granulation methods are used, the techniques being similar to those for iron- and steel-making slags.

Copper and nickel are produced from sulfide concentrates in a two-stage process. The concentrate is smelted to form matte and slag, then the matte is oxidised in a converting step to produce blister copper or impure nickel sulfide, which is subsequently leached to produce metallic nickel. Matte smelting slags may be discarded directly to landfill if their copper or nickel content is not high enough to warrant recovery, or they may be treated to recover metal values. They are often allowed to solidify, then ground to liberate the entrained copper matte. This is recovered by froth flotation and returned to the smelter. The tailings are stored in a tailings storage facility. Copper and nickel converter slags, being highly oxidised, contain high levels of dissolved copper and nickel and entrained metal droplets due to the high agitation of the contents in the converter. These slags are processed to recover value, usually in a reducing and settling furnace (a slag-cleaning furnace) or by slow-cooling the slag to enable copper or nickel-bearing phases to form. The latter are then liberated by grinding and recovered by froth flotation. The concentrate is returned to the matte smelter and the tailings are stored in a tailings facility. Slag from a slag-cleaning furnace can be discarded to landfill or, if water-quenched to retain its glassy structure and stabilise potentially leachable toxic elements (e.g. copper, chromium, manganese, nickel), can be used in aggregate applications, particularly as road base. A recent trend has been to use converter slag as a grinding medium in tower mills for grinding ore for flotation. In this manner, metal values are recovered from the slag during the subsequent flotation and the remainder of the slag is discarded in the flotation tailings.

12.2 MANAGEMENT OF LIQUID WASTES

The waters produced on mine sites can contain a wide range of contaminants; some of the more common, and their sources, are listed in Table 12.1. Not all will always be found at a particular mining operation. These, and any other impurities in water, have to be reduced in concentration to levels acceptable for the purposes for which the water may be used, such as those listed in Table 11.6.

12.2.1 Technologies for water treatment

Many technologies have been developed to make water safe to drink or release to the environment. These are increasingly being adopted for treating and recycling domestic urban and industrial water. Many are also used by the mining industry to treat on-site water for recycle or discharge. The treatment technology used at a particular site depends on the current

Table 12.1: Common contaminants of water on mining sites, and their sources

Contaminant	Typical source
Metals – Fe, Mn, Cr, Zn, Cu etc.	AMD from oxidation of pyrite in mines, waste rock piles, tailings dams
Sulfate ions	AMD from oxidation of pyrite in mines, waste rock piles, tailings dams
Acidity (H^+ ions)	AMD from oxidation of pyrite in mines, waste rock piles, tailings dams
Cyanide	Spillage from gold leaching operations; spillage and seepage from gold tailings dams
Suspended solids	Inadequate underground settling; run-off from surface, tailings dams, rock piles etc.
Sodium ions	From groundwater, particularly artesian water and land-locked inland lakes; addition of sodium-based reagents (e.g. for flotation, leaching of bauxite in Bayer process, neutralisation)
Chloride ions	From groundwater, as above
Nitrogen compounds	Wastes from explosives (e.g. AMFO); sewage and other domestic wastes
Phosphate ions	Sewage and other domestic wastes
Radionuclides	AMD attack on radionuclide-bearing rocks
Microbes	Human faecal contamination; run-off from livestock grazing

Source: Pulles *et al.* (1995); with modifications.

and required water quality and the quantity to be treated. In general, water treatment systems consist of three stages: pretreatment, primary treatment and post-treatment, or polishing. A brief description of present technologies is given below (Kuyucak, 2009; Lévy *et al.*, 2006; Cartwright, 2006).

Thickening, clarification and filtration

Thickening, clarification and filtration technologies are well known to the mining and mineral processing industries where they are used extensively to recover solids and liquids, or to thicken tailings. They are used as pretreatments to remove suspended solid particles. Coagulants and flocculants are usually added to improve the efficiency of separation. The most common coagulants are mineral salts (of aluminium and iron) and the most common flocculants are synthetic polyacrylamide polymers.

Precipitation

Chemical precipitation using alkaline compounds is the most common primary treatment method. It is used to remove dissolved heavy metals and to neutralise acid solutions. Alkaline compounds such as lime ($Ca(OH)_2$), calcium carbonate ($CaCO_3$) and magnesium hydroxide ($Mg(OH)_2$) are readily available for this purpose, are relatively cheap, and are easy to use. All elements that precipitate as hydroxides can be removed in this way. The processes are simple and require minimal equipment. This aspect is discussed further in Section 12.2.3. The main disadvantages are that they produce a large quantity of sludge that requires disposal, and that the pH of the treated water needs to be adjusted back to neutral values. Chemical precipitation as sulfides is also used for removing heavy metals by the addition of a soluble sulfide such as sodium sulfide (Na_2S). Sulfide precipitation can be used to remove Pb, Cu, Cr^{6+}, Ag, Cd, Zn, Hg, Ni, Tl, Sb and V.

Membrane technologies

Membranes are a layer of a substance which acts as a barrier to the flow of small inorganic and organic substances ranging in size from about 1 μm down to 0.0001 μm. Membrane processes include microfiltration, ultrafiltration, nanofiltration, reverse osmosis (RO) and electrodialysis (ED). All usually require pretreatment of water by conventional clarification or filtration. The first four are filtration processes which operate under increasingly higher pressures; some of their characteristics are listed in Table 12.2. Electrodialysis is an electrochemical process which produces water similar in quality to that produced by reverse osmosis. Membrane processes have improved

Table 12.2: Pressure-driven membrane processes

	Pressure required (Pa)	Pore size (m)	Components removed
Microfiltration	0.7×10^5	10^{-6}–10^{-7}	Suspended particles and large colloids
Ultrafiltration	$1.0–7.0 \times 10^5$	10^{-7}–10^{-9}	Colloids, proteins, microbiological contaminants and large organic molecules
Nanofiltration	$3.5–16 \times 10^5$	10^{-9}	Organic molecules, divalent and some monovalent ions (e.g. SO_4^{2-})
Reverse osmosis	$5.0–84 \times 10^5$	10^{-9}–10^{-10}	Monovalent ions (e.g. Na^-, Cl^-), micro-solutes

Source: Lévy *et al.* (2006).

greatly in recent years with the decrease in price of membranes, the development of higher-pressure vessels, recycling pumps and energy recovery devices. Microfiltration is used to remove larger particles, such as algae or clays, from water. Microfiltration and ultrafiltration are usually performed before nanofiltration to protect the membrane elements from fouling. Reverse osmosis and electrodialysis were developed to produce demineralised water for the pharmaceutical and food industries and to provide clean water for boilers for power stations. Water produced by these technologies has to be re-mineralised for drinking purposes. Reverse osmosis and electrodialysis remove up to 99% of all contaminants. The water produced by nanofiltration is less demineralised and the process is less energy-intensive. All these technologies are increasingly being used to treat industrial waste waters.

Ion exchange

Ion exchange technology is used to demineralise water (for boiler feed) and to recover specific elements. In the minerals industry, ion exchange technology was developed to selectively recover and concentrate metals in solution in hydrometallurgical processes, particularly in the recovery of uranium. Its use is limited by the salinity of the water. In the minerals industry, ion exchange has potential use as a 'polishing' water treatment process, to produce very clean water. Most ion exchange substrates are specially developed organic resins but some natural materials such as zeolite clays have ion exchange properties and can be used.

Biological processes

Biological systems can be useful in some water treatment applications in the minerals industry; for example, removing organic materials (e.g. oxalates in alumina production by the Bayer process) and nutrients from waste streams, removal of sulfate and neutralisation of acid mine drainage, and production of sulfides and subsequent precipitation of metals such as copper, nickel and arsenic from copper refinery electrolyte.

12.2.2 Management of cyanide solutions

Following the Baia Mare cyanide spill in Romania (Section 11.5.1), a cyanide management code was developed for the use of cyanide in the gold industry (International Cyanide Management Code, 2009). This is a voluntary code, administered by the International Cyanide Management Institute, to promote responsible management of cyanide used in gold mining, enhance the protection of human health, and reduce the potential for environmental impacts. Companies that become signatories to the code have their operations audited by a third party; the audit results are made public. Most large gold producers, and many smaller producers, as well as cyanide transportation companies, are signatories to the code.

There are four basic approaches to reducing cyanide levels in tailings (DITR, 2008):

- natural degradation;
- enhanced natural degradation;
- chemical and biological methods;
- recovery and recycling.

Natural and enhanced degradation and chemical and biological methods rely on the fact that cyanide ions in solution are relatively unstable and can be fairly easily decomposed into benign products. In natural degradation, at least seven mechanisms are responsible for the natural breakdown of cyanide (Lottermoser, 2003). Of these, volatilisation as gaseous hydrogen cyanide (HCN) is the most important. Cyanide lost to the atmosphere by volatilisation from waste waters, tailings and heap leach piles disperses to the natural background concentration and degrades to form ammonia and carbon dioxide. The overall reaction is:[25]

$$(CN^-) + 0.5\{O_2\} + 2H_2O = \{NH_3\} + (HCO_3^-)$$

At some mine sites, natural degradation is sufficient to meet requirements for release of water to the environment. Natural degradation can be enhanced, if necessary, by lowering the pH of the solution, which makes CN^- ions more available for reaction. Increasing the surface area of tailings dams increases absorption of carbon dioxide from the atmosphere and lowers the pH. Repeated washing of leach heap piles with water lowers the pH and enhances volatilisation. Aeration of cyanide-containing solutions has a similar effect. Chemical methods of degradation involve oxidising dissolved cyanide by adding appropriate reagents. The most commonly used are hydrogen peroxide (H_2O_2), Caro's acid (prepared on-site from sulfuric acid and hydrogen peroxide), chlorine gas, and a sulfur dioxide–air mixture. Precipitation of metal cyanides by addition of ferric chloride (or ferric sulfate) and lime is another method. Microorganisms have a natural capacity to decompose cyanide and biological degradation can be used on heap leach piles, in artificial wetlands and in bioreactor tanks. However, bacteria require a correct nutrient balance and nutrients may have to be added (Lottermoser, 2003). Cyanide can be recovered from solutions using various non-oxidative processes. In one process, tailings are contacted with high volumes of turbulent air in stripping columns and the volatilised HCN is absorbed in calcium hydroxide slurry in absorption columns. The recovered cyanide is then reused in the leaching operation as calcium cyanide (Stevenson *et al.*, 1995).

25 The conventions adopted in this and subsequent chemical equations to indicate the nature of the phases involved are summarised in Table 8.2.

12.2.3 Management of AMD

AMD is often the major source of contaminated water at a mine site. As discussed previously (Section 11.3.2), it forms from pyrite-containing solid materials in contact with air and water. These materials include waste rock, heap leach residues, tailings, and exposed rock in the pits of open cut mines and mined areas of underground mines. Since AMD frequently contains dissolved metals, degradation of AMD to benign products, as can be done with cyanide, is not possible. Management of AMD is approached from two directions simultaneously. The first aim is to minimise and control the formation of AMD, and its subsequent movement; the second is to treat AMD to allow reuse or safe discharge of the water to the environment, and safe storage or disposal of the captured metals. Minimisation is preferred over control, and control is preferred over treatment. Hence, these represent a hierarchy of strategies.

Minimisation of AMD

Oxidation of pyrite must be prevented, to stop the formation of AMD. This means that ingress of oxygen or water, or preferably both, to pyrite-bearing material must be prevented or, at least, minimised. Selectively placing potentially acid-forming (PAF) materials (rock waste and tailings) and encapsulating them in benign waste is the preferred management practice during mining operations (DITR, 2007b). Storage sites should have drainage channels on the surface and clean water from upstream should be diverted around or under the storage site (through lined, buried drains). Typically, waste rock piles are constructed by placing a layer of PAF rock waste on a base layer of benign waste rock, then covering it with benign rock waste. The top of the pile is then covered with material to limit ingress of rain and air. To minimise water content due to rainfall, piles should be constructed sequentially. Each pile should be constructed to its full height and covered before a new one is commenced. When a rock pile storage facility is closed, a spillway may need to be added to cope with future rainfall and run-off.

The oxidation of reactive tailings in surface storage facilities is due mainly to diffusion of oxygen through the dried surface of the tailings. The oxidation products may then be transported by tailings water, rainfall run-off or seepage from the facility. Seepage may be reduced by use of a liner of impermeable material, such as clay, on the base. Tailings should be deposited as dry as possible. Evaporative drying, by cycling the deposition of tailings between cells in the storage facility, should be used to limit the amount of seepage. A covering should be placed over waste piles and surface storage facilities for tailings when they are closed. This usually consists of a layer of topsoil on a layer of clay, to reduce rainfall infiltration. The soil is usually separated from the clay layer and the clay layer from the reactive waste below by layers of crushed benign rock.

Blending and co-disposal approaches are also used to stabilise PAF materials. They can be mixed with cement, or a mixture of cement and unreactive tailings, and placed in underground voids as backfill. The most effective way to prevent exposure of reactive wastes to oxygen is to deposit them permanently under water. This is possible only where there is an assured supply of water and where the topography is suitable. Surface storage requires a valley containment with a catchment of sufficient size to maintain a water cover. Another option is to backfill a pit with PAF waste rock to the top of the watertable, then fill the rest with benign waste. Flooded underground workings may also provide permanent storage sites for PAF wastes under water. However, these approaches may make it impossible to reopen the mine or to reprocess the waste rock.

Treatment of AMD

The second approach to managing AMD is to treat it after it has formed, to produce water for reuse on the mine site or release to the environment. Treatment of AMD can be a costly part of mining operations and potentially even more costly after mine closure. Hence, prevention is preferable. There are many possible treatment methods. Selection of the appropriate method depends on site-specific conditions, including the composition of the AMD and the purity targets. Both active and passive treatment systems are used. Passive systems are those that require little or no mechanical pumping of water or mechanical addition of chemical reagents. They are best suited to treating waters with low acidity (<800 mg L^{-1} $CaCO_3$) with steady flow rates. When these conditions are not met, for example by sudden surges in the volume of AMD or acidity, the systems tend to fail and collapse.

Active treatment systems. By far the commonest active treatment method is neutralisation and precipitation of metal hydroxides by addition of lime; other methods in use include ion exchange, membrane separation and bioreactors (DITR, 2007b). Lime treatment is preferred because it is cheap and effective. Lime treatment involves dissolving lime (in the form of quicklime or hydrated lime) in water to produce Ca^{2+} and hydroxyl (OH^-) ions. The calcium ions react with sulfate ions to form gypsum, which is practically insoluble and precipitates as a solid:

$$(Ca^{2+}) + (SO_4^{2-}) + 2H_2O = [CaSO_4.2H_2O]$$

and the hydroxyl ions combine with the dissolved metals to produce metal hydroxide precipitates:

$$(Al^{3+}) + 3(OH^-) = [Al(OH)_3]$$
$$(Cu^{2+}) + 2(OH^-) = [Cu(OH)_2]$$
$$(Fe^{2+}) + 2(OH^-) = [Fe(OH)_2]$$
$$(Fe^{3+}) + 3(OH^-) = [Fe(OH)_3]$$

The mixed precipitate of gypsum and metal hydroxides, called *sludge*, is then separated from the water to produce a clear effluent, which may meet internal recycle or discharge purity criteria or require further treatment. The sludge must be disposed of in an environmentally acceptable manner (Section 12.1.3). Iron, zinc and copper are precipitated at a pH of around 9.5 while nickel and cadmium are precipitated at a pH of around 10.5–11 (Aubé, 2004). Some contaminants, particularly arsenic and molybdenum, cannot be removed simply by controlling pH; these must be co-precipitated with other metals, usually ferric iron. For this reason, and the fact that ferrous hydroxides do not settle as well as ferric hydroxides, AMD solutions are often aerated to oxidise the ferrous iron to ferric iron.

Pond and pit treatment of AMD are the simplest and cheapest techniques. They involve adding lime in a stream and allowing the precipitates to settle in a pond or pit (Aubé, 2004). Two ponds are often used. The first serves to accumulate the sludge (and requires

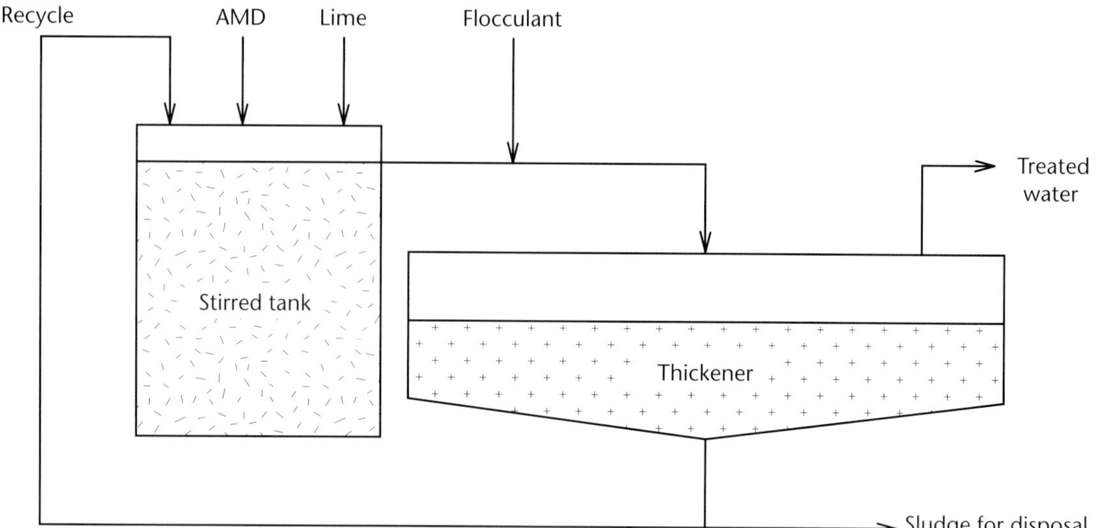

Figure 12.4: Simplified flowsheet for treating AMD, with recycle of some precipitate to improve efficiency (the high-density sludge process).

periodic dredging) and the second, larger pond with longer residence time, allows for final polishing of the effluent. Treating AMD in a pond or pit does not allow much control of the system. It is difficult to oxidise ferrous iron because the sludge does not settle if air sparging is used. Also, the efficiency of lime utilisation is low and much is wasted. An alternative for operating mines is co-deposition. Lime and AMD are added to tailings as they are pumped to the storage facility. This has the advantage of not requiring a separate sludge storage area. Typically, an excess of lime is added to ensure that the minimum pH is attained, thus co-precipitation is also fairly inefficient. There is a possibility that the sludge could redissolve if the tailings continue to oxidise. There is little information on the long-term effects of mixing sludge and tailings, though some studies have shown that sludge helps decrease the permeability of tailings and may therefore help maintain saturation and inhibit oxidation (Aubé, 2004). If the tailings are already acid, partial dissolution of the sludge is probable.

Purpose-built plants are also used to treat AMD. In a conventional plant, the AMD is neutralised in a mixing tank with addition of lime to give the desired pH. A flocculant is then added and the slurry is pumped to a thickener for solid–liquid separation. The sludge is removed from the bottom of the thickener and pumped to a storage area, or filtered to increase its density prior to transporting off-site. This process has a better utilisation efficiency of lime than do pond or pit treatment. The precipitates tend to form as particles less than 1 μm in diameter, which makes settling and separation of the sludge difficult. It is usual to recycle some of the sludge from the clarifier back to the mixing tank to provide 'seeds' onto which fresh gypsum and metal hydroxides can precipitate. This results in coarser particles, faster settling times and sludge with up to 20 wt% solids (compared with less than 5 wt% solids without recycle). Figure 12.4 shows a flowsheet of the process as it is widely practised. A further refinement is to mix lime and recycled sludge in a tank before they contact AMD in another tank. This causes the precipitation reactions to occur mainly on the surface of existing particles, thereby increasing their size and density. These processes are called high-density sludge processes.

Passive treatment systems. Passive treatment systems allow the naturally occurring chemical and biological reactions that aid in AMD treatment to occur in the controlled environment of the treatment system. In principle, passive systems have some advantages over active systems. The use of chemicals and energy (other than from the Sun) are practically eliminated and the operating and maintenance requirements are much less. Many types of passive systems have been proposed or adopted (PDEP, 2009) but, as noted earlier, with mixed success. The simplest passive system consists of open channels containing limestone through which AMD flows. The channels may be dug and lined with crushed limestone, or crushed limestone may be simply added along an existing stream. Impervious liners are sometimes placed under the limestone to prevent AMD seepage into the groundwater. Diversion wells are another simple system. AMD is conveyed by pipe to a downstream well which contains crushed limestone. Channels and diversion wells require periodic refilling with limestone, as it is consumed. A variation is a buried bed of limestone constructed to intercept subsurface AMD flows and prevent contact with oxygen in the air (an anoxic drain). This prevents oxidation of the metal ions, particularly ferrous ion, which helps to reduce formation of a coating of precipitate on the limestone. When the limestone becomes coated with precipitate it forms a barrier to further reaction. A sufficient flow rate and turbulence is required in all these systems to keep the precipitates in suspension. Anoxic drains are often used as a pretreatment step before the water is released to a wetland for further treatment.

AMD can be treated in wetland environments, which can be aerobic or anerobic (US EPA, 2006). Aerobic environments (in which oxygen is present) promote oxidation of metals, particularly manganese and iron. After oxidation, these will precipitate in neutral (or near-neutral) waters and potentially remove other contaminants, such as arsenic, by co-precipitation. Anaerobic environments (in which oxygen is excluded) contain microorganisms that promote the reduction of sulfate, nitrate, oxidised metals and metalloids (selenium, arsenic and antimony). A by-product of a number of anaerobic microorganisms is bicarbonate ions, which increase the pH and promote precipitation of insoluble metal hydroxides and carbonates. Sulfide ions produced from sulfate reduction promote the precipitation of metal sulfides (of Cu, Cd, Zn, Pb, Ni, Mo, As, Sb and Fe) under a wide range of chemical conditions. Under anaerobic conditions, Cr^{6+} and U^{6+} can be reduced by a number of microorganisms to Cr^{3+} and U^{4+}, respectively, and subsequently precipitated as hydroxides. The removal of some metals is facilitated by co-precipitation with aluminium and iron hydroxides.

Aerobic wetlands are shallow bodies of water (usually less than 0.7 m deep) in which plants grow. Anaerobic wetlands are subsurface bodies of water that support the growth of plants on the surface. Contaminated water is intercepted and diverted through the wetland system. The vegetation and sediments provide surfaces for the growth of bacteria. The construction of wetlands is relatively inexpensive compared with active treatment systems, and their operation and maintenance costs are lower. Wetlands are temperature-sensitive. Low temperatures reduce bacterial activity, and hence the rates of iron and manganese oxidation, and ice covers limit the rate of oxygen transfer to the wetland from the atmosphere. A larger area of wetland is required in colder climates than in more temperate regions. Highly variable flows may result in the re-suspension and flushing of collected precipitates; thus, subsequent polishing ponds are required. Collection of precipitates and sediments, and loss of permeability, leads to the need to rebuild the entire system periodically. The most successful applications of aerobic wetlands have been for AMD from coal mines, not from metal mines.

Management of AMD post mine closure

The volume of AMD requiring treatment can increase or decrease after a mine closes. The opportunities for use of water on-site decrease after closure, which may result in increased mine drainage volumes. On the other hand, completion of rehabilitation work after closure may decrease the ingress of water into old mining operations and result in decreased mine drainage. The long-term management and support for post-closure operation and maintenance of AMD treatment facilities may be quite limited and, where appropriate, passive treatment technologies are preferred after mines close. Mine planning needs to take into account the land requirements of post-closure water treatment in the design of tailings storage facilities and mine waste dumps to ensure that space is available and that post-closure water treatment does not become a major design constraint that forces the implementation of active treatment technologies after closure (Global Acid Rock Drainage Guide, 2009).

12.3 GASEOUS WASTES

Furnaces, roasters and other reactors used in thermal processing of mineral commodities are designed so that the gases produced during operation can be contained and collected by hoods and ducts. Fans are usually operated in-line in ducts and flues to maintain a pressure slightly below atmospheric pressure so gases will flow preferentially into the collection system. The gases, after treatment if required, are vented to the atmosphere through tall chimneys (stacks), which help create a draught and disperse the gases over a wide area. Process gases often contain fine solid particles (fume), which are removed prior to release or further processing of the gas. The gases are usually produced at high temperatures and therefore contain heat that may be able to be used elsewhere in the process or used to generate electricity (co-generation). If so, the gas is passed through heat exchangers to recover the heat. The basic functions of a gas handling system therefore include some or all of the following: containment and capture of process gas at the reactor exit; cooling the gas to a temperature suitable for subsequent handling or cleaning, or both; condensation of volatile components; separation of particulate matter (dusts and fumes); and transfer of the gas to gas cleaning, sulfur fixation or the stack. CO_2, H_2O and N_2 are not noxious gases and are released to the atmosphere, though the increasing concern about global warming may mean that in the future CO_2 will have to be captured and sequestered.

12.3.1 Gas cooling and heat recovery

A simple method of cooling gas is to dilute it with large volumes of air. Another is to pass the gas through ducts or a tower into which water is sprayed. This is effective, but it consumes large quantities of water through evaporation. Neither method allows heat in the gas to be recovered. A more sophisticated method is the use of water jackets around ducts to extract heat from the gas as it flows through the ducts. Water flowing through the jackets is heated; this heat can be used elsewhere and the water recycled. However, this is usually low-grade heat (less than 400°C). Waste heat boilers and fluidised bed heat exchangers provide the means to generate steam, under pressure at much higher temperatures, which can be used for production of electricity. Low-grade heat can also be recovered using recuperators and regenerators. Uses of low-grade heat include drying and preheating of raw materials, preheating of reaction air (used in the iron blast furnace), providing heat to other processes, heating of offices and accommodation, and providing heat for the local community. Heat recovery technologies are usually used only on continuous processes; batch processes, such as Peirce-Smith and BOF converting of copper and steel, respectively, generally do not have waste heat recovery systems.

Conventional *waste heat boilers* consist of a radiation section and a convection section (Paykan and Russell, 2010). In the radiation section, heat from the hot entry gas is transferred to pressurised water flowing through tubes (typically 40 mm diameter) in the roof and walls of a large rectangular chamber, cooling the gas to around 700°C. In the convection section, heat is transferred from the cooler gas to the water flowing through steel tubes suspended in the path of the gas in a long narrow chamber, cooling the gas to around 350–400°C. The radiation and convection sections may be arranged horizontally or vertically. Waste heat boilers can be designed to produce either saturated or superheated steam.[26] Superheated steam is usually used for power generation and saturated steam is used for other process applications.

Problems associated with the cooling of smelter gases include the condensation of volatile species, and the transition of molten material carried over in the gas through a 'sticky' stage before becoming solid. Surfaces therefore tend to become coated with solid material, which reduces the rate of heat conduction through the surfaces. Furnaces have to be shut down periodically to allow the surfaces of ducts and waste heat boilers to be cleaned. Key design challenges of gas cooling systems are minimising the quantity of material that adheres to surfaces, and making cleaning of the surfaces as simple as possible. When SOx are present in the gas, SO_3 can react with water vapour in the gas to form sulfuric acid. If the gas is cooled below

26 Saturated steam is steam at equilibrium with liquid water. When saturated steam is heated at constant pressure, its temperature rises, producing superheated steam.

the dew point[27] of sulfuric acid it will condense to liquid and cause corrosion problems in the steel ducts. The exit temperature of waste heat boilers in these cases should be kept above the dew point.

Fluidised bed reactors (Figure 8.6) have good heat exchange characteristics, and *circulating fluidised bed heat exchangers* have been developed to overcome some of the problems of conventional waste heat boilers (Dry and Beebie, 1997). The concept is illustrated in Figure 8.17, showing the heat recovery system on the ISASMELT™ copper smelter at Mt Isa in Queensland, Australia. The hot gas, containing sticky materials, is fed directly into the conical base of a distributor-free circulating fluidised bed reactor. Ground slag is used as the bed material. The gas meets the mass of cool particles in the bed and is quenched to about 450°C. Molten material in the gas freezes in-flight or on the colder particles rather than on the walls of the system. The quench also helps to minimise the amount of oxidation of SO_2 to SO_3. Heat is extracted by water-cooled tubes in the vertical riser section above the bed. The bed material leaves the furnace entrained in the gas. It is separated from the gas in two cyclones, operating in parallel, and returned to the bed.

Recuperators are gas-to-gas heat exchangers used in some metallurgical processes to recover low-grade heat from waste gases. They usually consist of a series of parallel tubes through which flows the gas that is to be heated or cooled. The tubes are housed within a vessel, or shell; the second gas flows from one end of the shell to the other over the tubes and provides the heat or absorbs the heat required. Usually the two gases flow counter-currently. These are referred to as shell-and-tube heat exchangers. Shell-and-tube heat exchangers can operate at atmospheric pressure or at elevated pressures, according to their design and type of construction materials. They have a low capital cost and can handle dusty gases. The heat is frequently recovered in the form of preheated air. However, the heat recovery savings are limited by the demand from the available end-users. *Regenerators* are heat exchangers which contain heat-storage materials which absorb heat from hot gases then give up that heat to preheat cold air. Thus, they serve a similar function as recuperators. In its simplest form, a recuperator consists of a chamber filled with a checkerwork of bricks laid in a pattern to create a multitude of vertical gas passages. The hot waste gas and cold air (or other gas to be preheated) are passed alternately through the chamber to heat then cool the bricks. To provide for a continuous flow of waste gas and preheated air, at least two chambers are required; while one is being cooled the other is being heated.

12.3.2 Gas cleaning

Figure 12.5 shows some key characteristics of particulate matter, and the types of gas cleaning methods which are applicable to different size particles. Several types of devices are used for separating solids from gas streams: settling chambers, cyclones (and other centrifugal separators), bag filters, scrubbers and electrostatic precipitators. These are illustrated schematically in Figure 12.6. These devices have progressively finer particle size recovery limits, as indicated in Figure 12.5. In practice, gases are often passed through a series of cleaning units which progressively remove finer particles; see, for example, Figure 8.17. The efficiency of gas–solid separation devices decreases as the particle size of the particulate decreases. Settling chambers, cyclones and electrostatic precipitators can operate with hot gas. Bag filters require relatively cool gas, to avoid damage to the cloth filter medium – hot gases must be cooled before they are filtered. Scrubbers can operate with hot or cool gases. The residues from gas cleaning and scrubbing processes contain potentially harmful impurities (see Tables 11.2 and 11.7), which need to be disposed of or stored appropriately.

Settling chambers are the simplest gas–solid separation device. They consist of a chamber in which the gas velocity is reduced by allowing it to expand into a larger volume, thereby allowing particles to settle out under the action of gravity. They generally consist of a long rectangular chamber with an inlet at one end and an outlet at the side or top of the other end. *Cyclones* operate on the same principle as hydrocyclones used for separating solid particles from water

27 The dew point is the temperature to which a gas must be cooled for a particular volatile component to condense. It is the temperature at which the vapour pressure of the pure component is equal to the partial pressure of the component in the gas.

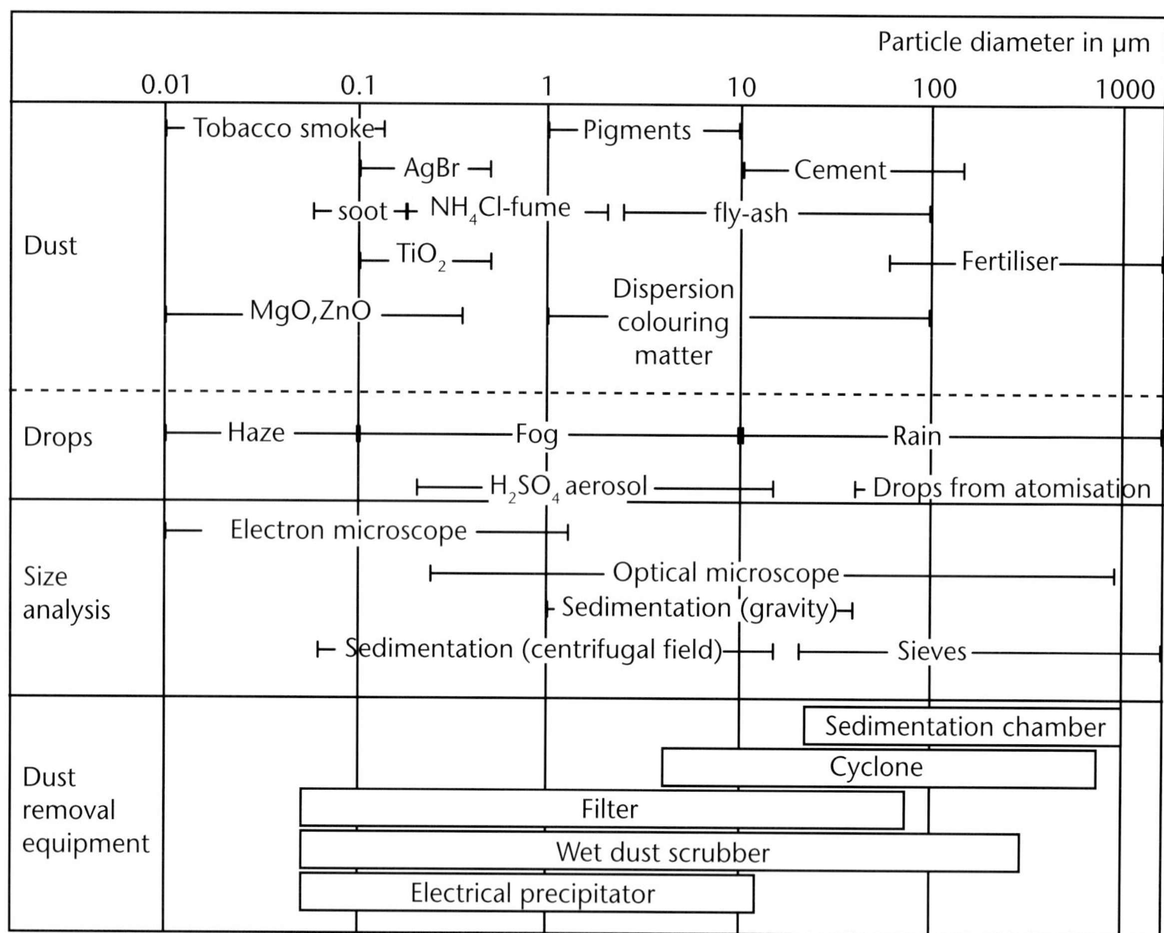

Figure 12.5: Characteristics of particles (after Bauer and Varma, 1981).

(Section 7.2.2). *Bag filters* are typically 100–200 mm in diameter and 2.5–5 m long, open at one end, and made of woven fabric or felt. Several bags are hung vertically, open end downwards, in a baghouse and the gas is drawn into the bags from the bottom. Dust collects on the inner surface of the bags and the cleaned air passes out at the top of the baghouse. The bags are periodically shaken by a mechanical device. The dust falls to the floor of the baghouse, from which it is regularly removed.

Scrubbers are a class of devices for removing particles or gaseous impurities from gas streams using droplets of water or solid particles. The simplest type is a spray tower in which water droplets, produced by spray nozzles, settle through a rising stream of gas. The droplets are typically 0.5–1 mm in diameter, which ensures they collect particles effectively and are heavy enough to settle against the upwards flow of gas, which is typically around 1 m per second. Cyclone spray scrubbers are more efficient since the difference in velocity of the droplets relative to the gas stream is greater. Tower or column scrubbers, in which a gas stream is passed upwards through a bed of inert particles through which liquid flows downwards, are another common type of wet scrubber. These are frequently used to remove reactive gases from a gas stream. For example, the sulfur dioxide content of low-concentration gas streams, such as coal-fired power station gas, sulfuric acid plant tail gas and smelter ventilation gases, can be reduced to less than 20 ppm SO_2 in wet scrubbers using sodium, calcium, magnesium or aluminium hydroxide solutions.

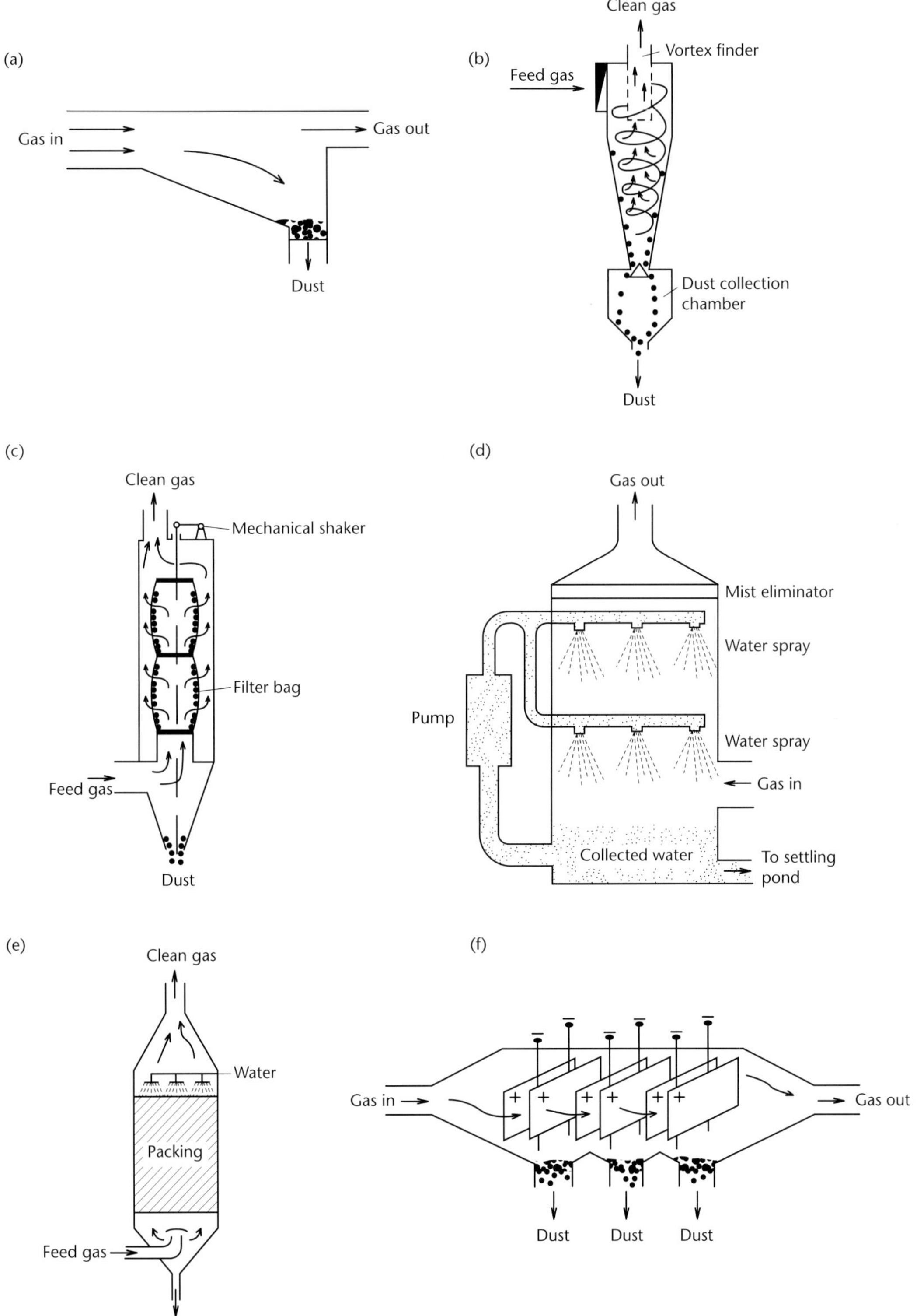

Figure 12.6: Large-scale devices for removing particulate matter from gas streams. (a) Settling chamber. (b) Cyclone. (c) Baghouse. (d) Wet scrubber – spray tower. (e) Wet scrubber – packed bed. (f) Electrostatic precipitator.

Dry scrubbers are used mainly to absorb specific impurity gases onto solid particles rather than to remove particulates from the gas. These utilise packed or fluidised beds through which the gas flows. Alternatively, the solid particles may be entrained in the gas stream long enough for reaction, then separated using, for example, cyclones or bag filters. A particularly important application of dry scrubbers is removing volatile fluorides from the gases produced during aluminium smelting. Some or all of the alumina feedstock for the reduction cells is passed through the dry-scrubbing system to absorb fluorides and other volatiles onto the surface of the alumina particles. Two systems are commonly used. Either alumina powder is injected into a gas duct, or the gas is passed through a fluidised bed of alumina powder. The alumina is then separated from the gas and recovered using bag filters. This is a particularly effective way of capturing fluorine (up to 99% recovery), which would otherwise escape with the flue gases, and returning it to the reduction cells. Dry scrubbing with alumina powder is also used to scrub organic volatiles from fumes produced during the baking of anodes for aluminium smelting.

Electrostatic precipitators operate on the same principles as electrostatic separators (Section 7.2.2). Electrostatic precipitators are designed to simultaneously:

- ionise the gas;
- charge the particulates, causing them to move towards the charged surface;
- allow sufficient residence time in the precipitator for the particles in the gas to move and attach to the surfaces;
- prevent re-entrainment of the captured particulates;
- allow the particles to be removed from the device during operation.

There are two main types: single-stage, in which ionisation and particle collection are performed in one chamber, and two-stage, in which ionisation is achieved in one section and collection in another.

12.3.3 Sulfur dioxide removal

Sulfur dioxide is a major component of gases produced during the roasting and smelting of sulfide concentrates, particularly during the production of copper, nickel, lead and zinc, and during the roasting of sulfidic gold-bearing concentrates to release the gold for leaching. Gas streams that contain greater than 5 vol% SO_2 can be used to produce sulfuric acid using the contact process, though the cost of production increases the lower the concentration of SO_2. The tail gas from sulfuric acid plants is released to the atmosphere, after further scrubbing. In acid production from smelter gases, the gas stream is cleaned to remove solid particles then cooled to about 400°C using a combination of waste heat boiler and gas cleaning stages (e.g. settling chambers, sprays, cyclones, electrostatic precipitators). Air is added if necessary, to ensure excess oxygen. The gas is then passed over vanadium pentoxide catalyst in a packed bed column reactor in a single or double contact process, in which sulfur dioxide is converted to sulfur trioxide gas:

$$\{SO_2\} + 0.5\{O_2\} = \{SO_3\}$$

In double contact plants, SO_3 is removed from the gas stream, which is then passed through a second reactor to achieve a higher overall conversion. About 97% sulfur conversion is achieved in single contact acid plants and more than 99.5% in double contact plants. For smelters with a single contact plant, conversion to a double contact plant has to be made or acid plant tail gas scrubbing is required to achieve an overall sulfur recovery of 99%. To achieve very high recoveries some smelters (e.g. in Japan) scrub the acid plant tail gases from double contact acid plants. The conversion reaction is exothermic and the temperature of the gas rises to about 600°C. The gas from the reactor is cooled in a heat exchanger then passed upwards through an absorption tower containing an inert packed bed down through which flows 98% concentrated sulfuric acid. The SO_3 is absorbed by the sulfuric acid to form oleum (H_2SO_7). The oleum leaving the base of the tower is diluted with water back to 98% sulfuric acid. Some of this is recycled to the tower, the remainder being the product for sale or internal use.

Sulfuric acid is highly corrosive and very hazardous. For this reason it is not transported great distances in large quantities. Where large quantities are required, it is manufactured locally. If a suitable waste

Table 12.3: GRI sustainability indicators for solid, liquid and gaseous wastes

EN16	Total direct and indirect greenhouse gas emissions by weight	Core
EN17	Other relevant indirect greenhouse gas emissions by weight	Core
EN18	Initiatives to reduce greenhouse gas emissions and reductions achieved	Voluntary
EN19	Emissions of ozone-depleting substances by weight	Core
EN20	NO, SO and other significant air emissions by type and weight	Core
EN21	Total water discharge by quality and destination	Core
EN22	Total weight of waste by type and disposal method	Core
EN23	Total number and volume of significant spills	Core
EN24	Weight of transported, imported, exported or treated waste deemed hazardous under the terms of the Basel Convention Annex I, II, III and VIII, and percentage of transported waste shipped internationally	Voluntary

Source: GRI (2006).

gas stream is not available, elemental sulfur is usually purchased and burned to produce the SO_2. Hence, acid production as a way of removing SO_2 from smelter gas streams is feasible only if there is a local use for the large quantities of acid. There are many examples of this. At Chuquicamata, in Chile, acid produced from the copper smelter gases is used to leach oxidised ores and low-grade waste heaps. At Mt Isa in Queensland, Australia, sulfuric acid produced from copper smelter gases is railed 150 km south to Phosphate Hill, where it is used in the manufacture of fertiliser from phosphate rock.

12.4 WASTE, EFFLUENT AND EMISSION SUSTAINABILITY INDICATORS

The Global Reporting Initiative has two indicators for solid wastes, two for liquid wastes and five for gaseous emissions against which companies are required to report performance. These are listed in Table 12.3. Seven of these are mandatory (core) and two are optional (voluntary). In the survey of 25 mining companies that produced sustainability reports in 2007, Mudd (2009) found reporting against the above indicators to be highly variable. Some companies interpreted EN22 as referring to non-mining waste, such as municipal and hazardous wastes (e.g. batteries, oils, spent chemicals and medical wastes). Rarely was the quantity of waste rock reported, even though waste rock is the largest quantity of solid waste produced at a mine. Mudd (2009) stated, 'given the massive scale of mine wastes now being generated annually ... this problem cannot be ignored, downplayed or left to over-confidence ... The problem of large volume mine wastes is also cumulative. For these fundamental reasons alone it should be absolutely compulsory for all mines to report annually on the quantities of mine waste they generate, as well as the location, conditions and management of all mine wastes'. The level of reporting of spills (EN23) was variable. Some companies reported the number of incidents, others gave volumes, and others even reported on potential environmental impacts and follow-up studies. All the companies reported greenhouse gas emissions. Ten reported just the totals, one reported just the direct emissions, and only 12 reported both direct and indirect emissions. Two companies provided variable reporting across their individual mines or business units. Mudd stated, 'to further improve the level and accuracy of the data, it is important for mines to release site data as well as separately reporting direct and indirect emissions', and that 'it is good practice to cite the report or basis upon which emissions were estimated ... to ensure accurate interpretation and analysis of reported emissions data'. Few companies reported against EN18. Although some reported on programs to improve energy efficiency, the evidence on reducing emissions did not present a clear picture.

The emissions of noxious air pollutants (EN20), such as NOx and SOx, are of variable relevance according to the nature of the operations. For smelters, such emissions are significant. Some companies reported SOx, though less emphasis was given to NOx, and some demonstrated major reductions in SOx emissions due to implementation of capture or treatment processes.

12.5 REFERENCES

Aubé B (2004) The science of treating acid mine drainage and smelter effluents. <www.enviraube.com>.

Bauer H and Varma YBG (1981) *Air Pollution Control Equipment*. Springer Verlag: Berlin.

Cartwright P (2006) Process water treatment – challenges and solutions. *Chemical Engineering* **March**: 50–56.

DITR (2007a) *Tailings Management*. Leading Practice Sustainable Development Program for the Mining Industry. Department of Industry, Tourism and Resources, Commonwealth of Australia: Canberra. <http://www.ret.gov.au/resources/resources_programs/lpsdp>.

DITR (2007b) *Managing Acid and Metalliferous Drainage*. Leading Practice Sustainable Development Program for the Mining Industry. Department of Industry, Tourism and Resources, Commonwealth of Australia: Canberra.

DITR (2008) *Cyanide Management*. Leading Practice Sustainable Development Program for the Mining Industry. Department of Industry, Tourism and Resources, Commonwealth of Australia: Canberra. <http://www.ret.gov.au/resources/resources_programs/lpsdp>.

Dry RJ and Beeby CJ (1997) Applications of CFB technology to gas-solid reactions. In *Circulating Fluidized Beds*. (Eds JR Grace, AA Avidan and TM Knowlton) pp. 441–465. Chapman and Hall: London.

Engels J and Dixon-Hardy D (2009) Tailings.info – The website all about tailings. <http://tailings.info/index.htm>.

Global Acid Rock Drainage Guide (2009) <http://www.gardguide.com>.

GRI (2006) *Sustainability Reporting Guidelines*. 3rd edn. Global Reporting Initiative. <http://www.globalreporting.org>.

International Cyanide Management Code (2009) <http://www.cyanidecode.org>.

Janke D, Savov L and Vogel ME (2006) Secondary materials in steel production and recycling. In *Sustainable Metals Management*. (Eds A von Gleich, RU Ayres and S Gößling-Reismann) pp. 313–334. Springer: Dordrecht, The Netherlands.

Kuyucak N (2006) Selecting suitable methods for treating mining effluents. *Water in Mining*, Brisbane, 14–16 November 2006. pp. 267–276. Australasian Institute of Mining and Metallurgy: Melbourne.

Lévy V, Fabre R, Goebel B and Hertle C (2006) Water use in the mining industry – threats and opportunities. In *Water in Mining 2006*, Brisbane, 14–16 November 2006. pp. 289–295. Australasian Institute of Mining and Metallurgy: Melbourne.

Lottermoser BG (2003) Cyanidation wastes of gold-silver ores. In *Mine Wastes: Characterization, Treatment and Environmental Impacts*. pp. 182–187. Springer-Verlag: Berlin.

Mudd GM (2009) Sustainability reporting and mining – an assessment of the state of play for environmental indicators. In *Sustainable Development Indicators in the Minerals Industry Conference (SDIMI 2009)*, 6–8 July 2009, Gold Coast, Queensland, Australia. pp. 377–391. Australasian Institute of Mining and Metallurgy: Melbourne.

Norman DK and Rathorth RL (1998) Innovations and trends in reclamation of metal mine tailings in Washington. *Washington Geology* **26**(2/3): 29–42.

Paykan S and Russell M (2010) Heat recovery and energy optimization in smelter gas cleaning. *Proceedings of Copper 2010*, 6–10 June 2010, Hamburg, Germany. pp. 497–515. GDMB: Clausthal-Zellerfeld, Germany.

PDEP (2009) *The Science of Acid Mine Drainage and Passive Treatment*. Pennsylvania Department of Environmental Protection, Bureau of Abandoned Mine Reclamation. <http://www.depweb.state.pa.us/abandonedminerec>.

Pulles W, Howie D, Otto D and Easton JA (1995) 'Manual on mine water treatment and management practices in South Africa'. Report no. TT80/95, June 1995. Water Research Commission: Pretoria, South Africa.

Starke L (Ed.) (2002) *Breaking New Ground: Mining, Minerals and Sustainable Development*. Earthscan Publications: London.

Stevenson J, Botz M, Mudder T, Wilder A and Richens R (1995) Cyanisorb recovers cyanide. *Mining Environmental Management* **June**: 9–10.

UNEP (2005) *Mining for Closure: Policies and Guidelines for Sustainable Mining Practice and Closure of Mines*. United Nations Environment Programme. <http://www.grida.no/_res/site/file/publications/envsec/mining-for-closure_src.pdf>.

US EPA (2006) *Management and Treatment of Water from Hard Rock Mines*. EPA/625/R-06/014, October 2006. United States Environmental Protection Agency.

Wills BA (1992) *Mineral Processing Technology*. 5th edn. Pergamon Press: Oxford, UK.

Wilmoth RC, Hubbard SJ, Burckle JO and Martin JF (1991) Production and processing of metals: their disposal and future risks. In *Metals and their Compounds in the Environment: Occurrence, Analysis and Biological Relevance*. (Ed. E Merian) pp. 19–65. VCH Verlagsgesellschaft: Weinheim, Germany.

12.6 USEFUL SOURCES OF INFORMATION

All about tailings. <http://tailings.info>.

An account of the Ok Tedi mine, its history and environmental and social impacts. <http://en.wikipedia.org/wiki/Ok_Tedi_Mine>.

Global Acid Rock Drainage Guide. <http://www.gardguide.com>.

Hazardous Materials Management (Draft). Leading Practice Sustainable Development Program for the Mining Industry. Department of Industry, Tourism and Resources, Commonwealth of Australia. July 2009. <http://www.ret.gov.au/resources/resources_programs/lpsdp/>.

Hutchison IPG and Ellison RD (Eds) (1992) *Mine Waste Management*. Lewis Publishers: Boca Raton, FL.

International Cyanide Management Code. <http://www.cyanidecode.org>.

International Cyanide Management Institute. <http://www.cyanidecode.org/whoicmi.php>.

International Network for Acid Prevention. <http://www.inap.com.au>.

Kelly EG and Spottiswood DJ (1982) *Introduction to Mineral Processing*. John Wiley and Sons: New York.

Lottermoser BG (2003) *Mine Wastes: Characterization, Treatment and Environmental Impacts*. Springer-Verlag: Berlin.

Managing Acid and Metalliferous Drainage. Leading Practice Sustainable Development Program for the Mining Industry. Department of Industry, Tourism and Resources, Commonwealth of Australia. 2007. <http://www.ret.gov.au/resources/resources_programs/lpsdp>.

Mining for Closure: Policies and Guidelines for Sustainable Mining Practice and Closure of Mines. United Nations Environment Programme, 2005. <http://www.grida.no/_res/site/file/publications/envsec/mining-for-closure_src.pdf>.

Tailings Management. Leading Practice Sustainable Development Program for the Mining Industry. Department of Industry, Tourism and Resources, Commonwealth of Australia. 2007. <http://www.ret.gov.au/resources/resources_programs/lpsdp>.

Technical Guidelines on Hazardous Wastes Physico-Chemical Treatment/Biological Treatment. Basel Convention series/SBC No. 02/09, October 2000. <http://www.basel.int/meetings/sbc/workdoc/techdocs.html>.

Williams RE (1975) *Waste Production and Disposal in Mining, Milling and Metallurgical Industries*. Miller Freeman: San Francisco.

13 Secondary materials and recycling

This chapter is concerned with the part of the materials cycle (Figure 2.2) relevant to products at the end of their useful life. It examines the options other than disposal of them as waste. In this context, all human-made objects with a function are considered to be products. These include items of infrastructure (roads, railways, buildings, electrical cables, dams), heavy industrial and commercial equipment (machines, trucks, trains, ships, aeroplanes), electrical and electronic equipment, cars and consumer goods. Of particular interest are the reclamation of materials from used products, so-called secondary materials, and the reuse of them in new products.

13.1 OPTIONS FOR END-OF-LIFE PRODUCTS

At the end of its life, a product can be disposed of (e.g. to landfill), burned to produce energy (if it is combustible), reused for a similar or related purpose, remanufactured and resold for a similar or different function, or its materials of construction can be recovered and recycled. Thus, disposal, recycling, reuse and remanufacturing are the main options available for a product at the end of its life. Figure 13.1, which is a form of the materials cycle, shows the flow of mineral and metallic commodities through the economy and the complex interrelationships between primary production, product design and manufacture, consumption and end-of-life management.

13.1.1 Recycling

Recycling is the reintroduction of recovered materials into the materials cycle. Metals and glass are examples of common mineral and metallic materials that are recycled. They are melted and reformed into new products, such as metal sheet or wire and glass containers. Recycling involves collecting, sorting and processing discarded materials to make them fit for use. The recovery, sorting and processing of materials together constitute *secondary production* of materials. This is distinct from primary production, which is the production of commodities from natural resources (mainly minerals, petroleum and timber). Figure 13.2 shows a generalised flowsheet for the secondary production of materials. Note that at each stage some of the material can be lost to the environment. A material that has been recycled may have the same characteristics and properties as the original material, or it may be of lower or higher quality. This depends on the process used in recycling the material and the nature of the material, in particular whether the intrinsic characteristics of the material are retained during the process.

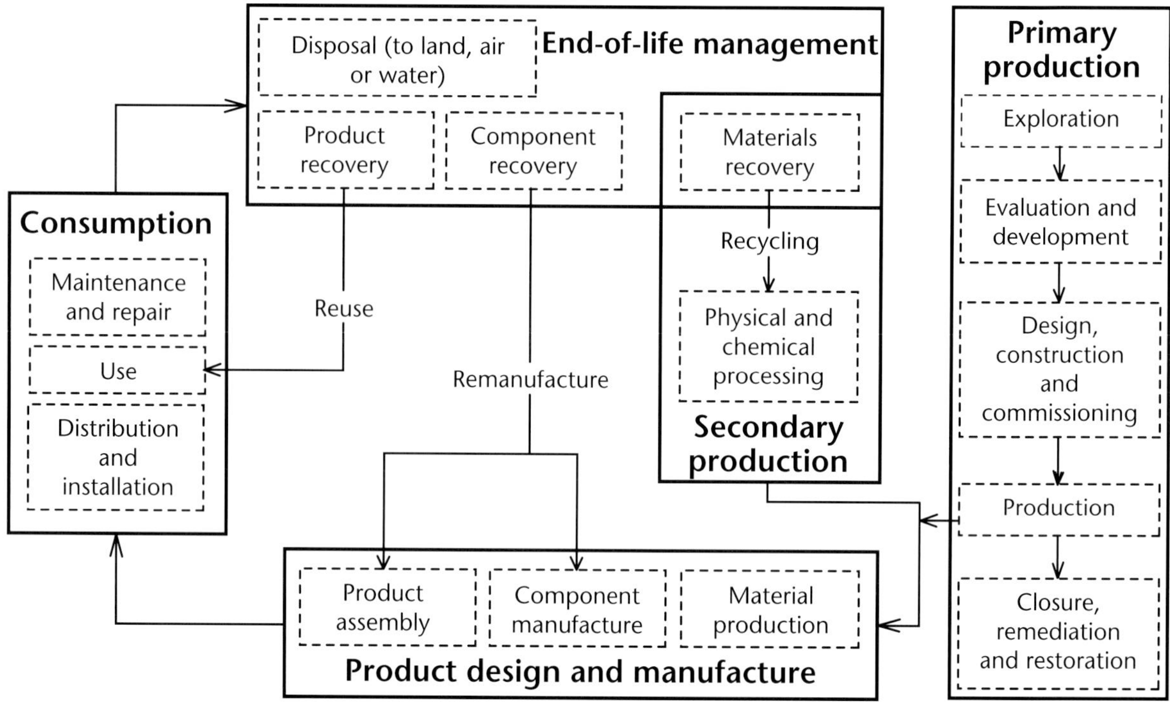

Figure 13.1: The flow of mineral and metallic commodities through the economy (after van Berkel, 2007; with modifications).

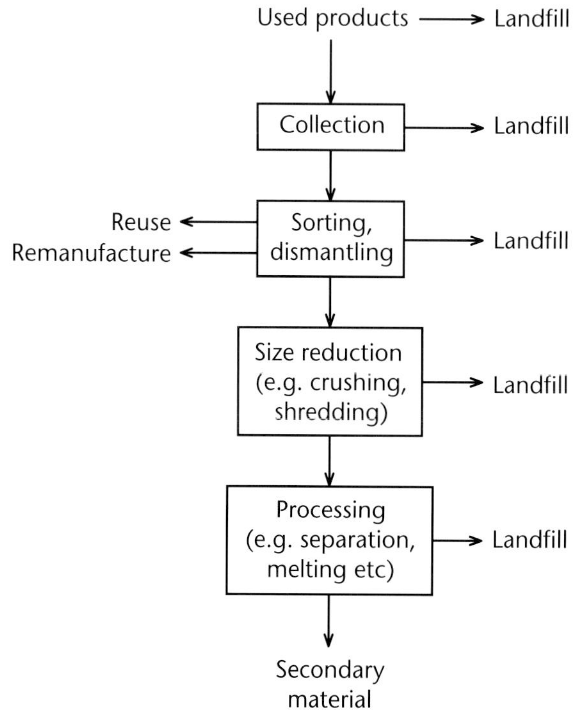

Figure 13.2: A generic flowsheet for recycling materials.

13.1.2 Reuse

Reuse is the use of a product or component of a product more than once, either for its original function or a different function. It is different from recycling in that recycling involves the breaking down of used products into separate materials which can be used to make new items, of the same type or different. Probably the best known example of reuse is the returnable glass soft drink or milk bottle which is collected, washed, refilled and resold. Other examples include the retreading of tyres for cars and trucks and the use of concrete as aggregate in construction projects.

13.1.3 Remanufacturing

Remanufacturing is the repair and overhaul of valuable parts, components, equipment and machines such as engines, office fittings, printer cartridges, aircraft hulls and cathode ray tubes, in a factory environment to meet the same or similar specifications as new. This can be a highly profitable activity, even in developed countries. In most jurisdictions it is legal to sell refurbished items under a new brand name but illegal to sell them under the original equipment manufacturer's brand unless they are sold as a used item.

13.2 DRIVERS OF RECYCLING, REUSE AND REMANUFACTURING

Historically, economics has been the driver for recycling, reuse and remanufacture. In the developing world this has led to very high levels of recycle and reuse where labour-intensive and often not very safe or environmentally friendly methods are used (e.g. Nilmani, 2005). Economics was a major driver in developed countries until after World War II. However, rising wages and the relatively low cost of landfill have made the reuse of low-value items, such as packaging and many consumer products, uneconomic in many countries. This move has been facilitated by corporate strategies aimed at increasing consumer consumption by creating disposable products with short life spans. This has often been aided, wittingly or otherwise, by governments and has resulted in the community bearing the costs of consumer waste. The huge quantities of domestic and consumer waste that are now produced in developed countries are a direct consequence of these changes.

Growing environmental awareness and the increasing cost of landfill disposal are gradually changing attitudes towards consumer wastes. In many developed countries government regulation of wastes lags behind public opinion, often due to industry resistance, though in others, particularly the European Union and Japan, government regulation is driving industry response. This is discussed further in Section 13.9. Some materials, particularly glass, concrete, metals and many electronic products, are sufficiently valuable that increasing quantities of them are recycled even in countries where recycling targets are not mandated. Some materials must be recycled because their toxicity makes their disposal too expensive.

13.3 THE BENEFITS AND LIMITATIONS OF RECYCLING

The potential environmental and other sustainability benefits of recycling are many.

- The consumption of new raw materials is reduced. As a result, the quantity of waste from production of primary materials is reduced and the life of mineral resources is extended.
- The energy required to produce secondary materials is much less than if they were made from primary sources.
- The dispersion of toxic elements into the environment (air, water and land) is reduced since the materials cycle is closed.
- The quantity of land required for landfill disposal of consumer and industrial wastes is reduced, an important consideration in countries and regions with high population densities.

However, recycling has some limitations. First, the quantities and proportions of recycled materials may not always match current demand profiles. Even if a material is completely recycled it may be insufficient to meet a growing demand. Alternatively, a material may become in excess if demand for it is declining due to new technologies, innovative design or changing fashions. Also, the combinations of metals in streams from recycling operations is usually quite different from the combinations that occur naturally. As a result, and with increasing reliance on recycled metals, deficiencies or surpluses of some minor metals may develop.

A second limitation of recycling is that recycled materials often have a lower quality than the original material. Ideally, recycling would produce a new supply of the same material; for example, used office paper would be recycled to make more office paper and used aluminium cans would be recycled to make new cans. This is called *closed-loop recycling*. However, this is often technically difficult and expensive compared to producing the same product from raw materials. Accordingly, recycling of many materials involves producing different materials or a similar material but of lower grade. This is called *open-loop recycling*. For example, newspaper and cardboard are made from higher-grade papers and cast aluminium products are made from recycled aluminium cans. The reason is that most materials degrade during recycling. Paper fibres become shorter with each recycle and must be used in progressively lower-grade products. Concrete cannot be reconstituted into cement and aggregate, so when concrete is recycled it is put to a lower-grade use as aggregate. Metals can often become contaminated with other elements during recycling, and removal of the impurities is often expensive. Thus the quality of the metal is lowered. At present, the problem of reduced quality is minimised by adding virgin or primary materials to recycled materials to dilute them and bring the composition or morphology to the desired level. As the quantity of a particular recycled material increases, however, the quantity of virgin material required for dilution will also increase. When the total demand for a material is growing rapidly, this may not become a problem. However, when growth in demand slows or decreases there will be a limit to the quantity of the material that can be recycled to make material of the same quality.

Third, recycling invariably produces wastes, both direct and indirect. The indirect wastes are mainly those associated with the generation of electricity used in collection, sorting and recycling. Direct wastes include CO_2 from trucks, trains and ships used to transport wastes, residual materials that cannot economically be recycled, materials that are toxic or have no market, and waste products formed during chemical processing to recycle. Though the volume of waste produced per unit of recycled material is small compared to that produced in primary production, secondary processing wastes are often potentially more toxic, both to humans and the environment.

Finally, recycling is an end-of-pipe solution and can be expensive because of the need to:

- collect products and materials;
- transport them to a central point;
- separate and sort them into streams suitable for reprocessing;
- perform the physical or chemical processing required.

This is discussed further in Section 13.6. In many situations, reuse and remanufacture have advantages over recycling because more of the energy and capital cost embodied in the product, and hence functionality, is conserved. Thus, as products become increasingly designed to facilitate reuse and remanufacture, the need for recycling may decrease. This could partially solve the dilemma posed by the need to dilute secondary materials with virgin material.

13.4 RECYCLING TERMINOLOGY

Several terms are frequently used (and sometimes misused!) to quantify aspects of recycling. Sometimes, different names are used for the same term. The most useful concepts and their most commonly used names are recovery rate, recycling rate and return rate. These terms are often used interchangeably; care must be exercised, when using published data, to understand which term is intended. According to Norgate *et al.* (2009), most data for recycling rates reported in the literature are actually return rates. Unfortunately, recycling data are often presented without an explanation of the terms used and in ways that are favourable to the presenter's position.

The *recovery rate* (also called the collection quota) is the quantity of a secondary material that is recovered by collection systems as a percentage of the total quantity of the material available (Equation 13.1, Box 13.1). The quantity of a material available is a difficult concept since availability of secondary materials for recycling is linked to the quantity consumed one product life cycle ago for the application in question (Norgate *et al.*, 2009). For example, the metal used in the production of a motor vehicle with a life span of, say, 10–15 years will not be available for return to the metals stream as scrap for that period whereas an aluminium can may return to the metals stream within months of leaving it. Table 13.1 lists the estimated life span of various metals in different applications. The quantity of a material available for recycle is not the current level of consumption of that material but the sum of the consumption of the material one product life cycle ago for each of the end-use categories. The estimation of the life span of a material in a particular product, or application, can be quite difficult. The *technical recovery rate* (technical recycling quota) is the quantity of material recycled as a percentage of the quantity of material collected (Equation 13.2, Box 13.1). The technical recovery rate is always smaller than the recovery rate because some

Box 13.1: Definitions of recycling terms

$$Recovery\ rate = \frac{quantity\ of\ collected\ material}{quantity\ of\ material\ available} \times 100\% \quad (13.1)$$

$$Technical\ recovery\ rate = \frac{quantity\ of\ recycled\ material}{quantity\ of\ collected\ material} \times 100\% \quad (13.2)$$

$$Recycling\ rate = \frac{quantity\ of\ recycled\ material}{quantity\ of\ material\ available} \times 100\% \quad (13.3)$$

$$Return\ rate = \frac{quantity\ of\ recycled\ material}{quantity\ of\ recycled\ material + quantity\ of\ virgin\ material} \times 100\% \quad (13.4)$$

of the collected material will be lost during processing. The technical recovery rate is therefore a measure of the efficiency (yield) of the technical processes used for recycling.

The *recycling rate* (recycling quota) is the quantity of a material that is recycled as a percentage of the total quantity of material available (Equation 13.3, Box 13.1). The recycling rate is equal to the product of the recovery rate and the technical recovery rate. It is also sometimes referred to as the resource-orientated recycling quota (Hoberg *et al.*, 1999). The recycling rate is always smaller than the recovery rate because not all the material available for recycling is collected. The total quantity of material that is lost by not being recycled is given by:

$$Amount\ lost = quantity\ of\ material\ available - quantity\ of\ recycled\ material$$

The *return rate* (recycled content or recycled quotient) is the quantity of a material that is recycled for use in manufacturing as a percentage of the total quantity of the material (primary plus secondary) used for manufacturing (Equation 13.4, Box 13.1). In other words, the return rate is the percentage of a particular material that is supplied from recycled sources, the balance being supplied from primary sources. Usually, the return rate for a material is smaller than its recycling rate because, when overall demand for a material is increasing, the quantity of virgin material exceeds the quantity of lost material.

13.5 RECOVERY, RECYCLING AND RETURN RATES FOR COMMON MATERIALS

Table 13.2 lists the recovery, recycling and return rates for some common materials, compiled from reliable sources. For most of the materials, secondary sources make an important contribution to their total production and the recycle rate is relatively high. Steel is the material which is recycled in greatest quantity. It also has very high recovery and recycling rates. Globally, about 90% of used steel is recovered and nearly 70% is recycled. Recycled steel makes up about 30% of total steel production, the balance coming from primary sources. Lead has the highest return rate, about 80% in the United States, higher in many other developed countries. This reflects the increasing concern about lead in the environment and the need to return used lead to the materials cycle rather than dispersing it in landfill. Collection is made easier by the fact that about 80% of all lead is used in lead-acid batteries (for vehicles) and most countries have voluntary or compulsory collection systems for these batteries. The quantity of concrete recycled as aggregate is very small, but growing.

13.6 THE ENERGY REQUIRED FOR RECYCLING

It is often claimed, and widely accepted, that the energy required for recycling of secondary materials is a small

Table 13.1: Estimated life span of metals in various applications

	Product/application	Lifespan (years)
Aluminium	Buildings and construction	40
	Transport – aerospace	30–40
	Transport – automobiles and light trucks	13–20
	Transport – trucks and buses	20
	Transport – rail	30
	Consumer durables	12–15
	Electrical	35
	Machinery and equipment	20–25
	Containers and packaging	0.25–1
Steel	Motor vehicles	5–15
	Buildings	20–60
	Major industrial	40
	Heavy industrial machinery	30
	Rails	25
	Consumer durables	7–15
	Cans	<1
Copper	Motor vehicles	8–10
	Small electric motors	10–12
	Cable	30–40
	Buildings	60–80
	Electrical plant and machinery	30
	Non-electrical machinery	15
	Housing	35
	Transport	10
Zinc	Dry batteries	1
	Building and plumbing	25
	Weighted average	13

Source: Norgate *et al.* (2009).

fraction of that required to make virgin material. This is generally true when only the actual processing step is considered. For example, the direct energy required to remelt aluminium for recycling is only about 5% of the direct energy required for primary production of aluminium from bauxite ores. Some materials, such as concrete, require only simple physical processing operations like crushing and screening; again, the energy requirement is relatively low compared to the energy required for primary production. However, beside remelting or crushing, a secondary material has to be collected and sorted; the energy required for this can be very high, depending on a number of factors. The further material has to be transported to the processing facility, the more energy is required per unit of material recycled and the more expensive it becomes. Also, as the return rate increases, the proportion of available material that has to be collected and sorted increases and the grade becomes progressively lower. This means that the quantity of products collected progressively becomes larger in relation to the valuable material content, since increasingly lower-grade sources must be included. Thus, the energy required for collection and sorting increases as the return rate increases. This is analogous to the effect of declining ore grade on the energy required to make a concentrate (Section 9.4).

Figure 13.3 illustrates these effects in a semi-quantitative way. The solid line shows the variation in energy required to produce a unit of a material as the proportion of secondary material used (the return rate) increases. This line is the sum of the three dotted lines which represent the energy required for primary production, the energy required for collection and sorting, and the energy required for reprocessing the sorted material, assuming a closed-loop recycling system. The energy for primary production decreases linearly as the return rate increases since the proportion of primary material in the total supply of the material decreases. The energy for reprocessing recovered material increases linearly with increasing return rate since the proportion of recovered material increases. The energy required for collection and sorting, however, increases in a manner similar to that in Figure 9.5 for the effect of declining ore grade on energy required for making a concentrate. This is because the grade, or quality, of the recovered material progressively decreases and the distance over which materials have to be transported to the processing centre increases. Summing the three dotted lines gives the total energy. Figure 13.3 shows that initially the total energy required to produce a unit of a material decreases with increasing proportion of recycled material but, with increasing influence of the energy for collection and sorting, the total energy at some point reaches a minimum then begins to rise and rises increasingly rapidly, eventually exceeding the

Table 13.2: Recovery, recycling and return rates for some common materials

Material	Location and year	Quantity recycled (tonnes)	Recovery rate (%)	Recycling rate (%)	Return rate (%)
Glass containers	EU (2007)	10 973 000[a]			62[a]
Glass – all types	US (2003)			18[b]	
	UK (2003)			35[b]	
	Japan (2003)			90[b]	
Paper and cardboard	US (2003)			48[b]	
	UK (2003)			52[b]	
	Japan (2003)			65[b]	
Concrete (for use as aggregate)	USA	100 000 000[c]			5[c]
Construction and demolition waste	EU (late 1990s)	50 400 000[d]		28[d]	
	UK (2005)	71 700 000[e]		80[e]	
Steel – all types	World (2007)		90[f]	68[f]	25[f], 32[g]
– all types	US (2007)	65 000 000[h]			53[h]
– construction	World (2007)		90[f]	85[f]	
– automotive	"		90[f]	85[f]	
– machinery	"		95[f]	90[f]	
– appliances	"		65[f]	50[f]	
– containers	"			69[f]	
Aluminium – all types	US (2007)	3 859 000[h]	54[i]		45[h]
– building	World (2000)		70[j]	60[j]	
– automobiles	"		75[j]	65[j]	
– aerospace	"		75[j]	65[j]	
– machinery	"		44[j]	38[j]	
– electrical	"		42[j]	36[j]	
– consumer durable	"		21[j]	18[j]	
– beverage cans	"		59[j]	32[j]	
Copper	US (2007)	925 000[h]			31[h]
Lead	"	1 180 000[h]			80[h]
Zinc	"	237 000[h]			20[h]
Magnesium	"	84 000[h]			52[h]
Tin	"	14 800[h]			34[h]

a) European Container Glass Federation (2009).
b) *The Economist* (2007).
c) Cleveland (2008).
d) Rao *et al.* (1997).
e) Delgado *et al.* (2009).
f) World Steel Association (2008).
g) Estimated from USGS data:[h] Global primary production of pig iron in 2007 was 940 000 t (all from iron ore); assuming a 95% yield this would produce 893 000 t steel. Total steel production (primary plus secondary) was 1 320 000 t. Therefore, the return rate is 100 × (1 320 000 – 893 000)/1 320 000 = 32.3%.
h) USGS (2007).
i) Aluminum Association; <http://www.aluminum.org>; retrieved 26 November 2009.
j) Bruggink and Martchek (2004).

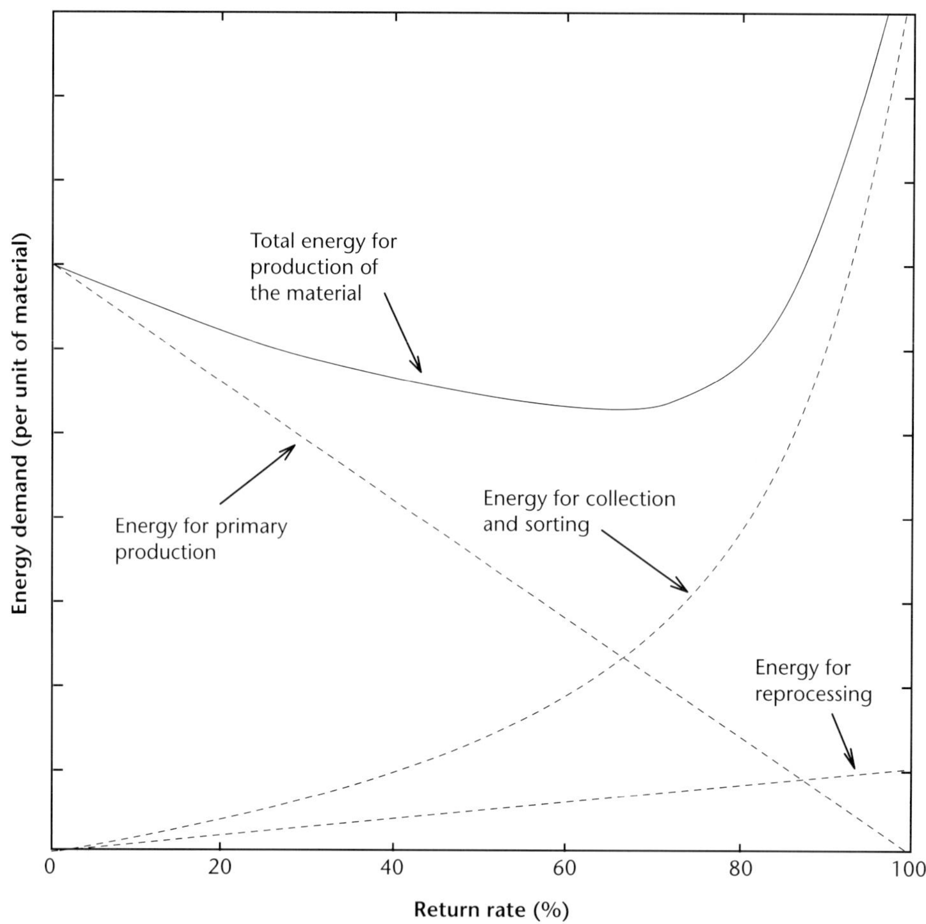

Figure 13.3: Qualitative representation of the total energy required for recycling.

energy required for primary production. The minimum point represents the optimum return rate in terms of energy consumption.

13.6.1 The Gross Energy Requirement for recycling

The Gross Energy Requirement (GER) for a recycled material comprises the direct and indirect energy required for collection, sorting, transportation, preparation and reprocessing to make a saleable material. Values of the energy required for recycling, or of the energy saved by recycling a material compared with producing it from raw materials, are frequently quoted in scientific and technical literature as well as in more general literature. The bases for these values, however, are frequently not given and it may not be clear whether collection and transportation are included. All values need to be treated with caution. Definitive energy data are difficult to obtain and even industry associations with an interest in recycling often present data only in processed or summary form. Table 13.3 lists the values of GER for recycling some common metals (Kusik and Kenahan, 1978). These values are based on scrap having zero embodied energy at the first point of collection, hence they do not include transport and other forms of energy required for collection. These values expressed as a percentage of the GER for primary production are also listed. Some values for metals, plastic, paper and glass commonly cited in the literature are listed for comparison. Generally, the GER for recycling of metals is significantly less than for primary production, particularly for metals and especially for aluminium, the primary production of which is very energy-intensive.

Table 13.3: Gross energy requirement for recycling some common commodities

		Embodied energy, or GER (GJ per tonne)		GER for recycling as % of GER for primary production		
Commodity	**Form**	**Secondary production[a,b]**	**Primary production (Table 9.5)**	**Calculated**	**Commonly quoted values (various sources)**	
Plastic						30[c]
Paper						60[c]
Glass						70[c]
Aluminium	Alloy	17.5	212	8.3	Al	5–10
	cans	10.1		4.8		
Copper	No. 1 scrap	4.4	33	13.3	Cu	15
	No. 2 scrap	20.1		60.9		
	Low-grade scrap	49.3		149		
Steel	Billets	9.7	23	42.7	Steel	20–40
Lead – soft	Batteries	9.4	20	47.0	Pb	35
Lead – hard	Batteries	11.2		56.0		
Nickel	Alloy scrap	12.9	114	11.3	Ni	10
Zinc	New scrap	3.8	48	7.9	Zn	25–40
	Slab	22.0		45.5		

a) Kusik and Kenahan (1978).
b) Does not include the energy required to transport goods to the first point of collection.
c) *The Economist* (2007).

13.6.2 The effect of repeated recycling

The more times a material is recycled the greater is the embodied energy content of the material but the less is the embodied energy per application of the material. For example, producing 1 kg of primary aluminium requires 212 MJ of energy (Table 9.5). If the aluminium is recycled once after its initial use, the additional energy required is 10 MJ, therefore the total embodied energy is 222 MJ kg^{-1}. Since this is spread over two applications of the material, the embodied energy in the aluminium per application is 111 MJ kg^{-1}. If the metal is recycled again, the embodied energy is $212 + (2 \times 10) = 232$ MJ kg^{-1}. The embodied energy per application is $232/3 = 77.3$ MJ kg^{-1}, and so on. If the return rate is less than 100%, the energy required to produce the primary aluminium that is added to the recycled aluminium, needs to be included. For the previous example, if the return rate is 20% (i.e. 80% of the metal is primary and 20% is from secondary sources), the energy required after two recycles is $(0.8 \times 212) + (0.2 \times 77.3) = 185$ MJ kg^{-1}. Figure 13.4, calculated as above, shows how the embodied energy of aluminium per application of the metal decreases as the number of times the metal is recycled increases, with the absolute values decreasing as the return rate increases. Similar trends will be exhibited by all materials that can be recycled. As noted previously, the number of times a material can be recycled is sometimes limited by deterioration in the quality of the material.

13.7 THE EFFECT OF RECYCLING ON RESOURCE LIFE

When materials are recycled, the draw on natural resources is reduced to the extent that the recycled material replaces material that would otherwise be produced from natural resources. In the case of minerals, which are non-renewable natural resources, recycling

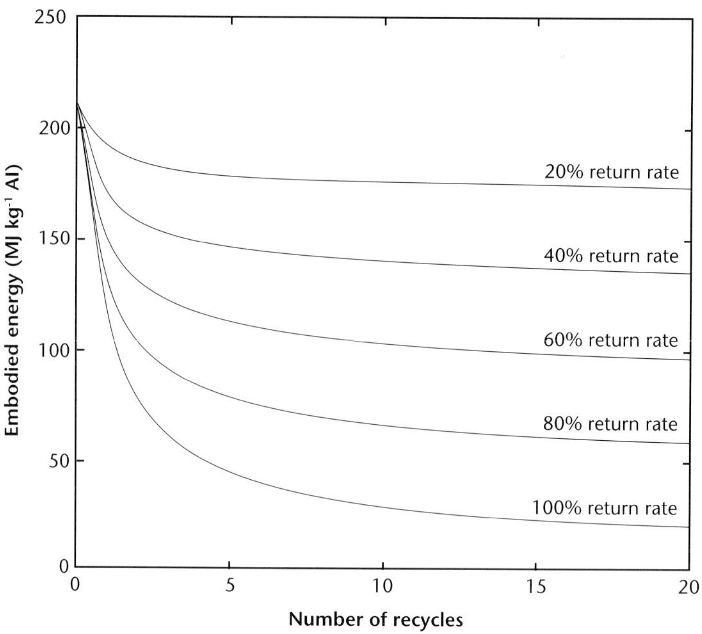

Figure 13.4: The effect of the number of recycles and return rate on the embodied energy of aluminium.

of material therefore extends the life of the material reserves. Table 6.4 shows the years of supply of common mineral and metal commodities based on 2007 consumption rates and the known reserve base. Figure 13.5 shows the calculated variation in years of supply for some commodities for various assumed return rates. To illustrate the method of calculation, consider copper and assume that the return rate is 40%. The total

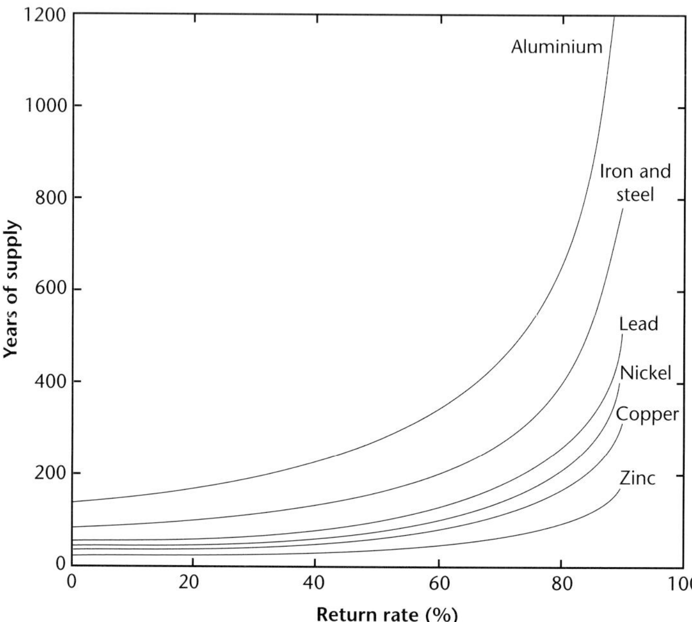

Figure 13.5: The effect of return rate on the years of supply of some common mineral and metal commodities, assuming 2007 values for production rates and reserves.

primary production of copper in 2007 was 15 600 000 tonnes. If 40% of this could be replaced by secondary copper, only 9 360 000 tonnes would have to be mined. Since the known reserves of copper in 2007 were 490 000 000 tonnes, this would have the effect of increasing the life of the reserves from 31 years to 52 years. Figure 13.5 shows that the life of reserves increases as the return rate increases. The curves rise rapidly at returns greater than around 70% and asymptote to infinity at return rates of 100% – if all of a material is supplied from recycled material no mining would be necessary and the reserves would have an infinite life. Figure 13.5 is intended only to demonstrate the effect of recycling on the life of mineral and metal reserves. In an environment in which the total quantity of most mineral and metal commodities consumed is growing annually (a situation which has existed for most of recorded history), it is not possible to meet the demand for commodities entirely, or in most cases largely, from secondary materials.

13.8 RECYCLING MATERIALS FROM SIMPLE PRODUCTS

13.8.1 Construction and demolition wastes

Construction and demolition wastes are those generated during the construction of buildings, dams, roads, bridges and other forms of infrastructure and during their demolition. They consist predominantly of concrete, rock and asphalt but often contain steel (structural or as reinforcing bars) and other metals as well as smaller quantities of other materials used in particular types of infrastructure. Construction and demolition waste can be disposed of directly in landfill, recycled or reused. As it is a low-value material and usually relatively environmentally benign, landfill is the major option, with recycle and return rates among the lowest of the common commodities (Table 13.2). However, the recycle rate is increasing due to the rising cost of landfill and encouragement by governments.

Concrete is reused by crushing to produce aggregate. This can be done at the demolition site using portable crushing equipment, in which case the aggregate can be reused on site, or the concrete can be transported to a recycling centre, usually nearby because of the cost of transport. Often two crushers are used in series, such as a jaw crusher followed by a cone crusher or roll crusher. Metal reinforcing bars and mesh are removed from the product of the first crusher by hand picking or magnetic separators. The crusher product is then passed over a series of screens. Undersize material is sent to landfill. Asphalt paving from demolition of roads is processed in a similar way. Most aggregate produced from concrete is used as road base; a small quantity is used in new concrete mixes and asphalt hot mixes. Concrete made from concrete aggregate generally has lower strength than conventional concrete, so the proportion of concrete aggregate in the mix must be limited or the concrete used in lower-grade applications. Recycled asphalt is used in the preparation of fresh asphalt mixtures or as road base. The proportion of reused asphalt in asphalt mixtures can be quite high, up to around 40% for roads and 100% for parking lots, but most recycled asphalt is used as aggregate for road base.

13.8.2 Glass

Glass containers are the most common form of glass to be recycled. Many countries have developed collection systems to facilitate the recycling of glass containers, through kerb-side collection along with other household wastes or through local collection centres. Window glass, electric light globes, mirrors and special glasses such as Pyrex® are less frequently recycled. Most glass is recycled by remelting it then reforming it into new containers (Section 8.4). Glass must be sorted to avoid cross-contamination. It is usually sorted into colourless, green and amber streams which are then processed separately. Contaminants (metals, stones, plastics and ceramics) are removed. Sorting is usually done by machines which use optical sensors to identify the different types of glass. The sorted glass is then crushed, the product being called *cullet*. The size of cullet varies according to the proposed use. Most cullet is sold to glass manufacturers; it is remelted directly or is mixed with new raw materials. Cullet melts at a lower temperature than that required to form glass from raw materials, hence the energy required for the melting step is less than that for glass made solely from raw materials. Cullet is a hard, granular material that can also be used as construction aggregate in concrete and asphalt. Finely ground cullet has similar properties to sand and can be used in concrete in place of natural sand. Cullet is an ingredient in ceramics and bricks and can be used as an abrasive.

13.8.3 Metals

Metals are the materials most recycled, due to their high value. This particularly applies for steel, aluminium and copper. Scrap metal originates from three main sources and is classified accordingly. *Home scrap* (in-house scrap) is scrap produced in primary and secondary plants as part of the metal production process. It has a known composition and is free of coatings and attachments. Home scrap is recycled simply by feeding it back into the production process. It is not usually included in recycling statistics. All other scrap is distinguished as new scrap or old scrap depending on whether it reaches the end of its life before or after the consumer has used it. *New scrap* (prompt scrap) is scrap generated during the manufacturing or fabrication process. It can sometimes be recycled directly within the manufacturing facility; otherwise, it is sent to a metal recycler. Since the composition of new scrap is known, in principle it does not need any pretreatment before it is remelted, although cutting to size might be necessary. *Old scrap* (post-consumer scrap) is scrap collected after a consumer cycle, either separately or mixed. Its composition is usually not known and it is often contaminated with other materials, depending on its origin and the collection system.

The processing steps in metal recycling are collection, sorting, shredding, sizing, separation and melting. Usually, not all steps are performed by the one company; several types of companies are involved (Figure 13.6). Scrap metal is collected separately or mixed, sorted in a scrap yard then sold to scrap treatment plants or sent directly to a refiner or remelter. At the scrap treatment plant, different types of metals are separated and prepared for shredding and sizing, which is often needed to achieve liberation of the different materials in products. After shredding and sizing, ferrous metals are extracted using magnetic separators, then the non-ferrous metals are separated from non-metallic materials using various separation technologies. Finally, the different non-ferrous metals are separated. If necessary, the metal is dried or further cleaned of contaminants such as oil, grease, lubricants, lacquers, and rubber and plastic coatings. This might be done at the scrap treatment plant; if thermal treatment is required it is more energy-efficient to do this at the metal smelter or remelter. At the smelter or remelter the scrap is charged to furnaces where it is melted, refined and tapped to produce metal ingots or billets.

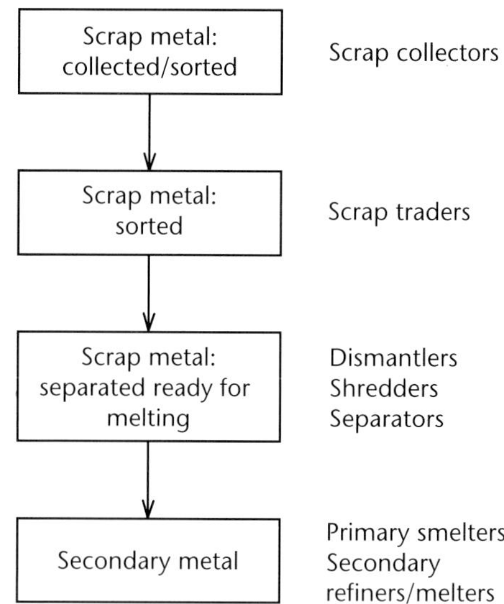

Figure 13.6: The typical structure of the scrap metal industry.

Scrap steel and aluminium

Ferrous scrap processing is part of the integrated steel industry. Up to 30 wt% of the charge to basic oxygen furnaces is scrap steel, the balance being hot metal from the blast furnace (World Steel Association, 2008). Scrap, or an equivalent such as DRI, is an essential part of the charge: it controls the temperature rise due to the exothermic reaction between the injected oxygen and the carbon and silicon in the hot metal. Scrap is also processed in electric arc furnaces independently of primary metal production. In the case of aluminium, all secondary aluminium production is performed independently of primary aluminium production, in remelting facilities, because there is no technical or economic benefit in recycling aluminium scrap to the Hall-Héroult process. Remelting metals produces considerable waste, typically off-gases and molten or semi-molten slags and drosses, some of which are hazardous.

Because aluminium is so reactive, typically around 5–10% of the input scrap oxidises during remelting and forms *dross*, a solid consisting of aluminium oxide (and some aluminium nitride) with up to 80 wt% metallic aluminium trapped within it. To minimise the loss of aluminium by oxidation, particularly when highly reactive, finely divided aluminium is remelted, and to improve the quality of the metal, a mixture of NaCl and KCl is added in some remelting processes. This melts and coats the aluminium, thereby reducing its exposure to oxidising furnace gases. The dross from these furnaces is called black dross, distinguishing it from white dross which is produced in salt-free remelting and molten metal handling processes. Dross is processed to recover the metal content, by heating it with salt flux in a rotary furnace. The salt helps to release the metal. This collects as a molten pool and results in formation of saltcake, which contains up to 10 wt% aluminium and 20–60 wt% aluminium oxide (Peterson and Newton, 2002). As much as 500 kg of saltcake can be generated per tonne of metal produced. Salt-free processes have been developed but not widely implemented. They are of two types – high-temperature melting processes and low-temperature separation processes. In the latter, dross is crushed, ground and screened to liberate the entrapped metal.

Landfilling of saltcake is banned in many countries though it is still common practice in others. Approximately 800 000 tonnes of saltcake were landfilled in the United States in 2005 (Zhang, 2006). Processing saltcake to reduce its environmental impact involves several steps (Prillhofer *et al.*, 2009). In the conventional process, saltcake is crushed and ground to liberate trapped aluminium metal, then leached in water to dissolve the salts. Noxious gases are emitted during this stage. The oxide residue is separated in a thickener and the salt is recovered, by evaporation and crystallisation, for reuse. The recovered metal is remelted and the thickener residue (mainly aluminium oxide) is used in cement production or sent to landfill.

When melting steel scrap in electric arc furnaces, around 100–150 kg of slag and 10–20 kg of dust per tonne of steel are generated. The dusts often contain high levels of zinc and lead from the coatings of galvanised steel, and chromium, nickel and molybdenum from stainless steel. The typical composition range of electric arc furnace dusts is Fe 20–30 wt%, Zn 19–21 wt%, Pb 2–4 wt%, Cr 0.3–0.4 wt%, Mn 2.5–4.6 wt% and Cu ≤0.5 wt% (Rao, 2006). In many less developed countries the dust is still disposed of in landfill, usually after stabilisation with cement, or is stockpiled. However, due to waste management regulations, the quantity of dust going to landfill is decreasing. More is being treated for recovery of its remaining metal content. In the United States about 60% of dust is recycled; the balance goes to landfill (Bruckard *et al.*, 2006). In Europe and Japan virtually 100% is recycled. There are numerous commercial processes for treating electric arc furnace dust including hydrometallurgical, pyrometallurgical, pyro-hydro hybrids and stabilisation processes (Zunkel, 1997; Liebman, 2000). The slag produced from remelting steel in electric arc furnaces can be used as aggregate in building and road construction.

The problem of tramp elements

Because recycling of metals involves melting the scrap, it is almost impossible to prevent contamination by other elements, particularly metals that are mixed with the scrap. In primary metal production the flowsheet has evolved so that impurities are progressively removed at each stage – mining, beneficiation, smelting, leaching/SX – and the remaining impurities are adjusted in the refining step. In recycling of metals, the potential contaminants are different, in nature and quantity, from those in ores. For example, copper from electrical cables is a common contaminant in scrap steel but is not a concern in production of steel from iron ores. Elements that are difficult to remove from metals, for example copper from iron and iron from aluminium, are called *tramp elements*. Tramp elements in steel and aluminium generally increase the strength of the metal but reduce its ductility; the application of these metals, once contaminated, is limited to lower-grade uses. Tramp elements are usually less chemically reactive than the host metal. If refining reagents are added the host metal tends to react preferentially, making it difficult or impossible to remove the tramp element without significant loss of the host metal. Removal of tramp elements requires application of highly selective physical or chemical processes which are technically difficult and expensive. This is particularly the case for aluminium, which is highly reactive. It is for this reason that, in the production of primary aluminium, the impurities present in the bauxite are removed in the Bayer process to produce highly pure aluminium oxide (alumina) which is then reduced electrolytically to metal with only a little recontamination from the molten salt electrolytes and the carbon-based anode and cathodes.

The major potential tramp elements in aluminium are iron and manganese (from unseparated steel), silicon and magnesium (from aluminium alloys), copper (from electrical components) and zinc (from unseparated galvanised steel). The major potential tramp elements in steel are copper (from electrical components, particularly in car bodies), tin (from tin plate) and chromium, nickel and molybdenum (from alloy steels). Other tramp elements include lead, antimony, bismuth and arsenic. Zinc (from galvanised steel products) is not considered a tramp element in steel. Being volatile at steel remelting temperatures, zinc concentrates in the fume in electric arc furnace scrap melting and is thus effectively separated. It is, however, a tramp element in aluminium because aluminium melts at a much lower temperature than steel (around 660°C compared with around 1500°C) and zinc dissolves in the aluminium rather than volatilises. There are four main ways of dealing with the problem of tramp elements in metals. These may be applied independently or in combination.

Separation. The various metals in scrap can be liberated then separated to remove as much as possible of the potential tramp elements. This can be quite effective for simple products and components. In complex products and components, such as electrical and electronic equipment, good liberation of the individual materials is difficult or impossible.

Dilution. Primary metal can be blended with recycled metals to dilute tramp elements to acceptable concentrations. This is a common approach but, as the return rate of metals increases, its effectiveness will decrease as the possible quantity of dilution decreases.

Refining. Where technically possible, specific refining technologies can be applied to scrap to lower the concentration of tramp elements. This is rarely done because it is expensive and technically difficult.

Downgrading. A lower-grade product, containing tramp elements, can be accepted and used in applications which require less demanding properties. Recycled aluminium from beverage cans is frequently downgraded to lower-quality cast or wrought alloys.

13.9 RECYCLING MATERIALS FROM COMPLEX PRODUCTS

Many modern manufactured products are complex and contain many different components made from a wide range of materials combined in varying proportions and forms. The challenge of recycling these products lies in liberating the materials by breaking the products into small enough pieces that effective separation of the different materials can be made. In many respects this is similar to the way an ore is treated to liberate and separate the individual minerals. The flowsheets and equipment used bear close resemblance, not surprisingly, to mineral processing flowsheets. Two important classes of complex products, both ubiquitous, are cars and electrical and electronic products. The approaches to recycling these are similar though the scale of operations for recycling vehicles is usually much larger. However, complete liberation of all materials is not possible for much electronic equipment because of its often small size and complex nature. Such equipment must be processed chemically to separate the various elements. This is similar to the strategy for processing ores when the valuable element is too finely dispersed to be able to be liberated (Section 5.7.2).

13.9.1 Cars

The main types of materials used in cars are listed in Table 13.4, which shows that metals comprise about 75 wt% and polymer materials about 16 wt%. Steel is used in by far the largest quantities. It and other materials are combined to form parts from which a car is assembled. A typical motor vehicle contains around 15 000 parts made from materials in various and often complex combinations of type and form

Table 13.4: The approximate composition of a modern car, in terms of types of material

Material	wt%
Steel	66
Zinc, copper, lead	2
Aluminium	6
Plastics	9
Rubber (tyres)	4
Adhesive, paints	3
Glass	3
Textiles	1
Fluids	1
Other	5

Source: DEH (2002).

(DEH, 2002). Cars that have reached the end of their useful life are referred to as end-of-life vehicles. The reuse of parts and the reclamation of materials from end-of-life vehicles is a long established industry which has continually evolved. Approaches vary widely depending largely on the cost of labour, the cost of landfill, and legislation regarding recycling. Typical modern practice in Europe is summarised in the flowsheet for mechanical processing of end-of-life vehicles in Figure 13.7.

The technology of vehicle recycling

The process of recycling a vehicle usually commences when a vehicle is sold to a dismantler (wrecker) who removes parts that can be sold for reuse (wheels, engines, transmissions, plastic bumpers) and the environmentally polluting materials (operating fluids, batteries), then sells the hulk to a shredding company. Shredding companies shred cars and other products such as white goods, demolition scrap and prompt scrap. These are fed into shredders, which are large, high-capacity mills that break the products into small pieces, the objective being to liberate the individual materials so they can be separated. Shredders consist of a casing and a housing in which a heavy horizontal rotor turns at around 500–600 rpm. Swinging hammers on the rotor pulverise metal objects against a breaker bar and drag the pieces around the housing until they become small enough to fall through grates at the base (Nijkerk and Dalmijn, 2001). The size of product from a shredder is typically less than 50–80 mm. The shredded material is next preconcentrated by air classification to remove dust and light non-metallic materials such as foam, textiles, foil and wood. This material is called Light Auto Shredder Residue (Light ASR) and is usually sent to landfill. Next, steel is recovered by magnetic separation (Dalmijn and De Jong, 2007). Since copper wiring is often not fully liberated, hand sorting of the separated steel fraction must be done if a copper content of less than 0.25 wt% is required.

The non-magnetic fraction consists of non-ferrous metals and stainless steel, as well as glass, stones, plastics and rubber. It is quite heterogeneous with respect to composition, size and shape. Separation is therefore difficult. The material is screened into several size fractions to facilitate subsequent operations (Dalmijn and De Jong, 2007); for example, <12 mm, 12–65 mm and >65 mm. The >65 mm material is hand-sorted. Separation of the other fractions begins with removal of the light components (plastic, textile, wood) using eddy current or rising current separators. The latter involves using a rising current of water to separate materials of different density. The former uses permanent magnets to induce eddy currents in moving, electrically conducting materials, such as non-ferrous metals. These currents give rise to secondary magnetic fields which interact with the field of the permanent magnets to move the conducting materials relative to the non-conducting materials and thus effect a separation. The non-ferrous fraction is subjected to a two-stage dense medium separation. The first is at about 2.3 g mL^{-1}, to recover magnesium, dense plastics and rubber. The second is at about 3.3 g mL^{-1}, to separate aluminium from the dense non-ferrous metals. The sink product of the second step consists mainly of copper, brass, lead, zinc and stainless steel; separation of these is difficult. Automated systems using colour recognition as the basis for sorting are constantly improving. The sink product is hand-sorted in countries where this is economical. Residues from this stream are called Heavy Automotive Shredder Residue (Heavy ASR), and are usually sent to landfill.

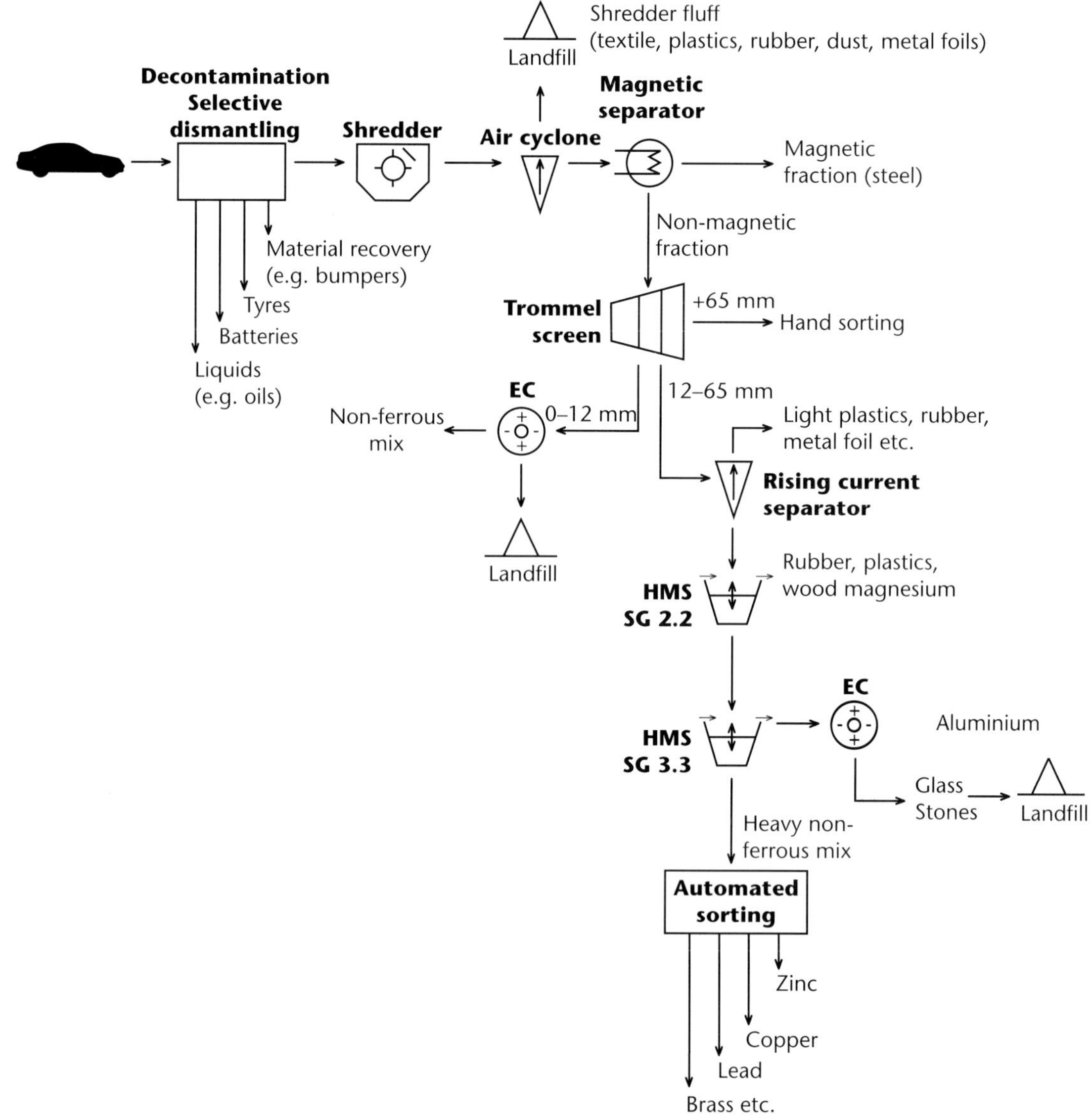

Figure 13.7: A generalised flowsheet for the mechanical processing of end-of-life vehicles. EC = eddy current separator, HMS = heavy medium separator. (Dalmijn and De Jong, 2007; reproduced with kind permission from Springer Science+Business Media).

Legislation relating to vehicle recycling

Many countries have no legislation requiring the recycling of vehicles and rely on market forces, particularly the cost of landfill and the value of scrap metal, to determine the recycle rate. Other countries have legislated certain requirements. The most advanced in this regard is the European Union. An end-of-life vehicle directive (Directive 2000/53/EC) enacted in 2002 (European Parliament, 2000) requires member states to:

- achieve targets of 85% reuse and recovery and 80% materials recycling by average weight per vehicle deposited for treatment by 1 January 2006, and 95% reuse and recovery and 85% materials recycling by 1 January 2015;
- ensure that all end-of-life vehicles are dismantled, treated and recovered by industry at no cost to the final holder of a vehicle and in a manner that does not cause environmental pollution;

- introduce systems whereby certificates of destruction are notified to the vehicle registration authorities on the deposit of end-of-life vehicles by their registered owners at authorised treatment facilities;
- minimise the use of specified hazardous substances in new vehicles.

It is likely that other countries will progressively adopt similar legislation.

13.9.2 Waste electrical and electronic equipment

Waste electrical and electronic equipment (WEEE) consists of all equipment which depends on electric currents or electromagnetic fields for its operation, and equipment for generating, transferring and measuring currents and fields. It includes household appliances, IT and telecommunications equipment, consumer equipment, lighting equipment, electrical and electronic tools, toys, leisure and sports equipment, medical devices, and monitoring and control instruments. There are two broad classes of WEEE: electronic scrap (e-scrap or e-waste) such as personal computers, mobile (cell) phones, televisions, radios, LCD and plasma monitors and cathode ray tubes (CRTs); and electrical scrap such as refrigerators, washing machines, dishwashers, air-conditioners and power tools.

Waste electrical equipment is often recycled, along with car bodies and other largely metallic products, as described in Section 13.8.3 or in purpose-built facilities using similar technologies. Refrigerators, washing machines, dishwashers, air-conditioners and similar electrical goods are extensively dismantled and sorted to remove components that may be reused or that are difficult to shred. For example, compressors and detachable parts such as shelves are removed from refrigerators, and oil and refrigerant are removed from the compressor. The main body of the refrigerator is then sent to the shredder.

Electronic equipment is generally much more complex than electrical equipment. Electronic goods can contain up to 60 different elements. Their manufacture involves a large share of the demand for some precious metals (particularly Ag, Au, Pd and Pt) and some rare metals, particularly Co, In and Sb (Meskers and Hagelüken, 2009). The elements and materials used in electronic goods are often combined in complex ways, both chemically and physically, and this makes recycling them difficult. The volume of electronic waste produced has been growing rapidly since the 1970s with the development of consumer electronic products and their rapid uptake in all countries. The volume of obsolete electronic equipment is already a serious global problem, particularly because the many toxic materials used in their manufacture (lead, beryllium, mercury, cadmium, chromium, brominated flame retardants in plastics) make disposal by landfill or incineration potentially hazardous. The options for electronic waste are direct disposal to landfill (which may lead to contaminated groundwater), stabilisation before disposal to landfill (which is expensive), incineration (which releases metals and other toxic substances to air and to ashes) and recycling. About 5–7 million tonnes of electronic waste is generated per year in the United States. Only about 10% is recycled, 30% is stored for future disposal, and the remainder is landfilled (Kaya and Sözeri, 2009). A similar quantity is produced in Europe. The volume is increasing at 3–5% per year (Kang and Schoenung, 2005).

Electronic wastes contain significant quantities of metal, some of which are very valuable. Recycling of electronic wastes not only reduces potential environmental and health-related effects due to dispersion of toxic elements, it also saves reserves by providing a source of many rare metals. The major types of materials in a mobile (cell) phone, for example, are listed in Table 13.5. A typical mobile phone weighs about 100 g. Each phone contains about 25 g of various metals, the largest quantity of which is copper. Some very expensive metals occur in only small quantities but, because of the volume of end-of-life mobile phones, the value of these metals is significant. Table 13.6 lists the value of the main metals in mobile phones. It was projected that about one billion mobile phones would be produced globally in 2009 (Sullivan, 2006). The total value of metal in those phones was about US$1.2 billion. There were predicted to be about 2.6 billion mobile phones in use at any one time by 2009 (Sullivan, 2006). These would contain metal to the value of about US$3.1 billion.

Table 13.5: Typical quantities of the major types of materials in a mobile phone

Material	wt%
Plastic	58
Metals	25
Ceramics	16
Flame retardant	1
Total	*100*

Source: Sullivan (2006).

The technology of recycling of electronic scrap

Electronic products generally consist of a large number of individually complex components, often produced by different manufacturers. Many valuable as well as hazardous substances are used in many combinations, often closely intermingled. The recovery of metals that occur at a low concentration, such as precious and rare metals in circuit boards or indium in LCD screens, is technically challenging. The accessibility of critical components like batteries and printed circuit boards is important, since easy access and removal can ensure they will be recycled rather than lost. The key operations in the recycling of electronic waste are collection, sorting and recovery, recycling, and disposal. Several types of collection methods are used (Kang and Schoenung, 2005). Kerbside collection consists of the collection of waste periodically (like municipal waste) or by request, and is the most convenient for the consumer. Collection of e-waste as part of an existing kerbside waste collection program substantially reduces the costs, but kerbside collection is usually more costly than other systems. Drop-offs are short (1–2 days) events during which consumers drop off unwanted electronic products at a central collection point. The location of the collection site is important. Permanent collection is essentially year-round collection; a municipal solid waste collection site is often used. This type of collection is often the most cost-effective. Point-of-purchase collection is a method whereby retailers serve as the collection agency; consumers return unwanted electronic equipment when they purchase new equipment. The active participation of the retailer is essential and it may need to be legally mandated to be effective, as has been done in the European Union and other countries.

A simplified flowsheet for materials recovery from waste electronic equipment is shown in Figure 13.8. Collected equipment is transported to a materials recovery facility, where it is sorted into streams. One stream is for refurbished equipment that can be sold or donated to secondary users, a second is for components that can be reclaimed, resold and reused, and the third is for salvaged and recycled materials. Although straightforward, examination and testing for reuse is time-consuming and labour-intensive. A plug-and-play test is used to identify equipment that is operational. Equipment that fails the test may be dismantled for component resale and reuse. Hazardous materials are removed at this stage. The material in the third stream is shredded to liberate the major types of materials (plastics, glass and ceramics,

Table 13.6: Mass and value of selected metals in mobile phones

		Per phone	1 billion sales per year (2009)		2.6 billion phones in use (2009)
	Price[a]	Mass	Mass	Value	Value
Metal	(US$ per tonne)	(grams)[b]	(tonnes)	(US$, millions)	(US$, millions)
Copper	7385	16	16 000	118	307
Silver	430 819	0.35	350	151	392
Gold	21 701 723	0.034	34	738	1920
Palladium	11 574 252	0.015	15	174	451
Platinum	40 509 882	0.00034	0.34	14	36
Total		*16.4*	*16 399*	*1194*	*3105*

a) Based on 2007 metal prices, Table 6.4.
b) Sullivan (2006).

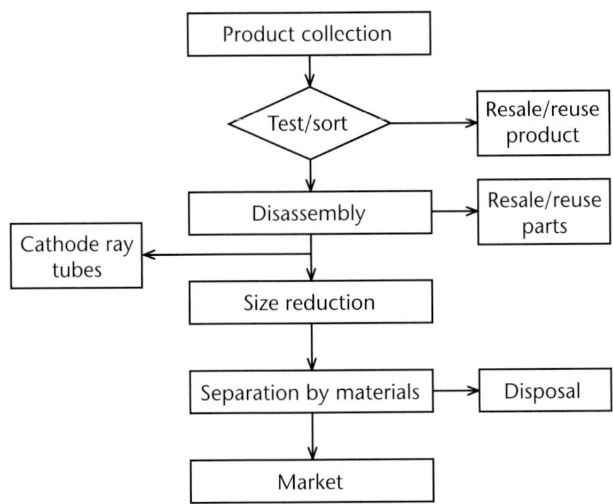

Figure 13.8: Simplified flowsheet for the recycling of electronic products (after Kang and Schoenung, 2005).

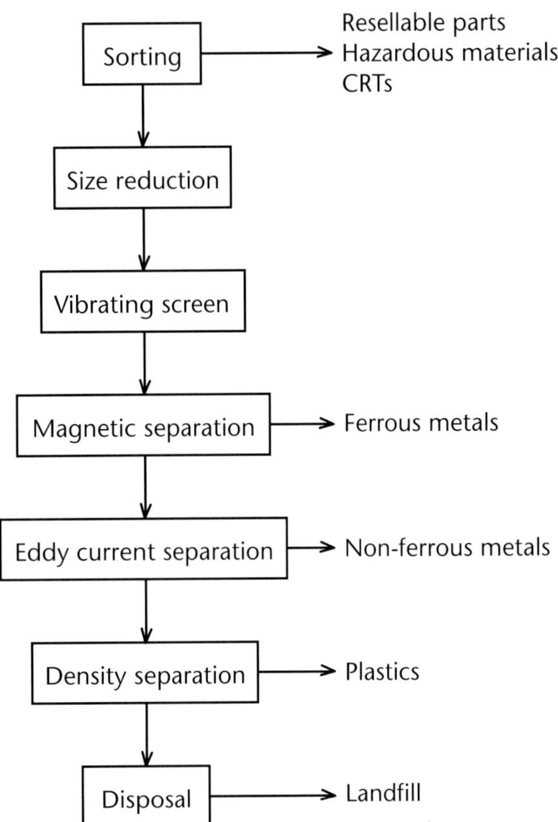

Figure 13.9: Simplified flowsheet of a materials recovery facility for recycling electronic scrap (after Kang and Schoenung, 2005).

ferrous and non-ferrous metals), which are then separated using various techniques (Figure 13.9). Complete or near-complete liberation of the non-ferrous metals from other materials is impossible because of their intimate association in printed circuit boards and integrated circuits.

The recovered plastics, glass and ceramics are treated in a variety of ways (Kang and Schoenung, 2005). Plastics can be recycled mechanically or chemically, or combusted to provide heat for power generation or for cement kilns. Glass from CRTs, because it contains lead, needs special caution if it is disposed to landfill. There are two technologies for CRT glass recycling: closed loop recycling to new CRT glass, and smelting the glass to remove the lead. The latter produces a lower-value glass product.

There are many options for treating the non-ferrous metals stream. These all involve chemical processing. The flowsheets are complex because of the number of elements to be separated and their widely differing properties. Most commercialised recycling processes involve both pyrometallurgical and hydrometallurgical stages. Pyrometallurgical processing is very effective in separating the common base metals (copper, lead, zinc), collecting precious metals in copper matte, copper metal or lead bullion, from which they can be recovered in subsequent operations, and collecting many of the rare elements in slag, from which they can be recovered later. Many base metal primary smelters also process significant quantities of electronic scrap. Smelters can tolerate incomplete liberation of non-ferrous metals from other materials. Glass, for example, will simply report to the slag, polymers (from printed circuit boards) will serve as reductant or as process fuel and metals will report to the metal, slag or matte phase according to their stability. Combustion of polymers can give rise to dioxins and furans, so off-gases from smelting operations that treat electronic waste need extra gas-cleaning steps. Hydrometallurgical processing is effective for making highly selective separations of elements from complex feed materials using chemical precipitation, solvent extraction, ion exchange and electro-refining. Leaching kinetics is generally slow so the size of the feed materials needs to be small, to provide a high surface area; high temperatures and pressures may also be required. Leaching and hydrometallurgical separation processes consume large quantities of acidic or alkaline reagents and produce toxic solutions and solid residues.

A primary base metal smelter which processes large quantities of electronic waste is Boliden's Rönnskär integrated base metal smelter in Sweden which produces copper, lead, zinc clinker, precious metals and other materials from purchased copper concentrates, lead concentrates, copper scrap, lead scrap, catalysts, electronic scrap, EAF dust and other waste materials. The Umicore refinery in Hoboken, Belgium, treats printed circuit boards, electronic components, metallic granules, plastics containing metals, metal-containing dusts and small WEEE devices after removal of batteries. It is a purpose-built recycling facility which combines pyrometallurgical and hydrometallurgical steps. The materials are fed to an ISASMELT™ furnace which produces metallic copper and lead-rich slag, which is reduced in a blast furnace to produce lead. The crude copper and lead are refined using hydrometallurgical and pyrometallurgical steps to remove and recover the precious and rare metals (Meskers *et al.*, 2009).

Legislation relating to the recycling of WEEE

Legislation regulating the disposal of e-waste is relatively undeveloped in many countries. European legislation is the most advanced. Two directives passed by the European Parliament in 2003 are particularly relevant (European Parliament, 2003a; 2003b).

Directive 2002/96 On waste electrical and electronic equipment. This directive requires producers to set up and operate collection, recycling and disposal systems, at no cost to consumers. It classifies WEEE into 10 categories and sets recovery, reuse and recycle targets for each (Table 13.7). It also requires labelling of e-waste to identify the different components and materials.

Directive 2002/95 On the restriction of the use of certain hazardous substances in electrical and electronic equipment (RoHS). This directive requires the phasing out of the use of hazardous substances in electrical and electronic waste by 2008. The RoHS Directive specifically restricts the use of lead, mercury, cadmium, hexavalent chromium, polybrominated biphenyls (PBB) and polybrominated diphenyl ethers (PBDE) in electronic equipment, the maximum permitted concentrations being 0.1 wt%, except for cadmium which is limited to 0.01 wt%. These limits apply to single materials that could, in principle, be mechanically separated, not to the final product or to individual components. Batteries are not included within the scope of RoHS but are

Table 13.7: European Union minimum target recovery, reuse and recycling rates, by average weight per appliance

Category		Recovery rate (%)	Reuse and recycling rate (%)
1	Large household appliances	80	75
2	Small household appliances	70	50
3	IT and telecommunications equipment	75	65
4	Consumer equipment	75	65
5	Lighting equipment	70	50
6	Electrical and electronic tools (except large-scale stationary industrial tools)	70	50
7	Toys, leisure and sports equipment	70	50
8	Medical devices (except all implanted and infected products)	70	50
9	Monitoring and control instruments	70	50
10	Automatic dispensers	80	75
	Gas discharge lamps		80

Source: European Parliament (2003a).

covered by a separate directive (European Parliament, 2006):

Directive 2006/66. On batteries and accumulators and waste batteries and accumulators and repealing Directive 91/157/EEC. Commonly known as the Battery Directive, this directive regulates the manufacture and disposal of batteries.

Japan, South Korea and Taiwan have extended producer responsibility laws which require manufacturers and importers to take back end-of-life electronic equipment for recycling and waste management. Recycling targets have been set and a fee on consumers helps to cover the costs of collection and transport. The United States, Canada and Australia do not have national laws regulating WEEE but state laws, particularly relating to landfill, influence the recycling of WEEE. In some states, landfilling of CRTs is banned. In California, electronic waste cannot be disposed of in landfill nor exported. California introduced an electronic waste recycling fee in 2004 on all new monitors and television sets to cover the cost of recycling. Maine, Maryland and Washington have extended producer responsibility schemes but no recycling fee. Several Canadian provinces have similar legislation. In 2005, 60–85% of all WEEE in the United States was deposited in landfill (US EPA, 2008). India and China do not have regulatory frameworks to manage WEEE generated domestically and WEEE that is imported for recycling.

WEEE recycling in developing countries

Though the Basel Convention bans international trade in hazardous wastes, large quantities of WEEE are shipped from developed countries to China, India, Pakistan and other countries which have cheap labour and low environmental standards. This is an area of growing international concern (Herat, 2010). It is estimated that 50–80% of the e-waste collected for recycling in the United States, which is not a signatory to the Basel Convention, is exported to developing countries for recycling. In those countries, WEEE recycling is performed largely by an unregulated, informal industry using labour-intensive, primitive technology which is unsafe and poses health risks to workers and the environment. Frequently, the wastes are disposed of in unregulated ways, into the atmosphere and rivers and to waste land. In China, around 700 000 people were employed in e-waste recycling in 2007, 98% of them in the informal economy. In India, over a million people are similarly employed.

13.10 DESIGN FOR THE ENVIRONMENT

Design for the Environment (DfE) is a general approach to product design that aims to reduce the overall environmental impact of a product across its life cycle. The general principles of DfE are applicable to all types of manufactured goods, from large items of infrastructure (buildings, bridges), to vehicles (cars, trucks, aeroplanes, ships) to domestic, consumer and other goods. There are four main areas of focus: the manufacturing process, materials selection (including packaging materials), product use, and end-of-life.

Manufacturing. This involves producing goods in ways that reduce the quantity of energy and water consumed, that produce minimal emissions and that result in the smallest possible quantity of leftover materials.

Materials. This involves the selection and use of materials in products. No unnecessary materials should be used, the materials should not have a harmful impact on the environment, and they should be easily recyclable. The variety of materials used in a product should be limited to make the product more easily processed at its end-of-life. Particular combinations of materials should be avoided if they are technically difficult to separate and liberate when intimately mixed. Figure 13.10 shows a preliminary attempt to classify the more common elements according to the difficulty of separating them by chemical processing. Incompatible elements should not be used in components if they cannot be easily liberated. The smallest possible quantity of material should be used to achieve the desired function. This conserves resources and reduces the mass of the product, making it less energy-intensive to transport.

Use. This involves designing a product for optimum use. The energy consumption of the product during use should be minimised by, for example, using the lowest energy-consuming components and insulating where appropriate to retain or exclude heat. The life of

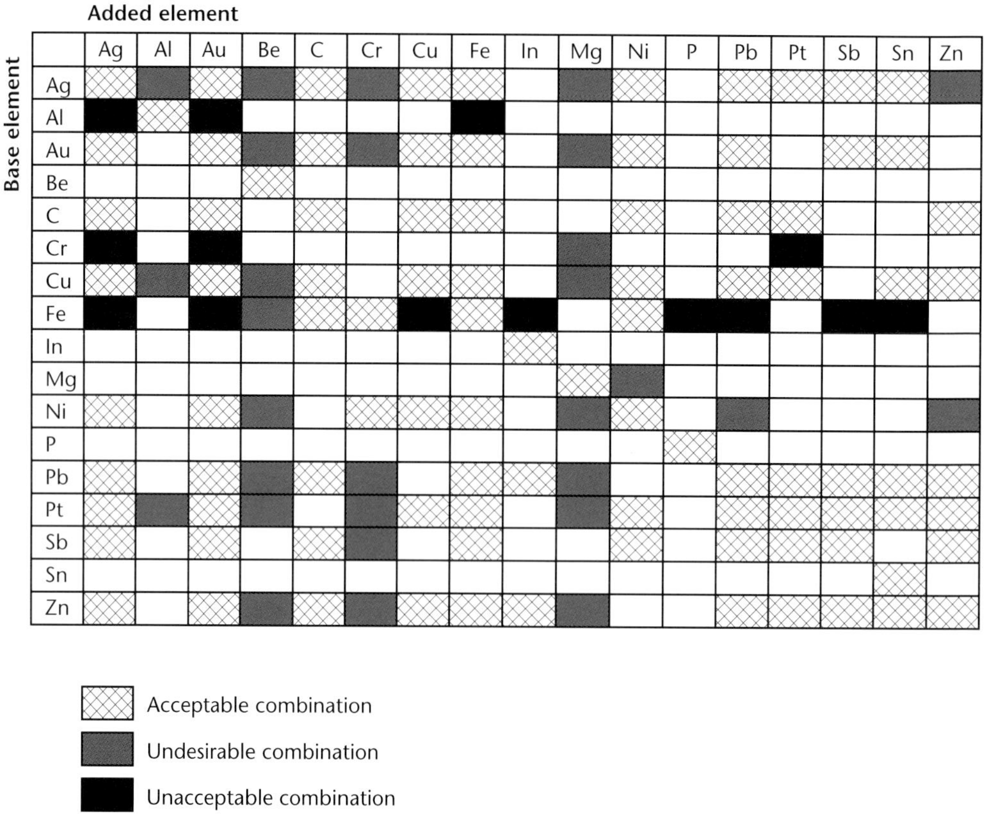

Figure 13.10: Element combinations classified according to the degree of difficulty in separating them by chemical processing (after Lehner and Henriksson, 2009; with modifications).

the product should be optimised by increasing its reliability and durability, designing for easy maintenance and repair, and avoiding designs with a technical life span which exceeds the aesthetic life span. Products should be designed so they can meet possible future needs or be easily modified to do so (e.g. bridges designed so extra traffic lanes can be added or converted to some other function).

Reuse and remanufacture. This involves designing products so they can be easily refurbished and reused for the same or a different purpose. Printer ink cartridges are a good example. They can be easily removed from the printer, returned to a manufacturer, refilled with fresh ink, and resold. Products should be designed for easy disassembly; for example by use of hierarchical and modular structures, use of detachable points and use of standardised joints. This improves maintainability of products and extends their useful life.

Design for the Environment is a complex process. All the above aspects, and others, need to be taken into consideration. They cannot be treated independently since they are interrelated – changes in one area will have implications in other areas. It may not be possible to satisfy all criteria. Hence, a life cycle approach is needed to guide the design of products. LCA methodology is frequently used to predict the impacts of alternative designs, in order to choose the most environmentally favourable.

13.11 REFERENCES

Bruckard W, Davey KJ and Woodcock JT (2006) Characterisation of Australian electric arc furnace (EAF) dusts and the application of simple physical separation techniques to upgrade them. In *Green Processing 2006*. pp. 43–48. Australasian Institute of Mining and Metallurgy: Melbourne.

Bruggink PR and Martchek KJ (2004) Worldwide recycled aluminium supply and environmental impact model. In *Light Metals*. pp. 907–911. Minerals, Metals and Materials Society: Warrendale, PA.

Cleveland CJ (Topic Ed.) (2008) Recycled aggregates. In *Encyclopedia of Earth*. (Ed. CJ Cleveland) Environmental Information Coalition, National Council for Science and the Environment: Washington DC. <http://www.eoEarth.org/article/Recycled_aggregates>.

Dalmijn WL and De Jong TPR (2007) The development of vehicle recycling in Europe: sorting, shredding and separation. *JOM* **59**(11): 52–56.

DEH (2002) *Environmental Impact of End-of-Life Vehicles: An Information Paper*. Department of the Environment and Heritage, Commonwealth of Australia: Canberra.

Delgado L, Catarino AS, Eder P, Litten D, Luo Z and Villanueva A (2009) *End-of-Waste Criteria*. Joint Research Centre, Institute for Prospective Technological Studies: Luxemburg. September, 2009. <http://ipts.jrc.ec.europa.eu/publications/pub.cfm?id=2619>.

European Container Glass Federation (2009) <www.feve.org>.

European Parliament (2000) Directive 2000/53/EC of the European Parliament and of the Council of 18 September 2000 on end-of-life vehicles. *Official Journal of the European Communities*, L 269, 21 October 2000.

European Parliament (2003a) European Parliament Directive 2002/95/EC of the European Parliament and of the Council of 27 January 2003 on the restriction of the use of certain hazardous substances in electrical and electronic equipment. *Official Journal of the European Union*, L 37/19, 13 February 2003.

European Parliament (2003b) Directive 2002/96/EC of the European Parliament and of the Council of 27 January 2003 on waste electrical and electronic equipment (WEEE). *Official Journal of the European Union*, L 37/24, 12 February 2003.

European Parliament (2006) Directive 2006/66/EC of the European Parliament and of the Council of 6 September 2006 on batteries and accumulators and waste batteries and accumulators and repealing Directive 91/157/EEC. *Official Journal of the European Union*, L 266/1, 26 September 2006.

Herat S (2010) Emerging issues, challenges and opportunities in environmentally sound management of e-waste. *International Consultative Meeting on Expanding Waste Management Services in Developing Countries*, 18–19 March 2010, Tokyo, Japan.

Hoberg H, Wolf S and Meier-Kortwig J (1999) Modelling the material flow of recycling processes for aluminium alloys by means of technical recycling quotas. In *Global Symposium on Recycling, Waste Treatment and Clean Technology, REWAS 99*. (Eds E Gaballah, J Hager and R Solozabl) pp. 1023–1033. San Sebastian, Spain.

Kang H-Y and Schoenung JM (2005) Electronic waste recycling: a review of US infrastructure and technology options. *Resources Conservation and Recycling* **45**: 368–400.

Kaya M and Sözeri A (2009) A review of electronic waste (e-waste) recycling technologies. Is e-waste an opportunity or threat? In *EDP Congress 2009*. (Ed. SM Howard) pp. 1055–1060. Minerals, Metals and Materials Society: Warrendale, PA.

Kusik CL and Kenahan CB (1978) Energy use patterns for metal recycling. *Information Circular 8781*. US Bureau of Mines: Washington DC.

Lehner T and Henriksson H (2009) Industrial recycling of electronic scrap at Boliden's Rönnskär smelter (oral presentation). *TMS Annual Meeting*, San Francisco, 16–18 February 2009.

Liebman M (2009) The current status of electric arc furnace dust recycling in North America. In *Recycling of Metals and Engineered Materials*. (Eds DL Stewart, JC Daley and R Stephens) pp. 237–250. Minerals, Metals and Materials Society: Warrendale, PA.

Meskers CEM and Hagelüken C (2009) Closed loop WEEE recycling? Challenges and opportunities for a global recycling society. In *EDP Congress 2009*. (Ed. SM Howard) pp. 1049–1054. Minerals, Metals and Materials Society: Warrendale, PA.

Meskers CEM, Hagelüken C and Van Damme G (2009) Green recycling of EEE: special and precious metal recovery from EEE. In *EDP Congress 2009*. (Ed. SM Howard) pp. 1131–1136. Minerals, Metals and Materials Society: Warrendale, PA.

Nijkerk AA and Dalmijn W (2001) *Handbook of Recycling Techniques.* 5th edn. Nijkerk Consultancy: The Hague, The Netherlands.

Nilmani M (2005) Zero waste – poverty driven. In *Sustainable Developments in Metals Processing.* (Eds M Nilmani and WJ Rankin) pp. 295–301. NCS Associates: Sydney.

Norgate TE, Jahanshahi S and Rankin WJ (2009) Metals and sustainable development. In *Environmental Impacts Assessments.* (Eds GT Halley and YT Fridian) pp. 355–395. Nova Science Publishers: Hauppauge, New York.

Peterson RD and Newton L (2002) Review of aluminum dross processing. In *Light Metals 2002.* (Ed. W Schneider) pp. 1029–1037. Minerals, Metals and Materials Society: Warrendale, PA.

Prillhofer R, Prillhofer B and Antrekowitsch H (2009) Treatment of residues during aluminum recycling. In *EDP Congress 2009.* (Ed. SM Howard) pp. 857–861. Minerals, Metals and Materials Society: Warrendale, PA.

Rao A, Jha KN and Misra S (1997) Use of aggregates from recycled construction and demolition waste in concrete. *Resources, Conservation and Recycling* **50**: 71–81.

Rao SR (2006) *Resource Recovery and Recycling from Metallurgical Wastes.* Elsevier: Amsterdam.

Sullivan DE (2006) Recycled cell phones – a treasure trove of valuable metals. United States Geological Survey, Fact Sheet 2006-3097, July 2006. <http://pubs.usgs.gov/fs/2006/3097>.

The Economist (2007) The price of virtue. **383**(8532), June 2007: 12.

US EPA (2008) Management of electronic waste in the United States. United States Environmental Protection Agency, Fact Sheet EPA530-F-08-014, July 2008.

USGS (2007) *Minerals Yearbook 2007.* United States Geological Survey. <http://minerals.usgs.gov/minerals/pubs/commodity/myb>.

van Berkel R (2007) Eco-efficiency in primary metal production: context, perspectives and methods. *Resources, Conservation and Recycling* **51**: 511–540.

World Steel Association (2008) *Fact Sheet – Energy.* October 2008. <http://www.worldsteel.org>.

Zhang L (2006) State of the art in aluminum recycling from aluminum dross. In *Light Metals 2006.* (Ed. TJ Galloway) pp. 931–936. Minerals, Metals and Materials Society: Warrendale, PA.

Zunkel AD (1997) Electric arc furnace dust management: a review of technologies. *Iron and Steel Engineer* **74**(3): 33–38.

13.12 USEFUL SOURCES OF INFORMATION

Green JAS (Ed.) (2007) *Aluminum Recycling and Processing for Energy Conservation and Sustainability.* ASM International: Materials Park, Ohio.

Henstock ME (1996) *The Recycling of Non-ferrous Metals.* International Council on Metals and the Environment: Ottawa, Canada. <http://www.icmm.com/page/1658/the-recycling-of-non-ferrous-metals>.

Nijkerk AA and Dalmijn W (2001) *Handbook of Recycling Techniques.* 5th edn. Nijkerk Consultancy: The Hague, The Netherlands.

Reuter MA *et al.* (2005) *The Metrics of Material and Metal Ecology: Harmonizing the Resource, Technology and Environmental Cycles.* Elsevier: Amsterdam.

Schluep M, Hagelueken C, Kuehr R, Magalini F, Maurer C, Meskers C *et al.* (2009) Recycling – from e-waste to resources. United Nations Environment Programme: Paris, France, July 2009. <http://www.unep.org/PDF/PressReleases/E-Waste_publication_screen_FINALVERSION-sml.pdf>.

14 The future availability of minerals and metals

There is a widespread belief that, materially at least, everything will continue to improve, that products will become better, cheaper and more plentiful. This has certainly been the general trend in the Western world for over 200 years and, particularly since the 1950s, in much of the rest of the world. It is important to examine whether this trend can and will continue. There appears to be no definitive or simple answer. The pessimistic view is that the world is rapidly running out of non-renewable mineral resources, that energy consumption to produce mineral and metal commodities will rise rapidly as grades fall, and that as prices rise the material standard of living must decline. The optimistic view is that there are no real limits, that new mineral discoveries, new sources of minerals and improved technologies will ensure the long-term supply of relatively low-cost minerals and metals, and that recycling, substitution and reduction in the intensity-of-use of materials (dematerialisation) will reduce the demand for primary minerals and metals. This chapter explores the future availability of minerals-based materials by looking beyond the present boundaries of what are considered to be mineral resources. This is the major geological and technical supply issue. Chapter 15 examines the concepts of substitution and dematerialisation, and explores the technical aspects of the long-term demand for mineral and metal commodities.

14.1 THE DETERMINANTS OF LONG-TERM SUPPLY

Advances in exploration technology can result in the discovery of new ore bodies and increase the reserve of particular minerals and metals. New technologies for mining and for processing ores can reduce the cost of producing commodities and make previously uneconomic mineral deposits profitable to mine, again resulting in an increase in mineral reserves. Changes to the economic and political systems of countries may increase the supply of mineral and metal commodities. For example, over the past four decades, financial deregulation, workforce reforms and the reduction or removal of tariffs and quotas in many countries have resulted in an increase in trade. This has enabled new mining and processing operations to commence, particularly in developing countries. All the above factors have, historically, increased the supply of minerals and metals. Thus, the main factors which affect the long-term supply of mineral and metal commodities can be summarised as follows (Sohn, 2006):

- the cost of their production;
- developments in technologies associated with primary production;
- the nature of the political and economic system.

It is widely believed that the future cost of producing minerals will increase as ore grades decline and deeper and more remote mineral deposits have to be exploited. As discussed in Section 6.3, this has not occurred to date, largely due to improved technologies for exploration, mining and processing. In fact, in real terms the production cost of many mineral commodities has been declining for many decades, if not centuries. How long can similar types of developments continue? The answer would tell us the long-term availability of minerals.

Exploration can only discover what is already there, and the cost of extraction and processing depends strongly on the grade of deposits. Hence, the fundamental technical question regarding long-term supply is the quantity and grade of a particular mineral or metal that is available to be extracted. This chapter restricts consideration of long-term supply to this question. It is not concerned with socio-political factors that may also significantly determine the future supply.

14.2 POTENTIAL SOURCES OF MINERALS

There are a surprising number of actual and potential sources of minerals and metals. The most important of these are summarised in Figure 14.1, where they are grouped as conventional and unconventional sources. They are arranged vertically by increasing cost of extracting elements from them, both financially and in terms of energy required. From left to right, the sources move from geologically known (for which the locations, quantities and compositions are reasonably well established) to increasingly geologically unknown. Some sources span the entire range, others only part. The figure is not drawn to scale and the areas do not indicate the relative sizes of the various sources.

The conventional sources of primary metals, industrial minerals and rocks are what we know as reserves. What we call resources are the future potential conventional sources. These lie almost totally within the continental crust, though a few minerals are obtained from seawater and inland waters (salts) and from the continental shelves. The area labelled hypothetical mineral resources represents undiscovered mineral deposits in the crust, at concentrations above which they could be extracted profitably in the future. This concentration is the so-called mineralogical barrier, discussed further in Section 14.3.2. The major unconventional, and at present mainly uneconomic, primary sources of mineral and metal commodities are:

- crustal rocks (excluding mineral deposits);
- seawater;
- the seabed;
- mining, milling, leaching and smelting wastes;
- concentrated brines from inland waters;
- extraterrestrial material such as from the Moon or asteroids.

Some consumer wastes, construction and demolition wastes, and industrial wastes are already used as

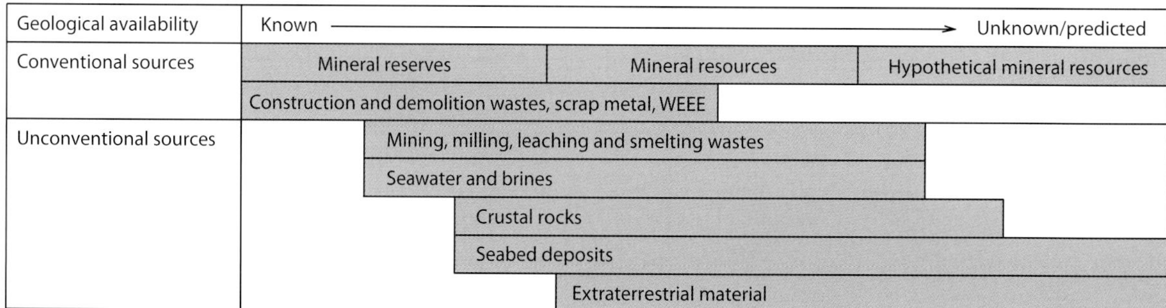

Figure 14.1: Actual and potential sources of mineral resources. Conventional sources are those being exploited. Unconventional sources represent possible future sources.

secondary sources of materials. For example, scrap metal is a conventional and important source of metals through recycling. Building and construction wastes and electronic wastes are increasingly being recovered and recycled. The wastes produced from mining, milling, leaching and smelting often contain useful minerals and elements at moderate or high concentrations. These could be a source for many metals and minerals, though little is done at present to exploit this opportunity. Doing so would also reduce the quantities of wastes from primary processing that need to be stored or otherwise disposed of. This is an important topic, explored in detail in Chapter 16. This chapter examines the potential for finding additional conventional mineral resources and the possibility of extracting mineral and metal commodities from crustal rocks, seawater and deposits on the seabed.

14.3 CRUSTAL RESOURCES

14.3.1 The distribution of the elements in the crust

As noted in Section 5.3.3, the elements of the crust fall into two quite distinct groups, the geochemically abundant elements (O, Si, Al, Fe, Ca, Mg, Na, K, Ti, P and Mn), which make up nearly 99.6 wt% of the crust, and the remaining, geochemically scarce elements. Their distribution in the crust determines their accessibility. It is generally accepted that the concentration distribution of the geochemically abundant elements in the crust has the form of a bell-shaped curve, as shown qualitatively in Figure 14.2. The mathematical form of this curve is not known for certain but there is evidence that it is described by a log-normal function (Ahrens, 1954). The curve is unique for each element, and the area under the curve is the measure of the total quantity of an element in the crust. The portion that has been mined lies to the right of the dotted line in Figure 14.2. High-grade deposits are mined first, so the boundary between mined and unmined material moves progressively towards the left in Figure 14.2. What is remarkable about this form of distribution is that, as the grade of ore decreases, the quantity of an element available from mineral deposits, at the mined grade, increases until the grade falls to that of the mean concentration of the element in the crust. Because the curve is nearly symmetrical, this commences when about half the total quantity of the element in the crust has been mined. At lower concentrations, the quantity available decreases progressively as the grade continues to fall.

For geological reasons, the distribution of the geochemically scarce elements is likely to be quite different. These elements rarely form separate minerals but occur most commonly in substitution at the atomic level within the crystal structure of rock-forming minerals. For example, lead and zinc occur in most common rocks by substituting for the abundant elements potassium and magnesium, respectively (Skinner, 1976). However, where there are localised high concentrations of geochemically scarce elements in geological anomalies, scarce elements form their own minerals and occur at much higher than average concentrations. These deposits are relatively rare and are our present source of mineral commodities for the scarce elements. Because of these two modes of occurrence, it is believed that scarce elements occur within the crust in the form of a bimodal distribution (Figure 14.3). The large peak represents the distribution of a scarce element in common rocks (in substitution) and the small peak represents deposits of the element produced by ore-forming processes. The latter contain discrete minerals of the scarce element; after mining, it is possible to concentrate these by liberation and separation processes and to process the concentrate to extract the element or a compound of the element.

14.3.2 The mineralogical barrier

The bimodal distribution of geochemically scarce elements, if correct, has two important implications. First, as material under the first curve is mined, starting with the richest material, there will be an initial period during which declining ore grades result in larger and larger reserves of ore becoming available, as in the case of the abundant elements. This situation has existed throughout human history. Eventually, the peak will be reached for individual scarce elements; further reduction in grade would result in decreasing reserves of the element. Second, as the ore grade falls, a point is reached beyond which a scarce element occurs only in substitution within rock-forming minerals in

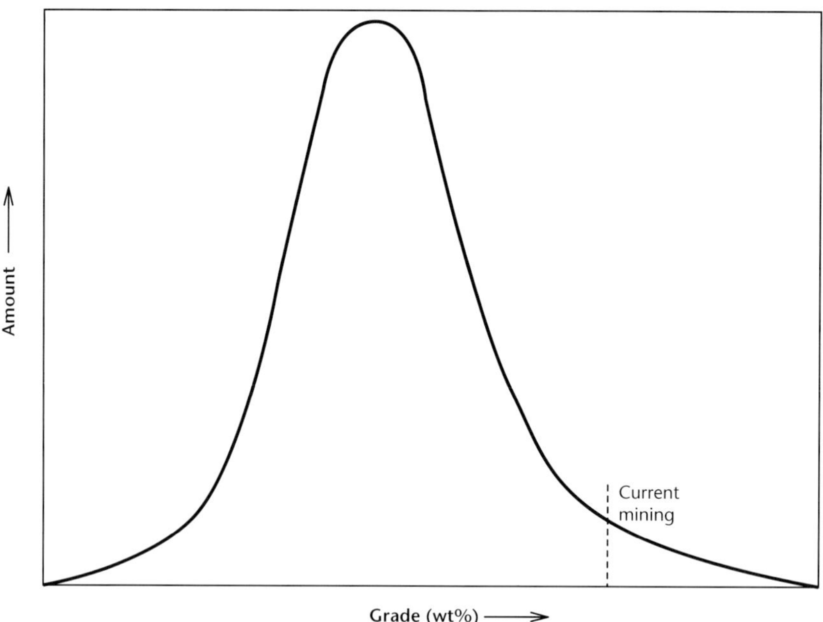

Figure 14.2: The probable distribution of a geochemically abundant element in the Earth's crust (Skinner, 1976).

the common rocks of the crust and is no longer amenable to concentration. This is referred to as the *mineralogical barrier*.

There is never likely to be a shortage of the geochemically abundant elements because they occur at relatively high concentrations in the crust in such large quantities. Even when the present rich deposits of bauxite and iron ore have been exhausted, it will be possible to obtain aluminium and iron by processing progressively lower-grade deposits. However, the cost

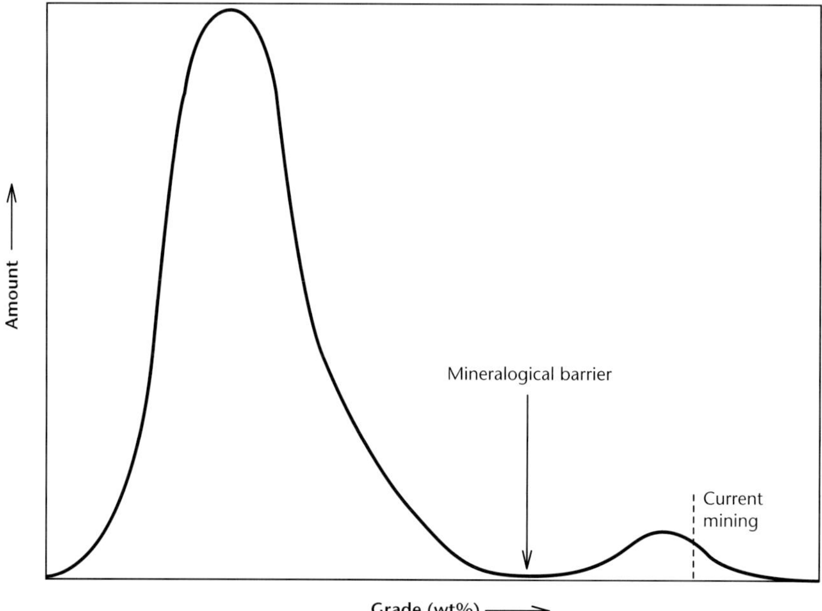

Figure 14.3: The probable distribution of a geochemically scarce element in the Earth's crust (Skinner, 1976).

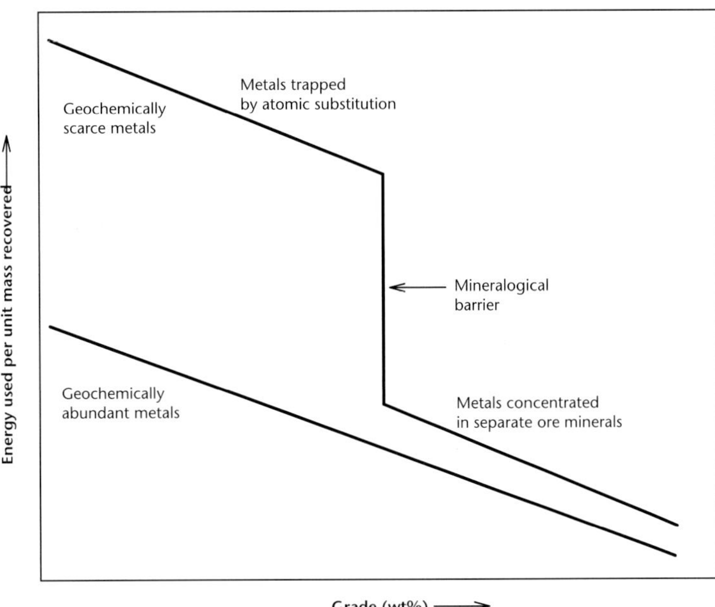

Figure 14.4: The qualitative relationship between element grade and the energy required for production per unit of element recovered, for geochemically abundant and scarce elements (after Skinner, 1976).

per unit of production would increase gradually since more energy would be required to mine and process the ores (Figure 14.4). Also, larger quantities of wastes would be produced. In contrast, the quantities of scarce elements in geological anomalies are relatively small and mineral deposits of scarce elements will eventually become exhausted. In principle, scarce elements could be extracted from common rock (the large peak sources) but, since they occur in substitution not as discrete minerals, it is not possible to concentrate them by liberation and separation. The entire rock would have to be chemically processed, by leaching or smelting. Because scarce elements occur at such low concentrations, very large quantities of rock would have to be processed and the energy required would be very much greater than that required for extracting abundant elements. This could be 100–1000 times that used for presently worked deposits of minerals (Skinner, 1976). Also, huge quantities of waste materials would be produced.

A steadily rising quantity of energy would be required to produce geochemically abundant elements as ore grade declines, but the quantity of energy required to produce scarce elements when the mineralogical barrier is reached would involve a change of several orders of magnitude (Figure 14.4). If large quantities of cheap energy (e.g. from nuclear fusion) become available in the future it would be possible to mine and process common rocks to produce all the elements humans need. However, the elements do not occur in common rocks in the ratios in which they are presently used, or are likely to be used in the future. If common rock were processed to obtain the base, precious and other scarce metals needed by humans, far more iron and aluminium, in particular, would be extracted than could possibly be used! Much of this would have to be returned to the Earth as waste.

14.3.3 Hubbert's curve and the concept of peak minerals

The curve in Figure 14.2 (and the smaller of the two curves in Figure 14.3) should not be confused with the similarly shaped and related Hubbert curve (Hubbert, 1956). That was developed to predict the rate of production of petroleum over time and to estimate the life span of petroleum reserves, but has since been applied to many other natural resources, including metals. Hubbert assumed that, after oil is discovered, production rises slowly then increasingly rapidly, as more oil fields are discovered and more efficient extraction technologies are developed, until peak output is reached and production begins declining, again slowly initially then increasingly rapidly. Since the Hubbert curve is approximately symmetrical, peak production is reached when about half the oil that will ultimately be produced, has been extracted.

The difference between the Hubbert curve and the curve in Figure 14.2 is that the latter refers to mineral resources in the ground, not the rate at which they are extracted. That can depend on many factors other than the quantity of the resource and its grade, in particular the demand for the commodity and the development of the technology for extracting it. Hubbert developed an equation to fit his scenario and predicted that peak production of oil in the United States would occur between 1965 and 1970. Peak oil production actually occurred in 1970. Hubbert's prediction was so disturbing to the petroleum industry that his methodology was heavily criticised, but no serious flaw in the logic has been found. Hubbert's approach can be applied to other non-renewable commodities but the data are less reliable. It remains unclear when peak production of particular metals and other elements will occur.

There are fundamental differences between oil and many minerals which makes application of Hubbert's approach problematical. Most elements (in the form of minerals) occur at a range of concentrations throughout the entire crust of the Earth and therefore will not run out, unlike oil, which occurs only in a finite (though unknown) number of discrete reservoirs. However, it is more expensive and energy-intensive to extract an element from the crust the lower the concentration at which it occurs. Many elements, particularly the metals, can be recycled and hence, in principle, can remain in use indefinitely. Oil, on the other hand, is consumed by burning (as in transport and heating applications) or converted into polymer materials. Physical depletion, therefore, is not the primary determinant of a mineral's availability. While physical depletion could become a concern at the level of national or regional economies, it is much more likely that economic depletion due to the rising cost of extracting a mineral as its grade falls and due to social and environmental constraints and impacts, will most influence which mineral deposits are exploited in the future.

14.3.4 Are many mineral deposits still to be discovered?

Another way of posing this question is to ask when peak production will occur for various minerals and metals required by humans, since at this time the discovery rate of exploitable mineral deposits starts to decline. As noted previously, peak production for minerals is not easy to predict using Hubbert's methodology. However, another approach is possible. It is fairly easy to estimate the total quantity of an element, present in the Earth's crust, that might be accessible. Assume initially that all continental crust to a depth of 10 km is potentially available to be mined should a suitable deposit be discovered. Then, since the radius of the Earth is about 6375 km and the continents occupy about 30% of the Earth's surface, the volume of rock available is:

$$\frac{4}{3}\pi(6375^3 - 6365^3) \times 0.3 = 1.530 \times 10^9 \text{ km}^3$$

The average density of the continental crust is 2.7 g mL^{-1}, therefore the mass of rock in this volume is 4.13×10^{18} tonnes. Using the average crustal abundance data in Table 5.2, the quantity of each element in the first 10 km of continental crust can be readily calculated. The values for some industrially important elements are listed in Table 14.1 (column 3). This calculation excludes oceanic crust, which makes up 70% of the total crust. This is not because it lies largely below present-day oceans and is difficult to access, though that could be a limiting factor, but because oceanic crust, being less than 200 million years old, is too young for geological processes to have formed ore deposits (Skinner, 1976). However, large areas of the ocean floors contain deposits of minerals formed from erosion processes on the continents and from tectonic processes in the oceanic crust; these are considered separately in detail later.

The total quantity of an element present in mineral anomalies is more important than the total quantity in the crust, but the former is much more difficult to estimate. Until the last deposit has been discovered and measured, the answer cannot be known precisely. However, several observations can provide a basis for

Table 14.1: Estimate of the quantities of elements in the continental crust to a depth of 3 km and 10 km, and the quantity of those elements that could be present in deposits at concentrations greater than the mineralogical limit

Element (1)	Crustal abundance[a] (2)	Total quantity in crust (tonnes) (3)	Abundance relative to copper (4)	Estimated quantity in mineral deposits (tonnes) (5) (6)		Identified resources or reserve base[b] (tonnes) (7)	Ratio of estimated resources to known resources (8) (9)	
				Minimum	Maximum		10 km depth	3 km depth (av. of range)
Al	8.3 wt%	3.43×10^{17}	1.22×10^{3}	3.43×10^{12}	3.43×10^{13}	1.72×10^{10}[c]	200–2000	330
Fe	6.2 wt%	2.56×10^{17}	9.12×10^{2}	2.56×10^{12}	2.56×10^{13}	$>3.20 \times 10^{11}$	8–80	13
Mg	2.76 wt%	1.14×10^{17}	4.06×10^{2}	1.14×10^{12}	1.14×10^{13}			
Ti	0.632 wt%	2.61×10^{16}	9.29×10	2.61×10^{11}	2.61×10^{12}			
P	0.112 wt%	4.63×10^{15}	1.65×10	4.63×10^{10}	4.63×10^{11}			
Mn	0.106 wt%	4.38×10^{15}	1.56×10	4.38×10^{10}	4.38×10^{11}	5.20×10^{9}	8.4–84	14
Zr	162 ppm	6.69×10^{14}	2.38	6.69×10^{9}	6.69×10^{10}			
V	136 ppm	5.62×10^{14}	2.00	5.63×10^{9}	5.62×10^{10}			
Cr	122 ppm	5.04×10^{14}	1.79	5.04×10^{9}	5.04×10^{10}			
Ni	99 ppm	4.09×10^{14}	1.46	4.09×10^{9}	4.09×10^{10}	1.5×10^{8}	27–270	45
Zn	76 ppm	3.14×10^{14}	1.12	3.14×10^{9}	3.14×10^{10}	1.90×10^{9}	1.7–17	3
Cu	68 ppm	2.81×10^{14}	1.00	2.81×10^{9}	2.81×10^{10}	$>3.00 \times 10^{9}$	>0.94–94	>1.5
Co	29 ppm	1.20×10^{14}	4.26×10^{-1}	1.20×10^{9}	1.20×10^{10}	1.50×10^{7}	80–800	130
Ga	19 ppm	7.85×10^{13}	2.79×10^{-1}	7.85×10^{8}	7.85×10^{9}			
Li	18 ppm	7.43×10^{13}	2.65×10^{-1}	7.43×10^{8}	7.43×10^{9}			
Pb	13 ppm	5.37×10^{12}	1.91×10^{-2}	5.37×10^{7}	5.37×10^{8}	$>1.5 \times 10^{9}$	>0.4–4	>0.6
U	2.3 ppm	9.50×10^{12}	3.38×10^{-2}	9.50×10^{7}	9.50×10^{8}			
Sn	2.1 ppm	8.67×10^{12}	3.09×10^{-2}	8.67×10^{7}	8.67×10^{9}	1.1×10^{7}	7.9–79	13
Be	2 ppm	8.26×10^{12}	2.94×10^{-2}	8.26×10^{7}	8.26×10^{8}			
Ge	1.5 ppm	6.20×10^{12}	2.21×10^{-2}	6.20×10^{7}	6.20×10^{8}			
W	1.2 ppm	4.96×10^{12}	1.76×10^{-2}	4.96×10^{7}	4.96×10^{8}			
Mo	1.2 ppm	4.96×10^{12}	1.76×10^{-2}	4.96×10^{7}	4.96×10^{8}			
In	0.24 ppm	9.91×10^{11}	3.53×10^{-3}	9.91×10^{6}	9.91×10^{7}			
Sb	0.2 ppm	8.26×10^{11}	2.94×10^{-3}	8.26×10^{6}	8.26×10^{7}			
Cd	0.16 ppm	6.61×10^{11}	2.35×10^{3}	6.61×10^{6}	6.61×10^{7}			
Ag	0.08 ppm	3.3×10^{11}	1.18×10^{-3}	3.30×10^{6}	3.30×10^{7}	5.70×10^{5}	5.8–58	10
Pt	0.01 ppm	4.13×10^{10}	1.47×10^{-4}	4.13×10^{5}	4.13×10^{6}	1.0×10^{5}[d]	4.1–41	7
Bi	0.008 ppm	3.30×10^{10}	1.18×10^{-4}	3.30×10^{5}	3.30×10^{6}			
Au	0.004 ppm	1.65×10^{10}	5.88×10^{-5}	1.65×10^{5}	1.65×10^{6}	9.0×10^{4}	1.8–18	3

a) Data from Table 5.2.
b) Data from Table 6.4.
c) Assuming bauxite contains 50 wt% Al_2O_3; i.e. 26.5 wt% Al.
d) Total platinum group metals.

estimating the total quantity of an element that may be present in geological anomalies.

1 The size of the largest known deposit of each scarce metal is proportional to the average crustal abundance of the element (Skinner, 1976).
2 The number of known deposits of over 1 million tonnes of a given metal is proportional to the average crustal abundance of the element (Skinner, 1976).
3 The known reserves of the elements are proportional to their crustal abundance (McKelvey, 1960). This is illustrated in Figure 14.5, which is a more recent version of McKelvey's graph.

Taken together, these relations strongly suggest that the combined size of ore deposits of a particular element is directly proportional to the crustal abundance of the element. It follows that, if the area under the small curve in Figure 14.3 can be estimated for one element, the area for all other elements can be calculated by using their crustal abundance to scale the area under the curve. Skinner (1976) proposed the following assumptions for copper.

Assumption 1. The mineralogical barrier is reached at 0.1 wt% copper in the crust (i.e. the quantity of energy required to liberate copper at concentrations below this would be prohibitive).

Assumption 2. The quantity of copper that occurs in deposits having grades of 0.1 wt% Cu or higher would be between 0.001% and 0.01% of the total quantity of copper in the volume of accessible crust.

The quantity of copper in mineral deposits could, therefore, be between 2.81×10^9 and 2.81×10^{10} tonnes (0.001% and 0.01% of 2.81×10^{14}, respectively). Using these values and the abundance of other elements relative to copper (column 4, Table 14.1), estimates of the quantities of other elements present in mineral deposits at concentrations above their own mineralogical limit can be calculated. The values are listed in Table 14.1 (columns 5 and 6). These values include material

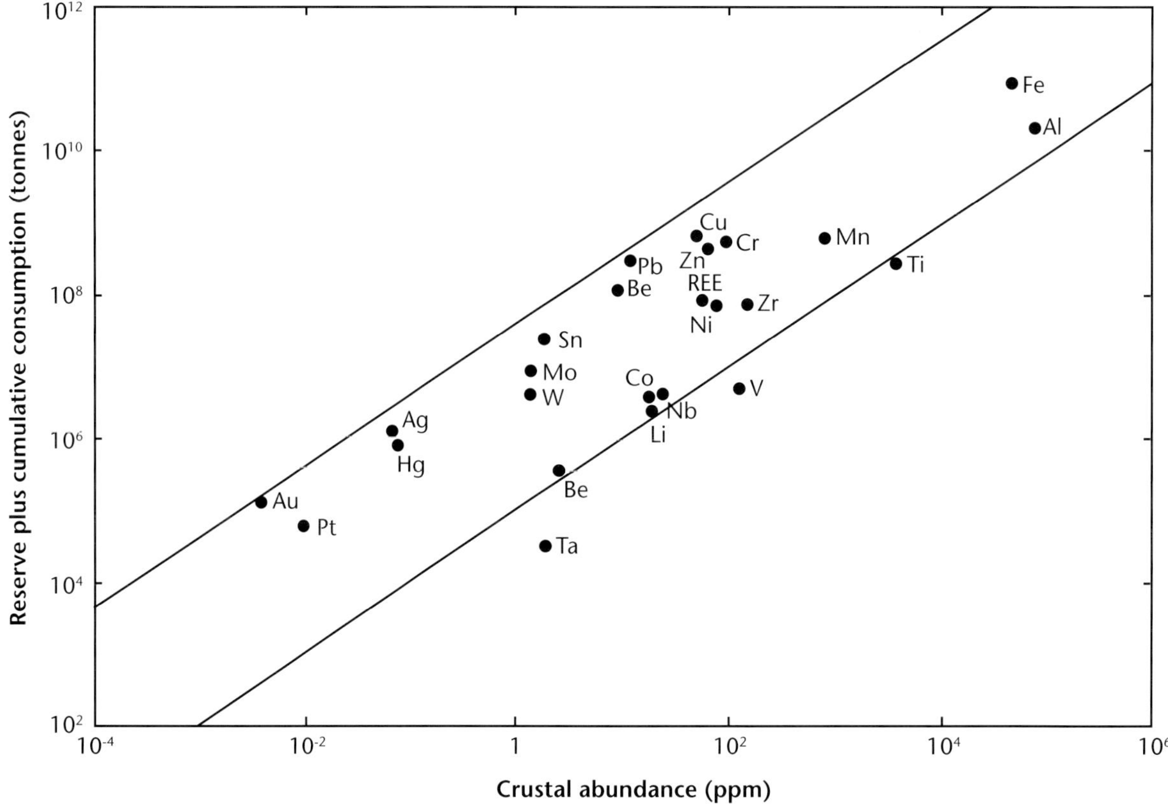

Figure 14.5: The known reserves of the elements as a function of their crustal abundance (after Nishiyama and Adachi, 1995).

that has already been mined (exclusively from mineral deposits, not from common rock), known reserves and resources, and undiscovered mineral deposits at concentrations above the mineralogical limit.

The quantities listed in column 3 of Table 14.1 are optimistic values – not all continental crust would be available for mining, and mining to a depth of 10 km is technically impossible at present (the deepest mines are less than 4 km deep). The calculated quantities of elements available in mineral deposits above their mineralogical barrier depend on the arbitrary assumption that, for copper, it lies between 0.001% and 0.01% of the total quantity of copper in the volume of accessible crust. With these serious limitations in mind, it is possible to compare the predicted quantities with the known, or identified, resources for some common metals. These are listed in columns 7 and 8 of Table 14.1. A more realistic assessment might be made if consideration is restricted to the first 3 km of crust, the depth to which mining is presently performed for high-value metals. These values are listed in column 9. According to this estimate there are large reserves of iron, aluminium and manganese still to be discovered. This is unsurprising, since they are geochemically abundant. Of the common base metals, there may be considerable quantities of nickel, cobalt, tin and zinc still to be discovered; this applies much less for copper and lead. The value for lead is clearly an underestimate since the predicted value is less than the known resources. Even so, of all the common base metals, lead may be the one with least resources still to be discovered. Figure 6.1 indicates that the increase in production rate of lead is slowing. Figure 6.5 indicates that the reserves of lead have remained relatively constant, while the reserves of most other metals have increased, in recent decades. These trends do not prove that peak lead production is approaching, since the usage pattern (and therefore demand) for lead has changed remarkably with the phasing out of lead additives in gasoline in most countries in recent decades, the high recycling rate of lead from batteries, and the banning of lead paints in earlier decades.

It is possible to do a sanity check on the values predicted in Table 14.1 by comparing them with an assessment of undiscovered deposits of gold, silver, copper, lead and zinc in the United States using a different methodology (USGS, 2000). In that study, areas were delineated according to the types of deposits permitted by the local geology. Next, the quantities of metal in typical deposits of those types were estimated using grade and tonnage models. The number of undiscovered deposits of each type was estimated using a variety of subjective methods. The exercise was conducted by 19 assessment teams comprising geologists, geochemists, geophysicists and resource analysts. Estimates of undiscovered resources were made to a depth of 1 km, the approximate depth to which present exploration techniques are effective, even though deposits once discovered may be explored and developed up to a depth of 3 km. The results of the study are summarised in Table 14.2. The ratio of the total quantity of each metal to the identified resource of the metal is compared with the estimates in Table 14.1. The values are of similar order of magnitude though

Table 14.2: Estimates of undiscovered deposits, identified resources, past production and discovered resources of zinc, copper, lead, silver and gold in the United States

	Undiscovered deposits (tonnes)	Resources ('000 t)[a]	Cumulative production to date ('000 t)[a]	Total quantity in ore bodies ('000 t)[a]	Ratio of total quantity to resources for US[a]	Ratio of total quantity to resources for entire continental crust[b]
Zinc	210 000	55 000	44 000	309 000	5.6	3
Copper	290 000	260 000	91 000	641 000	2.5	>1.5
Lead	85 000	51 000	41 000	177 000	3.5	>0.6
Silver	460 000	160 000	170 000	790 000	4.9	10
Gold	18 000	15 000	12 000	45 000	3.0	3

a) USGS (2000).
b) Estimate from Table 14.1.

the ratios for the United States are generally larger than the global values. This may be due to error in the estimation methods, or because the crust comprising the United States contains more or larger mineral deposits than the global average. Clearly, care must be exercised not to overinterpret the data of Tables 14.1 and 14.2 because of the gross assumptions made in deriving them. Nevertheless, they give some comfort that there remain significant resources of many metals yet to be discovered.

14.3.5 Crustal rocks as a source of scarce elements

In principle, all elements could be obtained from common crustal rock. There are many chemical processes that could be used, both hydrometallurgical and pyrometallurgical, though little research has been undertaken to determine optimum processes since there is no present incentive to extract elements from crustal rock. One of the few studies is that of Borg and Steen (2001), who carried out leaching tests on ground rocks to assess the technical possibility of extracting elements into solution, from which they could subsequently be recovered. They found that recoveries in excess of 50% could be obtained for most elements, with some common metals having much higher recoveries, when ground rock was reacted with 30% hydrochloric acid solution at 50°C for one hour. The rock samples were ground to mean particle sizes of 140 μm (granite), 28 μm (basalt) and 70 μm (granodiorite) for the tests and the acid consumption based on the results was 135, 255 and 165 litres of HCl per tonne of granite, basalt or granodiorite, respectively. The yield of aluminium was low and an alkaline leach (e.g. using NaOH) would be required in that case. The yield of some alkali earth metals was also low.

In subsequent work, Steen and Borg (2002) proposed a flowsheet (Figure 14.6) for producing metal concentrates from crustal rock. The aim was to produce sulfide and oxide concentrates which could

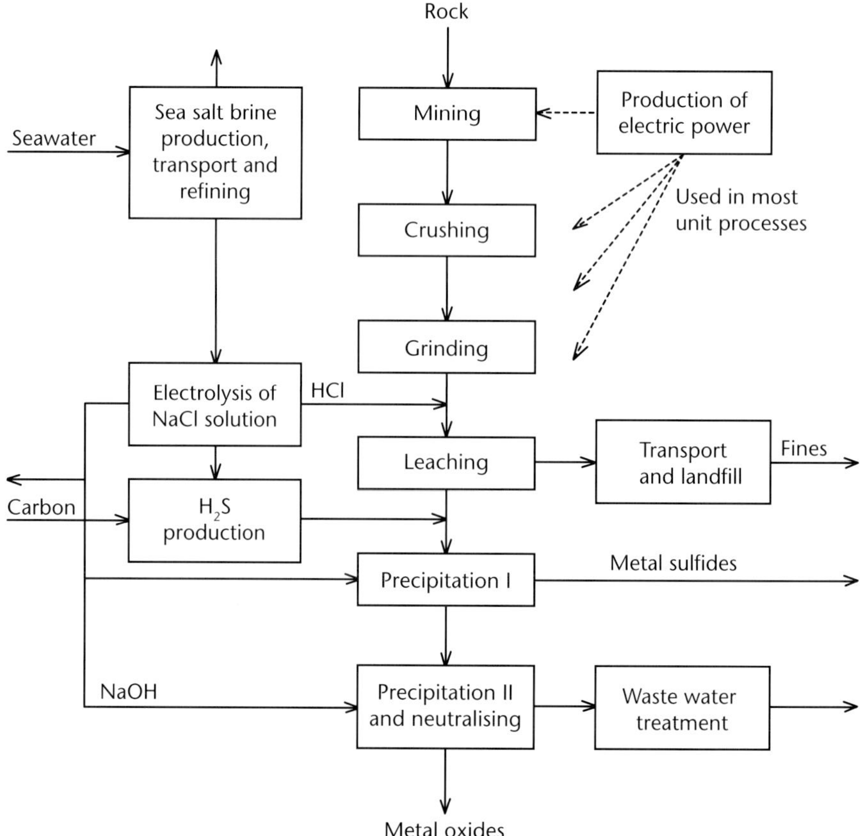

Figure 14.6: A possible flowsheet for producing metal concentrates from crustal rock (Steen and Borg, 2002).

be smelted or leached using conventional technologies. Instead of physical beneficiation (liberation and separation), as used to prepare concentrates from sulfide ore deposits, chemical beneficiation would be required for processing crustal rock. In the proposed flowsheet, rock would be mined by open cut methods, crushed and ground using conventional technology, then leached using hydrochloric acid. Hydrogen sulfide gas (H_2S) would then be used to precipitate most of the scarce metals as sulfides to produce metal sulfide concentrates. The remaining scarce metals would then be precipitated as hydroxides by addition of NaOH. The abundant elements would remain in solution and, with any small quantities of remaining scarce elements, be precipitated in the final, water treatment step. The HCl and NaOH could be manufactured by electrolysis of seawater, using the chlor-alkali process (Bommaraju *et al.*, 2004), and the H_2S from sulfates in seawater. Hydroelectric or solar sources of electricity were assumed. Metals could be produced virtually indefinitely using this flowsheet.

Steen and Borg determined the cost of producing concentrates from crustal rock using the flowsheet in Figure 14.6 by estimating the capital and operating costs. They included estimates for costs due to loss in human health values, loss of ecosystems, loss of non-renewable resources, loss of biodiversity and loss of recreational values (external costs). The results are summarised in Table 14.3, which also lists the price of metals in 2002 for comparison. Present-day prices do not include most of the external costs included by Steen and Borg, who estimated them at about 17% of the total average cost. Clearly, the cost of producing metal concentrates from crustal rock, even excluding the external costs, is in most cases one to two orders of magnitude greater than the cost of today's processes.

Table 14.3: Estimated production costs for metal concentrates from crustal rock compared to present-day prices of pure metals

Metal	Average production cost of concentrate from crustal rock (€ per kg metal)[a]	Price of metals (€ per kg metal)[a]
Cadmium	67 700	0.5–16
Cobalt	262	8
Chromium	141	8
Copper	149	1.7
Manganese	5.60	0.5
Nickel	254	6.8
Lead	392	0.6
Tin	1060	2.5
Tungsten	4780	0.03–0.2
Zinc	44.2	0.9

a) Based on prices in 2002.
Source: Steen and Borg (2002).

14.4 RESOURCES IN SEAWATER

Seawater contains every naturally occurring element. Most of the dissolved constituents of seawater originated from the continents. These were released from continental crustal rocks by weathering and carried to the oceans by run-off. Other constituents were leached directly from oceanic crust. Over time, the concentration of elements increased until equilibrium was obtained. Elemental analyses of fossilised organisms indicate that the composition of seawater has not changed significantly in the last 600 million years (Pidwirny, 2006). During this time, elements carried into the oceans by run-off have largely precipitated in solid forms onto the seafloor. These seafloor deposits are potentially very large and are discussed further in Section 14.5.

The relative abundance of the main dissolved elements in seawater is the same in all the oceans but the actual concentration varies slightly due to the different rates of water loss through evaporation and water gain through run-off and precipitation. Table 14.4 lists the average approximate concentration in seawater of the elements which occur at a concentration of 1 ppm (0.01 mg L^{-1}) or greater. It can be seen that hydrogen and oxygen, combined as water, make up about 96.7 wt% of the oceans; the remaining 3.3 wt% consists of dissolved salts, mainly in ionic form. Six elements (Cl, Na, Mg, S, Ca and K) make up nearly 99.9 wt% of the dissolved salts.

Seawater is an important source of some salts, particularly common salt (NaCl), magnesium and potassium chlorides ($MgCl_2$ and KCl) and magnesium and potassium sulfates ($MgSO_4$ and K_2SO_4). These are extracted by flooding seawater into large shallow

Table 14.4: Composition of seawater; elements at concentration greater than 0.01 ppm

Element	Principal species	Concentration (mg L^{-1} or ppm)	Wt%	Cumulative wt%	Wt% (excluding H_2O)	Cumulative wt% (excluding H_2O)
Oxygen	H_2O	857 000	85.90	85.90		
Hydrogen	H_2O	108 000	10.83	96.73		
Chlorine	Cl^-	19 000	1.90	98.63	58.24	58.24
Sodium	Na^+	10 500	1.05	99.69	32.196	90.43
Magnesium	Mg^{2+}	1350	0.14	99.82	4.148	94.57
Sulfur	SO_4^{2-}	885	0.09	99.91	2.71	97.28
Calcium	Ca^{2+}	400	4.01×10^{-2}	99.95	1.23	98.51
Potassium	K^+	380	3.81×10^{-2}	99.99	1.16	99.67
Bromine	Br^-	65	6.52×10^{-3}		0.199	99.87
Carbon	HCO_3^-, CC_3^{2-}, organic compounds	28	2.81×10^{-3}		0.086	99.96
Strontium	Sr^{2+}	8	8.02×10^{-4}		0.025	99.98
Boron	$B(OH)_3$, $B(OH)_2O^-$	4.6	4.61×10^{-4}		0.014	99.99
Silicon	$Si(OH)_4$, $S\ (OH)_3O^-$	3	3.01×10^{-4}		0.009	100.00
Fluorine	F^-	1.3	1.30×10^{-4}		0.004	
Argon	Ar	0.6	6.01×10^{-5}		0.002	
Nitrogen	NO_3^-, NO^{2-}, NH_4^+, N_2	0.5	5.01×10^{-5}		0.002	
Lithium	Li^+	0.17	1.70×10^{-5}		0.001	
Phosphorus	HPO_4^{2-}, $H_2PO_4^-$, PO_4^{3-}, H_3PO_4	0.07	7.02×10^{-6}			
Iodine	MoO_4^{2-}, I^-	0.06	6.01×10^{-6}			
Barium	Ba^{2+}	0.03	3.01×10^{-6}			
Molybdenum	MoO_4^{2-}	0.01	1.00×10^{-6}			
Zinc	Zn^{2+}	0.01	1.00×10^{-6}			
Iron	$Fe(OH)_3$	0.01	1.00×10^{-6}			

Source: Goldberg (1963).

ponds and allowing the water to evaporate. The solids left behind are mechanically removed and further purified. The process is relatively inexpensive since the Sun provides the energy required to evaporate the water. In principle, other elements could be extracted from seawater. The quantities potentially available are immense, as illustrated in Table 14.5 (column 4) for some important geochemically scarce metals. The values were calculated using the average concentration values and the total quantity of water in the oceans (Table 3.3). The quantity of many of these metals in the oceans is one to two orders of magnitude greater than those in known continental resources. The relative quantities of the metals that could be recovered, assuming 100% recovery, is listed in the second column of Table 14.5. The problem with extracting the metals, however, is the extreme dilution at which they occur and the resulting large volumes of water that would need to be pumped and processed to extract the elements. To extract 1 tonne of matter comprising the metals listed in Table 14.5, 35 billion L or 35 million tonnes of seawater would have to be processed.

Natural evaporation would be impractical for extracting dissolved metal salts because of the large quantity of water to be processed. The only suitable technology would be a membrane process, such as reverse osmosis, to concentrate the metals. This could be followed by solvent extraction or ion exchange to further concentrate and separate them into streams. The metals could then be recovered from their solutions by electro-winning. Energy would be required for pumping the water to the processing plant, pumping the water through the membrane to effect separation, for discharging the spent water, and for the extraction of metals from the solution. Desalination plants typically use 2–4 kW h per tonne of water treated for all operations. This would be the bulk of the energy required. If it is arbitrarily assumed that an additional 10% energy is required for SX/ion exchange and electro-winning, then extracting 1 tonne of metal would require 7.6×10^7 kW h to 1.5×10^8 kW h. If electrical energy is costed at US$0.05 per kW h (a typical commercial rate in developed countries), the cost of producing 1 tonne of metal would be $4–8 million. The value of the metal, at present-day prices, is a minute fraction of the cost of production. Even if the energy requirement could be halved through technological improvements (a challenging

Table 14.5. The quantities of some common metals in the oceans

Metal	Concentration[a] (mg L^{-1} or ppm)	Wt% (dry basis)	Total quantity in oceans (tonnes)	Continental resources + reserve base[b] (tonnes)
Zn	0.01	34.7	1.37×10^{10}	2.38×10^9
U	0.003	34.7	4.11×10^9	
Cu	0.003	10.4	1.23×10^9	$>3.94 \times 10^9$
Sn	0.003	6.9	1.11×10^9	
Ni	0.002	3.5	2.74×10^9	2.8×10^8
Mn	0.002	3.1	5.48×10^8	5.2×10^9
Al	0.001	2.8	1.37×10^8	
Co	0.0005	1.4	5.34×10^8	2.8×10^7
Ag	0.0003	1.4	3.80×10^8	5.7×10^5
Cr	0.00005	0.972	2.74×10^8	
Pb	0.00003	0.104	4.11×10^7	$>1.67 \times 10^9$
Au	0.000004	0.038	1.51×10^7	9.0×10^4
Total		*100.00*		

a) Goldberg (1963).
b) data from Table 6.4.

target), and the cost thereby roughly halved, the cost of production would still be prohibitive. Seawater is therefore unlikely to become a significant source of elements, other than hydrogen, chlorine, sodium, potassium, calcium, magnesium and sulfur, which occur at high or relatively high concentrations.

14.5 RESOURCES ON THE SEABED

Oceanic crust is unlikely to contain mineral deposits of the types found in continental crust, but the seafloor contains large quantities of mineral deposits. The quantities and locations of these are still largely unknown but the mechanisms by which they form are reasonably well understood (Rona, 2008). This knowledge, and the limited exploration of the seafloor, have established that there are very large resources of some important elements, particularly metals. These deposits are largely unexploited and extracting them poses great technological challenges. Their exploitation also poses potentially large environmental and social challenges, as the deep oceans remain one of the last areas relatively untouched by humans. The locations and nature of the major known marine mineral resources are illustrated in Figure 14.7. These can be classified according to their source as deposits originating from land sources, deposits from sources in ocean basins, and deposits from sources on both continents and in ocean basins (ISA, 2004).

14.5.1 Deposits originating from land sources

Mineral deposits from land sources consist of beach deposits, placer mineral deposits and deposits precipitated from seawater. Beach deposits consist of sediments and may lie at, above or below the present shoreline because the sea level has changed by up to 150 m over the past 18 000 years. They are not necessarily on present-day coastlines and may be covered in sedimentary deposits. These sediments generally contain silt, sand and gravel-sized particles consisting of quartz, derived by erosion of continental rocks, or calcium carbonate from broken shells or direct precipitation from seawater. Placers are deposits of high-density, inert metallic minerals and gemstones that have been eroded from source rocks on land and transported to oceans by rivers where they are sorted and concentrated by water motion. Principal metals in minerals of placer deposits are barium, chromium, gold, iron, rare earth elements, tin, titanium, thorium, tungsten and zirconium. The principal gemstone is diamond. Placer deposits forming parts of the present-day continents due to tectonic processes in the distant past are considered to be crustal deposits and are not considered here. The potential for placer deposits to occur on and within the sedimentary accumulations of continental margins (the continental shelf, slope and rise) associated with land sources of minerals is great. However, current knowledge of their occurrence is limited and only the most accessible deposits on or near the seafloor of small areas of inner portions of the continental shelves have been discovered. Lime, phosphorite and salt deposits occur on continental shelves. These were formed by precipitation from seawater. Phosphorite ($Ca_{10}(PO_4CO_3)_6F_{2-3}$), which occurs as nodules and crusts, contains at least 20 wt% P and is a potential source of phosphorus for fertiliser production. Lime ($CaCO_3$) may be precipitated in shallow water in subtropical and tropical climate zones or extracted from seawater by microscopic plants and animals and deposited as their remains (shells and other forms). Salt occurs as layers up to kilometres thick and as dome-shaped masses buried within the sediments that cover passive continental margins. These were deposited at an early stage of opening of an ocean basin by rifting of continental margins (as in the Red Sea) then buried beneath thick layers of sediment, as in the continental margins on both sides of the Atlantic Ocean (Rona, 1985).

Sand and gravel are extensively mined from beach deposits around the world and are the most utilised marine mineral commodities. They are dredged directly from a beach or from off-shore sandbars in shallow water close to the beach. Diamonds are mined from off-shore placer deposits in Namibia and South Africa. Quite a few off-shore placer deposits of gold are known (Alaska, Nova Scotia, Chile, China, India, South Korea, Fiji, Philippines, Russia, Sierra Leone, Solomon Islands, southern New Zealand) but only the Alaskan deposits have been mined (1987–90). Globally, there are 88 known occurrences of titanium-bearing placers, of which 26 are located off-shore, and

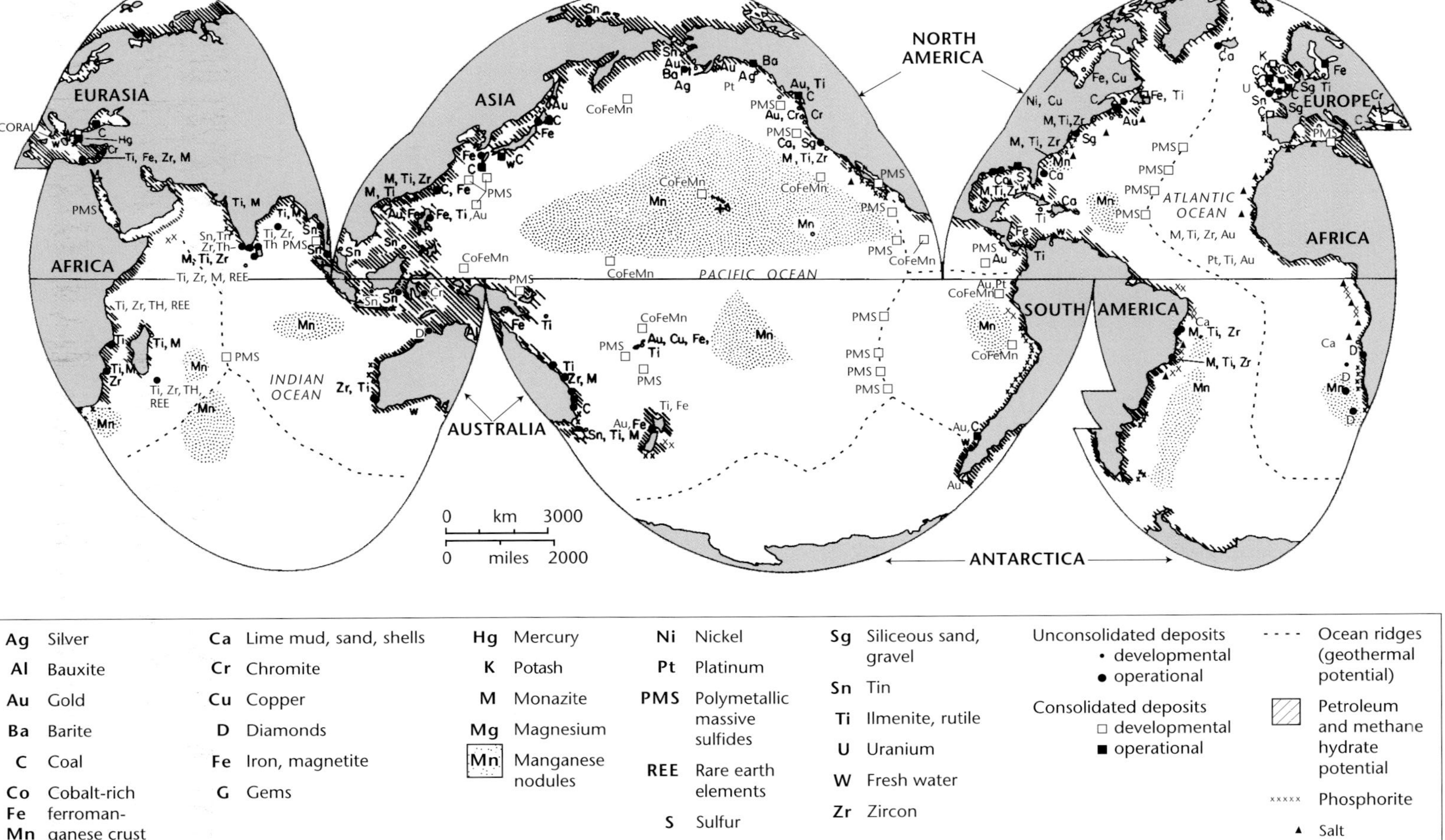

Figure 14.7: The global distribution of known marine mineral resources (Rona, 2003). Reproduced with permission from the American Association for the Advancement of Science.

there are 56 known occurrences of off-shore phosphate deposits (Lenoble *et al.*, 1995).

14.5.2 Deposits originating from sources in ocean basins

Mineral deposits from sources in ocean basins form as a result of tectonic activity and comprise polymetallic massive sulfides and metalliferous sediments formed by hydrothermal processes. The general mechanism for the formation of these deposits is reasonably well understood (Rona, 2002). The oceanic crust of the ocean basins is heavily fractured and highly permeable. Seawater extends from the bottom of the oceans several kilometres downwards in these fractures. It is believed the quantity of water in the mantle below the oceans is similar to that in the oceans themselves (Meade and Jeanloz, 1991). Where this water encounters hot or molten rocks it becomes heated, expands and rises (by thermal convection). This occurs at divergent plate boundaries, where hot magma upwells to form new oceanic crust, and at convergent plate boundaries where subduction occurs. As the heated water rises, it carries some magmatic fluids and dissolves metals, present as soluble chlorides that occur in low concentration in the rocks. When these liquids discharge at the surface of the seafloor, polymetallic massive sulfide deposits and metalliferous sediments may form.

Polymetallic massive sulfide deposits form when the upwelling liquid reacts with sulfur, mainly from sulfates in seawater, to precipitate metallic sulfides on the seafloor. These usually contain high concentrations of copper, iron, zinc and silver as sulfide minerals. Gold and other metals may also be present in lower concentrations. The hot water (up to 400°C) then discharges into the cold ocean water. The remaining dissolved metal precipitates out to form clouds of metallic mineral particles which, through buoyancy, may rise several hundred metres into the ocean above. The vents from which the clouds emerge are called 'black smokers'. Massive sulfide deposits occur at divergent plate boundaries, such as mid-ocean ridges, at or within kilometres of the spreading axis, and at intervals of tens of kilometres along the axis. They also occur at convergent plate boundaries with island arcs in calderas of seafloor volcanoes. The term massive in this context refers to the metallic sulfide content (>60%), not the size or form of the deposit. Individual deposits are actually quite small, with diameters up to several hundred metres.

Table 14.6: Likely metallic resources of the Red Sea

Resource	Grade (dry, salt-free basis)	Quantity (tonnes, dry, salt-free basis)
Metalliferous sediments		89 500 000
Zinc	2.06 wt%	1 838 000
Copper	0.45 wt%	402 000
Silver	38 g t^{-1}	3432

Source: Nawab (2001).

Metalliferous sediments form when the upwelling liquid flows through several kilometre thick deposits of salt, previously precipitated, thereby increasing the salinity by an order of magnitude, and deposits in a trough on the seafloor. The first deposit of metalliferous sediments discovered, and the largest to date, is in the northern Red Sea, in an area known as the Atlantis II Deep. The dense discharging fluids ponded in the Atlantis Deep trough; metallic sulfides, oxides and hydroxides precipitated to form layers of unconsolidated, jelly-like sediments up to 30 m thick in an area about 10 km in diameter. A number of smaller troughs occur along the axis of the northern Red Sea. The estimated grades and quantities of metalliferous sediments in the Red Sea are listed in Table 14.6.

14.5.3 Deposits originating from sources on continents and in ocean basins

The mineral deposits from sources on continents and in ocean basins comprise polymetallic manganese nodules and cobalt-rich ferromanganese crusts. Polymetallic manganese nodules were the only mineral resource known in the deep ocean prior to the advent of the theory of plate tectonics. *Polymetallic nodules* typically range in size from 20–100 mm. They are partially buried on the sedimentary surface of the plains which make up about 70% of the area of ocean basins. The upper portion of the nodules accumulates metals that are precipitated from seawater while the partly buried portion accumulates metals from pore-water in the underlying sediments. The growth rate of nodules is very slow, typically 1–4 mm per million years. This is very much slower than the sedimentation rate, therefore it would be expected that most nodules would be within, rather than on top of, the sediment layer. It is thought that local currents and the action of burrowing organisms keep the nodules near the surface of the sediment. Areas of commercial interest for polymetallic nodules are the eastern equatorial Pacific Ocean, between Hawaii and Central America (the Clarion-Clipperton zone) and the central equatorial Indian Ocean. These cover millions of square kilometres. The total quantity of polymetallic nodules lying on the seabed has been estimated as 500 billion tonnes (Archer, 1981, as cited by ISA, 2004). Like manganese nodules, *cobalt-rich ferromanganese crusts* are slowly precipitated from metals dissolved in seawater. The crusts contain a suite of metals, including iron, manganese, cobalt, nickel and platinum. Instead of accumulating as nodules on the sediment surface in the deep ocean, cobalt-rich ferromanganese crusts accumulate as layers directly on volcanic rock that forms submerged volcanic mountains and mountain ranges. The crusts accumulate slowly over millions of years and attain a thickness up to 250 mm. The mountains in the Pacific Ocean are likely locations for cobalt-rich ferromanganese crusts. It is estimated that in the central Pacific Ocean alone there are around 500 million tonnes of cobalt-rich ferromanganese crusts of economic value (Commeau *et al.*, 1984).

14.5.4 Recovery and processing of deep ocean deposits

The values of some deep ocean marine deposits, based on their metal content, an assumed recovery efficiency and the market price of metals, are listed in Table 14.7. To become economic, the cost of mining and processing these deposits to produce the indicated metals would need to be sufficiently less than the value of metal recovered, for a satisfactory return on investment. The production cost of many marine minerals is difficult to estimate. For massive sulfides and cobalt-rich ferromanganese crusts, mining and processing technologies are at the conceptual stage. For polymetallic nodules, technologies have been tested on small scales but are unproven. The key to commercial utilisation of deep sea deposits lies in mining technology since that is the most challenging from both the technological and environmental points of view. Processing the material to make metals may require new flowsheets but can make use of known chemical reactions and processing equipment; new technologies will be required for mining. The

Table 14.7: The average *in situ* value of marine resources 1960–2004, expressed in 2003 US dollars per tonne, based on metal content, assumed recoveries and average metal prices over the period

Type of deposit	Average value 1960–2004 (2003, US$ per tonne)
Polymetallic massive sulfides 10.9% Cu, 26.9% Zn, 1.7% Pb, 15 ppm Au, 230 ppm Ag	$807
2.1% Cu, 36.6% Zn, 6.1% Pb, 1.6 ppm Au, 260 ppm Ag (fractional recoveries in each case assumed to be 0.70 for Cu and Pb and 0.80 for Zn, Au and Ag)	$620
Polymetallic nodules 30% Mn, 1.37% Ni, 1.25% Cu, 0.25% Co (fractional recoveries assumed to be 0.85, 0.96, 0.95 and 0.94, respectively)	$547
Cobalt-rich ferromanganese crusts 1.2% Co, 0.6% Ni, 0.1% Cu (fractional recoveries assumed to be 0.70 for each metal)	$349

Source: ISA (2004).

major environmental issues related to the exploitation of deep ocean minerals are likely to be on the seafloor itself, where the deposits occur. There is insufficient knowledge for conclusions about the impact of mining on the reproductive capacity of organisms, and this is an area of active research.

The main ways that have been proposed for mining seabed deposits are dredgers and scoops, pneumatic and hydraulic systems, and independent, remote-controlled units (Avramov, 2005). Dredgers and scoop systems consist of buckets of 2.5 m by 5.0 m attached at 25 m intervals to a continuous 20 km cable, suspended between two ships in line. The cable is mechanically winched from the ships and the buckets scoop up material on the seabed. Early tests on manganese nodules found that the technique was not very efficient; at best, buckets were only a quarter-filled with nodules and many were filled entirely with mud (Earney, 1990). Pneumatic and hydraulic systems transfer nodules through vertical pipes using water as the transport medium; they are also relatively inefficient and have high energy consumption. Independent units are remotely controlled shuttles that move from the seafloor to the surface and back again. These are probably the most efficient of the methods proposed. The development of environmentally acceptable and efficient technologies will determine when, or if, deep sea mineral resources are exploited. Many methods have been proposed for processing material from deep sea deposits, particularly nodules (e.g. see reviews by Fursteneau and Han, 1983; Mukherjee *et al.*, 2004). These involve hydrometallurgical or pyrometallurgical steps or, commonly, a combination of both. Most have been tested only at the laboratory or small pilot plant scale. There are unlikely to be major technological challenges and the choice of processing route will be based on economics. It is likely that different processes, or variations of processes, will be used depending on the composition of particular deposits.

14.5.5 Legal aspects: the Convention of the Sea

The use of the oceans' natural resources is governed by the United Nations *Convention on the Law of the Sea* (UNCLOS), also called the Law of the Sea Convention. It defines the rights and responsibilities of nations in their use of the oceans and establishes guidelines for businesses, the environment, and the management of marine natural resources. It replaces the older 'freedom of the seas' concept, dating from the 17th century, in which nations' rights were limited to a band of water usually extending three nautical miles (~5.6 km) from a nation's coastline. By the early 20th century, some nations wanted to extend national claims to include mineral resources, to protect fish stocks and to provide the means to enforce pollution controls. In 1945 the United States extended control to all the natural resources of its continental shelf; other nations soon followed. The Law of the Sea Convention resulted from the third United Nations Conference on the Law of the Sea, which convened in 1973 and lasted until 1982. The Convention came into effect in 1994. To date, 158 countries and the European Union have ratified the Convention, the main exception being the United States. The International Seabed Authority (ISA) was established in 1994 as an autonomous international organisation through which nations party to the Convention organise and control activities in the oceans beyond the continental shelves, particularly with a view to administering the resources of the area. Its headquarters are in Kingston, Jamaica. The full text of the Convention is available on the ISA website (www.isa.com.jm).

Under the Convention, a coastal nation is entitled to a territorial sea, a contiguous zone, an exclusive economic zone and a continental shelf over which it has specific rights and jurisdiction. The Convention also specifies duties and obligations of coastal nations in each of these zones. Every nation has the right to a territorial sea which may extend up to 12 nautical miles from a baseline, usually the low-water mark. A coastal nation exercises sovereignty over its territorial sea, including living and non-living resources. Coastal nations may establish a contiguous zone not extending beyond 24 nautical miles from the baseline from which the territorial sea is measured. The rights of a coastal nation over the contiguous zone extend to prevention of infringement of customs, fiscal, immigration and sanitary laws. Beyond the territorial seas, nations may establish an exclusive economic zone extending not more than 200 nautical miles from the baseline. Within the exclusive economic zone, a coastal nation has rights for the purpose of exploring and exploiting, and for conserving and managing the natural resources of the waters, seabed and subsoil. In addition, the coastal nation has jurisdiction with regard to the establishment and use of artificial islands, installations and structures, marine scientific research and protection and preservation of the marine environment. The continental shelf of a coastal nation comprises the seabed and subsoil that extends beyond its territorial sea to the outer edge of the continental margin, or to a distance of 200 nautical miles if the outer edge of the continental margin does not extend to that distance. The continental margin consists of the seabed and subsoil of the shelf, the slope and the rise. It does not include the deep ocean floor with its oceanic ridges and subsoil. Coastal nations have the right to explore and exploit the natural resources of the continental shelf. These rights are exclusive; no one may undertake these activities without the consent of the nation.

The area beyond the continental shelf comprises the seabed, ocean floor and subsoil beyond the limits of national jurisdiction. The Convention provides that this area and its resources are the common heritage of mankind. Parties to the Convention organise and control activities in this area through the ISA, which is also empowered to engage in seabed mining in its own right. Since the early 1970s considerable investments have been made in research and prospecting for polymetallic nodules in deep-sea areas. In recognition of this, the ISA has registered seven pioneer investors and entered into 15-year exploration contracts with them.

14.6 SUMMARY AND CONCLUSIONS

There are many conventional and unconventional sources of minerals that have not been exploited. These include undiscovered mineral deposits in the continental crust and on the seafloor. Seawater and crustal rocks, though they contain large quantities of valuable elements, are unlikely to become important sources because most elements are present in such low concentrations that the energy required for extraction would be prohibitive. The quantities and concentrations of geochemically abundant elements in the crust are such that their long-term availability is unlikely to be an issue although the energy required, as a production cost, would be expected to increase gradually as deposit grades fall. The quantities of geochemically scarce elements at grades above their mineralogical barrier is very small compared with their overall abundance, although their availability at the global level is unlikely to be a problem in the medium term. There is little indication that, globally, conventional mineral resources for the elements used in large quantities will become exhausted in the short, or even medium, term. It is possible that the supply of some rare metals and of elements that occur exclusively in substitution, rather than as discrete minerals, may become constrained. Many are used in relatively small quantities but play an increasingly important role in evolving technologies such as electronics, energy storage and photovoltaics.

The rising cost of extracting a mineral as its grade falls, and social and environmental constraints and impacts, will influence which mineral deposits are exploited in the future. These factors will determine when sources of minerals other than mineral deposits in the continental crust – seafloor resources, mining wastes and secondary resources – will become more attractive. Practical exhaustion, or the inability to develop resources due to economic, social and environmental constraints, is of greater concern for the long-term supply of minerals than the physical exhaustion of specific minerals.

14.7 REFERENCES

Ahrens LH (1954) The log-normal distribution of the elements. *Geochim et Cosmochimica Acta* **5**: 49–73.

Avramov A (2005) The alternative: deep-water polymetallic nodules in the Pacific. *Journal of the University of Chemical Technology and Metallurgy* **40**(4): 275–286.

Bommaraju TV, Lüke B, O'Brien TF and Blackburn MC (2004) Chlorine. In *Kirk-Othmer Encyclopedia of Chemical Technology*. 4th edn, Vol. 6. (Eds J Kroschwitz and M Howe-Grant) pp. 130–211. John Wiley and Sons: New York.

Borg G and Steen B (2001) 'Availability of metals in the Earth's crust – leaching tests on silicate minerals'.

CPM Report 2001:12. Chalmers University of Technology: Gothenburg, Sweden.

Commeau R, Clark A, Manheim F, Aruscavage P, Johnson C and Lane C (1984) Ferromanganese crust resources in the Pacific and Atlantic Oceans. *Oceans* **16**: 421–430.

Earney FCF (1990) *Marine Mineral Resources*. Routledge: London.

Fursteneau DW and Han KN (1983) Metallurgy and processing of marine manganese nodules. *Mineral Processing and Technology Review* **1**: 1–83.

Goldberg ED (1963) The oceans as a chemical system. In *The Sea – Ideas and Observations on Progress in the Study of the Seas*. Vol. 2. (Ed. MN Hill) pp. 3–25. Interscience Publishers: New York.

Hubbert MK (1956) Nuclear energy and the fossil fuels. In *Drilling and Production Practices*, 7–9 March 1956, San Antonio, Texas. pp. 7–25. American Petroleum Institute.

ISA (2004) *Marine Mineral Resources – Scientific Advances and Economic Perspectives*. International Seabed Authority: Kingston, Jamaica. <http://www.isa.org.jm/en/documents/publications>.

Lenoble JP, Augris C, Cambon R and Saget P (1995) Marine mineral occurrences and deposits of the Economic Exclusive Zones. MARMIN: A data base. Ifremer: Brest, France.

McKelvey VE (1960) Relations of reserves of the elements to their crustal abundance. *American Journal of Science* **258**(A): 234–241.

Meade C and Jeanloz R (1991) Deep-focus earthquakes and recycling of water into the Earth's mantle. *Science* **252**: 68–72.

Mukherjee A, Raichur AM and Natarajan KA (2004) Recent developments in processing ocean manganese nodules – a critical review. *Mineral Processing and Extractive Metallurgy Review* **25**: 91–127.

Nawab Z (2001) Atlantis II Deep: a future deep sea mining site. In *Proposed Technologies for Mining Deep-Seabed Polymetallic Nodules*. Proceedings of International Seabed Authority's Workshop, Kingston, Jamaica, 3–6 August 1999. pp. 26–27.

Nishiyama T and Adachi T (1995) Resource depletion calculated by the ratio of the reserve plus cumulative consumption to crustal abundance for gold. *Natural Resources Research* **4**(3): 253–261.

Pidwirny M (2006) Seawater. In *Encyclopedia of Earth*. (Ed. CJ Cleveland). Environmental Information Coalition, National Council for Science and the Environment: Washington DC. <http://www.eoEarth.org/article/seawater>; retrieved 22 September 2009.

Rona PA (1985) Hydrothermal mineralization at slow-spreading centers: Red Sea, Atlantic Ocean, and Indian Ocean. *Marine Mining* **5**(2): 117–145.

Rona PA (2002) Marine resources for the 21st century. *Episodes* **25**(1): 2–12.

Rona PA (2003) Resources of the sea floor. *Science* **299**(5607): 673–674.

Rona PA (2008) The changing vision of marine minerals. *Ore Geology Reviews* **33**: 618–666.

Skinner BJ (1976) A second iron age ahead? *American Scientist* **64**: 258–269.

Sohn I (2006) Long-term projections of non-fuel minerals: we were wrong but why? *Resources Policy* **30**: 259–284.

Steen B and Borg G (2002) An estimation of the cost of sustainable production of metal concentrates from the Earth's crust. *Ecological Economics* **42**: 401–413.

USGS (2000) 1998 Assessment of undiscovered deposits of gold, silver, copper, lead and zinc in the United States. United States Geological Survey, Circular 1178, 2000.

14.8 USEFUL SOURCES OF INFORMATION

Bleischwitz R, Welfens PJJ and Zhang Z (Eds) (2009) *Sustainable Growth and Resource Productivity*. Greenleaf Publishing: Sheffield, UK.

Earney FCF (1990) *Marine Mineral Resources*. Routledge: London.

International Seabed Authority. <http://www.isa.org.jm/en/home>.

Polymetallic Massive Sulphides and Cobalt-Rich Ferromanganese Crusts: Status and Prospects. International Seabed Authority, United Nations, Kingston, Jamaica, 2002. <http://www.isa.org.jm/en/documents/publications>.

Tilton JE (2003) *On Borrowed Time? Assessing the Threat of Mineral Depletion*. Resources for the Future: Washington DC.

15 The future demand for minerals and metals

The demand for minerals is closely related to the demand for materials derived from minerals. The demand for particular materials is related to the demand for the goods and services that use those materials. There is no consumer demand for steel, for example, but there is a consumer demand for cars, trains and bridges which contain large quantities of steel. Hence, the demand for most mineral and metal commodities is a *derived demand*; these commodities are demanded only indirectly, through the patterns and levels of consumption of goods and services. This chapter examines the long-term demand for mineral and metal commodities, and the factors that influence that demand.

15.1 THE DETERMINANTS OF LONG-TERM DEMAND

The long-term demand for minerals and metals depends on the complex interaction of many factors. Historically, the demand for goods and services has increased as population grows and as the average level of income rises. This growth has resulted in an increased derived demand for mineral and metal commodities. The mix of goods and services has changed over time, as has the mix of raw materials used in their production. These changes reflect changes in consumer preferences, developments in materials, and developments in manufacturing and recycling technologies. The substitution of one material for another and the recycling rate of materials both affect the relative demand for primary mineral and metal commodities. Of particular interest is the substitution of more plentiful raw materials for scarcer raw materials, to achieve the same outcome. The less intensive use of materials to achieve the same outcome (dematerialisation), is another factor. Thus, the main factors which determine the long-term demand for mineral and metal commodities can be summarised as follows (Sohn, 2006):

- the population level, and its growth rate over time;
- the level of income, and its growth rate;
- the change in the mix of goods and services over time;
- the change in the quantity and mix of raw materials used in each product and service;
- the extent to which materials are recycled;
- the extent to which recycled materials could meet the total demand for materials;
- developments in technologies for manufacturing and recycling.

The long-term demand for metals, in particular, can also be affected by health, safety and environmental concerns associated with certain elements, for example, mercury, lead and chromium.

15.2 PROJECTIONS OF THE DEMAND FOR MINERAL COMMODITIES

Minerals companies, industry associations and financial institutions make projections of the future demand for mineral commodities for planning, finance-raising and other purposes. These are based on projected population trends, assumed rates of growth of gross domestic product (GDP) and historical usage patterns and intensities for specific commodities. Likely future changes in usage patterns and recycle rates are often taken into account. These are usually 'business-as-usual' projections which assume no major disruptions to growth of the world economy, either positive or negative, or to technology. They have some use, but historically it has proved problematic to forecast demand with any great accuracy because of the difficulty of predicting the likely values of the model parameters for more than a few years ahead (Sohn, 2006). Nevertheless, it is instructive to consider the major socio-economic trends that will in some way affect the future demand for minerals.

The world population is projected to continue growing until around the middle of the 21st century, with most growth occurring in China, India and other rapidly developing countries but little or negative growth in many developed countries. Some United Nations projections are shown in Figure 15.1. The global population is expected to increase from about 6.9 billion in 2010 to 9.1 billion in 2050 (United Nations, 2008). The global GDP has risen rapidly in recent decades, both in absolute terms and on a per capita basis, and is usually projected to continue rising. Figures 15.2 and 15.3 show an estimate of trends of GDP and per capita GDP, respectively, for a number of developed and developing countries. Historically, the demand for non-fuel mineral commodities and energy (which is strongly fossil fuel based) have increased as the per capita GDP of a country increases, at least during the rapid growth period (see Figures 6.2 and 6.3). It follows that business-as-usual

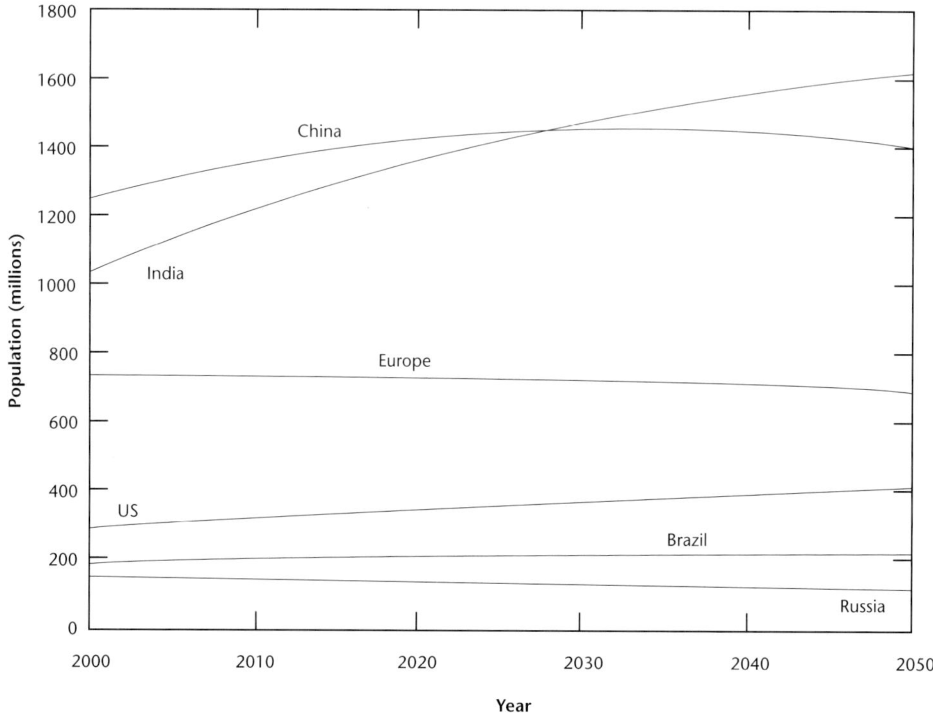

Figure 15.1: Population trends to 2050 (United Nations, 2008).

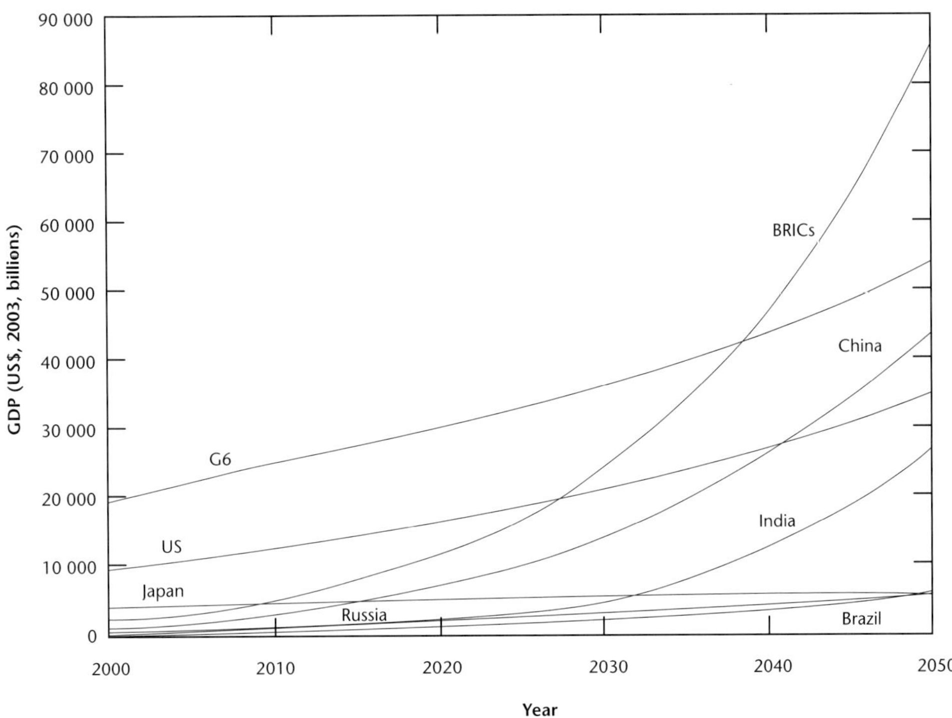

Figure 15.2: GDP trends to 2050 (Goldman Sachs, 2003). BRICs are Brazil, Russia, India and China.

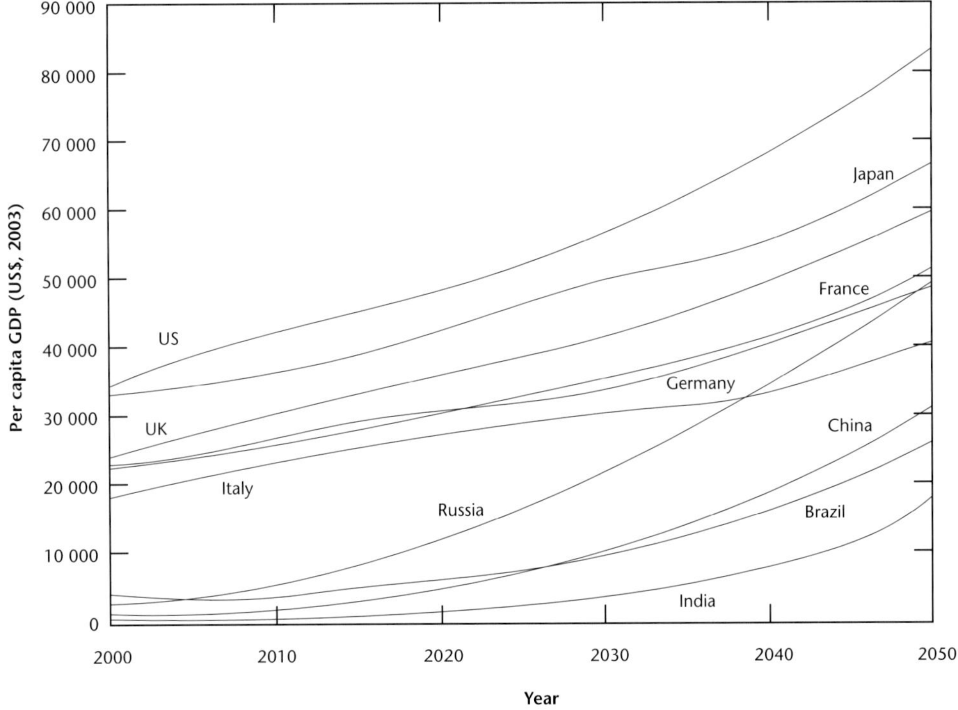

Figure 15.3: GDP per capita trends to 2050 (Goldman Sachs, 2003).

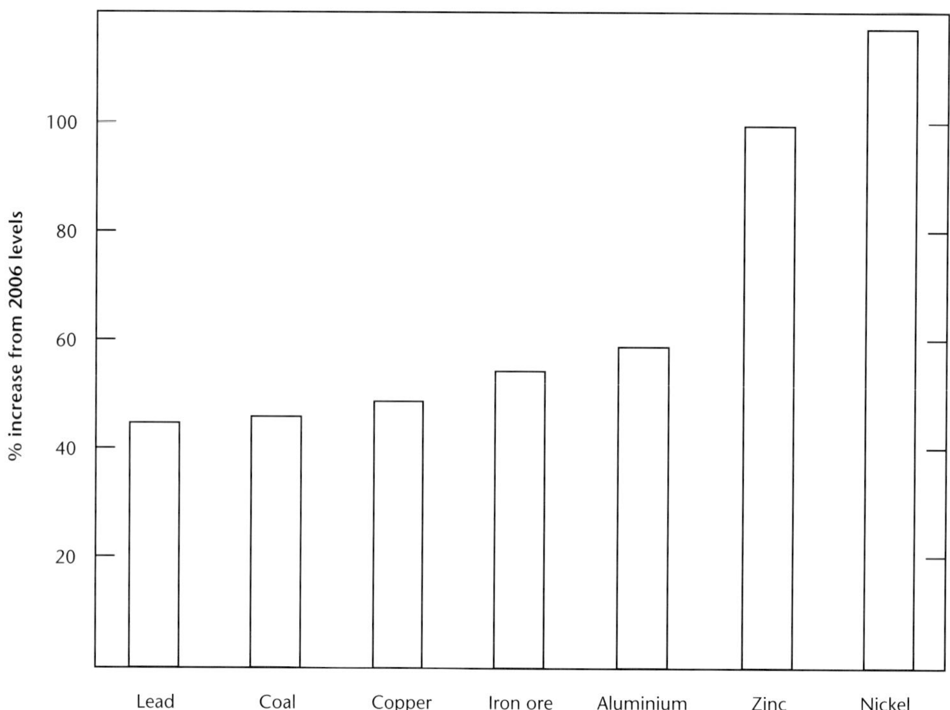

Figure 15.4: The increase in production required, relative to production in 2006, to meet projected demand in 2020 for some common mineral commodities (Access Economics, 2008).

projections invariably show rapidly increasing demand for mineral commodities, both fuel and non-fuel, well into the 21st century. Figure 15.4 shows a projection to 2020 for some common mineral commodities using modelling based on projected trends in population and productivity for key developing and developed countries. It indicates that demand for these commodities in 2020 will be 45–120% greater than in 2006.

The historical trends leading to continued growth in demand may well continue into the future. However, the increasing scarcity, and hence cost, of oil and the increased quantity of wastes (particularly mining wastes and CO_2 from fossil fuel combustion) and their environmental impact may make the relatively unrestrained growth of the past, as assumed in business-as-usual projections, no longer tenable. This could occur, for example, as a consequence of increased regulations restricting the disposal of wastes (including CO_2) and the introduction of financial instruments aimed at reducing the quantity of wastes (e.g. increased disposal charges, cap-and-trade schemes, a carbon tax). The situation is made more complex by the counteracting effects of substitution and dematerialisation. Predicting the long-term future demand for mineral and metal commodities is fraught with uncertainties.

15.3 MATERIALS AND TECHNOLOGICAL SUBSTITUTION

The term *substitution* is used in at least two different ways. Economists use the term to refer to substitution of the various forms of capital, as discussed in Section 4.3.2. A more technical use, which is relevant to the present discussion, relates to substitution of materials and technologies. This has several interrelated aspects – substitution of services for materials or goods, substitution of materials for other materials, and substitution of technologies by other technologies. Substitution is also closely related to the concept of dematerialisation, the meeting of people's needs and wants with a decreasing use of materials and the decreasing production of waste. Dematerialisation is a possible, but not inevitable, outcome of substitution. That substitution is occurring, and always has, is not in doubt. Its impact on dematerialisation, and indeed whether dematerialisation is actually occurring, is much more problematic.

The concept of materials substitution is well established; substitution has occurred throughout history. The Stone Age, Bronze Age and Iron Age can be viewed as periods of wide-scale materials substitution – bronze for stone, then iron for bronze. The idea of materials substitution can be extended to include technological substitution, the substitution of one technology for another. The substitution of steam engines, then internal combustion engines and electrical power, for manual and animal labour, and the replacement of wind power to propel ships with steam and later diesel engines, are examples. Technological substitutions are often enabled by developments in materials and *vice versa*. There are many examples of materials and technological substitution. The following are just a few well-known examples.

- In England, the price of charcoal used for smelting iron ore increased during the late 17th century because of the demand for timber for ship-building. In 1709 Abraham Darby developed coke, made from coal, as a substitute. Coke was then used in iron blast furnaces, a practice which continues to the present.
- Sperm whales, the source of oil for lamps and tallow for candles (the major sources of illumination prior to electricity), became increasingly scarce in the mid-19th century, due to overexploitation. This led to the replacement of whale oil and tallow by kerosene (also known as paraffin) which was made from petroleum, then a little-used resource. Ignacy Lukasiewicz, who established the world's first oil well in Bobkra in Poland in 1854, developed kerosene in 1852 and patented a kerosene lamp in 1853.
- Gasoline was a low-value by-product of kerosene production until the automobile became a common form of transport in the early 20th century. This cheap source of energy was the reason for the rapid expansion of automobiles as a means of transport, replacing horse-drawn transport in much of world by the 1950s.
- Natural rubber became impossible to obtain in the United Kingdom and United States during World War II due to the Japanese occupation of South-East Asia. Synthetic rubber, made from petroleum, was developed as a substitute and today it remains the major type of rubber used.
- Synthetic fertilisers, made from phosphate rock and nitrogen from the atmosphere (the Haber process), have largely replaced fertilisers obtained from natural deposits of guano and nitrates, which are largely exhausted.
- A more recent example is the introduction of lightweight components to replace components previously made of steel, reducing vehicle mass to achieve higher fuel efficiency. Figure 15.5 shows how the use of aluminium and magnesium to reduce the weight of motor vehicles, by replacing heavier steel components, significantly reduces

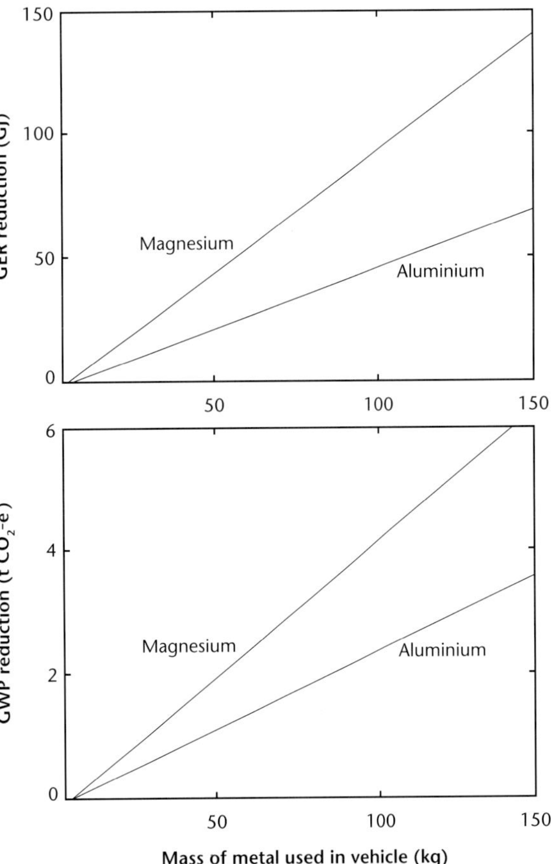

Figure 15.5: Calculated reduction in gross energy requirement and global warming potential as a result of substituting magnesium and aluminium for steel in motor vehicles, over the life cycle of the vehicle. GER and GWP of primary production of aluminium and magnesium are included (after Norgate *et al.*, 2004).

the gross energy requirement (GER) and global warming potential over the vehicle life span despite the much higher cradle-to-gate GERs of aluminium and magnesium compared to steel. The energy saving due to reduced fuel consumption over the life of the vehicle more than offsets the higher embodied energy in the aluminium and magnesium.

The substitution of services for goods is one of the outcomes of technological substitution. Using public transport rather than owning a car, using teleconferencing rather than travelling to a meeting, and reading an electronic copy of a book or newspaper rather than purchasing a paper copy are examples where services have substituted for goods. There has been a long term-trend in this direction for several decades.

In the above examples, substitution was made possible by technological developments but necessity was the driver of substitution. The necessity could result from legislation (e.g. restricting the use of charcoal for iron-making, and fuel efficiency of cars), increasing cost (kerosene replacing sperm oil), the unavailability of a material due to a temporary shortage (rubber during World War II) or the depletion of a resource (guano). Technological developments can facilitate substitution in several ways. They may make a material cheaper to produce than an existing material. They may enable development of a material with new or improved properties; it then substitutes for existing materials, thereby improving product quality. They may create entirely new uses and products (gasoline for cars). Finally, they may replace an existing technology (electric lights for kerosene lamps). These ideas are developed further in Table 15.1, which lists the types of changes in circumstances, arranged in a hierarchy, under which technological and materials substitution can occur. Starting at the lowest levels (levels 6 and 7) with changes in materials and materials processing, the hierarchy moves upwards to changes in components, subsystems, systems, strategies and finally changes in values. At the lower levels, supply forces drive substitution. Demand forces become stronger at higher levels. Substitution at any of the higher levels tends to drive substitution at the levels below. The frequency and rate of substitution are much faster at lower levels than at higher levels where political and social resistance to change are more influential. The rapid uptake of microwave ovens, personal computers and mobile phones (level 4), compared with the slowness of changes in personal values and development and implementation of strategies to address global warming and climate change (levels 1 and 2), illustrates this point well.

Table 15.1: The circumstances under which technological and materials substitution can occur, with examples

Level	Circumstance	Examples
1	A change in social or personal values or goals resulting in a change in demand	Better quality of life rather than more consumer goods
2	A change in strategy to achieve goals	Telecommunication instead of personal travel Mass transport (trains, buses) replacing individual personal transport (cars, bikes) (in response to climate change and energy self-sufficiency)
3	A change in systems through technical means to implement a strategy	Photovoltaic cells replacing fossil fuel generated electricity (in response to climate change) Individual personal transport (cars, bikes) replacing mass transport (trains, buses)
4	A change in subsystems within a system through design changes	Batteries and electric motors replacing internal combustion engines in cars
5	A change in components through design changes	Turbine engines replacing piston engines in aircraft
6	A change in materials for specific components	Use of aluminium instead of cast iron in engine blocks
7	A change in materials processing technology	Continuous casting to replace ingot casting of metals

Source: After Altenpohl (1980); with modifications.

Closely related to the concept of technological substitution is the concept of backstop technologies. A *backstop technology* is a technology that acts as a substitute for a finite resource and, in effect, gives the resource a quasi non-depletable characteristic (Nordhaus *et al.*, 1973). The substitution of fossil fuels with renewable energy technologies is a familiar example of a backstop technology. A backstop technology may be available but may not be implemented if it is uneconomic relative to continued use of the resource. However, as the resource becomes scarcer and its price increases, and/or as market-based instruments make the backstop technology more attractive, the backstop technology may become competitive. Usually, there would be a transition period when the new technology would co-exist with continued exploitation of the finite resource before taking over (Tsur and Zemel, 2003). The availability of a developed backstop technology may also have the effect of reducing the price of the resource since customers would know an alternative exists; this could accelerate the rate of depletion (Levy, 2000). The widespread adoption of electric cars, for example, would most likely result in a decrease in the price of gasoline. With increased social and environmental pressures on the mining industry, it is likely that backstop technologies that overcome issues arising from falling ore grades will become viable. In this context, materials substitution and recycling are backstop technologies for the mining of minerals. Both will increasingly influence the long-term demand for minerals.

15.3.1 Substitution limits and constraints

Views differ on the limits to which materials substitution can occur. It is clear that there are some limits. For example, some elements are essential for life (Table 11.9) and substitution for these is not possible. Most of these are recycled within ecosystems and will not run out. However, concern has been expressed about phosphorus because of its extensive use in fertilisers (Cordell *et al.*, 2009; Vaccari, 2009). Once mined and processed to make fertiliser, phosphorus is dissipated in the environment with much of it ending up in waterways and, ultimately, the oceans. Based on known resources of phosphorus, scarcity could become an issue in the long term since there is no substitute for it. Ayres (2005) gives other examples of specific substitution limits.

- Shelter and clothing are essential for humans and these require inputs from natural resources. Some could be provided by renewable resources but others cannot (e.g. glass, concrete and bricks). Substitution for these by other materials is unlikely.
- While there is some potential for substitution between transportation and communication (e.g. video-conferencing instead of travelling to a meeting) transportation of people and goods will always be required even if it could be greatly reduced by adopting more localised economies. Transportation requires containment and propulsion. Containment implies structural materials and these have to be derived largely from metals and plastics. Propulsion requires an energy source and storage. Other than human and animal power, energy sources (even flowing water and wind) and energy storage devices require construction materials, including metals.
- Copper is used in electrical applications, particularly in generators, electric motors, domestic wiring and transmission lines, because of its high conductivity and malleability (which enables it to be drawn into wire form). The only possible substitute for copper in electrical applications is aluminium, which is a poorer conductor. Consequently, its application is largely limited to high-voltage transmission lines.

Taking a long-term perspective, Goeller and Weinberg (1976) expounded the view that as society exhausts one raw material it will turn to lower-grade inexhaustible substitutes. They argued that it is possible to conceive of depletion and substitution in three stages. Stage 1 is a continuation of present patterns of use of non-renewable resources. During stage 2, society would still depend on coal but there would be little remaining oil and gas. There would be greater use of steel, aluminium, magnesium and titanium (geochemically abundant elements) and much less reliance on base metals such as copper, nickel, lead, zinc and tin (geochemically scarce elements). In stage 3, the age of substitutability, fossil fuels would be exhausted and society would be based on materials that have virtually unlimited availability – the abundant elements and crustal rocks. Energy for a stage 3 society could be provided by fast-breeder nuclear fission reactors, nuclear fusion (if feasible) or solar energy. Goeller and Weinberg proposed a principle of infinite substitutability, which can be summarised as follows.

- Society can exist on inexhaustible or nearly inexhaustible minerals with relatively little loss of living standard.
- Such a society would be based largely on glass, plastic, wood, cement, iron (steel), aluminium and magnesium.
- Whether that society would be like our present society would depend on how much energy will cost, both economically and environmentally.

Goeller and Weinberg based their research on the materials consumption pattern of the United States in the 1970s. With growing concern about climate change and the contributing role of fossil fuels, the use of fossil fuels may be phased out much earlier than they anticipated, but their argument would be essentially unchanged. Their perspective is long-term (several hundred years to reach stage 3) and they acknowledge that moving to the age of substitutability, while technically feasible, would be socially and politically difficult because the anticipated shortages in particular materials would create tensions between the 'haves' and 'have-nots'.

15.4 DEMATERIALISATION

In the 1960s it became apparent that the importance of materials in the United States economy was declining, and had been from at least the 1950s. By comparing input–output data for 1947 and 1958 for nine categories of materials, Carter (1966) found that there had been a decline in the quantity of materials and semi-finished goods used to produce the economic output of the United States over the 10-year period. Over the same time there had been a corresponding increase in non-material, or general, inputs (defined as energy, communications, trade, packaging, maintenance, real estate, finance, insurance, printing and publishing, and business machines and related information technologies). *Dematerialisation* is the term now used to describe this trend. It can be defined as the absolute or relative decline over time in the quantities of materials used, or the quantity of wastes generated, or both, in the production of a unit of economic output. Carter also found that the dominance of single types of materials (metals, stone, clay, glass, wood, fibres, rubber, plastics and so on) had given way by 1958 to increasing diversification of the materials consumed by each industry – products were becoming more complex. This was due to changes in the quality and types of materials and product design, driven by changes in end-use demand.

15.4.1 Intensity-of-use

The most widely used measure of dematerialisation is intensity-of-use. The intensity-of-use of a material (IU) is defined as the ratio of the quantity of materials used to the quantity of value added, which in economic terms is gross domestic product (GDP):

$$IU = \frac{X_i}{GDP} \qquad 15.1$$

where X_i is the annual consumption of a specific material, i, in terms of mass or volume. Intensity-of-use is usually expressed on a per capita basis. This definition includes only material flows directly used in the economy, and not the hidden material flows (as defined in Section 2.6.2). Progressive decrease in materials intensity-of-use over time is often called *weak dematerialisation*. Progressive decrease in the total materials use in absolute terms is called *strong dematerialisation* (de Bruyn and Opschoor, 1997).

The pioneering work on dematerialisation was done by Malenbaum (1978), who examined the intensity-of-use of 10 major metals and ores for 1950–75 for 10 regions. Malenbaum found that the intensity-of-use of a material in an economy (e.g. a country or region) initially increased with both time and with increasing GDP, reached a peak or plateau then started to decline (Figure 15.6). For example, in the United States, iron and steel, manganese, copper, zinc and tin reached a peak of intensity-of-use in the 1950s or earlier, nickel, cobalt and tungsten peaked in the 1960s, while chromium, aluminium and platinum peaked in the 1970s or later. Numerous other studies for different materials and different countries and

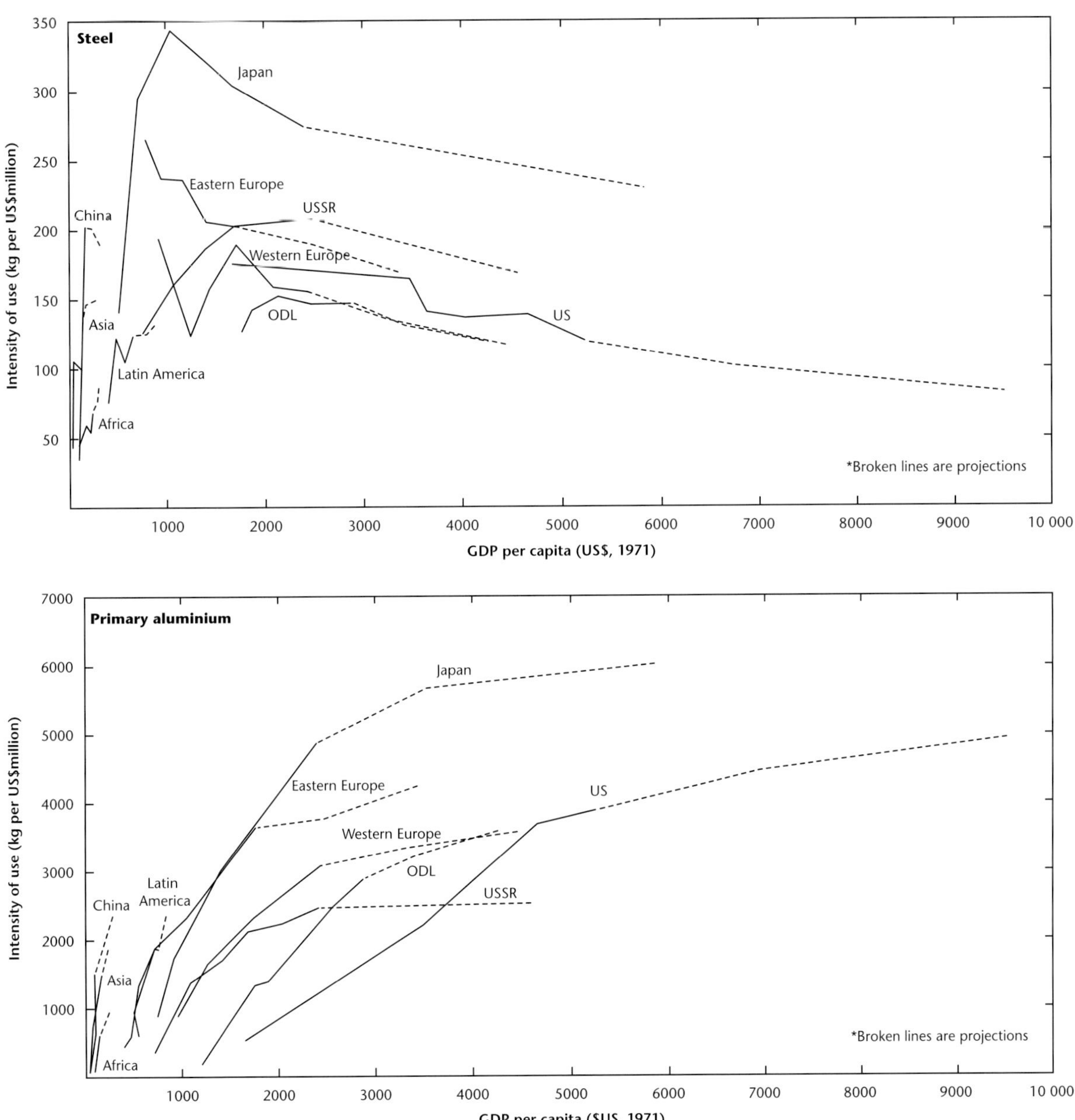

Figure 15.6: The trends in intensity-of-use of steel, aluminium, copper and platinum group metals in 10 world regions as a function of per capita GDP (referred to 1971 US$) (after Malenbaum, 1978). ODL is Other Developed Lands (Australia, Canada, Israel, New Zealand, South Africa)

regions have confirmed the general trends (Cleveland and Ruth, 1999). The characteristic inverted U-shaped curves of Figure 15.6 are now known as environmental Kuznets curves, after Simon Kuznets, who postulated a similar relation between income inequality and income levels (Kuznets, 1955).

At an economy-wide level, trends in direct material input (DMI) and total material requirement (TMR), as defined in Section 2.6.2, provide a guide to the linkage between GDP and dematerialisation. In many countries, material use has grown at a lower rate than the economy; i.e. there has been relative decoupling of

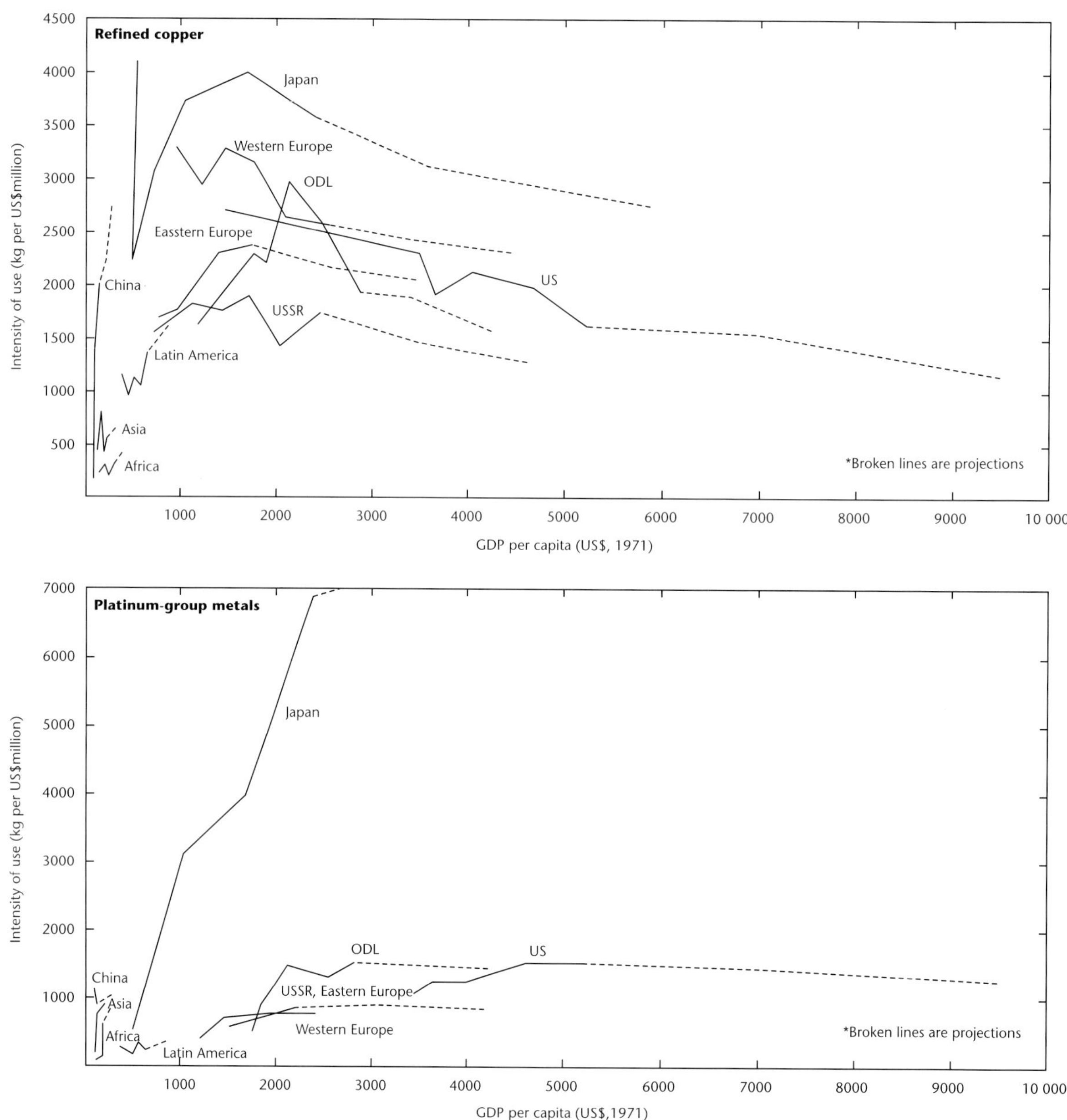

Figure 15.6 (continued)

DMI and GDP (weak dematerialisation). Although there is a high variability, material use seems to follow a non-linear trend in the course of economic growth (Bringezu *et al.*, 2009). For per capita GDPs below US$10 000 the rate of growth in DMI may be fast up to a value of 15–45 tonnes per person per year; for per capita GDPs above US$10 000 DMI continues to vary widely within this range (or higher for resource-exporting countries). Similarly, TMR increases with economic growth although at a slower rate, indicating a relative decoupling. The variation between countries is significantly higher than for DMI.

Malenbaum claimed that the relation between IU and GDP (or time) can be understood in terms of three factors. First, the types and quantities of materials used by an economy vary through its successive stages of development. In the early stages, when incomes are low, material requirements are low. With increasing industrialisation there is an increasing demand for materials to build infrastructure, such as roads, railways, factories and power grids. However, as development continues, the need for additional basic infrastructure decreases and there is increased consumer demand for services, which are usually less materials intensive. Second, technological developments alter the efficiency with which raw materials are discovered, extracted, processed, distributed and used in the production of goods. This is likely to result in a decrease in the consumption of raw materials. Finally, substitution among raw materials may occur if new materials are superior in price or functionality.

Malenbaum further postulated that the inverted U-shape curves should repeat in a similar manner for different regional or national economies independent of the time at which they start to industrialise. However, the intensity-of-use pattern would occur in successively shorter time intervals due to ongoing technological developments of which later-developing economies could take advantage. Larson (1991) developed this idea further and concluded that developing countries might not need to reach the same level of per capita consumption to attain an equivalent or higher standard of living; the maximum intensity-of-use of a material would occur at successively lower values in developing economies because they could take advantage of more recent technological developments. Bernadini and Galli (1993) summarised the work of Malenbaum and others as follows (Groenenberg *et al.*, 2005).

1 The intensity-of-use of a particular material follows the same pattern for all economies, at first increasing with per capita GDP, reaching a maximum at about the same GDP (in real terms), and eventually declining.
2 The maximum intensity-of-use declines the later it is attained by a given economy.

These trends are illustrated qualitatively in Figure 15.7. Figure 15.7a shows how countries 1 to 5 complete development in subsequent periods but at around the same value of per capita GDP. Figure 15.7b shows how the maximum intensity-of-use declines the later each country develops. An optimistic interpretation is that the future supply of materials will become decoupled from the demand for materials so that economic growth can continue while demand for materials continues to fall. This view is epitomised by the proponents of the Factor 10 approach (Schmidt-Bleek, 1997). Bernadini and Galli do not go that far, but state that, 'Taken together, the two postulates imply a declining trend in the rate of growth in materials and energy consumption at the world level and, depending on the rate of economic growth in the developing countries and in the world as a whole in the coming decades, they may even imply a decline in absolute terms in the rate of material consumption over the next half century or so'.

The trends noted by Bernadini and Galli are based on empirical evidence; they have no theoretical basis. Also, they apply to single materials not to the total materials intensity. There is abundant evidence of dematerialisation trends for particular materials, but this could be counterbalanced by substitution of alternative materials or by an increase in the number of products consumed. Further, the inverted U-shaped curves of Figure 15.7 are not the only trajectories for materials intensity-of-use in an economy (Groenenberg *et al.*, 2005). Heavy industry was dominant throughout the period of central planning in Eastern Europe and the former Soviet Union. This resulted in high rates of metal and concrete consumption, which fell only when economic restructuring began around 1990. The intensity-of-use has been historically low in the Middle East, due to the high national incomes from petroleum export, and in much of South America, where heavy industry has not developed for a variety of reasons. Some researchers have found that an N-shaped curve is possible; the inverted U-shape is followed by a period when material consumption again increases at a faster rate than GDP. For example, de Bruyn and Opschoor (1997) found that the curve for the transport sector in Western Europe began to rise after a period of dematerialisation in the 1960s and 1970s.

15.4.2 Drivers of dematerialisation

A number of social, economic, technological and environmental forces tend to drive dematerialisation

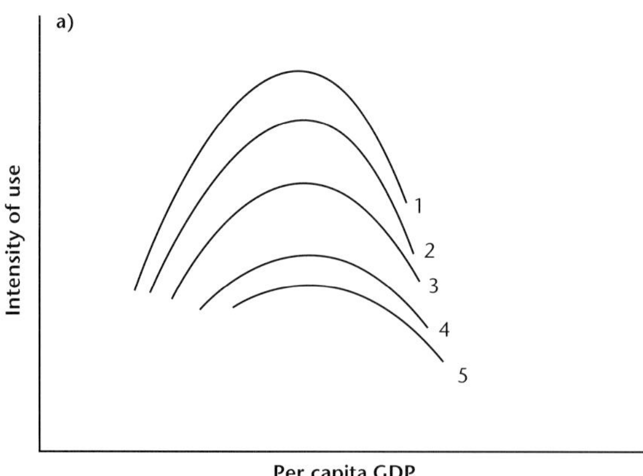

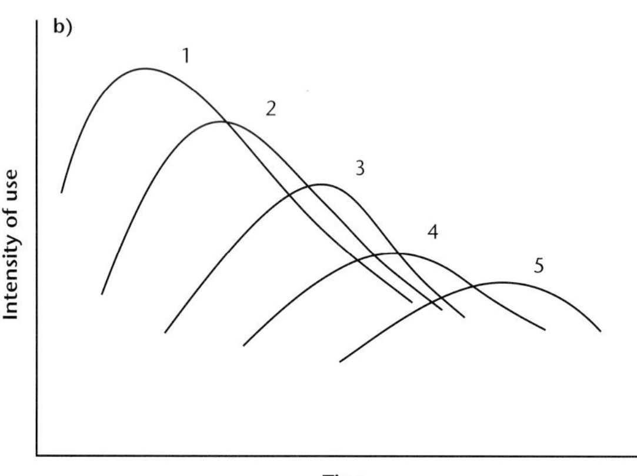

Figure 15.7: Qualitative representation of the environmental Kuznets curves for countries 1 to 5 (after Bernadini and Galli, 1993).

of the economy. Some of these were anticipated by Malenbaum.

Technological developments. There are numerous examples of the effects of technology on the decreasing intensity-of-use of materials. These have been due to new and improved materials, to incremental and novel developments in engineering sciences, and to improved design. The increase in power:weight ratio of engines is a dramatic example – from around 1 W kg^{-1} for steam engines in the early 19th century to over 600 W kg^{-1} for a modern car engine. The use of thinner steel in car bodies and white goods and of aluminium in cans, and the resulting reduction in materials use, resulted from the development of better alloys and fabrication processes. Miniaturisation of mechanical and electromechanical parts with electronic components, made possible by the development of transistors and the integrated circuit, means smaller and fewer parts and a resulting reduction in materials use. Better design and quality control has resulted in less waste per unit of production. Figure 15.8 shows qualitatively how the quantity of particular materials required in a product correlates with the knowledge embedded in the material and with the general trend towards more sophisticated materials. Also, products are increasingly being designed to increase their recyclability, reuse or remanufacture so there is less waste produced at their end-of-life.

Substitution. Materials substitution has always occurred. Usually new materials, with better functionality or price, substitute for old materials. For example, the steel content of automobiles in the United States decreased by 22% between 1976 and 1986. Most of it was replaced by aluminium, plastics and composites, resulting in an overall decrease in vehicle weight of about 16% (Bernadini and Galli, 1993). Optical fibres have replaced copper in many applications in the telecommunications industry, and in power transmission aluminium has replaced copper in some areas.

Changes in the structure of final demand. The mix of goods and services produced and consumed by an economy changes over time due to changes among sectors (the rise of the service sector) or changes within sectors (the increasing dominance of computers and other high-tech goods within the manufacturing sector) (Cleveland and Ruth, 1999). The shift towards services and high-technology products reduces the quantity of materials used per economic unit of output.

Saturation of markets for bulk goods. As economies mature there is less demand for new infrastructure. This should reduce the demand for steel, concrete and other basic materials as they are increasingly used only for replacement purposes.

Regulation. Governments are increasingly regulating the use of substances and materials, and their disposal. This has resulted in restricted uses of some substances, for environmental and health reasons, and increased recycling rates of many materials. For example, lead has been banned as a component of gasoline in most countries and lead from batteries is highly recycled. There has thus been a relative decline in demand for primary lead. Overall, recycling of materials has increased; while recycling does not of

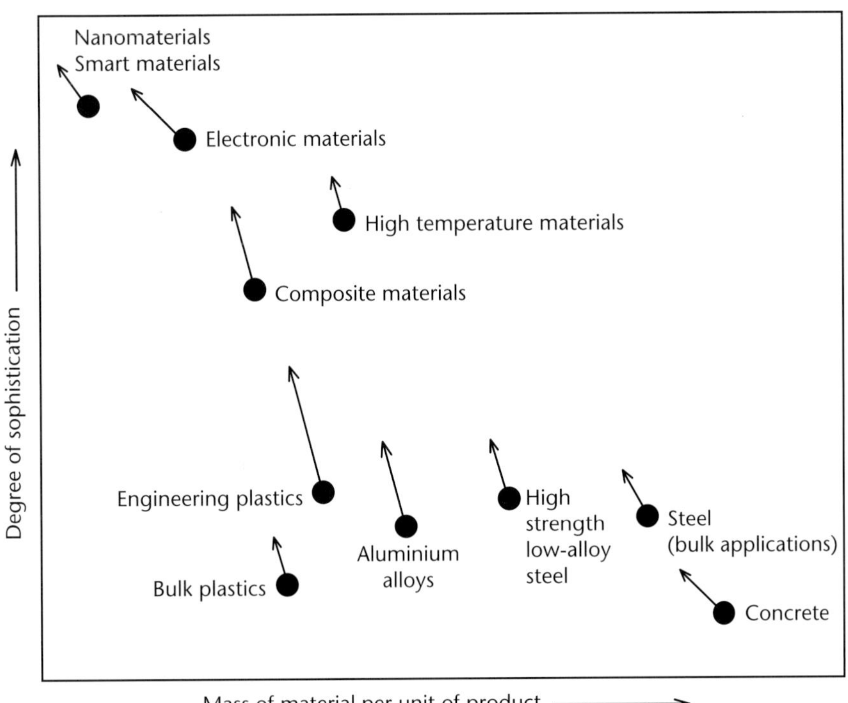

Figure 15.8: The relation between the quantity of material required in a product and its degree of sophistication or embedded knowledge. Arrows indicate the trends over recent decades (modified after Altenpohl, 1980).

itself reduce the rate of final consumption of materials, it reduces the energy used and the wastes created relative to the production of the materials from natural resources.

15.4.3 Counters to dematerialisation

Dematerialisation gains from substitution and technological developments may be offset by the growth in population and rising affluence (Wernick, 1994). This implies that dematerialisation is counteracted by the increase in the number of goods and services used. With the real cost of goods falling, due to technological and other improvements which have resulted in higher productivity, and with rising affluence, the demand for products and services has actually increased. In the case of products approaching demand saturation, the expenses forgone by consumers may well be spent on other goods and services. This is known as the *rebound effect*. There is much anecdotal evidence to support this. For example, a household in developed countries in the 1960s may have had one radio and one television set; nowadays it is common to have multiple sets within the one household. Similarly, in the 1950s and 1960s a household may have had one car; now it is not uncommon to have one car for every person of driving age in the household! The average house built today is much larger than the typical house of the 1960s, mainly due to the relative decline in the cost of materials and labour-saving efficiencies in construction techniques. The quantity of clothing owned by individuals, and its rate of turnover, in developed countries has increased enormously since the 1950s. The rebound effect is a modern expression of the *Jevons paradox*, a proposition that technological progress that increases the efficiency with which a resource is used tends to increase the consumption rate of the resource (Jevons, 1865). William Stanley Jevons, an English economist, observed that improvements in the efficiency of coal use led to increased consumption of coal in a wide range of industries. He argued that, contrary to intuition, technological improvements could not be relied upon to reduce fuel consumption.

Every substitution and technological improvement that changes the types and quantities of materials has a unique environmental impact. It is possible that 'less' from a materials perspective may not always be 'less' from an environmental perspective (Cleveland and Ruth, 1999). The substitution of aluminium for steel and plastic for timber are good examples. Bauxite processing has large environmental impacts (mainly as a result of red mud production) and aluminium smelting is highly energy-intensive, whereas steel production from iron ore has much less environmental impact and uses much less energy. Plastics are made from fossil fuels and have significant recycling and disposal problems (Frosch, 1995). There are many other examples. The increasing use of relatively small quantities of speciality metals in electronic goods may lead to greater dissipation of these in the environment because of difficulties in collection, separation and recycling (Ayres, 1989). The composites and plastics that have partially replaced steel in cars are difficult to recycle. The fuel saving due to the lighter weight may be more than offset environmentally by the additional waste created (Frosch and Gallopoulos, 1989). The introduction of personal computers and the internet has significantly increased the consumption of paper and printer inks rather than decreased it, as was predicted. As electronic chips have become smaller and the manufacturing process more complicated, the ratio of indirect materials consumption to materials actually embodied in the product has become very high. For example, a chip weighing 1 g requires processing in which hundreds or thousands of grams of reagents are consumed and discarded (Ayres, 2001).

15.4.4 A case study

Table 15.2 illustrates some dematerialisation issues, using Canadian economic data between 1981 and 2005 (Victor, 2009). The Canadian economy grew rapidly during that period, almost doubling in size (measured in constant dollars). The provision of services grew faster than the provision of goods. However, the overall composition of the economy changed very little, from 65.1% of GDP for services in 1981 to 68.4% in 2005. Part of this increase was due to the outsourcing of services that companies previously provided in-house. GDP grew by 93% and the associated increase in materials, energy and wastes flows far exceeded any beneficial effect from the small shift from goods to services. Personal expenditure increased by 57% from 1981 to 2005, reflecting rising affluence and the rebound effect. However, the shift to the purchasing of services was relatively small (from 48.8% in 1981 to 53.3% in 2005). Part of this was due to the outsourcing of functions that were previously performed by the household, such as cooking (eating out more often, consumption of fast foods), cleaning and gardening. Canada's trade in goods increased faster than its trade in services over the period, probably reflecting the rapid growth of the minerals industry. This was opposite to the overall movement of the economy towards a slightly higher proportion of services. Thus, there was no significant dematerialisation of the Canadian economy between 1981 and 2005. Cleveland and Ruth (1999), similarly, found no evidence that the United States economy had become decoupled from material inputs. From an examination of economy-wide material flow analysis data, Bringezu *et al.* (2009) concluded

Table 15.2: Changes in the Canadian economy 1981–2005, in constant dollar terms

	% increase (1981–2005)	1981 (%)	2005 (%)
Size of GDP	93		
– services	103		
– goods	75		
Composition of GDP			
– services		65.1	68.4
– goods		34.9	31.6
Size of personal expenditure	57		
Composition of personal expenditure			
– goods		51.2	46.7
– services		48.8	53.3
Composition of trade			
– goods exports		84.7	87.4
– goods imports		78.4	85.1
– services exports		15.3	12.6
– services imports		21.6	14.9

Source: Statistics Canada, as cited by Victor (2009).

that per capita material resource use does not decline as economies grow but that the rate of growth does decrease (weak dematerialisation).

15.5 THE IPAT EQUATION

The IPAT equation, named for the symbols comprising it, is frequently used to quantify the relationships between population, degree of affluence and environmental impact of technology for an economy. It focuses on the quantities of wastes produced rather than on the quantities of materials or products entering the economy. If *P* is the population of a region or country, *A* the degree of affluence of the population (per capita GDP) and *T* the environmental impact of a particular technology (e.g. tonnes of wastes produced per dollar of GDP), then the total environmental impact of the technology, *I* (tonnes of waste per year) is simply:

$$I = PAT \qquad 15.2$$

This is known as the IPAT equation. Equation 15.2 implies that the environmental impact of a particular technology increases as the population increases and as the degree of affluence increases. It shows that reducing the environmental impact of a technology, through technological developments and improvements, makes it possible for the per capita GDP to continue to grow without increasing the environmental impact if the reduction in *T* is sufficiently large. It could even be possible to reduce the environmental impact. However, growth in population could counter the gains and may overwhelm them. The IPAT equation does not provide any new insights but it does show the quantitative relation between environmental impact, population, affluence and technology.

The weakness of the IPAT equation is that it assumes that population, affluence and technology (*P*, *A* and *T*, respectively) are independent variables whereas in fact they are interrelated – population growth may influence affluence, affluence may influence technology, and so on. This means that it is usually not possible to independently change only one of *P*, *A* and *T* to reduce the impact. Also, *T* can change not only through the use of a new technology but as a result of a change in the composition of GDP without a change in technology. For example, an increase in the proportion of services relative to manufactured goods at the same level of GDP could result in a lower value of *T*. Thus, while impact calculated using the IPAT equation responds to changes in composition of GDP, it does not distinguish between changes in composition and changes in technology.

Equation 15.3 overcomes the latter weakness of the IPAT equation by separating goods and services and imports and exports (Victor, 2009):

$$I_D = (P \times A_g \times T_g) + (P \times A_s \times T_s) - X_i - X_e \qquad 15.3$$

where I_D is the total domestic environmental impact, *P* is the population, A_g is the goods components and A_s the services component of per capita GDP, T_g is the impact per dollar of goods, T_s is the impact per dollar of services, X_i is the impact of producing and transporting imported goods and services, and X_e is the impact of transporting and consuming exported goods and services. The impacts of importing and exporting goods and services are predominantly felt abroad; hence they are subtracted to give the domestic impact or are omitted to give the global impact:

$$I_G = (P \times A_g \times T_g) + (P \times A_s \times T_s) \qquad 15.4$$

where I_G is the total global environmental impact.

To illustrate Equation 15.3, assume that initially an economy has no imports or exports, so the last two terms will be zero. Assume also that the impact per dollar of services is less than the impact per dollar of goods ($T_s < T_g$). Then, if GDP remains constant but the proportion of services rises relative to the proportion of goods, the total environmental impact of the economy will decline. The environmental impact could still decline even if GDP grows, provided the shift from goods to services is sufficiently large. Now, assume the economy is open to trade. Equation 15.3 shows that impact on the domestic environment is decreased when services are exported and/or goods are imported. This is the situation in many developed countries, where many basic manufactured goods are imported from low-cost countries with lower environmental standards, and services are exported. In reality, the net effect is the export of environmental impacts!

15.6 SUMMARY AND CONCLUSIONS

Substitution and dematerialisation are relatively poorly understood concepts and it is difficult to draw definitive conclusions about their impact on demand for materials. Knowledge of the extent of, and mechanisms behind, the pattern of materials use is largely limited to individual materials or specific industries. Even less is known about the net environmental effects due to changes in materials use, except for a few important effects such as the reduction in carbon intensity (Cleveland and Ruth, 1999). Nevertheless, some general observations can be made.

- The ratio of the weight of materials used to the economic value created is decreasing for virtually all products. Products are becoming lighter per unit of cost, though not necessarily smaller. This is occurring through substitution and technological developments in materials, and manufacturing processes and construction techniques. There is a tendency to use more knowledge-based and artificially structured materials in products. Weak dematerialisation is therefore occurring.
- All industry sectors (particularly manufacturing and construction) are improving the efficiency of use of materials and are continually introducing improved materials. This is being driven largely by economics, although environmentally targeted regulations are increasingly important.
- There has been no net decrease in materials consumption at the individual consumer level and it is likely the reverse is occurring through the rebound effect. The growing population and rising affluence means that the total quantity of materials used will continue to increase for the foreseeable future. Strong dematerialisation, therefore, is not occurring.
- Dematerialisation does not inevitably lead to less waste and less environmental impact. The potential environmental impact of a new or improved material is unique and needs to be considered on a case-by-case basis.

The extent to which materials are recycled in future will increasingly influence the demand for primary mineral and metal commodities. Many products now contain a complex mix of materials, to achieve a lower mass and improved characteristics. Each material adds to a product's overall performance. At its end-of-life, however, this complexity adds to the difficulty of separating materials and reduces the economic feasibility of recycling materials.

An overall conclusion (based on Chapters 13, 14 and 15) is that we cannot know whether the availability of particular mineral resources will constrain economic development in the future. This unsatisfactory conclusion follows from at least four uncertainties:

1. the rate and extent to which technological developments will improve, first, the efficiency of extracting materials from mineral resources and, second, their use to produce more output from a given amount (dematerialisation);
2. the extent to which social and environmental factors will limit the exploitation of mineral and metal resources;
3. the future possibilities for materials substitution;
4. the extent to which recycling can meet future material needs.

Optimists believe that human inventiveness will solve any problems and that all material needs can be met indefinitely. Pessimists believe that just because humans have successfully solved supply and demand issues in the past this does not mean that they will be able to do so in the future, as the problems become more intractable.

15.7 REFERENCES

Access Economics (2008) *Global Commodity Demand Scenarios*. Access Economics, May 2008.

Altenpohl DG (1980) *Materials in World Perspective*. Springer-Verlag: Berlin.

Ayres RU (1989) Industrial metabolism. In *Technology and Environment*. (Eds JH Ausubel and HE Sladovich) pp. 23–49. National Academy Press: Washington DC.

Ayres, RU (2001) 'Resources, scarcity, growth and the environment.' Center for the Management of Environmental Resources, INSEAD: Paris. <http://europa.eu.int/comm/environment/enveco/waste/ayres.pdf>.

Ayres RU (2005) 'On the practical limits to substitution'. Interim Report IR-05-036. International Institute for Applied Systems Analysis, 28 July 2005.

Bernadini O and Galli R (1993) Dematerialization: long-term trends in the intensity of use of materials and energy. *Futures* **25**: 431–447.

Bringezu S, Schülz H, Saurat M, Moll S, Acosta-Fernández J and Steger S (2009) Europe's resource use: basic trends, global and sectoral patterns and environmental and socioeconomic impacts. In *Sustainable Resource Management: Global Trends, Visions and Policies.* (Eds S Bringezu and R Bleischwitz) pp. 52–154. Greenleaf Publishing: Sheffield, UK.

Carter AP (1966) The economics of technological change. *Scientific American* **214**(4): 25–31.

Cleveland CJ and Ruth M (1999) Indicators of materialization and materials intensity of use. *Journal of Industrial Ecology* **2**(3): 15–50.

Cordell D, Drangert J-O and White S (2009) The story of phosphorus: global food security and food for thought. *Global Environmental Change* **19**: 292–305.

de Bruyn SM and Opschoor JB (1997) Developments in the throughput–income relationship: theoretical and empirical observations. *Ecological Economics* **20**: 255–268.

Frosch RA (1995) Industrial technology: adapting technology for a sustainable world. *Environment* **37**(10): 16–37.

Frosch RA and Gallopoulos NE (1989) Strategies for manufacturing. *Scientific American* **September**: 144–152.

Goeller HE and Weinberg AM (1976) The age of substitutability. *Science* **191**: 683–689.

Goldman Sachs (2003) Global Economics. Paper No. 99, October 2003.

Groenenberg H, Blok K and van der Sluijs J (2005) Projection of energy-intensive material production for bottom-up scenario building. *Ecological Economics* **53**: 75–99.

Jevons WS (1865) *The Coal Question: An Inquiry Concerning the Progress of the Nation, and the Probable Exhaustion of Our Coal-Mines.* Macmillan: London. <http://www.econlib.org/library/YPDBooks/Jevons/jvnCQ.html>.

Kuznets S (1955) Economic growth and income inequality. *American Economic Review* **45**(1):1–28.

Larson ED (1991) Trends in the consumption of energy-intensive basic materials in the industrialized countries and implications for developing regions. In *International Symposium on Environmentally Sound Energy Technologies and their Transfer to Developing Countries and European Economies in Transition*, Milan, Italy, 21–25 October 1991.

Levy A (2000) From hotelling to backstop technology. *Middle East Business and Economic Review* **12**: 46–51.

Malenbaum W (1978) *World Demand for Raw Materials in 1985 and 2000.* McGraw-Hill: New York.

Nordhaus WD, Houthakker H and Solow R (1973) The allocation of energy resources. *Brookings Papers on Economic Activity* **3**: 529–576.

Norgate TE, Rajakumar V and Trang S (2004) Titanium and other light metals – technology pathways to sustainable development. In *Green Processing 2004.* pp. 105–112. Australasian Institute of Mining and Metallurgy: Melbourne.

Schmidt-Bleek F (1997) Statement to Government and Business Leaders. Wuppertal Institute: Wuppertal, Germany.

Sohn I (2006) Long-term projections of non-fuel minerals: we were wrong but why? *Resources Policy* **30**: 259–284.

Tsur Y and Zemel A (2003) Optimal transition to backstop substitutes for non-renewable resources. *Journal of Economic Dynamics and Control* **27**: 551–572.

United Nations (2008) World Population Prospects: The 2008 Revision. <http://esa.un.org/unpd/wpp2008/index.htm>.

Vaccari DA (2009) Phosphorus: a looming crisis. *Scientific American* **June**: 42–47.

Victor PA (2009) Scale, composition and technology. *Bulletin of Science, Technology and Society* **29**(5): 383–396.

Wernick IK (1994) Dematerialisation and secondary materials recovery: a long run perspective. *JOM* **46**(4): 39–42.

15.8 USEFUL SOURCES OF INFORMATION

Bleischwitz R, Welfens PJJ and Zhang Z (Eds) (2009) *Sustainable Growth and Resource Productivity.* Greenleaf Publishing: Sheffield, UK.

Tilton JE (2003) *On Borrowed Time? Assessing the Threat of Mineral Depletion.* Resources for the Future: Washington DC.

16 Towards zero waste

The types of wastes created in the primary production of mineral and metal commodities and how they are presently handled were examined in Chapters 11 and 12, respectively. Chapter 13 examined the recycling of materials from end-of-life products to reduce the quantities of wastes due to disposal and primary production (since lesser quantities of minerals need to be mined when materials are recycled). This chapter considers the reduction and elimination of wastes at their source, focusing on the wastes associated with the primary production of mineral and metal commodities. Hence the emphasis is on wastes from mining and processing, both direct and indirect, and solid, liquid and gaseous, of the types listed in Table 11.1.

16.1 THE WASTE HIERARCHY

Waste reduction and elimination are not new concepts and many strategies for them have long been known. Our ancestors rarely wasted anything. An animal slaughtered for food also provided fur and leather for clothing and shoes, and bones for implements of all varieties; little, if any, of the carcass was discarded. Many products were remanufactured or reused and materials were recycled before being ultimately discarded. The present attitude towards wastes is recent, and largely results from the Industrial Revolution and the increasing availability of energy and materials. There has been a renewed focus on wastes only since the ever-increasing quantities of waste from energy and materials production and consumption began to have wide-scale environmental impacts, in the latter half of the 20th century.

The conventional waste hierarchy, sometimes called the 3Rs, is:

- reduce;
- reuse;
- recycle.

This lists waste management strategies in decreasing order of desirability. Thus, the most desirable strategy is to reduce the quantity of materials and wastes associated with a product, and to use fewer products. The next most desirable strategy is to reuse a product, and only as a last resort recycle the materials comprising the product. An expanded form of the hierarchy, which includes disposal as the least desirable strategy, is: reduce, reuse, remanufacture, recycle, return (to environment). Yet another, Re^4, is: reduce, replace, recycle, restrict (Halada and Ijima, 2006). This includes restriction of the use or production of certain substances. These hierarchies have much in common.

They focus on the reduction or minimisation of waste rather than on the elimination of wastes, and seem to have been developed with manufactured products, building and construction products, and domestic waste in mind. Their focus is on things that have had a useful life rather than on things or by-products that have not had a previous use, such as mining and processing wastes, or other types of wastes.

The focus on minimising waste started shifting towards eliminating wastes at source as a result of the European Union Council Directive 91/156/EEC of 1991 (European Union, 1991) which substantially amended the original Waste Framework Directive (Directive 75/441/EEC) of 1975. Directive 91/156/EEC established the hierarchy:

- waste prevention;
- recovery;
- safe disposal.

Importantly, it addressed both classes of wastes – things that have had a useful life and by-products, such as mining and processing wastes, that have had no previous use. The directive requires member states to take a number of measures.

1 Waste production and its harmful effect should be prevented or reduced. This should be achieved in particular by:
   - the development of technologies which are more sparing in their use of natural resources;
   - the technical development and marketing of products designed to make no contribution or to make the smallest possible contribution (in their manufacture, use or final disposal) to increasing the quantity or harmful effects of waste and pollution hazards;
   - the development of appropriate techniques for the final disposal of dangerous substances contained in waste destined for recovery.
2 Recycling, reuse, reclamation or other processes should be used with a view to extracting secondary raw materials or using wastes as a source of energy.
3 Wastes should be recovered or disposed of without endangering human health and without using processes or methods which could harm the environment. This should be done without risk to water, air, soil and plants and animals; without causing a nuisance through noise or odours; and without adversely affecting the countryside or places of special interest.
4 The cost of disposing of waste must be borne by the producer of the waste.

After several further revisions, the Waste Framework Directive of 1975 was codified in 2006; this replaced previous versions. It was revised in 2008 as Directive 2008/98/EC (European Parliament, 2008) and established the following waste hierarchy:

- prevention;
- preparation for reuse;
- recycling;
- other recovery (e.g. energy recovery);
- disposal.

Directive 2008/98/EC excludes wastes resulting from the prospecting, extraction, treatment and storage of mineral resources and the working of quarries, which had been included in earlier versions, since these were covered specifically by Directive 2006/21/EC (European Parliament, 2006). The latter required the following strategy for mining-related wastes.

1 Prevent or reduce waste production and its harmfulness, in particular by considering:
   - waste management in the design phase and in the choice of the method for mineral extraction and treatment;
   - the changes that the extractive waste may undergo in relation to an increase in surface area and exposure to conditions above ground;
   - placing extractive waste back into the excavation void after extraction of the mineral, as far as technically and economically feasible and environmentally sound in accordance with existing environmental standards at community level and with the requirements of this directive where relevant;
   - putting topsoil back in place after the closure of the waste facility or, if this is not practically feasible, reusing topsoil elsewhere;
   - using less dangerous substances for the treatment of mineral resources.

2 Encourage the recovery of extractive waste by means of recycling, reusing or reclaiming such waste, where this is environmentally sound, in accordance with existing environmental standards at community level and with the requirements of this directive where relevant.
3 Ensure short- and long-term safe disposal of the extractive waste, in particular by considering, during the design phase, management during the operation and after closure of a waste facility and by choosing a design which:
   - requires minimal and, if possible, ultimately no monitoring, control and management of the closed waste facility;
   - prevents or at least minimises any long-term negative effects, for example attributable to migration of airborne or aquatic pollutants from the waste facility;
   - ensures the long-term geotechnical stability of any dams or heaps rising above the pre-existing ground surface.

This strategy is consistent with the waste hierarchy of Directive 2008/98/EC.

16.2 REDUCING AND ELIMINATING WASTES

It might be possible to significantly reduce the quantity of waste, or eliminate wastes entirely, if we change our concept of waste. The concept of waste is relative in at least two respects (Pongrácz, 2002). First, something becomes waste when it loses its primary function for the user; someone's waste could become someone else's raw material input. Second, the concept of waste is relative to the state of the technology for processing and transforming it and to the location of its generation and potential uses. Waste, therefore, is a dynamic concept. In this respect it is analogous to a mineral deposit, which can become an ore body when the economics and technologies available for extracting and processing it are viable. A common definition of waste is:

> *A waste is a thing or substance that has been discarded, or which will be discarded, and eventually sent to landfill or other disposal site.*

This definition, however, fails to recognise the dynamic nature of waste because it precludes the possibility of the thing or substance being useful again. A preferable definition is (Pongrácz, 2002):

> *A thing or substance is a waste when it has no purpose; has fulfilled its purpose; is not used or not usable for its purpose because its performance is inadequate; or its owners failed to use, or did not intend to use, it for its assigned purpose.*

This definition says nothing about the thing or substance being discarded and implies that waste is a temporary state. It allows for the possibility of waste being turned into a non-waste. This different understanding of the nature of waste provides the basis for developing waste minimisation and waste elimination strategies.

Strategies for minimising and eliminating wastes in the production of mineral and metal commodities can usefully be grouped as follows:

- cleaner production;
- use of waste as raw materials;
- waste reduction through process re-engineering;
- industrial ecology.

These are arranged in order of increasing capacity to minimise or eliminate wastes, and hence form a hierarchy. This order also correlates with increasing degree of integration into the business of a company and the economy at large. Thus, cleaner production can be implemented at a single operation, whereas industrial ecology requires integration across companies and across industry sectors. These strategies are not mutually exclusive alternatives. There is considerable overlap between them and several strategies may be pursued in parallel. Figure 16.1 shows the historical trend in approaches to addressing the environmental impact of wastes. In the move towards industrial ecology, company behaviour has moved from complying with regulations to corporate social responsibility, and now needs to move to an integrated sustainability strategy. The drivers for this change have moved from being exclusively profit to include regulations and stakeholders. In parallel, the materials cycle focus has shifted from a focus on products

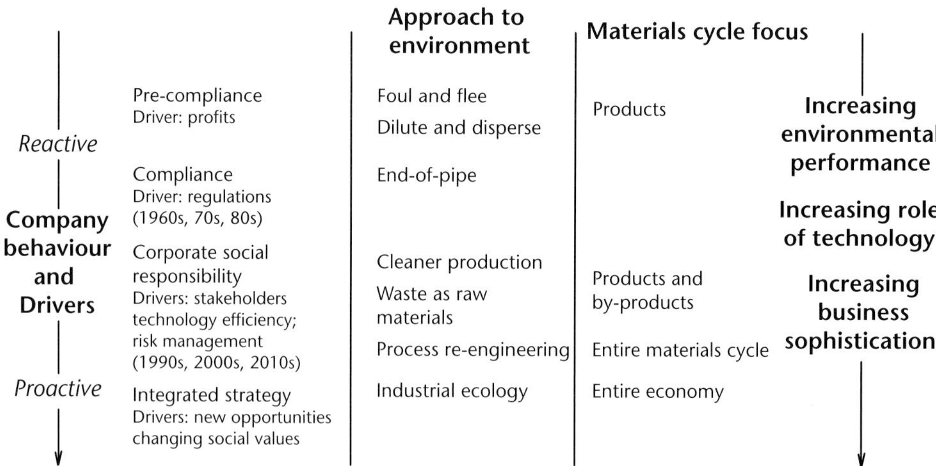

Figure 16.1: The historical trend in approaches to the environmental impact of wastes (based on a figure from Giurco and Petrie, 2007; with modifications).

only to by-products as well as products. It now needs to shift to the entire materials cycle and, ultimately, to the entire economy.

16.3 CLEANER PRODUCTION

The United Nations Environmental Programme defines cleaner production (CP) as follows (UNEP, 2010):

> *Cleaner production is the continuous application of an integrated preventive environmental strategy to processes, products, and services to increase overall efficiency, and reduce risks to humans and the environment.*

In some countries, the term pollution prevention (P2) is used interchangeably with cleaner production (Basu and van Zyl, 2006). Cleaner production can be applied to the processes used in any industry, to the products themselves and to various services provided in society. For production processes, cleaner production involves one or a combination of the following:

- conserving raw materials, water and energy;
- eliminating toxic and dangerous raw materials;
- reducing the quantity and toxicity of emissions and wastes at source during the production process.

Waste is considered as a product with negative economic value. Each action to reduce consumption of raw materials and energy, and prevent or reduce generation of waste, can increase productivity and bring financial benefits to enterprise. The similarities between eco-efficiency and clearer production are numerous. Like cleaner production, eco-efficiency (Section 4.4.2) goes beyond resource use and pollution reduction by emphasising value creation for the business and society at large. Eco-efficiency includes cleaner production concepts and captures the idea of pollution reduction through process change rather than end-of-pipe approaches. The terms cleaner production and eco-efficiency are often used interchangeably.

The key difference between pollution control and cleaner production is that pollution control is an after-the-event, react-and-treat approach while cleaner production looks forward and attempts to anticipate and prevent. Cleaner production aims to minimise or avoid practices such as waste treatment (including stabilisation, encapsulation and detoxification), waste dilution to comply with regulations (e.g. releasing contaminated water into rivers or streams during high flow periods, blending arsenic-containing fumes with flotation tailings), and transferring hazardous or toxic substances from one medium to another (e.g. wet-scrubbing gases then disposing of the contaminants as waste water).

Implementation of cleaner production requires a structured, holistic, common-sense approach using systems and people to both reduce environmental impact and improve the overall company perfor-

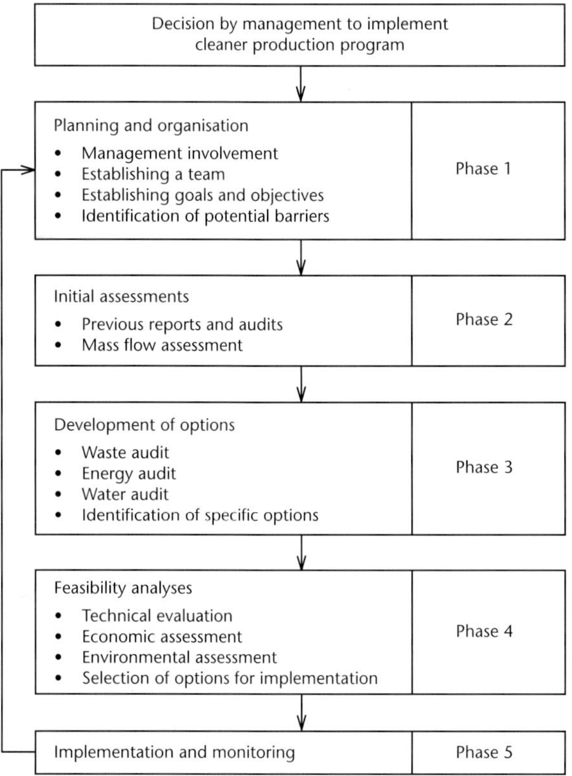

Figure 16.2: A generic model for developing and implementing a cleaner production program (after DEH, 2000).

mance (DEH, 2000). Figure 16.2 illustrates a generic approach for developing and implementing a cleaner production program. The audit process is usually time-consuming and may be quite challenging. It requires accurate knowledge of inputs, outputs and flows; these are often not known and have to be measured. When the information has been collected, material and energy balances are developed for parts or all of the operation. These form the basis for identifying waste reduction options based on knowledge gained from the literature, personal experience and expertise, discussions with suppliers, the example of other companies, or research and development.

Some obvious possibilities for improvement may be identified during the audit process but others may be identified only during a more rigorous analysis. A structured approach to analysis may be useful, for example dividing the analysis into several areas and considering each in turn, in an iterative manner. Some useful areas are input substitution, technology modifications, good housekeeping, product modifications and on-site recycling (UNEP, 1996). Input substitution refers to the use of less polluting raw materials and reagents and other consumables (refractories, lubricants, coolants, grinding media with a longer service life). Technology modifications include improved process automation, process optimisation, equipment redesign and process substitution. Good housekeeping refers to changes in operational procedures and management to eliminate the generation of wastes and emissions. Examples include management and personnel practices, training, material and inventory handling, loss prevention, waste segregation and production scheduling. Product modifications involve changing the product characteristics so the process is less polluting. There may not be much scope for this in processing operations, since mineral and metal commodities sold on the market must meet tight composition specifications. However, where a concentrate, for example, is used internally within a company there may be scope for optimising the mill-to-smelter flowsheet to produce less waste overall. On-site recycling is the useful application of waste materials or pollutants where they have been generated. This could take place through reuse as a raw material, recovery of materials or other useful application.

16.4 WASTES AS RAW MATERIALS

The consideration of wastes as raw materials follows naturally from the redefinition of waste in Section 16.2. Although processing of wastes could be considered an end-of-pipe environmental solution and therefore not particularly innovative, this is far from the truth. Many of the by-products from producing mineral and metal commodities, which are now considered wastes, contain much of value. Technologies for extracting that value are often technically sophisticated. There are several incentives for considering mining and processing waste as a raw material.

- A mining and processing waste is a 'free' raw material (since it has already been mined) and is often in a form suitable for further processing (since it may have been crushed and ground).

- Use of a waste reduces the demand for new mined material since the product produced replaces product that would otherwise have to be mined.
- Use of a waste to make a saleable product reduces the volume of waste that must be stored or disposed of. This saves on storage and disposal costs and could lead to a reduction in the environmental impact. This is particularly advantageous for large-volume wastes such as beneficiation tailings.
- Producing a saleable product adds another source of income for a company.

Use of mining and processing wastes as raw materials is not a new concept and some wastes from mining and processing are already treated to produce saleable products. Sulfur dioxide in smelter gases is used to make sulfuric acid. Slags from iron and steel production are used for making aggregate materials and as a raw material in cement manufacturing. However, mining and processing wastes are a largely untapped resource and there are many potential applications. An advance would be to optimise processes so that all output streams are useful products, rather than optimising processes around the principal product. In this way, all outputs would be optimised (in terms of composition and morphology) to maximise their effectiveness as inputs for further value-adding or, as a last resort, for disposal.

While utilisation rather than disposal of mining and processing wastes is preferable, it must be recognised that not all, or even most, mining-related wastes can be used productively. The quantities of mining wastes are so large that there are insufficient bulk applications, even in construction projects, to use significant quantities. Furthermore, many mines are located in remote and/or sparsely populated areas and the transportation of low-value construction products (sand, aggregate) to populated areas for use in infrastructure projects is uneconomic. Hence, the focus is necessarily on wastes from mines and processing operations close to populated areas and/or on higher-value products which can be transported economically over long distances. Some examples are given below to illustrate the possibilities.

Utilisation of red mud

As discussed, red mud from the production of alumina from bauxite is a solid waste produced in large quantities (Table 11.3). Increasing attention is being paid to finding uses for red mud because of the environmental problems associated with storing such a highly alkaline material. It is also a potentially valuable resource. A number of uses have been identified (Cooling, 2007):

- as a soil additive to help retain nutrients and adjust soil pH;
- as a neutralising agent for treating acid mine drainage and acid-producing soils;
- as a filtration medium to remove phosphorus and nitrogen from sewage effluent in domestic and industrial septic systems;
- as an additive to fertiliser to improve phosphorus retention in soils;
- as an additive to compost to aid the retention of trace metals;
- in brick and tile manufacture, both fired and non-fired;
- as a filler for plastics, to impart strength, resistance to UV, heat and chemicals, and colour;
- as road base, either using the sand fraction directly or the mud as a component of a composite with gypsum or fly ash;
- as raw material for the production of cement alternatives, such as mineral polymers and ceramics;
- as a pigment for a range of applications in coatings and materials manufacture.

The greatest potential for large-volume applications include use of the coarse sand fraction (>150 μm) as a general-purpose fill and construction medium and of the fine mud fraction as a soil additive. Separation of red mud into these fractions is relatively simple and cheap. Solar drying can be used in suitable locations. Early trials of the products have shown promising results (Cooling, 2007).

Chemical processing of red mud to recover elements of value is a distinct possibility but has not been adopted commercially. Red mud contains 30–60 wt% Fe_2O_3 and 10 wt%, or more, TiO_2 (Table 11.2). Figure 16.3 shows a flowsheet, tested at the laboratory scale, for recovering these and other substances of value. The process involves a number of beneficiation, pyrometallurgical and hydrometallurgical operations. A sand fraction is separated and the residue is melted to produce a slag which is cooled, crushed and leached to

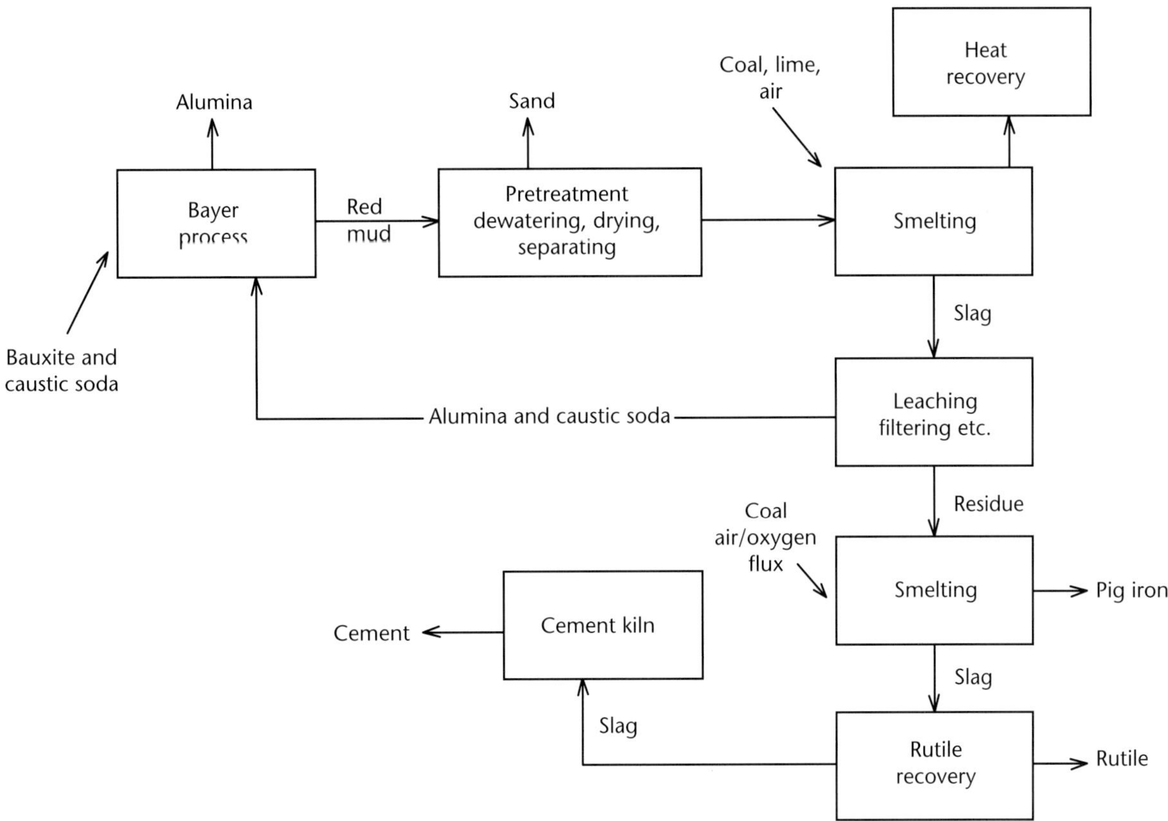

Figure 16.3: A conceptual flowsheet for value-recovery from red mud (after Jahanshahi *et al.*, 2007).

extract residual alumina and sodium hydroxide, which are returned to the Bayer plant. The leach residue is then smelted under reducing conditions to produce pig iron (which can be converted into steel) and a slag which contains the titanium. During solidification of the slag, after tapping, the titanium crystallises out as rutile (TiO_2), which is liberated by grinding then separated to form a rutile concentrate. The tailings are suitable as a raw material for cement manufacturing.

Utilisation of spent pot lining

Spent pot lining (SPL) from the smelting of aluminium is considered a hazardous waste in most countries because it contains significant quantities of fluorides and small quantities of cyanide (see Table 10.2). SPL is removed from aluminium smelting cells when they are periodically relined, typically every four to seven years. It consists of approximately equal quantities of carbon cathode material and refractories, with solidified bath material. Around 20–30 tonnes of SPL are generated for every 1000 tonnes of aluminium produced. A flowsheet for processing SPL (Figure 16.4) achieves the following objectives: destruction of the cyanide component of SPL, use of the carbon component as fuel, removal of sodium in slag, recycling of aluminium fluorides to the smelter, and production of a benign waste product which can be used as a sand substitute (Mansfield *et al.*, 2002). The flowsheet consists of a pyrometallurgical stage with related feed preparation and slag granulation, an off-gas cooling stage with particulate removal and recycling, and a stage for producing aluminium fluoride and final scrubbing of the outlet gas. An Ausmelt type furnace, which utilises a top submerged lance, is used for the pyrometallurgical stage. Carbon in the SPL provides most of the fuel requirement. A fluidised bed reactor is used for the aluminium fluoride recovery. A plant to process 12 000 tonnes per year of SPL was constructed at Alcoa's Portland aluminium smelter in Victoria, Australia, and ran successfully for a number of years.

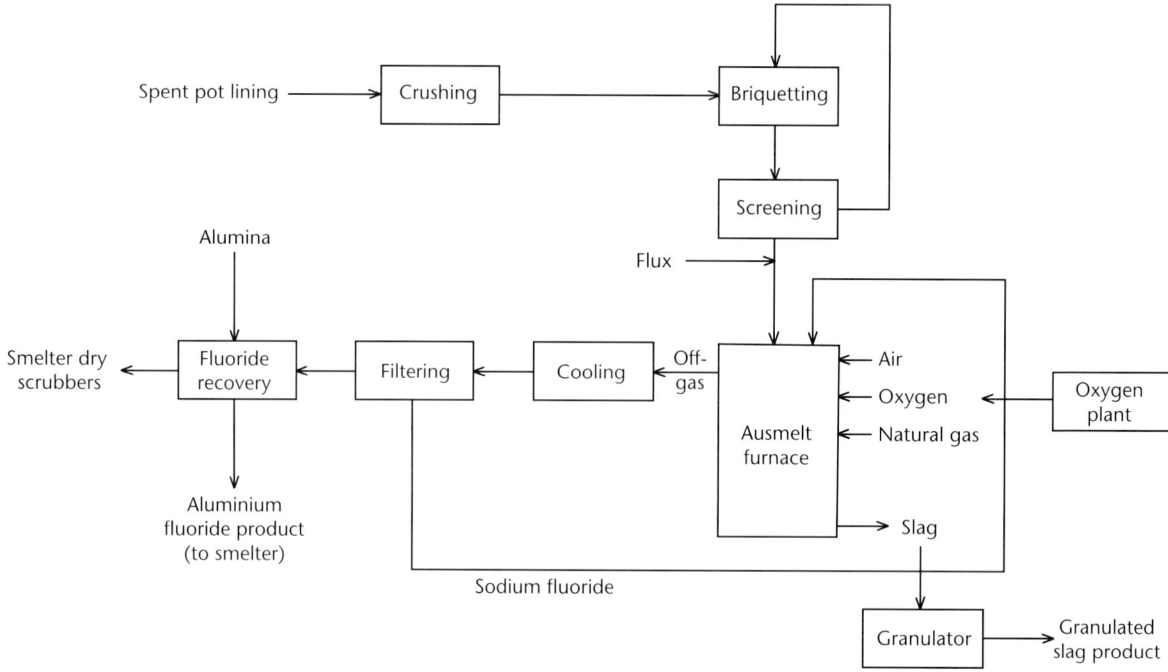

Figure 16.4: Flowsheet for the Alcoa Portland SPL process (after Mansfield *et al.*, 2002).

Utilisation of fly ash and slag

Ash is a product from coal-fired power stations and forms from the mineral matter contained in the coal. Bottom ash is the ash that remains in the furnace after combustion; fly ash is the ash that is carried over with the flue gas and is captured in baghouses. They have similar compositions, though bottom ash forms into lumps through sintering. Typically, about 90 wt% of the ash is in the form of fly ash. Depending on the type of coal burned, ash contains 15–60 wt% SiO_2, 5–25 wt% Al_2O_3, 5–40 wt% Fe_2O_3 and up to 30 wt% CaO, largely in the form of amorphous aluminosilicates. It may also contain small quantities of heavy metals. Globally, about 750 million tonnes of fly ash were produced in 2005 (Metha, 2009). About 40% of fly ash produced is presently utilised, and the remainder is stored in tailings dams or dry storage facilities. The major present uses of fly ash, in descending order of importance, are:

- for blending with ordinary Portland cement as a supplementary cementitious material (SCM);
- as an ingredient in structural fill;
- for stabilising wastes (by mixing with the wastes);
- as a raw material for cement clinkers;
- for stabilising soils;
- as road base;
- as aggregate;
- as a mineral filler in asphalt.

Some of these uses are possible due to the pozzolanic nature of fly ash, but only about 15% of fly ash is presently used in these applications. Bottom ash can be used as a sand or can be ground to produce a pozzolanic material with properties similar to those of fly ash.

The current global consumption of concrete is more than 5 billion tonnes per year. Cement is the key component of concrete, and binds the aggregate. In 1990, global cement production was about 1 billion tonnes; this increased to almost 2 billion tonnes by 2005. Despite the increased use of SCMs, demand is likely to increase to around 2.8 billion tonnes by 2020 (Mehta, 2009). Cement production results in about 0.9 tonne of CO_2 for every tonne of cement and, globally, contributes about 8% of the total CO_2 produced each year (Table 9.8).

Possible strategies to reduce the CO_2 production associated with cement manufacture are to reduce the

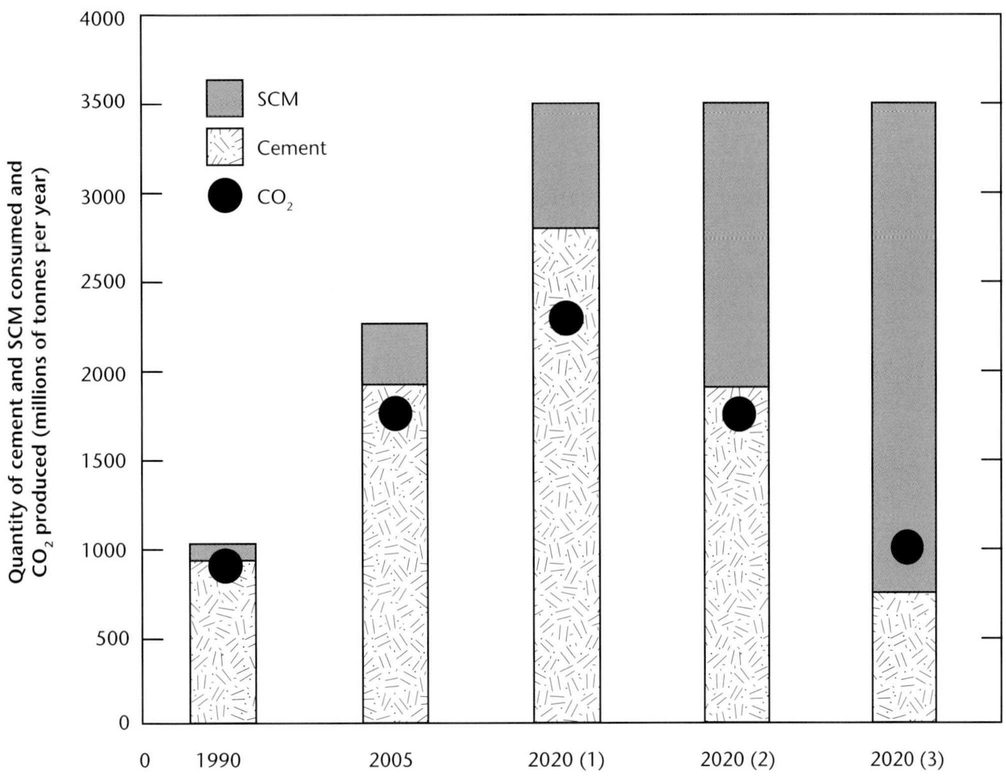

Figure 16.5: Projected global cement production demand and the impact of utilising SCMs (data for 1990 and 2005 are actual values; data for 2020 are projected values).
Case 1. Business as usual, 20 wt% SCM of binder.
Case 2. Calculated value of SCM if 2020 CO_2 levels are kept to 2005 levels.
Case 3. Calculated value of SCM if 2020 CO_2 levels are kept to 1990 levels.
(Mecasil Technology; http://www.mecasil.com/sustainability.html).

quantity of concrete consumed (through dematerialisation) and to reduce the cement content of concrete. One approach to the latter is to increase the proportion of SCM in the binder. Typically, SCMs make up about 20 wt% of the binder for standard concrete mixes. Fly ash and slag have the benefit of being by-products of other processes; they involve little extra emission of greenhouse gas. Figure 16.5 illustrates three possible scenarios. If the greenhouse gas emission level of 2005 were to be maintained while meeting the projected demand in 2020, the quantity of SCM in binder would have to be greater than 50 wt%. If 1990 CO_2 levels were to be achieved, the SCM proportion would need to be in excess of 75 wt%. The short-term challenge, therefore, is to raise the proportion of SCM in binder towards 50% for all concrete mixes without loss of performance. This can be facilitated by increasing the pozzolanic activity of fly ash by grinding it to increase the specific surface area. An alternative approach may be to replace cement, without loss of performance, with geopolymer binders which can utilise higher proportions of fly ash. This option is discussed below.

Geopolymer concrete

Geopolymers (also known as alkali activated materials, AAM) are a class of compounds formed by the reaction of an alkaline solution with an aluminosilicate material. The solid product has an amorphous three-dimensional structure similar to that of an aluminosilicate glass, which can act as a binder for aggregate materials. Though glass-like in structure, geopolymers are formed at low temperature and can be mixed with aggregate materials to form concrete. The reactants are a mixed alkali metal hydroxide – metal silicate solution (referred to as the activator) and an aluminosilicate binder with mean particle size in the range 1–30 μm. A significant proportion of the silicon and aluminium ions in the binder must be in an amorphous state. The most common activator is a mixture of water, sodium hydroxide and sodium silicate but other alkali metal systems can be used, as can any waste source of alkali, such as red mud. The most common binders are fly ash and ground granulated slags, but in theory any fine amorphous aluminosilicate material can be used.

During initial mixing the alkaline solution dissolves silicon and aluminium ions from the amorphous phases. These undergo a condensation reaction in which adjacent hydroxyl ions from near neighbours condense to form an oxygen bond linking the molecules, and a free molecule of water:

$$OH^- + OH^- = O^{2-} + H_2O$$

The water produced by the condensation reaction is physically expelled during the curing and drying process leaving closed, discontinuous nano-sized pores in the solid product. This is different from Portland cement in which the water added to the mix chemically combines during curing and is incorporated in the molecular structure. Each oxygen bond formed as a result of a condensation reaction bonds the neighbouring silicon or aluminium tetrahedron. Three monomers are commonly formed (Figure 16.6). The application of mild heat (typically up to 90°C) causes the monomers and other silicon and aluminium hydroxide molecules to polymerise, to form rigid chains or networks of oxygen-bonded tetrahedra. It is theoretically possible for every silicon ion to be bonded via an oxygen bond to four neighbouring silicon or aluminium ions to form a rigid polymer network. Aluminium ions, being trivalent, require an associated alkali metal cation (usually Na) to achieve charge balance. The resultant product consists of a rigid network of geopolymer containing a solution within the pores consisting of water, excess alkali metal ions and unreacted silicon hydroxide.

A potential benefit of geopolymers, in addition to providing a large market for fly ash and slag, is a reduction in the embodied CO_2 in concrete. The global warming potential of geopolymer concrete (kg CO_2-e per tonne) is highly dependent on

```
      |    |
      O    O
      |    |
 – O – Si – O – Al⁻ – O –
      |    |
      O    O
      |    |

      |    |    |
      O    O    O
      |    |    |
 – O – Si – O – Al⁻ – O – Si – O –
      |    |    |
      O    O    O
      |    |    |

      |    |    |    |
      O    O    O    O
      |    |    |    |
 – O – Si – O – Al⁻ – O – Si – O – Si – O –
      |    |    |    |
      O    O    O    O
      |    |    |    |
```

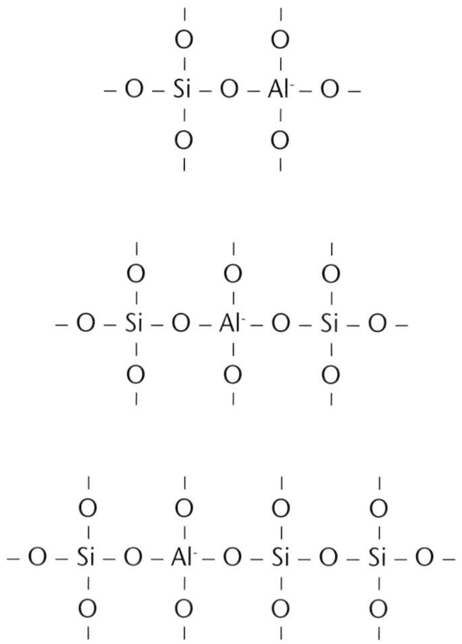

Figure 16.6: The monomer units which form the three-dimensional network of geopolymer materials.

the activator content since the energy embodied in manufacturing the alkali is several orders of magnitude greater than that for fly ash or slag. The chlor-alkali process for producing sodium hydroxide consumes about 3820 kW h (equivalent) per tonne of NaOH (Bommaraju *et al.*, 2004). It also produces chlorine and hydrogen gases (from electrolysis of sodium chloride in solution):

$$2NaCl + 2H_2O = 2NaOH + Cl_2 + H_2$$

If this energy is co-allocated to NaOH, chlorine and hydrogen in proportion to their stoichiometric amounts (52.3 wt%, 46.4 wt% and 1.3 wt%, respectively) it translates, for coal-fired electrical energy, to around 2000 kW h per tonne of NaOH. Using the conversion factor of 1 tonne CO_2 per kilowatt hour for black coal generated electricity (Table 9.9), this is equivalent to 2 tonnes of CO_2 per tonne of NaOH. Therefore, for a typical geopolymer formulation with 30 wt% NaOH in the binder and 25 wt% binder in concrete, the embodied CO_2 is about 0.15 tonnes of CO_2 per tonne of concrete. This can be compared with about 0.9/4 = 0.23 tonnes of CO_2 per tonne of concrete with the same percentage of ordinary Portland cement. This is about a 35% reduction. However, if use of geopolymers becomes widespread then NaOH manufacture must be increased to meet the demand. This means an excess of chlorine, a highly hazardous gas, would be produced and would have to be disposed of safely. It would mean allocating virtually the entire 3820 tonnes of CO_2 per tonne of NaOH to the NaOH, thereby doubling the embodied CO_2 to around 0.3 tonnes of CO_2 per tonne of concrete. In many cases, this would make it higher than for ordinary Portland cement. This would limit the quantity of geopolymer concrete that could substitute for OPC concrete.

Commercialisation of geopolymers, other than in niche applications, has been slow even though they have been known since the 1950s. Reasons for this include:

- cement and concrete manufacturers have well established processes and are reluctant to change to a radically new technology;
- the product is largely untested, certainly for long-term stability, hence potential users are reluctant to use it for major infrastructure projects for liability reasons;
- most standards for concrete are based on formulas which specify the ingredients and proportions to be used rather than the performance standards to be met. The latter would permit introduction of a new material if it met the performance standard.

Nevertheless, industry interest is increasing, due mainly to the potential greenhouse gas benefits and the increasing likelihood of the introduction of a carbon tax or trading scheme. Commercialisation is predominantly at the demonstration stage for small-scale applications. For example, in Australia geopolymer concrete paths have been cast at the Perth campus of Curtin University using a mix produced by the university's geopolymer research group, and a small ready-mix demonstration plant has been established in Melbourne (www.zeobond.com) to produce batches of up to 200 m^3 of geopolymer concrete per day for small applications. There are numerous other groups around the world developing and trialling geopolymer applications.

Utilisation of sulfide flotation tailings

The composition of tailings from flotation of metallic sulfide ores is highly variable but most tailings contain valuable minerals as well as low-value bulk minerals. The sulfide minerals present in the tailings can potentially lead to acid and metalliferous drainage (AMD) and its associated problems (Section 11.5.1). Separation of the sulfide minerals into a low-volume fraction would represent a major step forward in the environmental management of sulfide tailings. The fraction could be stored, thus requiring less space and less management of a potentially harmful material. If it contained valuable metals, it could be further processed to recover those before being stored or otherwise used. The non-sulfide fraction could be used to produce building and construction materials, used as a raw material for extracting useful elements (particularly iron) or stored as a benign waste without the potential to form AMD. This can be illustrated using the results of tests performed on four sulfide tailings to assess their value-adding potential (Bruckard and McCallum, 2007; McCallum and Bruckard, 2009). The tailings examined are listed in Table 16.1. The tailings were characterised by chemical analysis to determine their elemental composition and by X-ray diffraction to determine their mineralogical composition. These results were used to design four simple separation operations aimed at producing two or more fractions which could be used as feed to further processing (to produce value-added products) or stored safely as relatively benign materials. The separation operations tested were screening, gravity concentration, magnetic separation and flotation. The test procedure is summarised in Figure 16.7. The results are outlined in Table 16.2, from which the authors drew a number of conclusions.

- Simple screening is a useful method for concentrating silicates into coarse (low sulfide) sand fractions.
- De-sliming is a useful technique for producing bulk sulfide concentrates (sulfides of several metals), for further processing to recover valuable metals.
- Gravity separation can be used to produce high-grade silicate sand products (which are low in sulfides) and bulk iron sulfide products (containing pyrite and pyrrhotite).

Table 16.1: Sulfide tailings used to assess potential for making value-added products

Type of tailing	Sulfide content (wt% S)	Annual mine production of tailings (Mt per year)	Quantity of tailings stored (Mt)
Copper concentrate tailings			
Magnetite-rich sulfide tailing from a sulfide copper deposit with significant gold mineralisation	3.37	10.5	120
Nickel concentrate tailings			
Typical of pyrite and pyrrhotite rich tailings from a nickel deposit comprising bands of disseminated and massive sulfides	6.08	0.86	2.6
Cu-Ni concentrate tailings			
Typical of monzonite porphyry silicate/alumino-silicate rich tailings from a massive low-grade gold-copper sulfide deposit	0.29	22.0	177
Ag-Pb-Zn concentrate tailings			
Typical of lead-zinc concentrator tailings from a silver-lead-zinc deposit	1.44	1.5	–

Source: Bruckard and McCallum (2007).

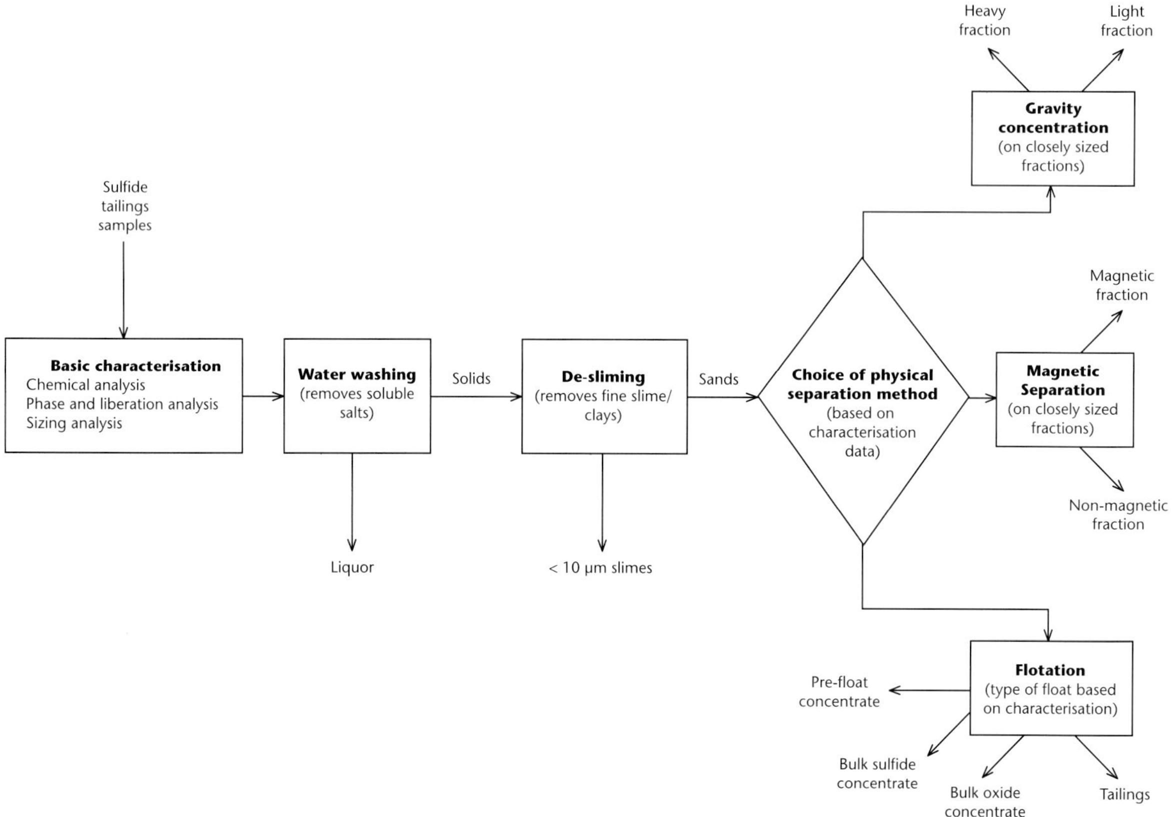

Figure 16.7: Protocol for assessing and testing sulfide tailings (after Bruckard and McCallum, 2007).

Table 16.2: Prospective products obtained from sulfide tailings samples by various physical separation methods

Screening	Gravity concentration	Magnetic separation	Flotation
Copper concentrate tailings			
Coarse (>212 μm) sand product, high in silica (>50 wt% SiO_2) and low in S (1 wt% S) but containing 57% of the Cu	A silicate sand product, high in silica (45 wt% SiO_2) and low in S (0.36 wt% S) Crude iron sulfide heavy fraction	High-grade magnetite fraction (60 wt% Fe) with low S (0.3 w% S) and silica (9.7 wt% SiO_2) content	Pre-float product high in Mo (1.29 wt% MoO_3) Bulk sulfide product high in S (28 wt%S) and Cu (0.56 wt% Cu) in a low-volume stream Possible benign float tail
Nickel concentrate tailings			
	A silicate sand product, high in silica (67 wt% SiO_2) and low in S (0.3 wt% S) Crude iron sulfide heavy fraction	High-grade pyrrhotite fraction (43 wt% Fe) with low silica (13.2 wt% SiO_2) content	Pre-float product high in talc (22 wt% MgO) Bulk iron sulfide (pyrrhotite) product (44 wt% Fe and 25 wt% S) also high in Ni (1.22 wt% NiO) in a low-volume stream Possible benign float tail
Cu-Ni concentrate tailings			
Crude split between sulfides and non-sulfides possible by de-sliming	A silicate sand product, high in silica (69 wt% SiO_2) and low in S (0.08 wt% S) Concentration of iron sulfides in slimes product		Bulk sulfide product concentrating S and Cu in a very low-volume stream (6.6 wt% S and 1.7 wt% Cu) Possible benign float tail
Ag-Pb-Zn concentrate tailings			
Coarse sand product (>212 μm), high in silica (>78 wt% SiO_2) with low sulfur level ($<$0.6 wt% S) Crude bulk sulfide product from slimes fraction	A silicate sand product, very high in silica (91 wt% SiO_2) and low in S (0.4 wt% S)	A silicate sand product very high in silica (85 wt% SiO_2) and relatively low in S (0.44 wt% S)	Bulk sulfide product concentrating Pb, Zn and S (61.2 wt% S, 1.95 wt% Zn, 3.7 wt% Pb) in a low-volume stream. Possible benign float tail

Source: Bruckard and McCallum (2007).

- Magnetic separation can be used to produce high-grade iron products (with low silicate content) containing magnetite or pyrrhotite, according to the mineralogy of the tailings.
- Froth flotation is useful for separating sulfides from non-sulfides. Most of the sulfides can be recovered in a low-volume fraction, leaving a relatively benign flotation tailings fraction.

Of particular note is the high-grade magnetite fraction obtained from the copper sulfide tailing by magnetic separation. This fraction contained 60 wt% iron and therefore could be a high-grade source of iron, depending on the types and quantities of impurities. The various coarse sand products would be suitable for use in the construction industry and in road making. The bulk iron sulfide concentrates could be processed to make elemental sulfur or sulfuric acid and the iron oxide product could make a suitable feed for steel-making. Various higher-value industrial minerals could also be produced in smaller quantities.

16.5 WASTE REDUCTION THROUGH PROCESS RE-ENGINEERING

The strategy of waste reduction through re-engineering aims to minimise the quantity of waste produced or to produce a by-product in a form that can be used more readily. This involves some process modification; often it may be necessary to completely redesign the flowsheet. There are three broad approaches:

- flowsheet simplification;
- use of novel equipment;
- use of novel processing conditions.

Flowsheet simplification involves the removal or combination of stages to reduce the overall number of stages required to produce a mineral or metal commodity. This reduces the amount of transport, handling and physical processing of material and can potentially reduce the amount of chemical processing. This saves energy (CO_2 emissions) and reduces the quantities of other wastes. The use of novel reactors or other equipment involves utilising unique characteristics of a reactor or other item of equipment to do something that was previously not possible, or was very difficult to do. The use of novel processing conditions involves utilising relatively standard reactors and flowsheet configurations with different reagents or processing conditions, such as temperature, pressure or concentration. While one of these approaches often predominates, a technological development usually combines aspects of two or all three.

16.5.1 Examples of flowsheet simplification

Heap leaching

Heap leaching is an established technology suitable for some ores and was discussed in Section 8.2.4. Heap leaching removes the need to grind ore to achieve liberation and to float it to make a concentrate. It thereby removes a number of energy-intensive steps and avoids the formation of fine, reactive tailings. Its main disadvantages are that recovery of metal is generally lower than when a concentrate is produced, and its application is limited to specific types of ore deposit and to specific metals.

The Finex iron-making process

This process is based on the direct use of iron ore fines and non-coking coal. It does not need the sintering and coking steps of the conventional blast furnace route. Because of this, SOx and NOx emissions are significantly lower compared to the blast furnace route, as are dust emissions because of the closed loop process. No dioxins are generated, unlike in sintering. The process has been implemented at a commercial scale at the Posco steelworks in South Korea.

A simplified flowsheet is shown in Figure 16.8. Iron ore fines and fluxes are fed into a series of fluidised bed reactors. They move downwards through the reactors, where they are progressively heated and reduced to metal by a counter-current stream of gas. The gas is generated from the gasification of coal in the melter-gasifier reactor. On leaving the last fluidised bed reactor, the hot direct-reduced iron is briquetted, transferred to a charging bin above the melter-gasifier, then fed into the melter-gasifier where

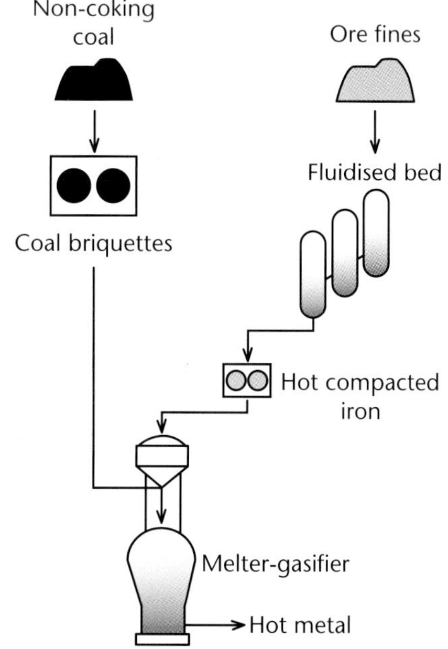

Figure 16.8: A simplified flowsheet of the Finex iron-making process.

it melts and separates into hot metal and slag. Briquetted coal fines and/or lump coal are charged at the top of the melter-gasifier and pulverised coal is injected into the vessel with oxygen. The coal is gasified and a hot reducing gas is generated, comprised mainly of CO and H_2. This hot gas is ducted to the fluidised bed reactor system to heat and reduce the iron ore. Part of the gas which leaves the top of the fluidised bed reactor is recycled, after CO_2 is removed, to achieve a higher utilisation. The heat generated from the gasification of the coal with oxygen provides the energy for melting the iron and forming the slag.

Production of titanium metal

All titanium metal is currently produced using the Kroll process, invented in 1940. This is a labour-intensive batch process in which $TiCl_4$, an intermediate product created by chlorination of rutile, is reduced with magnesium at about 900°C according to the overall reaction:

$$TiCl_4 + 2Mg = Ti + 2MgCl_2$$

The $MgCl_2$ is recovered and electrolysed to produce chlorine gas and magnesium metal, which is recycled to the Kroll reactor. The chlorine is recycled to the rutile chlorination reactor. The titanium metal is recovered in the form of a sponge. The refining of the sponge can involve up to 17 separate operations. It is first cleaned by acid leaching or vacuum distillation. The sponge is then melted to form ingots, further refined in at least two melting steps, then forged, hot rolled and annealed to produce semi-finished products such as plate and sheet. The major cost in producing titanium metal is not the sponge-forming step but the purification, melting, forging and rolling, which together account for over 60% of the cost, largely due to the high energy consumption of these operations (USDE, 2004). This, and the fact that demand for titanium would very likely grow if its cost of production could be significantly reduced, has led to the identification of a number of potential alternative, lower-energy routes (USDE, 2004).

Titanium is amenable to powder metallurgy operations and near net shape manufacturing techniques. These offer the potential for significant cost savings in many applications, provided a low-cost form of titanium powder can be produced, since the need for melting, forging and rolling is removed. Powder metallurgy is a forming and fabrication technology in which metal powder is injected into a mould, or passed through a die, to produce a weakly cohesive object with shape and dimensions very close to those desired for the final product. High pressures are used, typically 100–300 MPa. The object is then heated for a long period below the melting point of the metal. Solid state sintering of the powder particles occurs. Powder metallurgy fabrication produces very little scrap metal and can be used to make complex shapes with high tolerances. The key to reducing energy consumption and, therefore, the cost of titanium components lies in the provision of low-cost titanium powder and compatible powder metallurgy fabrication processes.

One approach to producing titanium powder is the TiRO™ process (Doblin and Wellwood, 2008). A simplified flowsheet is shown in Figure 16.9. The TiRO™ process utilises the same reaction chemistry as the Kroll process but is novel in that the reaction is

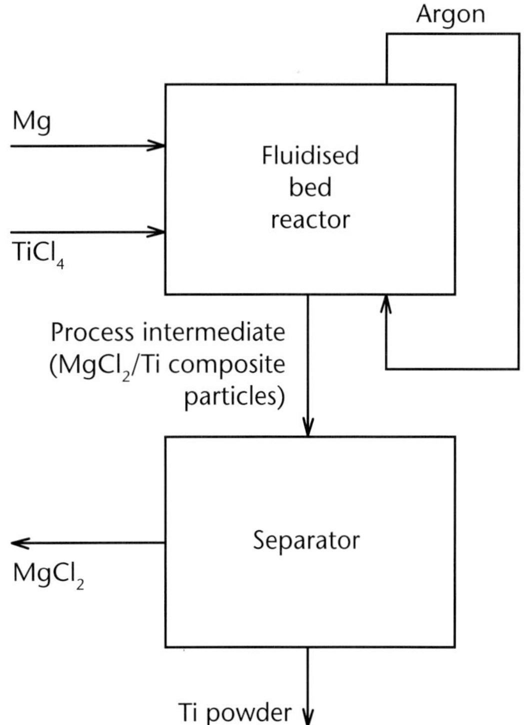

Figure 16.9: A simplified flowsheet of the TiRO™ process for producing titanium powder (after Doblin and Wellwood, 2008).

performed at a lower temperature and continuously in an argon fluidised bed. The reaction is conducted at a temperature above the melting point of magnesium (650°C) but below the melting point of $MgCl_2$ (712°C). The product formed in the bed is granules consisting of the reaction products, $MgCl_2$ and titanium particles. The second coupled stage involves separation of these two phases. There are several separation strategies, including vacuum distillation as used in the Kroll process. In the latter case, the product is friable granules consisting of loosely sintered titanium particles that are readily transformed into a powder. The titanium is reasonably pure and the $MgCl_2$ is recovered in an anhydrous form suitable for electrolysis to regenerate magnesium and chlorine.

16.5.2 Examples of novel equipment

Ore sorting

Hand sorting mined material based on visual observation has long been used in mineral processing. This is a simple form of ore sorting. However, except in very small-scale operations, ore sorting is now an automated operation. Ore sorting reduces the quantity of material that needs to be crushed and ground to achieve liberation. This saves energy and results in less process waste. If sorting is done in the mine, the waste can be left in the mine and in a form closer to its natural, unliberated state, which is more environmentally stable. Ore sorting involves evaluating the mineral content of individual rocks as they pass a sorter and separating them into accept and reject fractions, based on predetermined selection criteria. Four interactive subprocesses are involved: ore presentation, ore examination, data analysis and ore separation (Cutmore and Eberhardt, 2006). Most automatic ore sorting systems operate on particle sizes in the range 10–300 mm. Most current systems use custom conveyors for particle presentation and air blast ejectors for separation. These technologies are well established. Non-contact techniques used for ore examination are based on optical characteristics, magnetic susceptibility, X-ray diffraction, radiometric emission or absorption, or other techniques. Major advances in ore examination and data analysis technologies were made between 1965 and 1990 due to the increase in computing speed, developments in signal processing, decreases in computing costs and developments in sensing technologies. Despite these developments, automated ore sorting has not been very widely adopted, though its advantages are clear.

Underground and in-pit pre-concentration

The placing of primary crushing and pre-concentration operations underground reduces the quantity of material brought to the surface and provides a source of coarse backfill material. Crushing and pre-concentration can also be done in the pits of open cut mines, with the waste going directly to waste rock storage sites rather than to tailings storage facilities after liberation and separation. Underground primary crushing is quite common. Pre-concentration is not a new concept and has sometimes been performed on the surface; performing it underground is a relatively new idea. Potential benefits include (Bamber *et al.*, 2008):

- reduced mining costs through reducing the selectivity required during mining;
- reduced material handling requirements due to reduced quantities of ore to be hauled, hoisted and processed;
- increased metal recovery in the mill, particularly in flotation;
- less capacity required in surface transport, and mill and tailings disposal;
- improvements in fill quality through the use of coarse rejects;
- increased resource utilisation through a reduction in cut-off grade and increase in reserves;
- reduced surface impacts of operations.

Heavy- or dense-medium separation and jigs are technologies suitable for subsurface pre-concentration. The former are suitable for coarser lumps (15–40 mm) and the latter are suitable for finer material. Conventional technologies rely on water as the medium of transport and separation of particles, requiring water to be pumped to and from the surface. Dry processes are preferable. Ore sorting and dry separation (O'Connor *et al.*, 2006) are technologies which could be used.

Figure 16.10 shows an example of a flowsheet for an underground processing plant for gold which

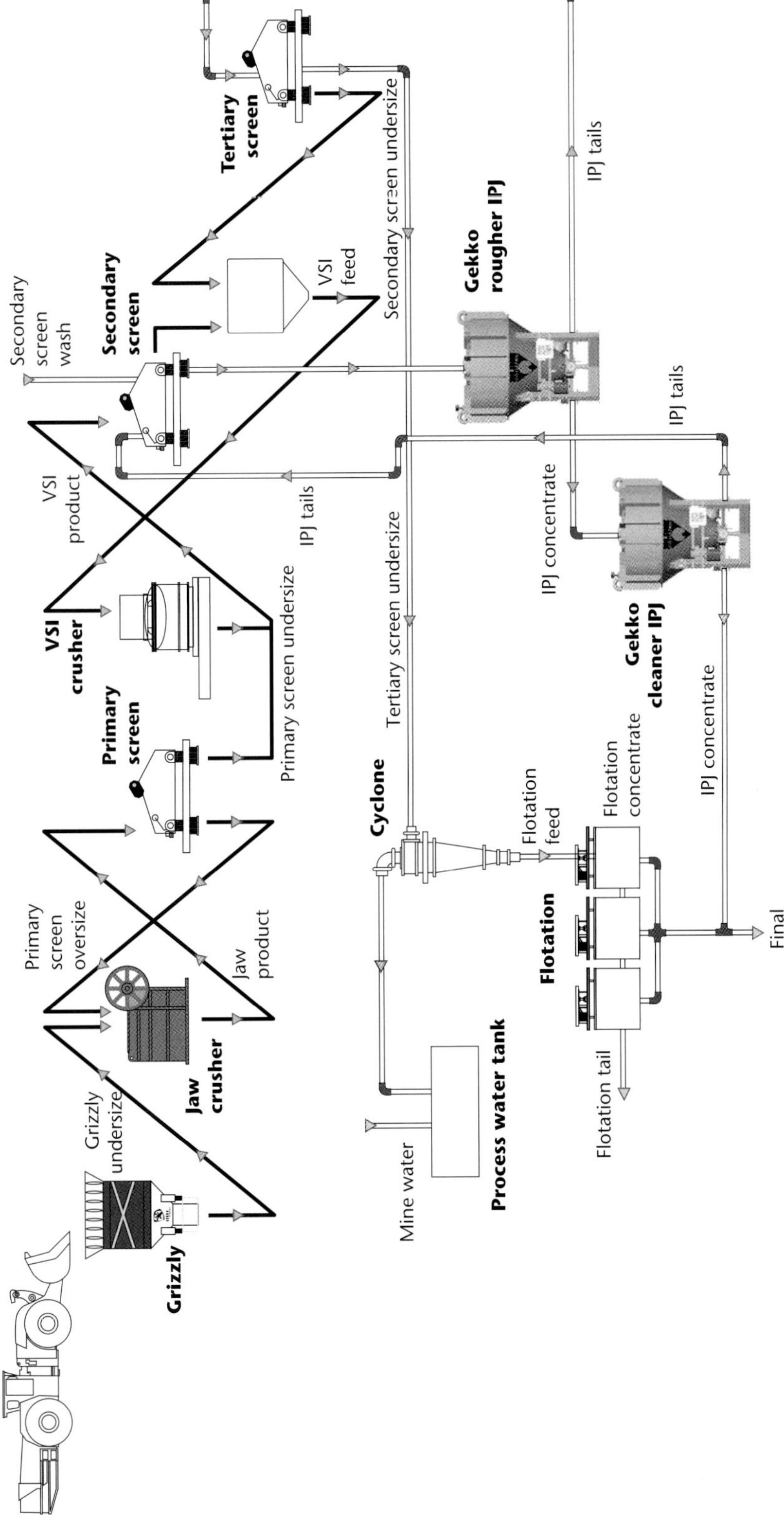

Figure 16.10: Process flowsheet for underground pre-concentration of gold ores (Hughes and Cormack, 2008). Reprinted with the permission of the Australasian Institute of Mining and Metallurgy.

incorporates primary and secondary crushing, a rougher jig, cleaner jig and flotation cells to produce a concentrate for further processing at the surface. The plant is operated to reject 65 wt% of the ore as tailings since 65 wt% bulk density is frequently used for backfilling. This relatively high concentrate:tailings ratio also ensures high recoveries. The flotation tailings are pumped to the surface or dewatered and used as backfill. The plant is 68 m long, 2 m wide and 4.8 m high and is designed to be installed in a 5 m by 5 m underground drive. It can process 20 tonnes of ore per hour. The first commercial system was installed by Central Rand Gold South Africa at the Central Rand Goldfields, south of Johannesburg, in 2008.

Continuous casting and thin-strip casting of metal

Continuous casting in the steel industry has replaced ingot casting but billets still need to be rolled to form plate and sheet. New technology eliminates the need to roll steel to produce thin strip, by using a twin-roll casting system in which molten steel is poured into a well between two counter-rotating steel rolls (Wechsler and Ferriola, 2002) (Figure 16.11). Depending on the gap between the rolls, steel may be cast to thicknesses of 0.7–2.0 mm. Greenhouse gas reductions of 70% compared with conventional technology (thin slab casting followed by rolling in a hot strip mill) have been claimed. The first commercial plant was installed and commissioned at the Nucor Steel plant in Crawfordsville, Indiana, in 2002 (Sosinsky *et al.*, 2008).

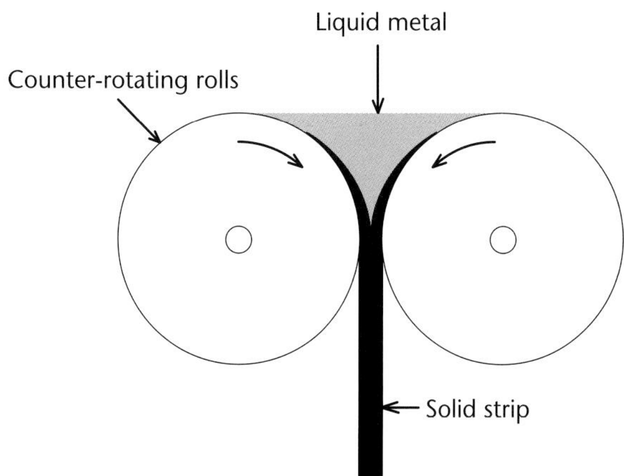

Figure 16.11: Schematic diagram of the twin-roll strip casting process (after Wechsler and Ferriola, 2002).

Dry granulation of slag

Blast furnace slag is routinely granulated using water to produce pozzolanic material suitable for incorporating into cement, but this process is wasteful of water and energy and produces noxious fumes (Section 12.1.4). An alternative approach, under development, is to quench atomised molten slag in air to produce fine particles and use the recovered heat for power generation (Xie and Jahanshahi, 2008; Xie *et al.*, 2010). This technology would provide a low CO_2 source of electrical energy, use less water, produce less emissions and produce a product suitable as a supplementary cementitious material or as feed for cement clinker. A two-stage process is used. In the first stage, molten slag is atomised and quenched by pouring a stream of slag onto a spinning disc and passing air through the reactor to cool the ejected slag droplets. The air leaves the reactor at 600°C and the solidified slag particles leave at below 900°C. In the second stage, the slag particles are cooled to ambient temperature by air in a counter-current packed bed heat exchanger. This air exits at 600°C, is mixed with the air from the granulator, then used for drying, preheating or steam generation.

Solar production of aluminium and other metals

A number of novel devices that concentrate solar energy to achieve temperatures up to 2400°C have been proposed (Haueter *et al.*, 1999). Of most interest is their potential application for the carbothermic production of aluminium (Murray, 1999, 2001; Halmann *et al.*, 2007), which requires temperatures in excess of 1900°C. Other metallurgical applications have also been considered (Murray *et al.*, 1995; Neelameggham, 2008). This technology has the advantage of using renewable energy directly to produce energy-intensive metals but there are many issues to be solved before it can be commercialised. Of particular concern are the cyclical and unreliable nature of solar energy and the severe materials problems of operating at such high temperatures.

16.5.3 Examples of novel processing conditions

Electrolytic production of metals

In principle, all metals, including iron (the metal used globally in greatest quantities) can be made by electrochemical reduction (Section 8.2.3). If the electrical energy were provided from renewable sources (wind, solar, hydro, tidal, wave) it would be a low carbon footprint method for producing metals. However, electrochemical methods for producing most metals are at a very early stage of development (Sadoway, 2000; Cox and Fray, 2009; Xiao and Fray, 2010).

Low carbon dioxide steel-making

Steel-making is not energy-intensive compared with the production of most other metals (Table 9.6). However, of all the metal-producing industries the steel industry is the greatest contributor to CO_2 emissions because of the huge volume of steel produced annually (Table 9.8). Accordingly, the industry has devoted much effort to develop low CO_2-impact technologies. The ultra-low carbon dioxide steel-making (ULCOS) program illustrates the types of approaches being examined. ULCOS is a consortium of 48 European companies and organisations from 15 European countries, formed in 2004 to undertake a six-year cooperative research and development program to reduce CO_2 emissions from steel production (ULCOS, 2010). The consortium consists of the major EU steel companies, energy and engineering companies, research institutes and universities, and is supported by the European Commission. The aim is to reduce the carbon dioxide emissions of best process routes by at least 50%. It is the largest program within the steel industry which seeks ways to reduce CO_2 emissions. In the first stage, about 70 process routes were examined. A small number were selected for further investigation. A second and third stage are envisaged, to implement the best solutions over 15–20 years. The technologies selected for further development are briefly described below.

Top gas recycling. In the top gas recycling blast furnace, off-gases are separated and the carbon monoxide is recycled to the furnace for use as a reducing agent. This reduces the quantity of coke needed. Oxygen is injected into the tuyeres rather than preheated, oxygen-enriched air as in the conventional blast furnace, removing unwanted nitrogen from the off-gas. This facilitates CO_2 capture and storage and further reduces the quantity of coke required, since the heat needed to heat the nitrogen to the off-gas temperature is saved. It is estimated that carbon consumption could be reduced by 100 kg per tonne of hot metal (from around 400 to 300 kg) and that this would reduce CO_2 emissions by 15%, or by 60% if the CO_2 were also captured and stored (Birat and Borlée, 2008). The concept was tested successfully in 2007 on LKAB's experimental blast furnace in Luleå, Sweden. The next step is to test it on a commercial-scale blast furnace in the second stage of the project.

HiSarna smelting reduction process. This process has three stages: preheating of iron ore and partial pyrolysis of coal, melting the ore in a cyclone reactor, and reduction of the ore in a bath smelter. As in the Finex process, the HiSarna process utilises iron ore fines and coal directly and avoids the need for sintering and coke-making. The process requires significantly less carbon than the conventional blast furnace and thus reduces the quantity of carbon dioxide produced. It is potentially more flexible and could allow partial substitution of coal by biomass, natural gas or hydrogen. The three stages have been demonstrated independently on a small scale and the next step is to construct and commission an integrated pilot plant to test the process up to a capacity of 65 000 tonnes per year. If successful, the plant will be extended in the second stage of the project to a semi-industrial scale of 700 000 tonnes per year.

ULCORED direct reduction process. Direct-reduced iron (DRI) is conventionally produced by reducing iron ore lumps or pellets with a reducing gas (produced from natural gas) at temperatures at which the products are solid. The reduced iron is then melted in an electric arc furnace (EAF) to produce steel. Conventional DRI processes are more expensive and energy-intensive than the conventional blast furnace–BOF route and require higher-quality iron ore. Their advantage is that there is no need to make coke and therefore no need for coke ovens. The main objective of the ULCOS program is to reduce natural gas consumption by replacing steam-reforming of natural gas by partial oxidation. This should reduce capital costs and make it possible to capture CO_2. There would be only a single source of CO_2 and it should be sufficiently concentrated to be sequestered geologically. A pilot plant to test the concept will be commissioned in the next phase.

Electro-winning. Electrolysis of iron ore is the least developed of the process routes being studied. Electro-winning would allow the transformation of iron ore into metal and gaseous oxygen using electrical energy. Coke ovens and blast furnaces would no longer be required, thus eliminating CO_2 production. Electrolysis is an established technology at the industrial scale for producing aluminium, zinc, copper and nickel but is not used commercially for iron. Aqueous-based electrolytes (both acid and alkaline) and molten salt and molten oxide electrolytes for electro-winning of iron from iron ore have been investigated at the laboratory scale (Birat and Borlée, 2008). Aqueous alkaline (the ULCOWIN process) and molten oxide electrolytes (the ULCOLYSIS process) have the most advantages and will be developed further in the next stage. The source of electricity for electrolytic processes is critical. For low CO_2 steel-making by electrolytic processes, the electricity must come from renewable or nuclear sources.

Use of biomass (Birat and Borlée, 2008). The early iron and steel industry used biomass in the form of charcoal as reductant and fuel before it was replaced in the 18th century by coke. The global warming challenge could bring a return to biomass, since it could lead to near carbon neutrality. Three biomass products have been investigated in the ULCOS project: charcoal made from dedicated plantations or agricultural residues, bio-oil, and syngas made from biomass. Bio-oil is probably better used in the transportation sector since it is expensive compared to other biomass products. Syngas could be made from agricultural residues and used, for example, as a substitute for natural gas in direct reduction processes. However, it may not be available in sufficient quantity or quality. The most obvious and practical form of biomass in the steel industry is charcoal made from sustainable plantations of eucalyptus trees grown in tropical countries (e.g. Brazil and Congo). The charcoal would be made on-site then shipped to iron-making plants. Pulverised coal injection can be substituted fully by charcoal in large blast furnaces (replacing up to 40% of the carbon input) while smelting reduction processes (e.g. HiSarna and Finex) can utilise up to 100% charcoal. Plantation forestry is mature and well developed in Brazil. Technological progress has yet to be made on charcoal production in continuous, high-productivity reactors.

16.6 INDUSTRIAL ECOLOGY

The term industrial ecology refers to an industrial system that operates much like a natural ecosystem in which materials circulate continuously in a complex web of interactions. While ecosystems produce some wastes (substances that are not recycled), such as fossil fuels and limestone and phosphate deposits, they are largely self-contained and self-sustaining through the constant input of energy from the Sun. In a similar fashion, industrial ecology involves focusing less on the impacts of each industrial activity in isolation and more on the overall impact of all such activities. This means recognising that the industrial system consists of much more than separate stages of extraction, manufacture and disposal, and that the stages are linked across time, distance and economic sectors (Frosch, 1995). The concept grew out of the work of Robert Ayres on industrial metabolism (Ayres, 1989) and was first popularised in an article in *Scientific American* by Frosch and Gallopoulos (1989). They asked, 'Why would not our industrial system behave like an ecosystem, where the wastes of a species may be resource to another species? Why would not the outputs of an industry be the inputs of another, thus reducing use of raw materials, pollution, and saving on waste treatment?' A useful definition of industrial ecology is (Graedel and Allenby, 2003):

> *The concept of industrial ecology is that an industrial system should be viewed not in isolation from its surrounding systems, but in concert with them. It is a system view in which one seeks to optimise the total materials cycle from virgin material, to finished material, to product, to waste product, and to ultimate disposal. Factors to be optimised include resources, energy and capital.*

An appreciation of the concept of industrial ecology can be gained by considering the simple models of industrial systems in Figure 16.12. Figure 16.12a shows the familiar flow-through, or open loop, system. Industry takes in new materials and processes

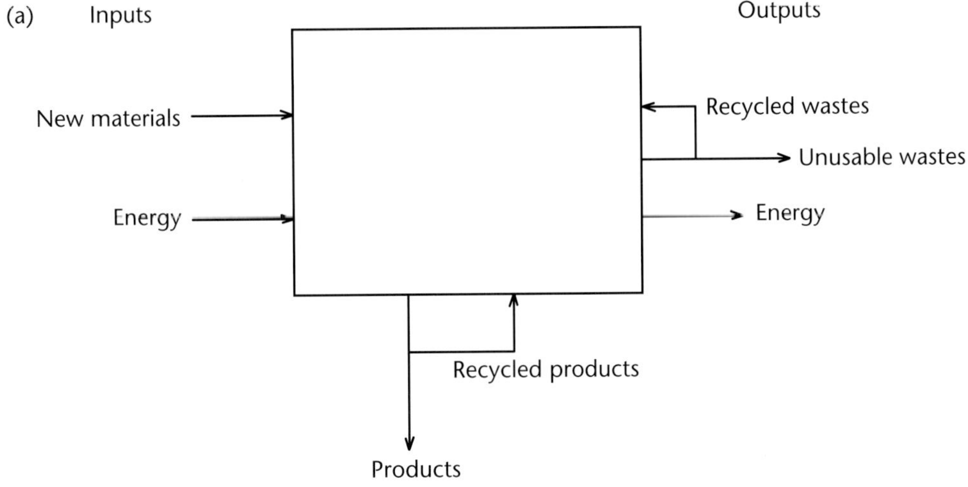

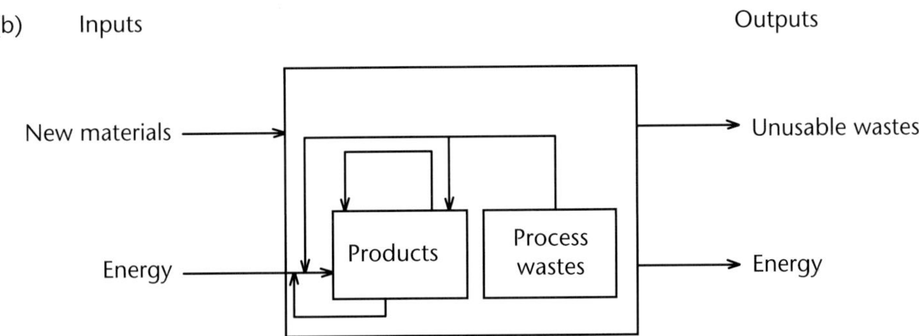

Figure 16.12: (a) Open and (b) closed material flow systems (after Frosch, 1995).

them using energy, and generates products and wastes. Both the products and the wastes are external to the boundary of the system. They are considered as externalities and their impacts are borne largely by society as a whole. New materials and energy come from outside the system and the impacts of their production are borne largely outside it. Some recycling may take place through recycling end-of-life products into the manufacturing system. Figure 16.12b shows an industrial ecosystem. This is not quite a closed loop. New materials and energy still come from outside the system and certain wastes still leave it, but products and process wastes remain within the system. Responsibility for products and for process wastes, and for the impacts of their use, is borne within the system. The unusable wastes which leave the system are of three main types:

- wastes generated during the extraction of new materials (e.g. overburden and waste rock from mining);
- wastes that escape from the recycling loop (since there are losses inherent in recovering and recycling materials);
- wastes lost through the use of products (e.g. by being discarded to landfill or incinerated).

Products that are currently in use or being held for recycling, and industrial wastes and other materials which will be reused at some time, constitute a reservoir (or stock) of materials available for use in the future.

Industrial ecology approaches environmental problems with the hypothesis that, by using principles similar to those that apply to natural systems,

Table 16.3: A comparison of the characteristics of natural ecosystems and industrial ecosystems

Natural ecosystems	Industrial ecosystems
The environment	The market
Organisms	Companies
Natural products	Industrial products
Natural selection	Competition
Ecological niches	Market niches
Anabolism and catabolism (metabolism)	Manufacturing and waste management (closed loop)
Mutation and selection (evolution)	Design for the Environment
Biological succession	Redevelopment
Adaptation	Innovation
Food web	Materials cycle

industrial systems can be improved to reduce their impact on the natural environment. The implementation of industrial ecology therefore implies shifting industrial processing from linear (open loop) systems, in which materials move through the system to become waste, to closed loop systems in which wastes become inputs for new processes. The analogy between industrial ecosystems and natural ecosystems is strong, as illustrated by the comparisons made in Table 16.3. However, the analogy with nature is strongly metaphorical and should not be overemphasised (Ehrenfeld, 2002).

An essential element of industrial ecosystems is industrial symbiosis, which can be defined as follows (Ehrenfeld and Gertler, 1997):

> *Industrial symbiosis is the use of practices such as reuse, remanufacture and recycling to create linkages between companies to increase the efficiency, measured at the scale of the system as a whole, of material and energy flows through the entire cluster of processes.*

The cascading use of energy, which involves using the residual energy in liquids or steam emanating from one process to provide heating, cooling or pressure for another process, and the use of industrial by-products as feedstocks for processes other than the ones that created them, are characteristics of industrial symbiosis. Some individual companies in a system, when viewed in isolation, may be inefficient by conventional measures of environmental performance but the overall environmental performance of the group of companies may be superior because of the linkages. The evolution, or planned implementation, of a set of interrelated symbiotic links among groups of firms gives rise to an industrial ecosystem. When these symbiotic links occur within a relatively defined area the links are often referred to as *regional synergies* and the area as an *eco-industrial park*. At the other extreme, the links may occur over long distances to form a sort of virtual industrial ecosystem.

There are many eco-industrial parks around the world and they are rapidly growing in number and complexity. Kalundborg (Denmark), Humberside

Table 16.4: Chronology of the development of the Kalundborg industrial ecosystem

Year	Development
1959	Asnaes power station commissioned
1961	Statoil refinery commissioned; water piped from Lake Tissø
1972	Plasterboard plant built; gas piped from Statoil
1973	Asnaes expands; draws water from pipeline
1976	Novo Nordisk begins shipping sludge to farmers
1979	Asnaes begins to sell fly ash to cement producers
1981	Asnaes produces heat for Kalundborg Kommune
1982	Asnaes delivers steam to Statoil and Novo Nordisk
1987	Statoil pipes cooling water to Asnaes
1989	Novo Nordisk switches from Lake Tissø to wells
1990	Statoil sells molten sulfur to Kemira in Jutland
1991	Statoil sends treated waste water to Asnaes for utility use
1992	Statoil sends desulfurised waste gas to Asnaes
1993	Asnaes supplies gypsum to plasterboard manufacturer

Source: Ehrenfeld and Gertler (1997).

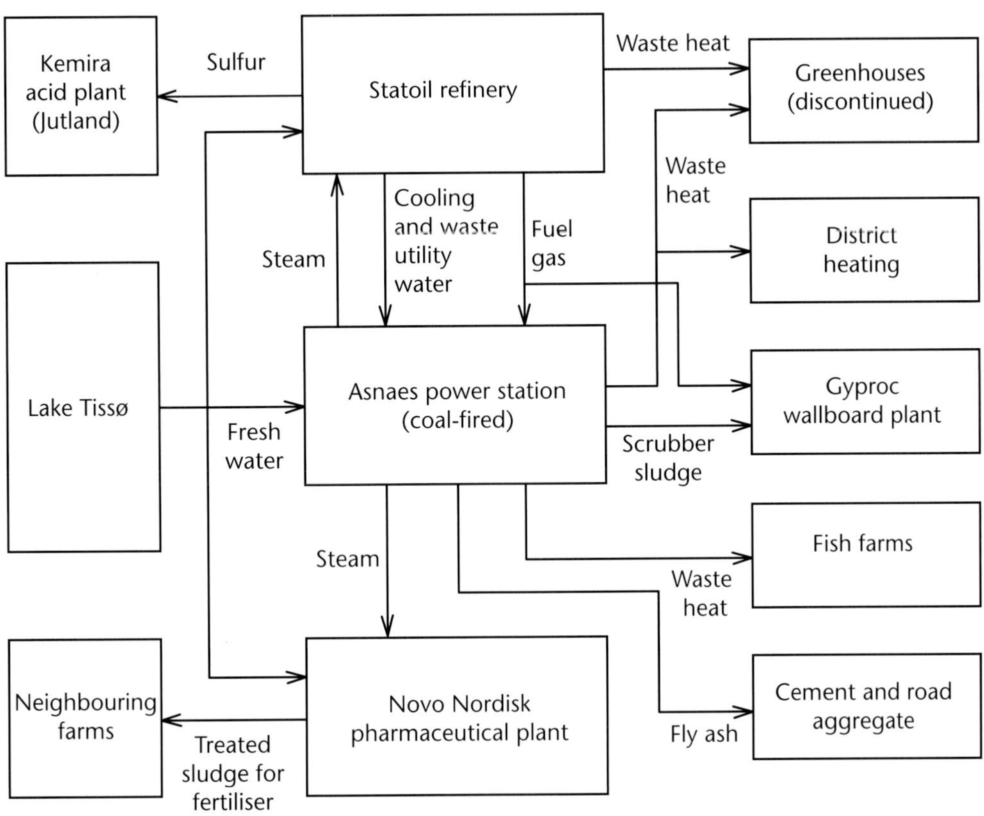

Figure 16.13: The Kalundborg industrial ecosystem (after Ehrenfeld and Gertler, 1997).

(United Kingdom), Moerdijk and Rotterdam (the Netherlands) and Kwinana (Australia) are frequently cited examples (Kurip, 2007). Kalundborg, an industrial coastal town in Denmark, is the location of a highly evolved eco-industrial park (Ehrenfeld and Gertler, 1997). It is probably the best known example of the implementation of industrial ecology principles. The key linkages in Kalundborg are shown in Figure 16.13. Four main industries – a 1500 MW coal-fired power station, a large oil refinery, a pharmaceutical company and a plasterboard manufacturer – and several other users, buy and sell wastes and energy and use them to turn by-products into raw materials. Companies outside the area also participate. The symbioses evolved gradually over about 35 years (Table 16.4) as companies sought to minimise the cost of complying with increasingly stringent environmental regulations.

An eco-industrial park based largely on resource processing is the Kwinana Industrial Area (KIA) in south-western Australia, a region which combines major resource processing operations with manufacturing, agriculture, aquaculture and recreational activities. Because of its resource base, this example is particularly relevant. The KIA is a coastal strip about 10 km long and 2 km wide about 40 km south of Perth (Western Australia) on the shores of Cockburn Sound, a sensitive marine environment. It was established in 1952 to cater for the construction of an oil refinery on the shores of Cockburn Sound. By 2002 the total economic output of the industries in the area was over A$4.3 billion per annum (Kwinana Industries Council, 2002). A number of diverse, non-competing processing industries coexist in the area (Figure 16.14). These include alumina production, nickel refining, oil refining, chemical production, cement manufacturing, fertiliser production and power generation. Agricultural and recreational areas adjoin the KIA. The existing regional synergies are arguably more diverse and significant than those

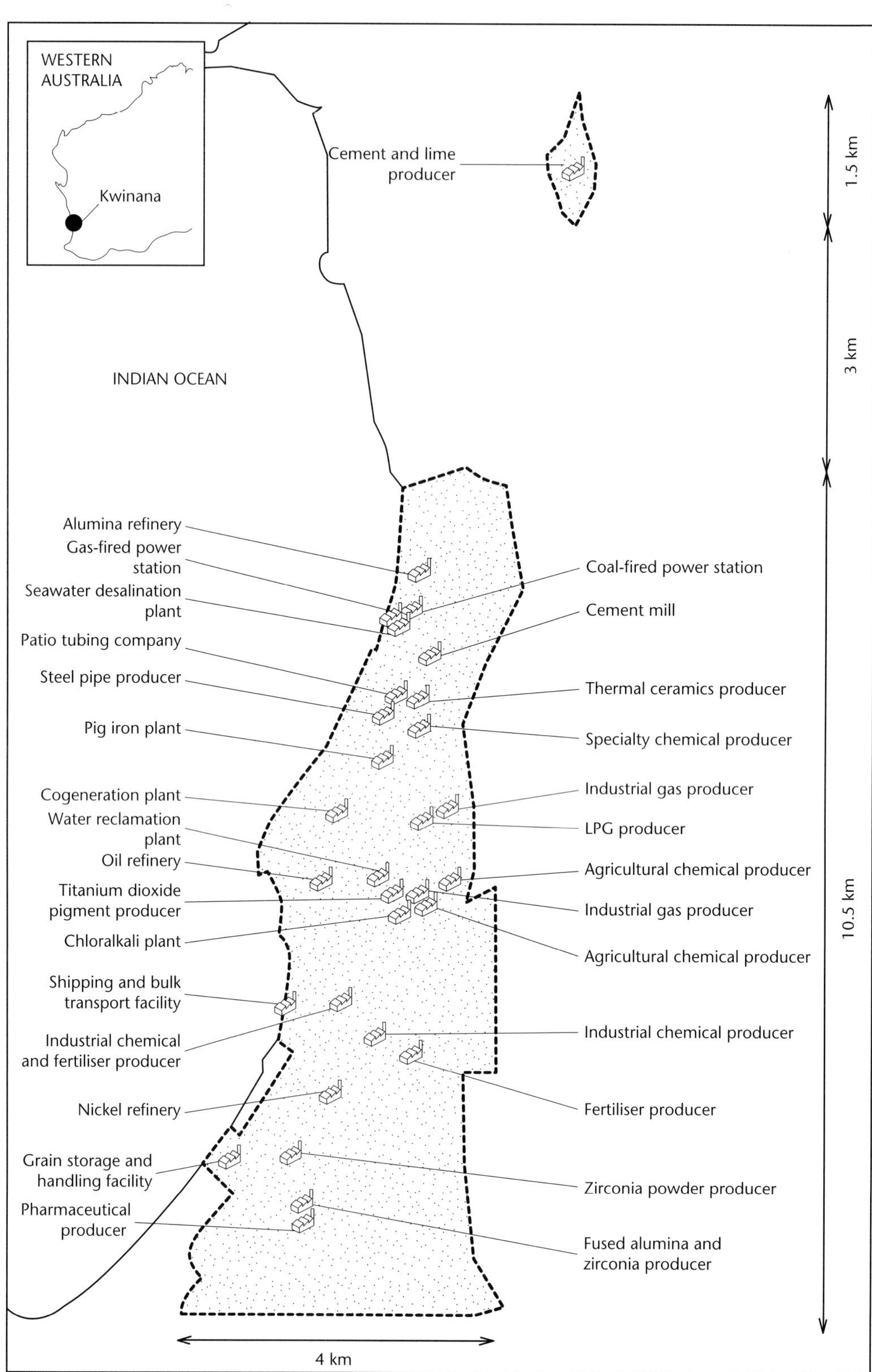

Figure 16.14: Location of the Kwinana Industrial Area, Western Australia, and its major industries (after van Beers, 2007).

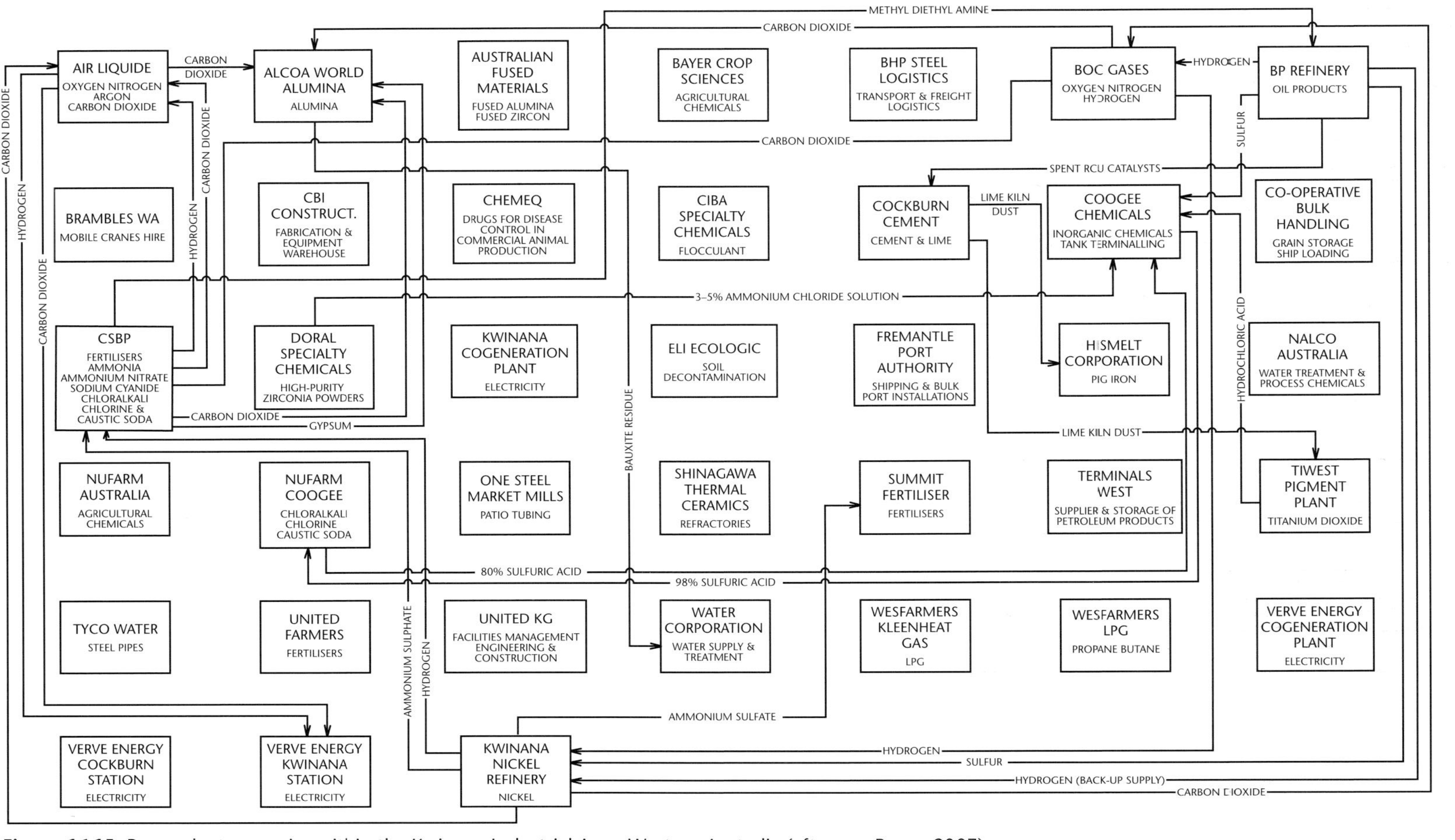

Figure 16.15: By-product synergies within the Kwinana Industrial Area, Western Australia (after van Beers, 2007).

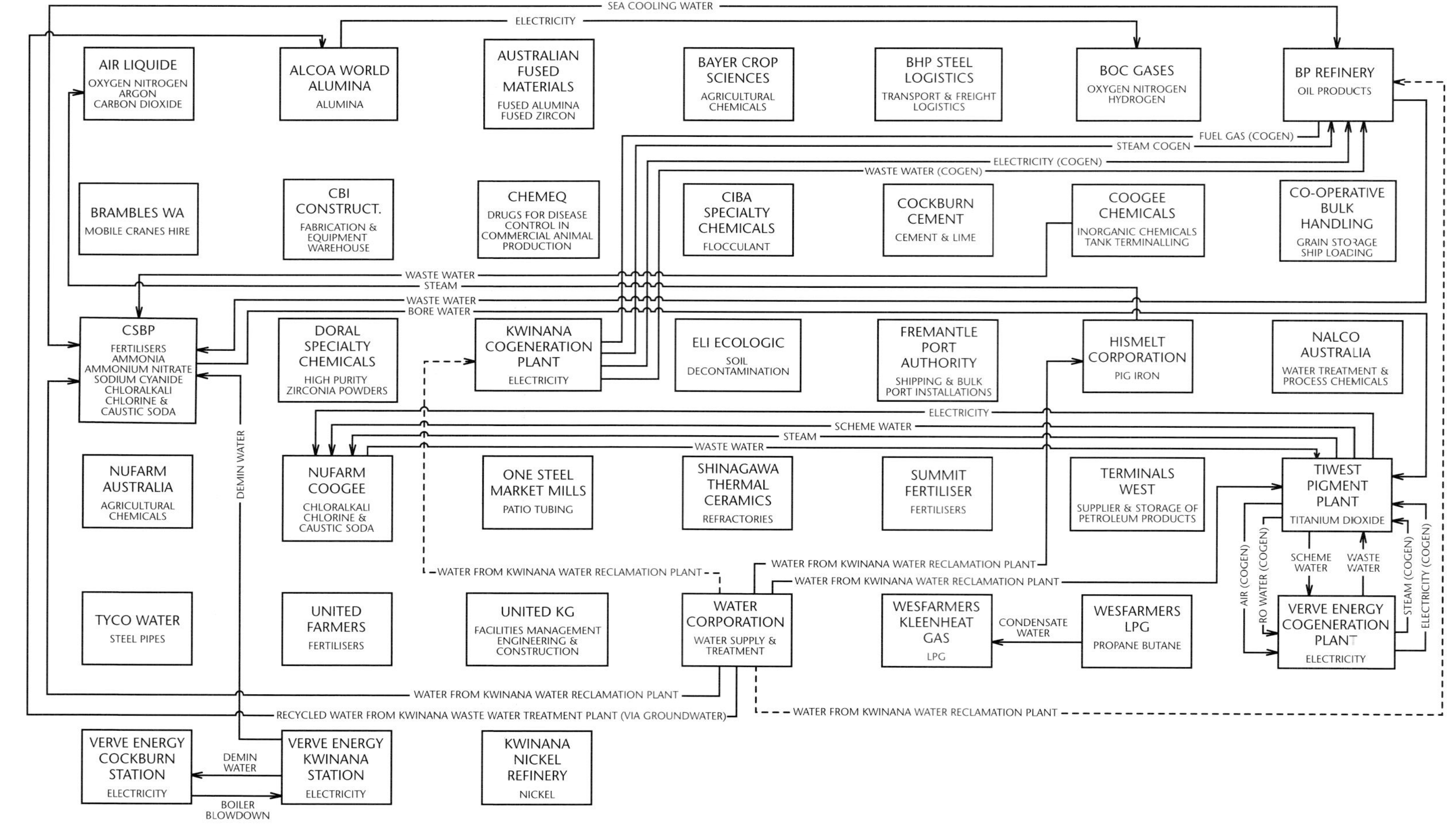

Figure 16.16: Utility synergies within the Kwinana Industrial Area, Western Australia (after van Beers, 2007).

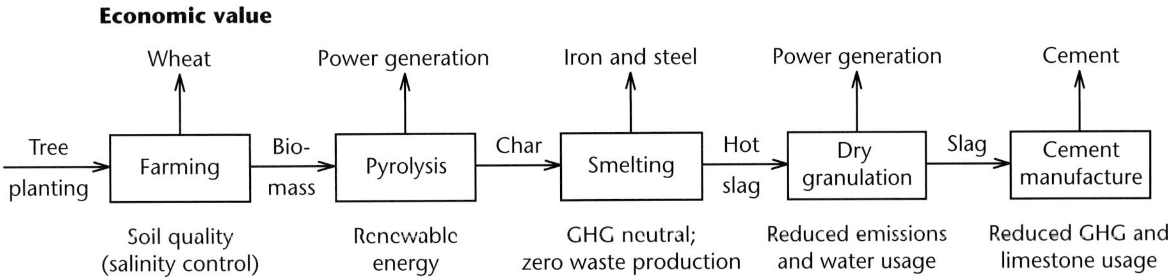

Figure 16.17: Concept for linking the KIA to agricultural activities (CSIRO Minerals Down Under Flagship; reproduced by permission).

reported for other heavily industrialised areas (van Beers, 2007). Forty-seven regional synergies have been identified. Thirty-two of these are by-product synergies and 15 involve shared use of utilities. The present synergies are illustrated in Figure 16.15 (for by-products) and Figure 16.16 (for utilities). These initially developed in a largely unplanned way in response to perceived business opportunities and environmental and resource efficiency considerations. A more coordinated approach to identifying and developing linkages was adopted with the formation in 1991 of the Kwinana Industries Council (KIC), an incorporated business association with membership drawn from the KIA.

Further possible synergies could link the KIA into the agricultural industry of the wheat belt to the east. This land was covered in forests before it was cleared for agriculture. The trees, through the action of transpiration, kept the watertable well below the surface; when the land was cleared and planted with shallow-rooting crops the watertable rose and gradually brought dissolved salts to the surface. Much of the soil is now contaminated with salt, and agricultural productivity has declined dramatically in recent decades. It is estimated that 2 million hectares of agricultural land in Western Australia has already been lost as a result of salinity and a further 100 000 hectares is being lost each year. One solution involves planting oil mallees, a hardy species of native eucalypt, in belts across fields. Crops could be planted between the tree belts. Every three to four years the trees would be cropped but they would continue to grow. Three products would be produced from the biomass: eucalyptus oil, charcoal/activated carbon and the residue leaves and twigs. The last would be burned to generate renewable electricity and the char could be used as a soil conditioner and CO_2 sequestering agent or as a reductant in metallurgical processes. A 20 000 tonne per year demonstration charcoal plant has been constructed at Narrogin, about 150 km east of Kwinana. Figure 16.17 shows how this could be linked into the KIA through the use of charcoal to replace coal or coke for iron- and steel-making.

16.7 MAKING IT HAPPEN

A number of barriers to the implementation of cleaner production (Moors *et al.*, 2005) and industrial ecology (Frosch, 1995) have been identified through case studies. The types of barriers are generally similar. They can be categorised as economic, technical, systemic, informational, organisational, and legal and regulatory.

Economic barriers. In the minerals sector, very large investments are needed for major changes in technologies. Established technologies have been refined over many years to be reliable and robust, and operations usually give good financial returns long after the capital costs have been depreciated. Introduction of new technologies introduces production risks which can, and often do, prove very costly. As discussed in Section 6.5.6, minerals companies are reluctant to introduce new technologies unless it can be done in an incremental way with minimum risk to overall production. Furthermore, the relatively low cost of disposal of mining and mineral processing wastes in most mineral resource rich countries is a disincentive to do anything other than discard them.

Technical barriers. The technical barriers are usually less important than financial and other barriers. Frequently, cleaner production technical solutions are available or are relatively easy to develop and implement, but are too costly, risky or difficult to implement for systemic, organisational or regulatory reasons. There are no shortages of ideas and possible approaches even for radically new technologies, many of which would, if pursued, lead to more efficient or cleaner processes. But developing the right technology at the right time and in such a way that it can be introduced with minimum risk to production has proved a challenge for both technologists and business strategists in the minerals industry.

Systemic barriers. There is a large degree of entwinement between minerals companies and other industry sectors such as power generation, infrastructure (roads, rail, ports) and suppliers of reagents and other consumables. Technological changes in one area have implications that flow through the entire system. The co-production of multiple products (due to the complex nature of many mineral deposits), and the need to sell these to different markets with differing and changing demand cycles, adds another layer of complication. The range of solid, liquid and gaseous wastes produced and the technologies for managing them is another complication. These combine to make mineral and metal production companies technologically complex. This constrains the changes that can be made easily, cheaply and with little risk. Moors *et al.* (2005) quoted a steel industry executive as saying, 'the most important barrier for implementation of new technology is the difficulty getting alternative new technologies within the existing infrastructure'.

Informational barriers. Most companies understand the technology and the markets for their primary products (e.g. concentrate or metal) but have little understanding of the technologies which could be used for adding value to their wastes or the potential markets for the by-products. An alumina producer understands the Bayer process and the market for alumina. However, if it wished to process red mud to produce materials for aggregate, for soil conditioning or to extract the iron to make steel, it generally lacks knowledge of the technologies needed and the markets for the products. It may not understand the regulatory requirements for materials produced from wastes. For a new product, such as geopolymer concrete or soil additives made from red mud, there has to be a first user who can accept the liability issues. One approach to overcome this is partnering with another company which has the necessary knowledge and expertise. Some steel companies do this in relation to slag. Coal-fired power stations are focused on producing energy and as a rule consider their secondary products (fly ash, bottom ash and heat) as waste – they view the effort to process and market these co-products a distraction from the core business. If addressed as a separate business, the co-products could be converted into useful products.

Organisational barriers. Organisational barriers are primarily cultural. Companies often perceive themselves as in the business of making steel or aluminium or copper. All other materials created in making their product are seen as wastes to be disposed of as cheaply as possible. Changing the culture of a company so that it perceives the resource in its entirety, not just part of it, as its greatest asset is a challenge which no minerals company has yet come near to tackling. Internal reward systems and the compartmentalisation of company activities discourage the interactions necessary to achieve the change.

Legal and regulatory barriers. The main regulatory issue in most jurisdictions is that regulations fail to promote closed loop systems and may actually discourage or prevent it. Of particular concern are regulations relating to the use of wastes or by-products as substitutes for virgin materials, and the assignment of liabilities. In many countries, defining a material as a waste or secondary raw material has consequences for what uses are permitted, what administrative procedures apply to its transport, export and processing, and what costs will be incurred (Pongrácz 2002). In some jurisdictions, a company that sells material classified as waste remains liable for any damages that may result from its use, even if it has been reused several times before the damage occurs (Frosch, 1995).

Governments have a leading role in helping and encouraging companies to overcome market failure barriers. They can do this by creating an environment that encourages adoption of cleaner production principles and that facilitates formation of industrial synergies, for example by:

- developing more appropriate regulations concerning wastes;
- entering into voluntary agreements with companies or industry associations on targets to achieve;
- applying market-based financial instruments such as tax concessions, taxes on emissions, emissions trading schemes and special purpose grants.

It is well established that connection (Boons, 2004; Rosenthal, 2000), communication (Wilderer, 2002) and collaboration (Chertow, 2000) between companies and other organisations enhance the likelihood that industrial symbioses will be established and maintained. The roles of the various stakeholders in establishing connection, communication and collaboration are crucial for success (Kurip, 2007). Particularly important stakeholders are local governments, environmental agencies, port authorities, regional and national government policy-makers, and community groups. Van Beers (2007) identified a number of drivers, barriers and triggers that were important in the creation of regional synergies in the Kwinana Industrial Area. These are listed in Table 16.5. Importantly, synergies can be triggered by a change of some kind, such as a proposed major new development or redevelopment, introduction of new environmental regulations, a study which identified an opportunity, or the actions of a champion who saw an opportunity. The possibility of gaining secure access to a vital resource can also be a trigger. Creation of triggers is therefore an important consideration in accelerating the move towards zero waste. Market-based financial instruments are likely to become triggers for major changes in the future.

Table 16.5: Drivers, barriers and triggers for regional synergies

Category	Drivers	Barriers	Triggers
Economics	Increased revenue through lower operational costs Reduced risks and liability	Relatively low price for utility resources Relatively low costs for waste disposal	Secure availability and access to vital process resources
Information availability	Local industry organisation Staff mobility	Confidentiality and commercial issues	Local and regional studies
Corporate citizenship and business strategy	Corporate sustainability focus Community engagement and perception	Core business focus	Industry champion
Region-specific issues	New company entering industrial area Geographic isolation	Distance between companies	Major new project developments
Regulation	Existing environmental regulations (e.g. air and water quality requirements and reporting)	Existing environmental regulations (intensive approval procedure for by-product reuse) Existing water and energy utility regulations	New pollutant targeted regulations (e.g. carbon tax and mandatory energy audits)
Technical issues	Research and technology developments Technical obsolescence of existing process equipment	Availability of (reliable) recovery technologies	Major brownfield development

Source: van Beers (2007)

Incremental progress on the reduction of wastes has been occurring for many years and it will continue under existing economic and regulatory regimes. However, the dramatic decrease required will not be achieved unless new approaches are developed. Two broad approaches are available to governments to influence the quantities of wastes produced and released into the environment:

- use of regulations (also known as command and control);
- use of market-based instruments.

Historically, governments have responded to community expectations for better environmental outcomes through regulatory responses. The regulatory approach often prescribes conditions for resource access and use. Mining regulations often specify the maximum allowable level of pollution, minimum requirements for mine-site rehabilitation and the type of management processes that should be used to reduce environmental damage. However, in many situations the regulatory approach has failed to achieve the goals or has proved very expensive. Market-based instruments (MBIs) for environmental management are relatively new but are increasingly used for the management of natural resources and the environment. They are financial instruments that encourage behaviour through market signals rather than through explicit directives (Stavins, 2001). MBIs have been used successfully to control NOx and SOx emissions in the United States. They are appropriate where regulatory approaches have failed to stop ongoing degradation or where the cost is prohibitive. The focus in applying MBIs is on achieving outcomes through the self-interest of companies and individuals. MBIs have two potential financial advantages over more traditional instruments (Whitten *et al.*, 2004):

- MBIs allow different companies to make different adjustments in response to their unique business structures and opportunities;
- MBIs provide companies with an incentive to discover cheaper ways to achieve outcomes.

Table 16.6 lists the available types of MBIs as price-based, rights-based and market friction. Price-based instruments alter the prices of goods or services to reflect their relative environmental impact. They provide certainty to industry, but the environmental outcome is less certain. A carbon tax is an example of a price-based MBI. Rights-based instruments can be used to control the quantity of an environmental good or service to a desired level. They provide certainty with respect to the environmental outcome but not with respect to the cost to industry. A cap-and-trade scheme for CO_2 emissions is an example of a rights-based MBI. Market friction instruments aim to stimulate a market to produce a desired environmental outcome through improving the workings of existing markets by reducing transaction costs or improving information flows. This has been the most commonly used MBI and is the easiest, politically, to introduce. There is often high resistance to the introduction of price-based and rights-based MBIs from interested industry sectors. The coal mining industry and its

Table 16.6: Types of market-based financial instruments

Price-based	Rights-based	Market friction
Emission charges User charges Product charges Performance bonds Non-compliance fees Subsidies (materials and financial) Removal of perverse subsidies/taxes Deposit–refund systems	Tradable permits, rights or quotas (cap-and-trade schemes) Offset schemes	Reducing market barriers Extension/education programs Research programs designed to facilitate market exchanges Labelling of products (eco-labelling) Information disclosure

Source: Whitten *et al.* (2004).

industry associations, for example, have lobbied extensively and successfully against carbon taxes and cap-and-trade schemes. Market-based approaches to transitioning to sustainability, and their potential limitations, are discussed further in Section 17.2.

16.8 REFERENCES

Ayres RA (1989) Industrial metabolism. In *Technology and Environment.* (Eds JH Ausubeland and HE Sladovich) pp. 23–49. National Academy Press: Washington DC.

Bamber AS, Klein B, Pakalnis RC and Scoble MJ (2008) Integrated mining, processing and waste disposal systems for reduced energy and operating costs at Xstrata Nickel's Sudbury operations. *Mining Technology* **117**(3): 142–153.

Basu AJ and van Zyl DJA (2006) Industrial ecology framework for achieving cleaner production in the mining and minerals industry. *Journal of Cleaner Production* **14**: 299–304.

Birat JP and Borlée J (2008) ULCOS: the European steel industry's effort to find breakthrough technologies to cut its CO_2 emissions significantly. In *Carbon Dioxide Reduction Metallurgy.* (Eds NR Neelameggham and RG Reddy) pp. 59–69. Minerals, Metals and Materials Society: Warrendale, PA.

Bommaraju TV, Lüke B, O'Brien TF and Blackburn MC (2004) Chlorine. In *Kirk-Othmer Encyclopedia of Chemical Technology.* 4th edn, Vol. 6. (Eds J Kroschwitz and M Howe-Grant) pp. 130–211. John Wiley and Sons: New York.

Boons F (2004) Connecting levels: a systems view on stakeholder dialogue for sustainability. *Progress in Industrial Ecology: An International Journal* **1**(4): 385–396.

Bruckard WJ and McCallum DA (2007) Treatment of sulfide tailings from base metal and gold operations – a source of saleable by-products and sustainable waste management. In *World Gold Conference*, Cairns, 22–24 November 2007. pp. 85–91. Australasian Institute of Mining and Metallurgy: Melbourne.

Chertow MR (2000) Industrial symbiosis: literature and taxonomy. *Annual Review of Energy and the Environment* **25**: 313–337.

Cooling D (2007) Improving the sustainability of residue management practices – Alcoa World Alumina Australia. In *Paste 2007.* (Eds A Fourie and RJ Newell) pp. 3–16. Australian Centre for Geomechanics: Perth.

Cox A and Fray DJ (2009) Electrolytic reduction of ferric oxide to yield iron and oxygen. In *Energy Technology Perspectives.* (Eds G Reddy, CK Belt and EE Vidal) pp. 77–85. Minerals, Metals and Materials Society: Warrendale, PA.

Cutmore NG and Eberhardt JE (2006) The future of ore sorting in sustainable processing. In *Green Processing 2006*. pp. 287–289. Australasian Institute of Mining and Metallurgy: Melbourne.

DEH (2000) Cleaner production. Department of Environment and Heritage, Commonwealth of Australia: Canberra, June 2000.

Doblin C and Wellwood GA (2008) TiRo™ – the development of a new process to produce titanium. In *CHEMECA 2007: Academia and Industry Strengthening the Profession.* pp. 280–286. Curran Associates: Red Hook, New York.

Ehrenfeld JR (2002) Industrial ecology: coming of age. *Environmental Science and Technology* **36**(13): 281–285.

Ehrenfeld JR and Gertler N (1997) Industrial ecology in practice: the evolution of interdependence at Kalundborg. *Journal of Industrial Ecology* **1**(1): 67–79.

European Parliament (2006) Directive 2006/21/EC of 15 March 2006 on the management of waste from extractive industries and amending Directive 2004/35/EC. *Official Journal of the European Union*, L 102/15, 11 April 2006.

European Parliament (2008) Directive 2008/98/EC of 19 November 2008 on waste and repealing certain Directives. *Official Journal of the European Union*, L 312/3, 22 November 2008.

European Union (1991) Directive 91/156/EEC of 18 March 1991 amending Directive 75/442/EEC on waste. *Official Journal of the European Union*, L 078, 26 March 1991.

Frosch RA (1995) Industrial ecology: adapting technology for a sustainable world. *Environment* **37**(10): 16–37.

Frosch RA and Gallopoulos NE (1989) Strategies for manufacturing. *Scientific American* **261**(3): 144–152.

Giurco D and Petrie JG (2007) Strategies for reducing the carbon footprint of copper: new technologies, more recycling or demand management? *Minerals Engineering* **20**: 842–853.

Graedel TE and Allenby BR (2003) *Industrial Ecology*. 2nd edn. Prentice Hall: New Jersey.

Halada K and Ijima K (2006) Hidden material flows by extraction of natural resources. In *Workshop on Material Flows and Environmental Impacts associated with Massive Consumption of Natural Resources and Products*, Tsukiba, Japan, 17 November, pp. 15–27.

Halmann M, Frei A and Steinfeld A (2007) Carbothermal reduction of alumina: thermochemical equilibrium calculations and experimental investigation. *Energy* **32**(12): 2420–2427.

Haueter P, Seitz T and Steinfeld A (1999) A new high-flux solar furnace for high-temperature thermochemical research. *Journal of Solar Energy Engineering – Transactions of the ASME* **121**(1): 77–80.

Hughes T and Cormack G (2008) Potential benefits of underground processing for the gold sector – conceptual process design and cost benefits. In *Future Mining 2008*. (Ed. S Saydam) pp. 135–142. Publication Series No. 10/2008. Australasian Institute for Mining and Metallurgy: Melbourne.

Jahanshahi S, Bruckard WJ and Somerville MA (2007) Towards zero waste and sustainable resource processing. In *International Conference on Processing and Disposal of Mineral Industry Waste 2007 (PDMIW'07)*, 14–15 June 2007, Falmouth, UK. CD-ROM, pp. 1–15.

Kurip B (2007) Methodology for capturing environmental, social and economic implications of industrial symbiosis in heavy industrial areas. PhD thesis. Curtin University of Technology: Perth, WA, December 2007. <http://www.kic.org.au/files/biji_kurupthesis08.pdf>; accessed 15 March 2010.

Kwinana Industries Council (2002) 'Kwinana industrial area economic impact study: an example of industrial interaction.' Kwinana Industries Council: Perth, WA. <http://www.kic.org.au>.

Mansfield K, Swayn G and Harpley J (2002) The spent pot lining treatment and fluoride recycling project. In *Green Processing 2002*. pp. 307–314. Australasian Institute of Mining and Metallurgy: Melbourne.

McCallum DA and Bruckard WJ (2009) The development of a diagnostic protocol for the assessment and testing of sulphide tailings. In *Enviromine 2009*. (Eds J Wietz and C Moran). GECAMIN: Santiago, Chile.

Mehta PK (2009) Global concrete industry sustainability. *Concrete International* **February**: 45–48.

Moors EHM, Mulder KF and Vergragt PJ (2005) Towards cleaner production: barriers and strategies in the base metals producing industry. *Journal of Cleaner Production* **13**: 657–668.

Murray JP (1999) Aluminum production using high-temperature solar process heat. *Solar Energy* **66**(2): 133–142.

Murray JP (2001) Solar production of aluminum by direct reduction: preliminary results of two processes. *Journal of Solar Energy Engineering* **123**: 125–132.

Murray JP, Steinfeld A and Fletcher EA (1995) Metals, nitrides and carbides via solar carbothermal reduction of metal-oxides. *Energy* **20**(7): 695–704.

Neelameggham NR (2008) Solar pyrometallurgy – an historical review. *JOM* **February**: 48–50.

O'Connor MT, Clout JMF and Nicholson RJ (2006) Dry particle separation – a contribution to sustainable processing. In *Green Processing 2006*. pp. 291–297. Australasian Institute of Mining and Metallurgy: Melbourne.

Pongrácz E (2002) Re-defining the concepts of waste and waste management: evolving the theory of waste management. Doctoral dissertation. University of Oulu, Department of Process and Environmental Engineering: Oulu, Finland. <http://herkules.oulu.fi/isbn9514268210>.

Rosenthal EC (2000) A walk on the human side of industrial ecology. *American Behavioral Scientist* **44**(2): 245–264.

Sadoway DR (2000) Electrochemical processing in molten salts: from green metals extraction to lunar colonization. In *Sixth International Conference on Molten Slags, Fluxes and Salts*, Stockholm and Helsinki, 12–17 June 2000.

Sosinsky DJ, Campbell P, Mahapatra R, Blejde W and Fisher F (2008) The Castrip® process – recent developments at Nucor Steel's commercial strip casting plant. *Metallurgist* **52**(11–12): 691–699.

Stavins RN (2001) Experience with market-based environmental policy instruments. Discussion Paper 01-58. Resources for the Future: Washington DC, November 2001.

ULCOS (2010) <http://www.ulcos.org/en/about_ulcos/home.php>.

UNEP (1996) Cleaner production – a training resource package. United Nations Environment Programme, March 1996. <http://www.unep.fr/shared/publications/pdf/WEBx0029xPA-CPtraining.pdf>; accessed 28 January 2010.

UNEP (2010) 'Understanding cleaner production.' United Nations Environment Programme. <http://www.unep.fr/scp/cp/understanding>.

USDE (2004) Summary of emerging titanium cost reduction technologies: a study performed for US Department of Energy and Oak Ridge National Laboratory. EHKTechnologies: Vancouver, Washington, January 2004. <www.ehktechnologies.com>.

van Beers D (2007) 'Capturing regional synergies in the Kwinana industrial area: 2007 status report'. Centre for Sustainable Resource Processing: Perth, Western Australia. <http://www.kic.org.au/files/70724_csrp__capturing_regional_synergies_in_the_kia__2007_report_final.pdf>.

Wechsler R and Ferriola J (2002) The Castrip® process for twin-roll casting of steel strip. *AISE Steel Technology* **79**(9): 69–74.

Whitten S, van Bueren M and Collins D (2004) An overview of market-based instruments and environmental policy in Australia. In *Market-based Tools for Environmental Management: Proceedings of the 6th Annual AARES National Symposium.* (Eds S Whitten, M Carter and G Stoneham). <http://www.ecosystemservicesproject.org/html/publications/docs/MBIs_overview.pdf>.

Wilderer MZ (2002) Eco-industrial parks and economic growth: analysis of a case study in India. *International Journal of Ecology and Environmental Sciences* **28**: 17–26.

Xiao YY and Fray DJ (2010) Molten salt electrolysis for sustainable metals extraction and materials processing – a review. In *Electrolysis: Theory, Types and Applications.* (Eds S Kuai and J Meng) pp. 255–301. Nova Science Publishers: New York.

Xie D and Jahanshahi J (2008) Waste heat recovery from molten slags. In *4th International Congress on the Science and Technology of Steelmaking (ICS2008)*, Gifu, Japan, October 2008. pp. 674–677. Iron and Steel Institute of Japan.

Xie D, Jahanshahi J and Norgate T (2010) Dry granulation to provide a sustainable option for slag treatment. In *Sustainable Mining 2010.* pp. 22–28. Australasian Institute for Mining and Metallurgy: Melbourne.

16.9 USEFUL SOURCES OF INFORMATION

Geopolymer Alliance. <http://www.geopolymers.com.au>.

Geopolymer Institute. <http://www.geopolymer.org>.

Provis JL and van Deventer JSJ (Eds) (2009) *Geopolymers: Structures, Processing, Properties and Industrial Applications.* CRC Press: New York.

Rao SR (2006) *Resource Recovery and Recycling from Metallurgical Wastes.* Elsevier: Amsterdam.

Waste Processing and Recycling in Mineral and Metallurgical Industries. A series of proceedings of conferences organised by the Canadian Institute of Mining, Metallurgy and Petroleum, Montreal, 1992, 1995, 1998, 2001, 2004, 2007.

17 Towards sustainability

We shall not cease from exploration
And the end of all our exploring
Will be to arrive where we started
And know the place for the first time.

T.S. Eliot, Little Gidding
(No. 4 of 'Four Quartets')

Materials produced from minerals are essential for society and will be essential during the transition to sustainability. What a sustainable world might be like is open to speculation; however, it is hard to imagine any acceptable form of sustainability in which mineral-based materials would not have an ongoing and important role. In the transition to sustainability, it will be necessary to close the materials cycle in ways that progressively reduce the demand for virgin minerals and reduce the quantity of wastes produced. How this can be done, and the mineral industry's role in achieving it, is the focus of this chapter. The chapter draws together a number of ideas developed in earlier chapters. The first part introduces the concept of stewardship and discusses its application to the transition to sustainability. The second part examines possible future directions, and the implications of those 'futures' for mineral resources and the minerals and metals sector of the global economy.

17.1 CLOSING THE MATERIALS CYCLE

The materials cycle (Figure 2.2) can be viewed as consisting of three major interrelated components:

- the resource from which materials are obtained;
- the materials themselves;
- the goods, products and infrastructure that contain materials.

This is illustrated in Figure 17.1. At any time there is a stock of each of these, which may increase or decrease. The *resource stock* consists of all the things and substances from which materials are derived. These are present in or on the Earth; for example, trees, coal, oil, gas, rocks and minerals. The *material stock* consists of all the materials used to make goods or provide services. Materials are produced from the resource stock; for example, timber from trees, metals from ore minerals and plastics from oil. The *product stock* consists of the goods and services used by society. These may be in present use or be obsolete or disused products. The material stock is present mainly within products that are in use but it also includes primary and secondary materials and materials in obsolete and disused products.

The sustainability challenge posed by non-renewable resources can only be addressed through

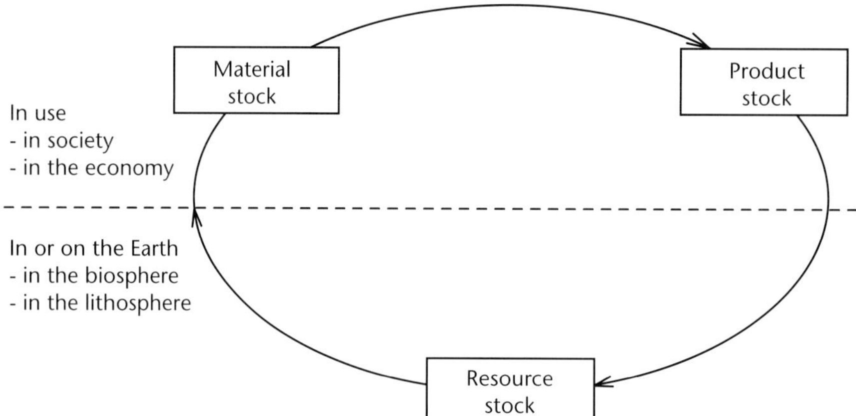

Figure 17.1: The interrelationship of the resource stock, material stock and product stock.

an integrated strategy for managing these stocks. Sometimes this is referred to as stewardship. *Stewardship* is the care and management of something throughout its life cycle. It has three aspects. What is being entrusted to care? To whom is it being entrusted? How will it be taken care of? In the context of the materials cycle, the stocks of resources, materials and goods are being entrusted to care. Hence, there are three types of stewardship to consider:

- resource stewardship;
- material stewardship;
- product stewardship.

Because of the different interpretations of sustainability, proposals for taking care of these stocks, and who should be responsible, vary widely. The Medveçka and Bangerter (2007) model of sustainability, summarised in Table 4.6, provides a framework for examining these proposals. Two widely known mineral industry stewardship models are the ICMM (ICMM, 2009) and Five Winds models (Five Winds, 2001). We start by examining these models, then consider a more integrated strategy.

17.1.1 The ICCM stewardship model

The ICMM stewardship model (ICMM, 2009) falls in the Compliers column of Table 4.6. The model uses the term materials stewardship as an all-encompassing term to include activities and actions which improve both the upstream processes that support the production of a material, and the downstream products that it goes into. It involves caring for and managing materials use along the entire materials cycle to:

- maximise benefit;
- minimise losses and risks;
- promote reuse and recycling;
- encourage the conservation of resources.

The model requires a life cycle management approach and involves risk assessment and risk management at all points in the materials cycle. It recognises that the mineral industry's capacity to act as stewards of the materials it produces varies along the materials cycle from aspects over which it has direct control to those which it does not manage directly. In the latter parts of the cycle, the ICMM proposes that stewardship be exercised through partnerships, provision of information and other routes of influence and support. Thus, the ICMM model requires the acceptance of complementary responsibility and the recognition of stewardship roles among all stakeholders.

Figure 17.2 illustrates the ICMM's approach, which divides materials stewardship into two aspects.

Process stewardship. This encompasses exploration, mining, processing, smelting and refining. These are the areas where the primary stewardship activities should be focused on efficiency, productivity of resources and minimising environmental, health and safety risks. This aspect of materials stewardship is where most mining companies have greatest direct control.

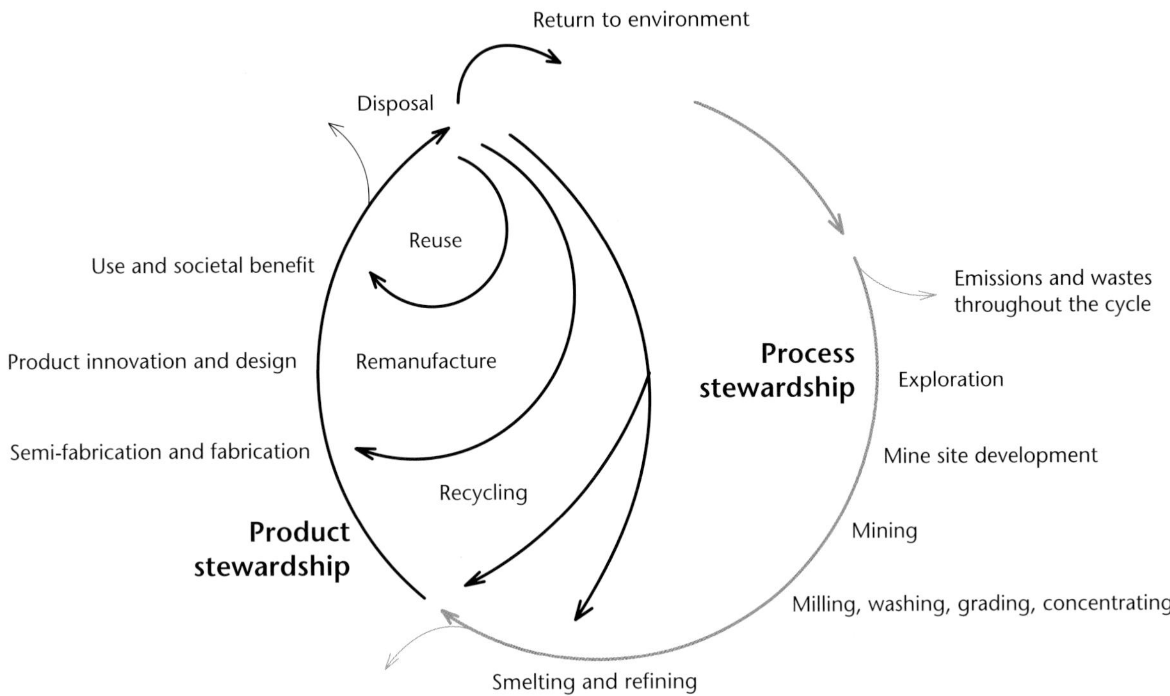

Figure 17.2: The process stewardship and product stewardship components of the materials cycle (after ICMM, 2006).

Product stewardship. As minerals and metals make their way into a wide variety of uses, stewardship responsibilities expand. These include supporting appropriate applications of minerals and metals and facilitating efforts towards recovery and reuse. This aspect of materials stewardship involves many more stakeholders. For example, choices made by product designers and engineers (e.g. selection of materials and processing technology) can significantly influence a product's overall environmental impact; these stakeholders therefore have an important role in implementing materials stewardship strategies.

Individual companies have varying degrees of control or influence at different points and this will affect their materials stewardship activities. A vertically integrated company involved in fabrication and product development as well as exploration, extraction and refining, will have many more opportunities to influence the design of products (e.g. to enhance materials recovery and recycling) than a company that sells its products as mineral or metallic commodities. The characteristics of specific minerals and metals will determine where materials stewardship activities need to be undertaken. A mineral or metal with hazardous properties will require more coordinated and structured stewardship activities than a more benign substance.

ICMM stewardship guidelines (ICMM, 2006) have been developed 'to support an industry facing the need to demonstrate its commitment to responsible performance of the whole materials cycle'. The guidelines are organised into four themes, each with a set of activities.

Taking a system perspective. This encompasses activities to gain a better understanding of material flows and life cycle impacts and benefits.

Building new, and strengthening existing, relationships. This involves activities to enhance interaction with other stakeholders in the materials cycle.

Optimising the production and application of minerals and metals. This involves activities to improve eco-efficiency at the process and product level.

Contributing to a robust, accessible base of information to support decision-making. This involves sharing and reporting data and information to identify and improve materials stewardship.

Adoption of the guidelines by ICMM member companies is voluntary and no activities are prescribed.

17.1.2 The Five Winds stewardship model

The Five Winds model falls predominantly in the Innovators column of Table 4.6. Its strategy is based on a consideration of the contributions of each major type of material to eco-efficiency by posing two related questions (Five Winds, 2001).

- How should the resource stock, material stock and product cycle be managed?
- How can the maximum value be achieved for each material type?

The principal types of commodity materials considered in the model are wood and paper, metals, and plastics. The proposed strategies for each are summarised in Table 17.1. The approach defines a primary strategy and supporting strategies for each material type, depending on the characteristics of the material. The primary strategy is anchored on the basic value associated with each resource–material–product system and focuses on the potential maximum value over time for each type of material. The primary strategy provides guidance on the actions needed to maximise the value of the resource–material–product system, while promoting eco-efficiency.

For metals, the primary strategy is to maximise the utility of the metallic element. The rationale is that metals, being elemental, are indestructible. In non-dissipative applications, they are potentially indefinitely recyclable if they do not become contaminated. This leads to a stewardship strategy aimed at maximising the value of the material stock of metals through reuse and recycling. The strategy recognises that the material stock is produced from a finite resource stock and relies on proper management of product cycles to ensure continual recovery and preservation of the value of metal. For wood and paper, the primary strategy is to maintain the integrity of the living resource

Table 17.1: Five Winds management strategy for the resource, material and product stock for the principal material groups (primary strategy in italics)

Material type	Resource stock	Material stock	Product stock
Wood and paper	*Manage stock to maintain integrity of the living resource stock* Extract without compromising the sustainability of the resource stock and promote replenishment	Recover and recycle, where appropriate, to achieve multiple lives and maximum value from each fibre Recover energy value or biodegrade to recover biological resource	Be efficient Return end-of-life products for nutrient recovery in support of the resource stock Manage risks associated with the product cycles
Metals	Extract metal resources wisely and efficiently Extract and process efficiently, to maximise material value	*Manage stock to maximise utility of the metal element* Encourage material recycling to conserve material stocks for future generations	Prevent metal contamination and dispersion Design product and systems for maximum value, and efficient material recovery Manage risks associated with the product cycles
Plastics	Extract wisely, based on need Extract and process efficiently, to maximise resource value Preserve oil and gas stocks to meet the needs of future generations	Maximise the utility and versatility of molecules for the product cycle Recover and recycle, where appropriate, to achieve multiple lives and maximum value from each molecule Recover energy where appropriate	*Manage stock to maximise value of the product* Ensure efficient, safe and responsible use of the product Design products for efficiency Manage risks associated with the product cycles

Source: Five Winds (2001).

stock since it is a renewable resource and wood and paper progressively degrade through repeated recycling. For plastics, the primary strategy is to maximise product value by taking advantage of their versatility in application in different product cycles. This recognises the need to ensure that products are used efficiently and responsibly because the resource from which they are made is finite. The primary strategy for each material type focuses on one of the three stocks; secondary strategies support the primary strategy. For example, recycling of metals depends on product recovery; a supporting strategy is to design products to ensure easy and economic recovery of the material for continued use in the material stock.

17.1.3 An integrated strategy for the minerals and metals sector

Adopting stewardship strategies for each of the three forms of stock, as in the Five Winds model, is a starting point for developing an integrated strategy. There is already a degree of integration between the three forms, which needs to be recognised and built on. Many primary metal smelters treat wastes of various kinds, including scrap metals and more complex materials, and combinations of materials (e.g. waste electrical and electronic equipment). Cement production is often highly integrated with other industry sectors; it utilises not only slag from the steel industry but a wide range of secondary materials, both as fuel and as raw material inputs. Greater integration, both within the minerals industry and between the minerals industry and other industry sectors, is possible and the potential sustainability benefits are great.

By building on the potential to integrate across the three forms of stock, a sustainability strategy which aligns more closely with the Marketeers column of Table 4.6 can be developed. The overall objectives of such a strategy would be to:

- maximise resource use efficiency (since the resources are finite) through extracting maximum value from mined products and causing no long-term environmental impact;
- maximise the materials use efficiency through substitution, reuse and recycling;
- maximise the end-use efficiency through Design for the Environment.

Table 17.2 lists the important strategies for each stock which assist in achieving these objectives. This creates a framework for many of the ideas and strategies identified in earlier chapters. Most of the concepts will already be familiar.

Table 17.2 refers to the natural resources hierarchy, a concept not previously discussed. It is illustrated in Figure 17.3, which shows mineral and fossil fuel resources arranged in a hierarchy which reflects their preferred order of use. Wastes from the beneficiation or combustion of higher-value resources are at the base of the hierarchy. The next level consists of construction materials which (from the geological point of view) are virtually unlimited in abundance, and those that are present in practically unlimited quantities in the oceans (e.g. magnesium and potassium). Non-energy resources which are extracted from deposits formed by natural enrichment processes, such as metals and some industrial minerals, follow these. The energy resources (coal, oil, gas and uranium), which are the most valuable, are at the top. To increase resource use efficiency, the lowest-value resource possible should be used in any application. For example, fly ash from coal-fired power stations should be used in cement and concrete as a substitute for primary cementitious raw materials from the second and third levels. Use of non-energy resources in place of energy resources may seem impossible, but it can be achieved by using non-energy resources in products to reduce their energy consumption. For example, replacing steel components in vehicles with lighter materials (aluminium, magnesium and plastics) can reduce the overall consumption of energy on a life cycle basis and conserve energy resources (Figure 15.5). Other examples are the double- or triple-glazing of windows and double-wall construction techniques to reduce energy demand for heating and cooling. Glass, bricks and concrete for these purposes are manufactured from quartz, limestone and clays, which are resources from the third level. Although additional energy would be needed to manufacture the extra glass and construction materials, the long-term energy saving would be considerable.

Integration of the strategies across the three forms of stock to achieve a better overall outcome can be achieved by using technologies and business strategies

Table 17.2: An integrated stewardship strategy for minerals and metals

Resource stock	Material stock	Product stock
Objectives		
Maximise resource use efficiency through extracting maximum value from mined products and causing no long-term environmental impact.	Maximise materials use efficiency through substitution, recycling and reuse	Maximise end use efficiency through Design for the Environment
Strategies		
Mine the least quantity of material and extract the most value from it as products and co-products (including from tailings, leach residues and slags). It follows that the smallest possible quantity of material should be discarded. This implies a high degree of mine site processing. Disturb land and the environment during mining to the least extent possible and restore land to its natural state when mining ceases Minimise the direct and indirect usage of energy and water, and the production of wastes, through technical innovation and the application of eco-efficiency and industrial ecology principles. Prevent environmental dispersion of toxic substances from mining and processing operations (toxic elements in ore, toxic reagents and toxic products) so there is no systematic build-up in the biosphere. Close every mine so it can be safely reopened, if potentially valuable minerals remain unmined, or close it in such a manner that it will never become an environmental liability.	Use each material in applications only where its characteristics make it the most appropriate material bearing in mind its eco- and human-toxicity effects, relative scarcity and recyclability. Use materials according to the natural resources hierarchy (Figure 17.3) Reuse and recycle materials in closed loop systems that maintain their quality (this keeps materials in circulation indefinitely and prevents dispersion of hazardous materials and contamination of the environment) Reuse and recycle materials in ways that prevent their loss and the dispersion of toxic elements	Design products which have the minimum materials content required to achieve their function (dematerialisation) Substitute services for goods wherever possible (dematerialisation) Design products to maximise the scope for remanufacture, reuse and recycling Provide systems for: – collecting all used products; – decommissioning all infrastructure; – separating and pre-sorting materials before reusing or recycling Landfill and incineration should be used as a last resort, and only for benign substances
Integrating strategies		
Industrial ecology; management of toxic elements along value chain; mine site processing; vertical integration of companies; recycling of materials; dematerialisation; Design for the Environment		
Tools and methodologies		
Material and substance flow analysis; life cycle assessment		
Drivers		
Shifts in societal values; national and international regulations concerning wastes; market-based instruments aimed at internalising environmental and social costs associated with the materials cycle		

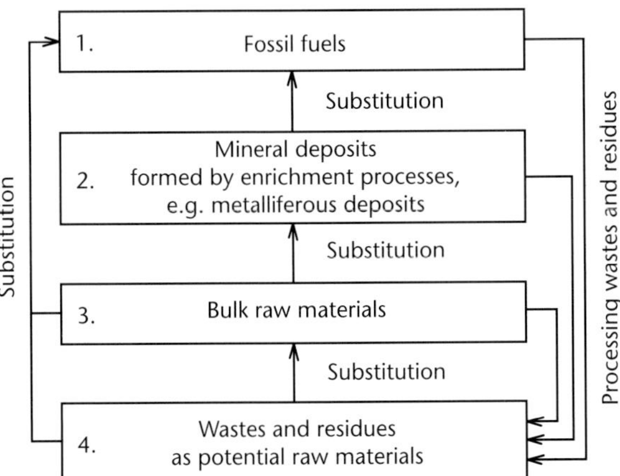

Figure 17.3: The natural resources hierarchy (after Wellmer and Becker-Platen, 2001).

that link aspects of two or more stocks in complementary ways. Some of these are described below.

Industrial ecology. This is potentially the most powerful linking strategy. Industrial ecology can integrate across industry sectors, commodities, products, wastes etc. Examples have been given previously.

Recycling. Mention has been made of the integration of primary smelters with recycling of metals. Cement kilns in urban areas have an invaluable role in consuming or recycling many secondary and waste materials that would otherwise be sent to landfill.

Vertical integration within companies. Vertically integrated companies have much scope for managing the use of primary metals and other commodities through the materials cycle and for optimising processes along their value-adding chain (ICMM, 2009). This was formerly common among large mining companies, but the advantage has been lost by most multinationals through the modern business strategy of focusing on the so-called 'core business' of mining. As a result, many large multinational companies have become less able to respond to the demands of the modern world, to move into areas of technological growth and to identify new business opportunities.

Mine site processing. The integration of mining, milling and smelting/leaching by optimising the combined value chain rather than optimising each step independently is easier when the three stages are performed at the one site. Wastes can be more easily managed and opportunities for new industries, based on extracting value from tailings, slags and other co-products, can be better realised. As with vertical integration, there has been a trend away from this in recent decades. Most of the large mining companies now prefer to do only the minimum processing needed to produce a tradable commodity (usually a concentrate).

Management of toxic elements. The integration of mining, milling and smelting/leaching provides the opportunity to determine the best stage or stages in the value chain at which toxic elements should be removed or managed in order to prevent their release and dispersion into the environment. For example, arsenic is an impurity in most base metal sulfide mineral deposits. Because specifications for traded concentrates require a relatively low upper limit for arsenic, arsenic usually has to be rejected at the concentrator. Rejection of arsenic-containing copper minerals (enargite, Cu_3AsS_4) in tailings, however, reduces the overall recovery of copper. A better approach might be to produce a high arsenic concentrate, with a resulting higher copper recovery, and remove the arsenic as fume during smelting. This would have the added advantage of eliminating most of the arsenic in one highly concentrated stream, which could be stabilised and returned to the mine. This method would be possible only if the smelter and concentrator were co-located.

Design for the Environment. DfE is a methodology for product design that aims to reduce the overall environmental impact of a product across its life cycle (Section 13.10).

Because of the high degree of integration across the columns in Table 17.2, the identification of opportunities and development of approaches for particular companies, commodities and industry sectors requires the use of tools and methodologies that cover the entire materials cycle, not just parts of it. These include material flow analysis, life cycle assessment, the ecological footprint, the water footprint, hidden material flows and others.

17.1.4 Drivers of stewardship

The ICMM lists the following as some reasons for the minerals and metals industry to adopt a form of materials stewardship (ICMM, 2006). These clearly fall within the Compliers column of Table 4.6.

Public commitments. The ICMM and its members have committed to a sustainable development framework. It is claimed that meeting this commitment will help secure credibility with external and internal stakeholders and ensure continued access to capital, land and mineral resources (the social licence to operate).

Competition and market access. A growing number of companies in the electronics, automotive, heavy manufacturing, and building and construction sectors are trying to avoid selecting materials that create undue environmental or social costs and risks. In some cases, companies seek materials that contribute to the environmental performance of their products or projects through their life cycle. The evaluation of environmental performance of materials will increasingly take into consideration the source of the materials (e.g. practices at the site where the material was produced).

Regulation. Global regulatory initiatives (the Basel Convention, REACH) aim to decrease the impacts of mining operations, reduce the quantity of material going to landfill, and direct the management and recycling of materials. This focus will increase.

Cost savings. Materials stewardship can improve resource productivity, increase mine efficiency, inform risk management, improve processes and eco-efficiency and contribute to other cost-saving measures.

Market development. Future markets will favour materials that are harvested, extracted, produced, used, recovered and reused with minimal negative impacts on the environment and society.

It is widely accepted that a combination of increased regulation and use of market-based financial instruments (cap-and-trade schemes, carbon taxes) will drive stewardship initiatives in the next decades. Already, extended producer responsibility, integrated product policy and other product-focused policies are being used to assign responsibility to companies along the materials cycle to manage environmental impacts and optimise resource recovery and recycling. Schemes such as the REACH Directive in Europe seek to limit the impacts of toxic materials on human health and the environment. Regulatory controls on end-of-life vehicles and waste electrical and electronic equipment have been introduced in a number of countries. This pressures manufacturers and material suppliers to consider how materials selection will facilitate easier recycling. Directives originating in the European Union are becoming the standard for the development of legislation in other countries. The directives also have a broad impact on global product design, as manufacturers recognise that the trends being established in Europe are spreading around the world. Will these trends be sufficient to achieve the required transition to sustainability? This is examined in the next section.

17.2 MARKET- AND POLICY-BASED APPROACHES TO TRANSITIONING TO SUSTAINABILITY[28]

Neo-liberal economics has been the dominant ideology of influential international institutions, politicians and intellectuals since the 1970s. It envisages a globally integrated, free market created by removing trade barriers, building market-enabling institutions and spreading the western model of development. The risk in continuing to pursue this ideology is that the globalised market may succumb to the large-scale degradation of the environment and collapse of ecosystems, economic instability, social polarisation and cultural conflict that seem inherent in the neo-liberal economic system. It is widely believed that corrections to a free market occur through the guidance of Adam Smith's 'hidden hand' (Smith, 1776). For example, environmental scarcity will result in higher prices that reduce demand and in business opportunities that promote technological innovation and materials substitution. However, to believe that self-correcting mechanisms will provide adjustments sufficiently quickly and on a sufficient scale is a matter of faith, unsupported by empirical analysis or historical experience.

Another widely held belief is that free-market forces alone can provide the social basis for sustainability and that economic growth will reduce poverty, improve international equity and reduce conflict. Again, the empirical evidence for this is weak. The experience in developed countries over the past two centuries is that income redistribution and social welfare programs are needed to ameliorate the dislocations and impoverish-

28 The ideas presented in this section are based on the work of Raskin *et al.* (2002).

ment caused by market-driven development. It is more likely that global poverty would persist as population grows and skewed income distributions counteract the poverty-reducing effect of the growth in average income. Even if market forces were able to deliver a stable global economic system, there is little reason to believe it would result in a sustainable world or dramatically reduce human deprivation. The economic and social polarisation is more likely to weaken social cohesion and make liberal democratic institutions more fragile. Resource demand and environmental degradation are likely to increase domestic and international tensions. Thus, market forces could, in the long run, undermine economic stability by weakening ecological resilience and social cohesion.

The alternative to an unrestrained free market is policy reform. Policy reform proponents seek a transition to sustainability by constraining the free market within politically imposed social and environmental targets. Policy reform is based on the belief that:

- environmental degradation is not a necessary outcome of development and that it can be mitigated by better choices of technology, resources and production processes;
- the cumulative effects of targeted incremental adjustments can make a substantial difference;
- poverty and extreme inequality are not inevitable, but result from social policy choices.

The policy reform path can only succeed if there is an unprecedented global commitment by governments to achieving sustainability, expressed through effective and comprehensive economic, social and institutional initiatives. There is no evidence of the political will for such reform. This was powerfully demonstrated at the Copenhagen Climate Change Conference of 7–19 December 2009. The conference brought together 120 heads of state and government and involved the participation of 193 countries. It failed, despite years of preparation and negotiation, to reach a significant and binding agreement on CO_2 reduction targets. The accumulation of wealth and the concentration of power erode the political basis needed for a transition to sustainability. The values of consumerism and individualism tend to undermine support for policies that promote long-term environmental benefits and social well-being. When the dominant interests of major constituencies and influential power-brokers are short-term, politicians remain focused on the next election rather than on long-term issues.

Market force and policy reform, alone or in combination, may be insufficient to bring about the scale and rate of change that are needed to transition to sustainability. This is explored in the next section.

17.3 WHAT DOES THE FUTURE HOLD?

It is not possible to predict the future, but it is possible to identify a range of futures, or scenarios, which can be used to explore the implications of decisions and actions (or lack thereof) and to help identify desirable goals and pathways to them. Two major exercises in scenario planning are reported in the 'Great Transition' and 'Vision 2050' reports, the former from an environmental organisation and the latter from the World Business Council for Sustainable Development. While quite different in approach, their conclusions are not dissimilar – a sign that some common ground is emerging. Both studies conclude that fundamental changes in governance structures, economic frameworks, and business and human behaviour are necessary to achieve the transition to sustainability.

17.3.1 The 'Great Transition' scenario

The 'Great Transition' (Raskin *et al.*, 2002), a report by the Global Scenario Group,[29] examined a number of scenarios for the future based on historical trends. It identified three periods of human history during which fundamental transformation of society occurred. They were the transition from a Stone Age culture to early civilisation around 10 000 years ago, the transition from early civilisation to the modern era over the past 1000 years, and a third transition, in progress, from the modern era to the planetary phase. The study argued that these transitions exhibit a pattern of development through sequences of quasi-stability,

29 The Global Scenario Group was convened jointly by the Stockholm Environment Institute and Tellus Institute to engage in scenario development for transitioning towards sustainability. It was succeeded by the Great Transition Initiative (http://www.gtinitiative.org).

rapid chaotic change, and restabilisation. It examined how features of past transitions have evolved, and projected the trends to describe the emerging properties of the planetary phase. Social complexity, technological complexity, spatial connectedness and pace of change have increased from one transition to the next and are leading to the globalisation of the world economy, global governance, and an information and communication revolution. The 'Great Transition' examined three classes of global 'futures'.

Conventional worlds. It is assumed that the global system in the 21st century will evolve without major surprises, sharp discontinuity, or fundamental transformation in the basis of human civilisation, that the dominant forces and values currently driving globalisation will shape the future, and that incremental market and policy adjustments will be able to cope with social, economic and environmental problems.

Barbarisation. It is assumed that social, economic and environmental problems will not be appropriately managed. Instead, they will cascade into self-amplifying crises that will overwhelm the coping capacity of conventional institutions. As a result, civilisation will descend into anarchy or tyranny.

Great transitions. It is assumed that transformations will occur in the fundamental values and organising principles of society. New values and development paradigms will emphasise quality of life and material sufficiency, human solidarity and global equity, and affinity with nature and environmental sustainability.

There are two variants of each scenario, as summarised in Table 17.3. The *Conventional worlds* variants are *Market Forces* and *Policy Reform*. In Market Forces, competitive, open and integrated global markets drive world development. Social and environmental concerns are secondary. In contrast, Policy Reform assumes that comprehensive and coordinated government action is initiated for poverty reduction and environmental sustainability. The *Barbarisation* variants are *Breakdown* and *Fortress World*. In Breakdown, conflict and crises spiral out of control and institutions collapse. Fortress World is an authoritar-

Table 17.3: Archetypal world views

World view	Antecedents	Philosophy	Motto
Conventional worlds			
Market Forces	Smith (1776)	Market optimism, hidden and enlightened hand	Don't worry, be happy
Policy Reform	Keynes (1936), UN (1987)	Policy stewardship	Growth, environment, equity through better technology and management
Barbarisation			
Breakdown	Malthus (1798)	Existential gloom, population/resource catastrophe	The end is coming
Fortress World	Hobbes (1651)	Social chaos, nasty nature of man	Order through strong leaders
Great transitions			
Eco-communalism	William Morris and social utopians (19th century), Schumacher (1972), Gandhi (1993)	Pastoral romance, human goodness, evil of industrialism	Small is beautiful
New Sustainability Paradigm	Mill (1848)	Sustainability as a progressive global social evolution	Human solidarity, new values, the art of living

Source: Raskin *et al.* (2002); with modifications.

ian response to the threat of breakdown, as the world divides. The elite would live in interconnected, protected enclaves and the impoverished majority would live outside. The two *Great transition* variants are *Eco-communalism* and *New Sustainability Paradigm*. Eco-communalism is a vision of bio-regionalism, localism, face-to-face democracy and economic independence. It corresponds approximately to the Localisers of the Medveçka and Bangerter (2007) model (Table 4.6). The New Sustainability Paradigm would see a change in the character of global civilisation rather than a retreat into localism. It would embrace global solidarity, cultural cross-fertilisation and economic connectedness while seeking a liberating, humanistic and ecological transition to sustainability. Figure 17.4 shows qualitatively the likely trends of selected varia-

Table 17.4: Implications for the minerals and metals sector of the global scenarios proposed by Raskin *et al.* (2002)

Conventional worlds	
Market Forces	Commodity markets grow according to business-as-usual predictions, with market forces determining supply and demand
	Increasing consumption generates further throughput of natural resources and associated environmental impacts
	Environmental scarcity would increase prices and market demand for businesses supporting technological innovation, resource efficiency and resource substitution
Policy Reform	Commodity markets may grow or shrink (e.g. for commodities such as fossil fuels), as regulation and market-based instruments impose full cost accounting and internalisation of environmental and social costs
	Wise policy on resource efficiency, renewable resources and environmental protection mitigate social and environmental impacts
	Interventionist public policies (e.g. industry development, protectionism) may drive a shift towards dematerialisation, backstop technologies and strengthened local and regional markets relative to global markets
Barbarisation	
Breakdown	Trade in commodities breaks down and local and global markets collapse
Fortress World	Commodity production and exchange systems break down as capacity to produce and trade declines, and a permanent siege and conflict environment decreases reliable access to resources
Great transition	
Eco-communalism	Commodity production decreases as scale of production is geared to local rather than global demand
	Global markets give way to multiple local markets
	Shift towards appropriate technology decreases demand for minerals and metals, especially rare metals and energy-intensive metals
New Sustainability Paradigm	Commodity production and resource consumption decrease to fit local and global bio-capacity (guided by adoption of sustainability principles and shift to industrial ecology)
	Rapid diffusion of environmentally benign technology and a shift to less materially intensive life styles
	A transition shaped by new values
	Energy transition prompts an age of renewable technology
	Materials transition instigates a reduction in resource throughput and phasing out of toxic materials
	Agricultural transition instigates increased reliance on ecological farming

Source: Giurco *et al.* (2009).

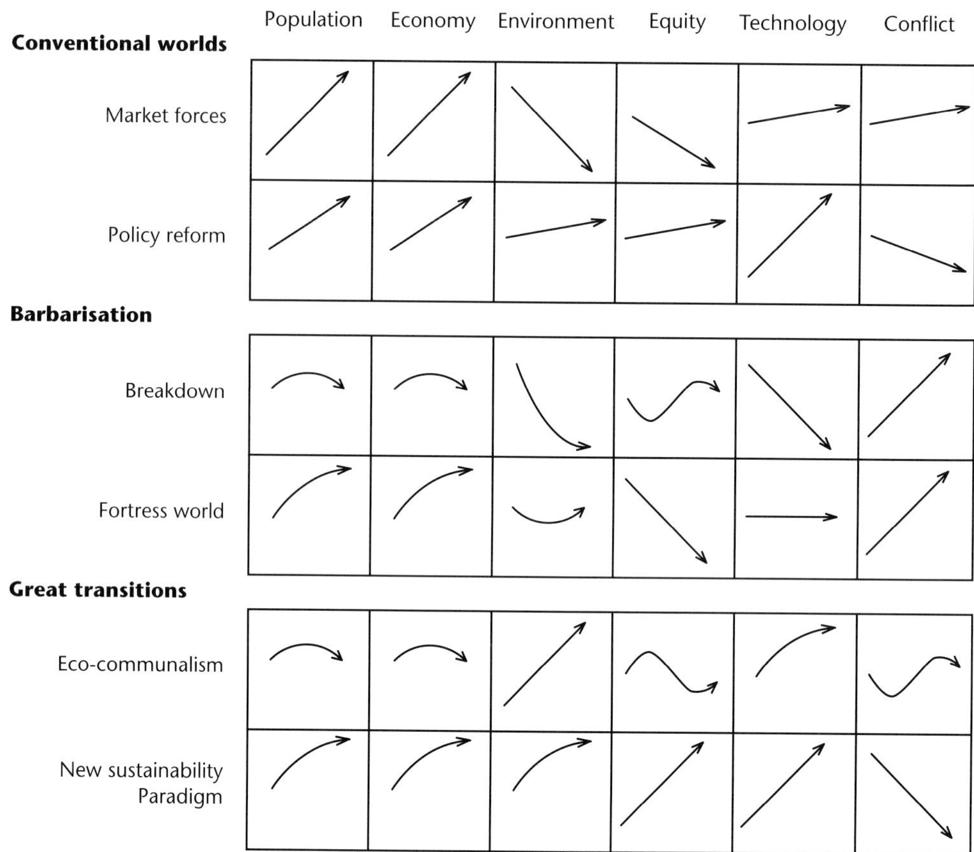

Figure 17.4: The six global scenarios proposed by Raskin *et al.* (2002). Arrows signify trends.

bles for each of the six scenarios. The potential impact of the scenarios on the minerals and metals sector has been examined by Giurco *et al.* (2009); their conclusions are summarised in Table 17.4. The impact on the minerals and metals production part of the materials cycle varies considerably – from growth in production and consumption along business-as-usual lines to decrease in demand (to varying degrees) and breakdown of the global market for higher-value mineral and metal commodities.

The Great Transition identified the New Sustainability Paradigm as the preferred and likely path to the future. It builds on the wealth-generating features of Market Forces and the technological changes of Policy Reform, and aligns with the Marketeers column of Table 4.6. However, it holds that these alone are not sufficient. It adds a third driver, a values-led shift towards an alternative global vision in which civil society and non-governmental organisations will play a leading role. A number of social trends which may already be driving the transition to sustainability include:

- anxiety about the future *versus* the promise of security and stability;
- concerns that policy adjustments are insufficient to avoid crises;
- the fear of loss of freedom and choice *versus* the opportunity to participate in community, political and cultural life;
- stressful life styles *versus* the desire for time for personal endeavours and stronger connection to nature.

17.3.2 The World Business Council for Sustainable Development scenario

There is a growing understanding, by some industry sector leaders, of the urgent need to transition to sustainability and industry's role in this. 'Vision 2050', the report of a project of the World Business Council for Sustainable Development (WBCSD, 2010) exemplifies this. The project examined three questions. What does a sustainable world look like? How can we realise it? What are the roles business can play in ensuring more rapid progress toward that world? Twenty-nine WBCSD companies contributed to develop a vision of a world on the way to sustainability by 2050, and a pathway leading to that world. The vision is for 'a planet of around nine billion people, all living well – with enough food, clean water, sanitation, shelter, mobility, education and health to make for wellness – within the limits of what this small, fragile planet can supply and renew, every day'. The pathway involves fundamental changes in governance structures, economic frameworks, and business and human behaviour. These are considered to be necessary to achieve the transition, to be feasible, and to provide business opportunities for companies that put sustainability principles into practice. The pathway involves the following strategies:

1 addressing the development needs of billions of people, enabling education and economic empowerment, particularly of women, and developing radically more eco-efficient solutions, life styles and behaviour;
2 incorporating the cost of externalities, starting with carbon, ecosystem services and water;
3 doubling agricultural output without increasing the amount of land or water used;
4 halting deforestation and increasing yields from planted forests;
5 halving carbon emissions worldwide (based on 2005 levels) by 2050, with greenhouse gas emissions peaking around 2020, through a shift to low-carbon energy systems and highly improved demand-side energy efficiency;
6 providing universal access to low-carbon forms of transportation;
7 delivering a four- to 10-fold improvement in the use of resources and materials.

Vision 2050 and the pathway are consistent with the Innovators column of the Medveçka and Bangerter (2007) model of sustainability (Table 4.6). Strategies 5 and 7 clearly have major implications for the minerals industry. Strategy 7, in particular, needs to be addressed by the approaches proposed in Table 17.2.

17.4 SUMMARY AND CONCLUSIONS

The challenges posed by non-renewable mineral resources can be addressed within the broader sustainability context through an integrated stewardship strategy for managing the stocks of resources from which materials are obtained, the materials themselves, and the goods, products and infrastructure that contain materials. However, because of the widely differing interpretations of sustainability (Table 4.6), there are wide variations in the proposed approaches for taking care of these stocks, and for who should be responsible. The ICMM stewardship model falls in the Compliers column of Table 4.6 while the Five Winds model falls predominantly in the Innovators column. The latter strategy is based on consideration of the contributions of each major type of material to eco-efficiency. A sustainability strategy which aligns more closely with the Marketeers column of Table 4.6 would involve greater integration across the three forms of stock. Such a strategy would involve:

- maximising resource use efficiency (since the resources are finite) through extracting maximum value from mined products and causing no long-term environmental impact;
- maximising materials use efficiency through substitution, reuse and recycling;
- maximising end-use efficiency through Design for the Environment.

Integration of the strategies can be achieved through application of technologies and business models that link aspects of two, or all three, forms of stock in complementary ways. These include wide-scale adoption of industrial ecology concepts, recycling of materials, vertical integration within companies, mine site processing, management of toxic elements along the value chain, and application of Design for the Environment principles.

Understanding within some industry sectors of the need to transition to sustainability and the role of industry is growing, but much activity remains at the level of rhetoric. Vision 2050, developed by the World

Business Council for Sustainable Development, represents a major step forward. It envisages by 2050 'a planet of around nine billion people, all living well – with enough food, clean water, sanitation, shelter, mobility, education and health to make for wellness – within the limits of what this small, fragile planet can supply and renew, every day'. The proposed pathway to achieve this vision involves fundamental changes in governance structures, economic frameworks, and business and human behaviour. It involves incorporating the cost of externalities (carbon, ecosystem services, water), halving carbon emissions worldwide (based on 2005 levels), and achieving a four- to 10-fold improvement in the use of resources and materials. The vision and pathway are consistent with the Innovators column of Table 4.6.

The minerals industry, while adopting the language of sustainability and the principles of corporate social responsibility, remains largely in the Compliers column of Table 4.6. The industry is yet to take the next major steps, to fully incorporate sustainability thinking within its business models at all levels and move to the Innovators column and, ultimately, to the Marketeers column. It is in the industry's long-term interest to identify social, political and economic trends and capitalise on them. The inevitable closing of the materials cycle and transition to sustainability will create new opportunities for companies prepared to adopt new business models. Minerals companies can help the transition to sustainability by working proactively with their stakeholders, particularly government and government agencies, non-governmental organisations and other business sectors, to implement the types of strategies listed in Table 17.2.

17.5 REFERENCES

Five Winds (2001) *Eco-Efficiency and Materials.* Prepared by Five Winds International for the International Council on Metals and the Environment: Ottawa, Canada, April 2001. <http://www.icmm.com/library>.

Gandhi M (1993) *The Essential Writings of Mahatma Gandhi.* Oxford University Press: New York.

Giurco D, Evans G, Cooper C, Mason L and Franks D (2009) Mineral futures discussion paper: sustainability issues, challenges and opportunities. Prepared for CSIRO Minerals Down Under Flagship. Institute for Sustainable Futures (University of Technology, Sydney) and Centre for Social Responsibility in Mining, Sustainable Minerals Institute (University of Queensland).

Hobbes T (1651) *The Leviathan.* Penguin: New York, 1977 edn. <http://www.publicliterature.org/books/leviathan>.

ICMM (2006) Maximizing value – guidance on implementing materials stewardship in the minerals and metals value chain. International Council on Mining and Metals: London, September 2006. <http://www.icmm.com/library>.

ICMM (2009) Minerals and Metals Management 2020 – responsible and integrative chemicals management in the mining and metals supply chain. International Council on Mining and Metals: London, May 2009. <http://www.icmm.com/library>.

Keynes JM (1936) *The General Theory of Employment, Interest, and Money.* Macmillan: London. <http://www.marxists.org/reference/subject/economics/keynes/general-theory/index.htm>.

Malthus T (1798) *An Essay on the Principle of Population.* Penguin: London, 1983 edn. <http://www.econlib.org/library/Malthus/malPop.html>.

Medveçka J and Bangerter P (2007) Engineering sustainable development into industry: unlocking institutional barriers. In *Cu2007. Vol. 6 – Sustainable Development, HS&E and Recycling.* (Eds D Rodier and W Adams) pp. 13–26. Canadian Institute of Mining, Metallurgy and Petroleum: Toronto, Canada.

Mill JS (1848) *Principles of Political Economy.* Oxford University Press: Oxford, UK, 1998 edn. <http://www.econlib.org/library/Mill/mlP.html>.

Raskin P, Banuri T, Gallopín G, Gutman P, Hammond A, Kates R and Swart R (2002) Great transition: the promise and lure of the times ahead. Stockholm Institute: Boston. <http://www.gtinitiative.org/documents/Great_Transitions.pdf>.

Schumacher EF (1972) *Small is Beautiful.* Blond and Briggs: London.

Smith A (1776) *The Wealth of Nations.* Amherst, Prometheus: New York, 1991 edn. <http://www.econlib.org/library/Smith/smWN.html>.

UN (1987) *Our Common Future.* Oxford University Press: Oxford, UK.

Wellmer F-W and Becker-Platen JD (2001) World natural resources policy (with focus on mineral resources) In *Our Fragile World – Challenges and Opportunities for Sustainable Development.* Vol. 1. (Ed. MK Tolba) pp. 183–207. EOLSS Publishers: Oxford, UK.

WBCSD (2010) *Vision 2050: The new agenda for business.* World Business Council for Sustainable Development: Conches-Geneva, Switzerland, February 2010. <http://www.wbcsd.org/web/projects/BZrole/Vision2050-FullReport_Final.pdf>.

Appendix I: A note on units and quantities

INTERNATIONAL SYSTEM OF UNITS

The International System of Units is the system of measurement used internationally by the scientific community and is the common system of measurement in all countries except the United States, Myanmar and Liberia. The International System was established in 1960 by the 11th General Congress on Weights and Measures (CGPM) and is abbreviated as SI (from the French *Système International d'unités*). SI units were developed from the old metre–kilogram–second (MKS) system. There are seven base quantities, as listed in Table I.1. All other SI units are derived from these. Derived units are expressed algebraically in terms of base units or other derived units and are obtained by the mathematical operations of multiplication and division. For example, velocity is distance divided by time, therefore it has the units m/s, or m s^{-1}. Some common examples are listed in Table I.1. A few derived units have been given special names and symbols; those used in this book, and defined as needed, are also listed. Some units that are not part of the SI are essential and used so widely that they are accepted by the CGPM. The most common are also listed.

A number of prefixes are used with SI units so that very large or very small numerical values can be avoided. For example, distances between cities are measured in kilometres rather than metres. A prefix name is attached directly to the name of a unit, and a prefix symbol is attached directly to the symbol for a unit. For example, one kilometre (1 km) is equal to one thousand metres (1000 m) and one millimetre (1 mm) is equal to one thousandth of a metre (0.001 m). The most common prefixes are listed in Table I.2. Note that the power represented by each prefix is 1000 times (10^3) larger than the prefix preceding it; for example, 1 km is 1000 times greater than 1 m, which is 1000 times greater than 1 mm, which is 1000 times greater than 1 μm, and so on. The terms million (1 000 000 or 10^6) and billion (1 000 000 000 or 10^9) are not SI terms but are used in everyday conversation and some technical literature. They are also used very commonly in the financial sector.

SCIENTIFIC NOTATION, SIGNIFICANT FIGURES AND ORDER OF MAGNITUDE

A way of writing very large or very small numbers is to use powers of 10 notation. For example, the distance 1 234 567 m is equal to 1.234567 m multiplied by 1 000 000, or 10^6, and therefore can be written as 1.234567×10^6 m. This is equivalent to 1.234567 Mm

Table I.1: The seven SI base units from which all other SI units are derived, and common examples of SI derived units, SI derived units with special names and symbols and widely accepted non-SI units

SI base units				
Quantity	**Name**		**Symbol**	
Length	metre		m	
Mass	kilogram		kg	
Time	second		s	
Electric current	ampere		A	
Temperature	kelvin		K	
Quantity of substance	mole		mol	
Luminous intensity	candela		cd	
Some derived units				
Quantity	**Name**		**Symbol**	
Area	square metre		m^2	
Volume	cubic metre		m^3	
Speed, velocity	metre per second		$m\ s^{-1}$	
Acceleration	metre per second squared		$m\ s^{-2}$	
Wave number	reciprocal metre		m^{-1}	
Density, mass density	kilogram per cubic metre		$kg\ m^{-3}$	
Specific volume	cubic metre per kilogram		$m^3\ kg^{-1}$	
Concentration	mole per cubic metre		$mol\ m^{-3}$	
Derived units with special names and symbols				
Derived quantity	**Special name**	**Special symbol**	**Expression in terms of other SI units**	**Expression in terms of SI base units**
Frequency	hertz	Hz		s^{-1}
Force	newton	N		$m\ kg\ s^{-2}$
Pressure, stress	pascal	Pa	$N\ m^{-2}$	$m^{-1}\ kg\ s^{-2}$
Energy, work, quantity of heat	joule	J	N m	$m^2\ kg\ s^{-2}$
Power, radiant flux	watt	W	$J\ s^{-1}$	$m^2\ kg\ s^{-3}$
Electric charge	coulomb	C		s A
Electrical potential difference	volt	V	$W\ A^{-1}$	$m^2\ kg\ s^{-3}\ A^{-1}$
Celsius temperature	degree Celsius	°C		K
Common non-SI units accepted for use with the SI				
	Symbol		**Value in SI units**	
minute	min		1 min = 60 s	
hour	h		1 h = 60 min = 3600 s	
day	d		1 d = 24 h = 86 400 s	
degree	°		$1° = (\pi/180)$ rad	
minute	′		$1' = (1/60)° = (\pi/10\ 800)$ rad	
second	″		$1'' = (1/60)' = (\pi/648\ 000)$ rad	
hectare	ha		$1\ ha = 1\ hm^2 = 10^4\ m^2$	
litre	L		$1\ L = 1\ dm^3 = 10^{-3}\ m^3$	
tonne (or metric ton)	t		$1\ t = 10^3\ kg$	
kilowatt-hour	kW h		$1\ kW\ h = 3.6 \times 10^6\ J$	

Table I.2: Commonly used SI prefixes

Factor	Prefix	Symbol
10^{15}	peta	P
10^{12}	tera	T
10^{9}	giga	G
10^{6}	mega	M
10^{3}	kilo	k
10^{2}	hecto	h
10^{1}	deka	da
10^{-1}	deci	d
10^{-2}	centi	c
10^{-3}	milli	m
10^{-6}	micro	μ
10^{-9}	nano	n
10^{-12}	pico	p

or 1.234567×10^3 km. This way of writing numbers is called *scientific notation.*

When we write the distance 1 234 567 m we are implying that the distance is known accurately to within 1 m. If the distance is known accurately only to within 10 m, we should write the distance as 1 234 570 m. The units have been rounded to the nearest 10. If we are confident that the distance is known only to within the nearest kilometre we would write the distance as 1 235 000 m or 1235 km. The number of *significant figures* in a measurement is the number of digits that are known with confidence plus the last digit, which is an estimate because it is formed by rounding off the subsequent digits. Thus, the measurement 1 235 000 m or 1235 km has four significant figures.

The phrase *order of magnitude* refers to the smallest power of 10 needed to represent a quantity. Order of magnitude is a useful concept for comparing quantities to get a feel for their relative magnitudes. For example, the average height of a human being is around 1.7 m which, rounded off to the nearest power of 10, is 1 m (or 10^0 m). This doesn't mean that that the average height of a human is 1 m but that the average height is closer to 1 m than it is to 10 m (or 10^1 m). If we say something is one order of magnitude greater (or smaller) than something else we mean it is 10 times greater (or smaller); similarly for two and three orders of magnitude, and so on:

	Greater than	Less than
One order of magnitude	$\times 10$	$\times \frac{1}{10}$
Two orders of magnitude	$\times 100$	$\times \frac{1}{100}$
Three orders of magnitude	$\times 1000$	$\times \frac{1}{1000}$

Table I.3 shows interesting order of magnitude comparisons of the mass and length of some well known things.

Table I.3: Order of magnitude comparisons of the mass and length of some common things

	Mass (g)		Radius or distance (m)
Electron	10^{-27}	Radius of proton	10^{-15}
Proton	10^{-24}	Radius of atom	10^{-10}
Virus	10^{-16}	Radius of virus	10^{-7}
Amoeba	10^{-5}	Radius of amoeba	10^{-4}
Raindrop	10^{-3}	Height of human being	10^{0}
Ant	10^{0}	Radius of Earth	10^{7}
Human being	10^{5}	Radius of Sun	10^{9}
Great Pyramid	10^{13}	Earth–Sun distance	10^{11}
Earth	10^{27}	Radius of solar system	10^{13}
Sun	10^{33}	Distance of Sun to nearest star	10^{16}
Milky Way galaxy	10^{44}	Radius of Milky Way galaxy	10^{21}
Known Universe	10^{55}	Radius of visible Universe	10^{26}

Appendix II: A review of some important scientific concepts

II.1 THE NATURE OF MATTER

All substances are composed of matter. This book is about matter derived from the Earth's crust and the materials made from it. Matter is made of atoms. These may exist individually as elements or in combination with other elements as compounds. There are 94 naturally occurring elements. These are shown in the periodic table in Figure II.1, where they are arranged in increasing atomic number (the number of protons in the nucleus) and in vertical groups and horizontal periods which demonstrate the periodic nature of the properties of the elements. The various classes of elements are indicated in the figure. The nuclei of elements with atomic numbers 83 and higher are unstable. They undergo radioactive decay, releasing energy in the process and transforming into an atom of different type with a smaller number of protons and/or neutrons in the nucleus. The mass lost as a result is converted into energy. Of the first 82 elements, 80 have stable isotopes. Elements 43 and 61 do not have stable isotopes and occur on Earth only in trace quantities. The elements from 83 to 94, though unstable, occur on Earth either as remnants from the original formation of the elements or as short-lived daughter-isotopes in the natural decay of uranium and thorium.

Atoms bond to form compounds when electrons in the outermost shells interact in such a way as to reduce the overall energy state of the atoms; i.e. the drive to form compounds is to minimise the energy of a system containing atoms. Thus compounds are the natural state for most elements. There are three common forms of bonding of atoms.

Covalent bonding. When the atoms in a compound form a stable, electrically neutral group of at least two atoms the unit is called a molecule. In these compounds the bonding is achieved by the sharing of electrons of the outermost shell to achieve a lower overall energy state. In solids and liquids composed of covalent molecules, the molecules themselves are loosely bound to other molecules of the same kind by weaker intermolecular bonds, including van der Waal bonds, to create the bulk substance. Covalent solids thus tend to be weak and soft. They usually have low melting and boiling points and are not very soluble in water. The gases hydrogen, nitrogen and oxygen occur in nature as covalently bonded diatomic molecules, represented by H_2, N_2 and O_2. Since these contain atoms of only one type, they are not considered to be compounds.

Ionic bonding. Bonding can also occur through electrons being donated by one atom to another to achieve an overall lower energy state. This is the case

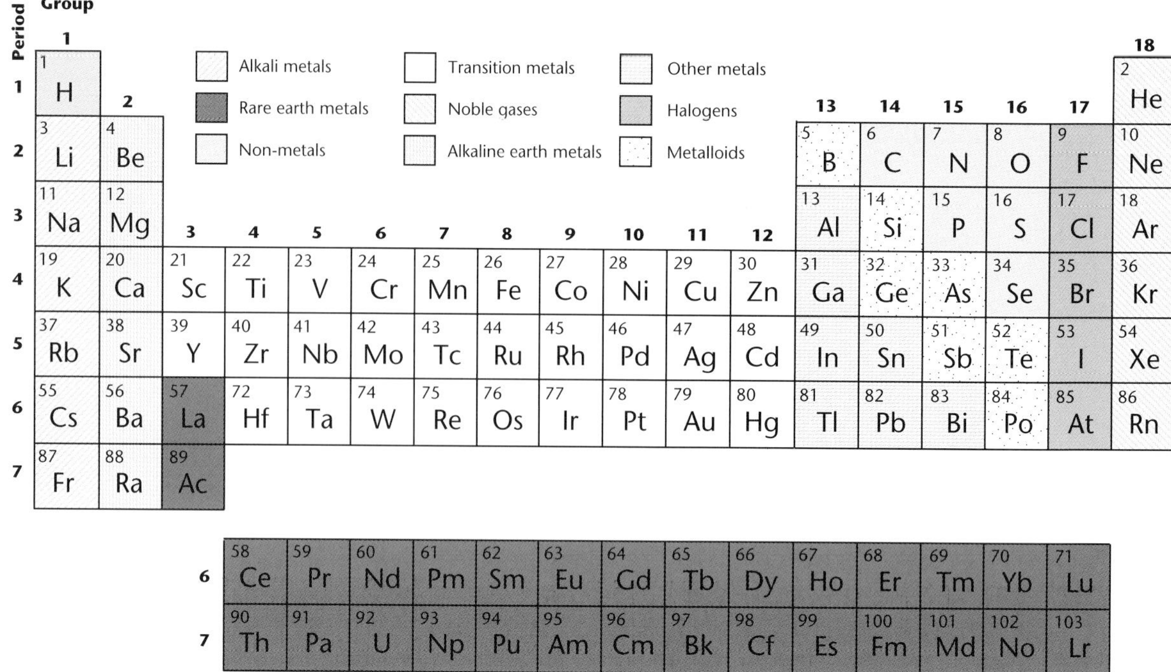

Figure II.1: The periodic table of the elements and their common groupings.

with salt (sodium chloride). The sodium atoms donate an electron to the chlorine atoms so each atom becomes more stable. In doing so, the sodium atoms become positively charged and the chlorine atoms become negatively charged. Atoms which have lost or gained one or more electrons are called ions. Negatively charged ions are called anions and positively charged ions are called cations. The anions and cations are held together by the electrostatic force between oppositely charged bodies in a regular lattice structure to create a bulk structure that overall produces the lowest energy state. Substances composed of positively and negatively charged ions are called ionic compounds. Ionic compounds are usually hard and brittle and have high melting points and high solubility in water. At high temperatures and in the molten state, they are good conductors of heat and electricity.

Metallic bonding. A third kind of bonding is metallic bonding, which is the type of bonding in solid metals. In metals, the electrons in the outermost shell are not associated with any single atom (as in ionic compounds) or small group of atoms (as in covalent compounds) but are mobile. The atoms form a regular array within a sea of electrons which are free to move. It is this characteristic which gives metals their high electrical and thermal conductivity.

Compounds may be classified as organic or inorganic. Organic compounds are those that contain carbon as a constituent element, although for historical reasons a few simple carbon-containing substances are considered to be inorganic (carbon monoxide, carbon dioxide, carbonates, cyanides, cyanates, carbides, thyocyanates). Most of the matter in the Earth and the universe is inorganic. This includes the minerals which make up the crust of the Earth from which many materials are obtained. The extraction of metals from ores relies on inorganic chemistry, as does the manufacture of glass, ceramics, artificial fertilisers and cement. Most organic compounds consist primarily of carbon and hydrogen but often contain other elements, particularly nitrogen, oxygen and the halogens as well as phosphorus, silicon and sulfur. Organic compounds with molecules containing multiple carbon atoms (usually arranged in chains and/or rings) occur in extremely large variety and the range of properties and their application is wide. Though the amount of organic matter in the Earth and the Universe is very small, the number of organic compounds is orders of magnitude larger than the number of inorganic compounds because of the large number of ways in which carbon and hydrogen atoms can be arranged in molecules. The molecules of organic compounds can range in size from a few atoms to tens of atoms, to tens of thousands of atoms. Organic compounds form the basis of all life on Earth. They are also important constituents of many useful products, including paints, plastics, food, explosives, drugs and petrochemicals.

II.2 CONSERVATION OF MATTER

Atoms are indestructible, except in nuclear processes. A consequence is the law of conservation of matter, which states:

> *Matter cannot be created or destroyed but it can be rearranged.*

The law implies that, for any chemical process in a closed system, the mass of the reactants will always be equal to the mass of the products. In science, a system is a part of the Universe in which we have a particular interest; for example, a reaction vessel, an engine, an area of land or the entire Earth. The rest of the Universe is called the surroundings.[30] The law of conservation of matter is a fundamental law of nature. Today it seems unremarkable, but the principle was not generally accepted until the early 1800s. The law of conservation of matter has profound implications. It provides the basis for determining the quantities of reagents required to make a certain quantity of a substance (or the quantity of products that can be extracted from a substance) and for establishing procedures for accounting for the quantity and distribution of elements in processes or the environment. The law is embodied in every equation written to describe a chemical reaction. The number of atoms of each type involved in a reaction is the same on the reactants side and the products side of the equation.

The only exception to the law is nuclear reactions, in which matter is converted into energy through the destruction of protons and neutrons. As a result, the original atoms are converted into different atoms with different numbers of protons and/or neutrons. This is the major process in the Sun and other stars, whereby hydrogen atoms combine through nuclear fusion to form helium atoms. It also occurs naturally on Earth through the nuclear fission (or splitting) of naturally occurring radioactive elements. During fission and fusion reactions, mass is converted into energy. The relationship is given by Einstein's equation:

$$E = mc^2 \qquad \text{II.1}$$

where m is the mass loss (kg), E is its energy equivalent (J) and c is a constant equal to the speed of light (299 792 458 m s^{-1}). Human-initiated nuclear reactions are the source of energy in nuclear reactors for generating electricity (by fission of uranium atoms) and in hydrogen bombs (by fusion of hydrogen atoms).

II.3 ENERGY, HEAT AND THE LAWS OF THERMODYNAMICS

The *energy* of a system is its capacity to do work. The unit of energy is the Joule (J), which is defined as the energy expended in applying a force of one Newton through a distance of one metre. The Newton (N) is defined as the force required to accelerate a mass of one kilogram at a rate of one metre per second per second. *Power* is the rate of expending energy (or doing work) and has the unit Joules per second, called the Watt. Energy can exist in various forms, such as potential, kinetic, thermal, electrical, chemical and nuclear. Potential energy is the energy stored in a body or a system due to its position in a force field or its configuration. For example, when a spring is stretched in one direction, it exerts a force in the opposite direction so as to return to its original state. Similarly, when a mass is raised vertically, the force of gravity acts to bring it back to its original position. The action of stretching the spring or lifting the mass requires energy. The energy that went into lifting the mass is stored in its position in the gravitational field and the energy

30 An open system is one in which matter and energy can move across the boundary between the system and surroundings. A closed system is one in which energy can move across the boundary but matter cannot. An isolated system is one in which neither matter nor energy can move across the boundary.

required to stretch the spring is stored in the metal. Kinetic energy is the energy of an object due to its motion. It is equal to the work required to accelerate a body of given mass from rest to its current velocity. The body maintains this kinetic energy unless, or until, its velocity changes. The same amount of work is required to decelerate the body from its current velocity to a state of rest. The kinetic energy of a non-rotating body of mass m travelling at a velocity v is equal to ½ mv^2. Energy can be converted from one form to another in various ways. A vehicle rolling down a hill gains kinetic energy as its velocity increases but loses potential energy as its height decreases – its potential energy is being converted into kinetic energy. Electrical energy can be produced by many kinds of devices which convert other forms of energy into electrical energy; for example, photovoltaic cells (thermal energy), turbines driven by running (kinetic energy) or falling water (gravitational potential energy), fuel-burning internal combustion engines and gas turbines (chemical energy), and batteries and fuel cells (chemical energy). The kilowatt-hour (kW h) is the unit of energy most commonly used for electrical energy consumption. It is the billing unit for energy delivered to consumers by electric utilities – 1 kW h is equal to $1000 \times 60 \times 60$ Joules, or 3.6 MJ.

The energy of a system can be increased or decreased by the transfer of energy as heat. Heat can be thought of as energy being transferred from one object to another, because of a difference in temperature between them – the energy flows from the hotter body to the cooler body. *Heat* is the macroscopic manifestation of the total kinetic energy of all the atoms and molecules making up a body or a system. Atoms and molecules are in constant motion in all substances. In gases, the atoms or molecules are free to move independently to occupy the volume available and to collide with each other and the walls of the container. In liquids, the atoms and molecules are free to move relative to one another to the extent that liquids take the shape of their container but retain their own volume. In solids, the atoms or molecules are held sufficiently tightly that solids retain their own shape and volume but their atoms and molecules vibrate about a mean position. The sum of the kinetic energies of all the atoms and molecules making up a system determines the thermal energy (or heat content) of the system. *Temperature* is the degree of hotness of a system. It is the macroscopic manifestation of the average kinetic energy of the individual atoms and molecules in the system. The faster these move, the greater their average kinetic energy and the higher the temperature. Thus, heat and temperature are related but different concepts. All motion of atoms and molecules ceases at the temperature of –273.15°C, which is called absolute zero and given the value of zero in the Kelvin temperature scale (written as 0 K). The Kelvin scale has the same temperature unit as the Celsius scale; to convert Celsius temperatures to absolute temperatures simply add 273.15. For example, the boiling point of water is 100°C or 373.15 K.

The first law of thermodynamics

Energy can be converted from one form to another, but the total quantity of energy always remains the same; in any physical or chemical process energy is neither created nor destroyed. This is the *law of conservation of energy*. A more precise statement is:

> *Whenever a quantity of one kind of energy is produced, an equivalent quantity of another kind (or other kinds) is used up.*

The first law of thermodynamics is an expression of the law of conservation of energy. It is usually formulated by stating that the change in the internal energy of a system is equal to the amount of heat supplied to the system minus the amount of work performed by the system on its surroundings. This law, like the law of conservation of matter, is an unremarkable concept today but it had a revolutionary impact on science and engineering during the 19th century as its implications became better understood. As with the law of conservation of matter, the only exception to the law of conservation of energy is nuclear reactions, in which energy is created through conversion of matter into energy.

The second law of thermodynamics

The second law of thermodynamics is equally important but more difficult to understand. It is based on the concept of *entropy*, which is the measure of randomness of the atoms or molecules in a system. *Spontaneous changes*, which are changes that occur of their own accord without any externally applied energy, result in an overall increase in entropy. Everyday examples of spontaneous changes are sugar dissolving in water, a hot cup of tea cooling to the temperature of its surroundings, and rust forming on a piece of steel. Spontaneous changes tend to smooth out differences in temperature, pressure, density and chemical composition that may be present in a system. Entropy is a measure of how far this smoothing-out process has progressed. In systems at constant temperature, the change in entropy, ΔS, is given by the relation:

$$\Delta S = \frac{Q}{T} \qquad \text{II.2}$$

where Q (J) is the quantity of heat absorbed by the system in a reversible process[31] and T (K) is the absolute temperature at which the process occurs. The unit of entropy is J K^{-1}. For chemical reactions, Q is equal to the enthalpy change of the reaction, therefore:

$$\Delta s = \frac{\Delta H}{T} \qquad \text{II.3}$$

There is no single, simple statement of the second law that conveys its full significance and it is often better to state it in several different ways. One statement is:

> *In an isolated system, a change can occur only if it increases the entropy of the system.*

An isolated system is one in which neither matter nor energy can cross the boundary between the system and its surroundings. Thus, an isolated system will either stay the same (at internal equilibrium) or undergo some change that increases entropy, i.e. a spontaneous change. Changes that decrease the total entropy of a system cannot occur without the application of energy from outside the system. An implication of this statement is that, as a result of the natural processes occurring throughout the Universe, the total entropy of the Universe is increasing with time and will continue to increase until all variations of temperature, composition, density etc. have been removed and the Universe is entirely at internal equilibrium. Of course, we don't know the ultimate fate of the Universe, or even if the second law of thermodynamics applies to the Universe as a whole, but this powerfully illustrates the significance of the second law.

Another statement of the second law (which in mathematical terms can be shown to be equivalent to the earlier statement) is:

> *Heat will not flow spontaneously from a substance at lower temperature to a substance at higher temperature.*

Put simply, this means heat doesn't flow from cold to hot without external energy being added. This is known from everyday experience. In a system where heat flows from cold to hot, entropy has to be decreasing. This can occur only if more entropy is created somewhere else, so that the total entropy of the system and its surroundings increases. For example, in a refrigerator (the system), heat flows from cold to hot (across the boundary between the system and its surroundings) but only under the action of a compressor, which is driven by a motor. The electrical energy consumed by the motor is converted to heat and results in an overall net increase in entropy of the Universe.

A third statement of the second law (also mathematically equivalent to the other statements), is:

> *It is not possible to convert a quantity of heat completely into work.*

This means that, for any device that converts heat into work, there will always be a quantity of remaining heat that cannot be converted into work. For example, in a car engine or gas turbine, hot gases are expelled but still contain thermal energy. Some of this energy is not converted to work due to inefficiencies in the engine but there is a theoretical limit beyond which efficiency cannot be increased, which is given by:

$$\varepsilon = 1 - \frac{T_c}{T_h} \qquad \text{II.4}$$

where T_c is the temperature of the heat sink and T_h is the temperature of the heat source (with temperatures expressed in Kelvin). This limit is never achieved in

31 In thermodynamics, a reversible process is one that can be reversed by means of infinitesimally small changes so the system remains in thermodynamic equilibrium throughout the process. Since it would take an infinite time to carry out a reversible process, perfectly reversible processes are impossible in practice. However, they are an essential concept in thermodynamic theory.

practice. If steam enters a turbine at 500°C and leaves it at 80°C, the maximum efficiency of conversion of heat into mechanical energy (or electrical energy if the turbine drives a generator) is 0.54, or 54%. If the steam is generated by burning fuel, if the flame temperature is 1400°C and if the combustion gases exit the boiler at 100°C, the maximum efficiency of conversion of the chemical energy in the fuel into thermal energy is 0.78, or 78%. In this case, the maximum overall efficiency of conversion of the chemical energy in the fuel into electrical energy is $0.54 \times 0.78 = 0.42$, or 42%.

II.4 ELECTROMAGNETIC RADIATION

All objects at temperatures above absolute zero radiate energy to their surroundings in the form of electromagnetic waves. These travel freely through space (or vacuum) and do not require a medium for their passage, unlike sound waves, for example. They also pass through low-density materials, their penetrating power depending on their frequency. In vacuum and in space they travel at the speed of light (299 792 458, or approximately 3.0×10^8 m s^{-1}); they are slowed slightly in other media. All waves, not just electromagnetic waves, are governed by the relation:

$$v = f\lambda \qquad \text{II.5}$$

where v is the velocity of the wave (m s^{-1}), λ is its wavelength, the distance between peaks (m), and f is the frequency of the wave, the number of times the wave pattern recurs in a given time period (s^{-1}). There are many types of electromagnetic radiation and each is defined by its own wavelength and frequency, as illustrated in Figure II.2. The radiation from the Sun spans wavelengths from approximately 0.1×10^{-6} m to 4.0×10^{-6} m. Visible light is that part of the spectrum perceived by our eyes as the colours red, orange, yellow, green, blue, indigo and violet. Light has wavelengths ranging from 0.40×10^{-6} m (violet) to 0.71×10^{-6} m (red). Wavelengths from 0.1×10^{-6} to 0.4×10^{-6} m lie in the ultraviolet radiation band and constitute about 7% of the energy from the Sun. Wavelengths between 0.71×10^{-6} and 4.0×10^{-6} m lie in the infrared radiation band and constitute about 48% of the energy from the

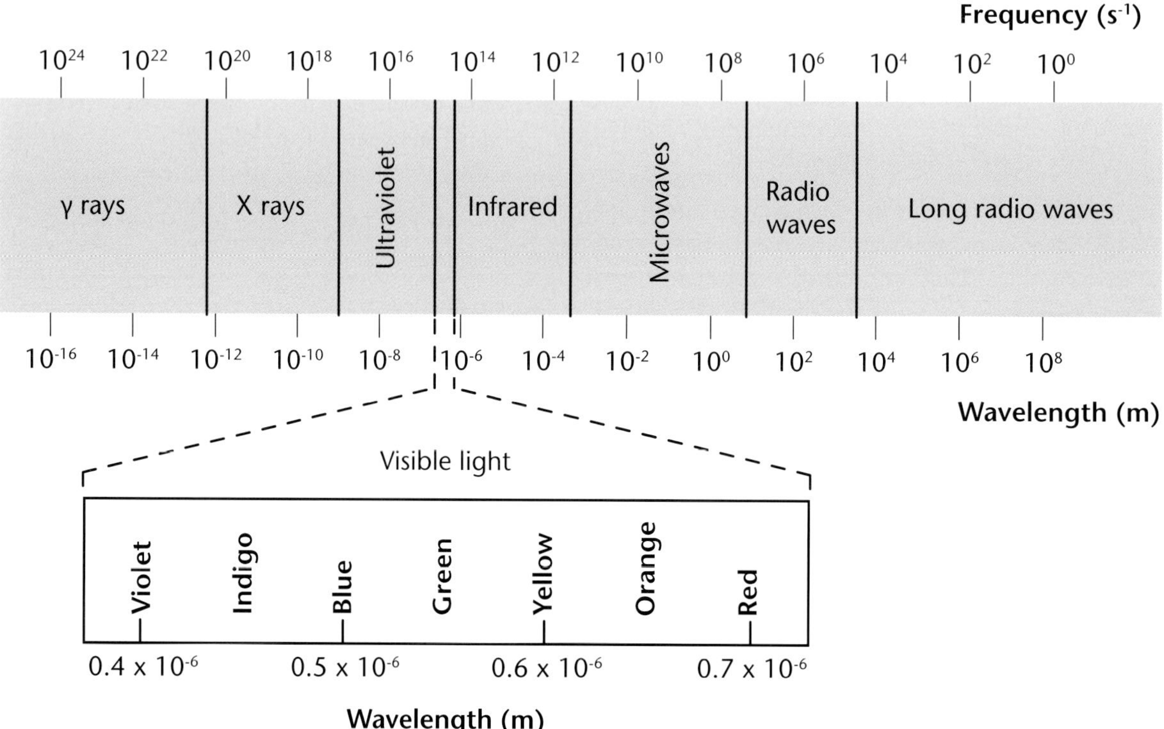

Figure II.2: The electromagnetic spectrum.

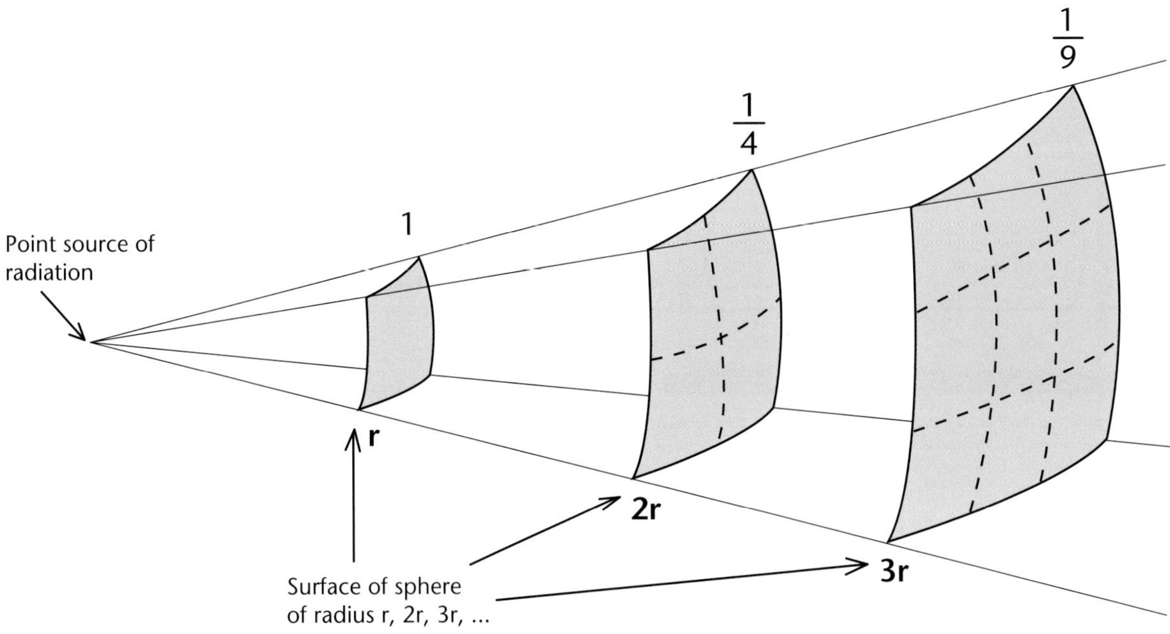

Figure II.3: The geometrical basis of the inverse square law.

Sun. The balance of the energy from the Sun is predominantly in the visible light band.

The intensity of radiation decreases progressively the farther away from its source, according to the inverse square law. This law is a consequence of simple geometry. It applies to any phenomenon which spreads its influence equally from a point radially in all directions, such as gravitational force, electric fields, electromagnetic radiation and sound. The principle is illustrated in Figure II.3 and may be expressed mathematically as:

$$I_2 = \frac{I_1}{\left(\frac{r_2}{r_1}\right)^2} \qquad \text{II.6}$$

where I_1 and I_2 are the intensities at radii r_1 and r_2, respectively, from the point source. To illustrate the inverse square law, suppose the intensity of radiation at radius *r* from the centre of a source is 100 000 units, then at two radii (2*r*) from the source the intensity will be $100\,000/2^2 = 25\,000$ units, at three radii (3*r*) it will be $100\,000/3^2 = 11\,111$ units, at four radii it will be $100\,000/4^2 = 6250$ units, and so on. At 10 radii, the intensity will be only 1000 units and at 100 radii it will be 10 units. It can be seen that the intensity decreases rapidly at first, then more slowly as distance increases.

When electromagnetic radiation encounters a medium it may be transmitted, reflected, scattered or absorbed. A red object appears red because it transmits, reflects and scatters the red part of the visible spectrum but absorbs the yellow, green and blue parts. Absorption is the process by which energy is taken up by an atom in the medium through valence electrons moving to a higher energy level. The absorbed energy may be re-emitted immediately as radiant energy or transformed into thermal energy and re-emitted at a lower wavelength. The amount of absorption by a body depends on its albedo, or surface reflectivity, and the wavelength of the radiation.

II.5 HEAT TRANSFER

As a form of energy, heat can be converted into other forms of energy. But heat may also be transferred from one place to another without changing form. This is achieved by three basic mechanisms: conduction, convection and radiation. Radiation has been discussed above. *Conduction* involves the transfer of thermal energy from atom to atom in a substance (solid, liquid or gas) without bulk movement of the atoms. Conduction results in the continuous flow of

Figure II.4: Convection currents in a pot of water being heated on a hotplate.

heat through a substance from regions of higher temperature to regions of lower temperature. The heating of a cup when hot water is poured into it results from heat being conducted through the wall of the cup from the inside (hotter) to the outside (cooler). *Convection* involves the transfer of heat by bulk movement of substance. Convection can occur only when the bulk material can flow. It is thus an important mechanism of heat transfer in liquids and gases and in substances which are plastic (able to flow under an applied force), such as in the Earth's interior. Convection depends on the fact that most substances expand slightly when heated, and therefore become less dense. Consider a pot of cold water placed on a hotplate (Figure II.4). The pot is heated at the base by conduction of heat through the base. The layer of water adjacent to the base is heated by conduction and expands slightly. Now being less dense, this water rises towards the surface due to buoyancy effects and displaces water from above, which being denser sinks towards the bottom of the pot where in turn it becomes heated and rises. Circulating convection currents are established, and eventually bring the entire contents of the pot to a uniform temperature. Convection and conduction of heat depend on the presence of matter but radiation does not require matter. Convection also results in the transfer of matter.

Appendix III: GRI Sustainability Indicators

Core indicators must be reported; others are optional.

Economic performance indicators		
Economic performance		
EC1	Direct economic value generated and distributed, including revenues, operating costs, employee compensation, donations and other community investments, retained earnings, and payments to capital providers and governments.	Core
EC2	Financial implications and other risks and opportunities for the organisation's activities due to climate change.	Core
EC3	Coverage of the organisation's defined benefit plan obligations.	Core
EC4	Significant financial assistance received from government.	Core
Market presence		
EC5	Range of ratios of standard entry-level wage compared to local minimum wage at significant locations of operation.	
EC6	Policy, practices and proportion of spending on locally based suppliers at significant locations of operation.	Core
EC7	Procedures for local hiring and proportion of senior management hired from the local community at locations of significant operation.	Core
Indirect economic impacts		
EC8	Development and impact of infrastructure investments and services provided primarily for public benefit through commercial, in-kind or *pro bono* engagement.	Core
EC9	Understanding and describing significant indirect economic impacts, including the extent of impacts.	

Environmental performance indicators		
Materials		
EN1	Materials used by weight or volume.	Core
EN2	Percentage of materials used that are recycled input materials.	Core
Energy		
EN3	Direct energy consumption by primary energy source.	Core
EN4	Indirect energy consumption by primary source.	Core
EN5	Energy saved due to conservation and efficiency improvements.	
EN6	Initiatives to provide energy-efficient or renewable energy based products and services, and reductions in energy requirements as a result of these initiatives.	
EN7	Initiatives to reduce indirect energy consumption and reductions achieved.	
Water		
EN8	Total water withdrawal by source.	Core
EN9	Water sources significantly affected by withdrawal of water.	
EN10	Percentage and total volume of water recycled and reused.	
Biodiversity		
EN11	Location and size of land owned, leased, managed in or adjacent to protected areas and areas of high biodiversity value outside protected areas.	Core
EN12	Description of significant impacts of activities, products and services on biodiversity in protected areas and areas of high biodiversity value outside protected areas.	Core
EN13	Habitats protected or restored.	
EN14	Strategies, current actions and future plans for managing impacts on biodiversity.	
EN15	Number of IUCN Red List species and national conservation list species with habitats in areas affected by operations, by level of extinction risk.	
Emissions, effluents and waste		
EN16	Total direct and indirect greenhouse gas emissions by weight.	Core
EN17	Other relevant indirect greenhouse gas emissions by weight.	Core
EN18	Initiatives to reduce greenhouse gas emissions and reductions achieved.	
EN19	Emissions of ozone-depleting substances by weight.	Core
EN20	NO, SO and other significant air emissions by type and weight.	Core
EN21	Total water discharge by quality and destination.	Core
EN22	Total weight of waste by type and disposal method.	Core
EN23	Total number and volume of significant spills.	Core
EN24	Weight of transported, imported, exported or treated waste deemed hazardous under the terms of the Basel Convention Annex I, II, III and VIII, and percentage of transported waste shipped internationally.	
EN25	Identity, size, protected status and biodiversity value of water bodies and related habitats significantly affected by the reporting organisation's discharges of water and run-off.	
Products and services		
EN26	Initiatives to mitigate environmental impacts of products and services, and extent of impact mitigation.	Core

EN27	Percentage of products sold and their packaging materials that are reclaimed by category.	Core
Compliance		
EN28	Monetary value of significant fines and total number of non-monetary sanctions for non-compliance with environmental laws and regulations.	Core
Transport		
EN29	Significant environmental impacts of transporting products and other goods and materials used for the organisation's operations, and transporting members of the workforce.	
Overall		
EN30	Total environmental protection expenditures and investments by type.	
Labour practices and decent work		
Employment		
LA1	Total workforce by employment type, employment contract and region.	Core
LA2	Total number and rate of employee turnover by age group, gender and region.	Core
LA3	Benefits provided to full-time employees that are not provided to temporary or part-time employees, by major operations.	
Labour/management relations		
LA4	Percentage of employees covered by collective bargaining agreements.	Core
LA5	Minimum notice period(s) regarding operational changes, including whether it is specified in collective agreements.	Core
Occupational health and safety		
LA6	Percentage of total workforce represented in formal joint management–worker health and safety committees that help monitor and advise on occupational health and safety programs.	
LA7	Rates of injury, occupational diseases, lost days and absenteeism, and number of work-related fatalities by region.	Core
LA8	Education, training, counselling, prevention and risk-control programs in place to assist workforce members, their families or community members regarding serious diseases.	Core
LA9	Health and safety topics covered in formal agreements with trade unions.	
Training and education		
LA10	Average hours of training per year per employee, by employee category.	Core
LA11	Programs for skills management and lifelong learning that support the continued employability of employees and assist them in managing career endings.	
LA12	Percentage of employees receiving regular performance and career development reviews.	
Diversity and equal opportunity		
LA13	Composition of governance bodies and breakdown of employees per category according to gender, age group, minority group membership and other indicators of diversity.	Core
LA14	Ratio of basic salary of men to women, by employee category.	Core
Human rights performance indicators		
Investment and procurement practices		
HR1	Percentage and total number of significant investment agreements that include human rights clauses or that have undergone human rights screening.	Core
HR2	Percentage of significant suppliers and contractors that have undergone screening on human rights and actions taken.	Core

HR3	Total hours of employee training on policies and procedures concerning aspects of human rights that are relevant to operations, including the percentage of employees trained.	
Non-discrimination		
HR4	Total number of incidents of discrimination and actions taken.	Core
Freedom of association and collective bargaining		
HR5	Operations identified in which the right to exercise freedom of association and collective bargaining may be at significant risk, and actions taken to support these rights.	Core
Child labour		
HR6	Operations identified as having significant risk for incidents of child labour, and measures taken to contribute to the elimination of child labour.	Core
Forced and compulsory labour		
HR7	Operations identified as having significant risk for incidents of forced or compulsory labour, and measures to contribute to the elimination of forced or compulsory labour.	Core
Security practices		
HR8	Percentage of security personnel trained in the organisation's policies or procedures concerning aspects of human rights that are relevant to operations.	
Indigenous rights		
HR9	Total number of incidents of violations involving rights of indigenous people, and actions taken.	
	Society performance indicators	
Community		
SO1	Nature, scope and effectiveness of any programs and practices that assess and manage the impacts of operations on communities, including entering, operating and exiting.	Core
Corruption		
SO2	Percentage and total number of business units analysed for risks related to corruption.	Core
SO3	Percentage of employees trained in organisation's anti-corruption policies and procedures.	Core
SO4	Actions taken in response to incidents of corruption.	Core
Public policy		
SO5	Public policy positions and participation in public policy development and lobbying.	Core
SO6	Total value of financial and in-kind contributions to political parties, politicians and related institutions by country.	
Anti-competitive behaviour		
SO7	Total number of legal actions for anti-competitive behaviour, anti-trust and monopoly practices, and their outcomes.	
Compliance		
SO8	Monetary value of significant fines and total number of non-monetary sanctions for non-compliance with laws and regulations.	Core
	Product responsibility performance indicators	
Customer health and safety		
PR1	Life cycle stages in which health and safety impacts of products and services are assessed for improvement, and percentage of significant products and services categories subject to such procedures.	Core

PR2	Total number of incidents of non-compliance with regulations and voluntary codes concerning health and safety impacts of products and services during their life cycle, by type of outcome.	
Product and service labelling		
PR3	Type of product and service information required by procedures, and percentage of significant products and services subject to such information requirements.	Core
PR4	Total number of incidents of non-compliance with regulations and voluntary codes concerning product and service information and labelling, by type of outcome.	
PR5	Practices related to customer satisfaction, including results of surveys measuring customer satisfaction.	
Marketing communications		
PR6	Programs for adherence to laws, standards and voluntary codes related to marketing communications, including advertising, promotion and sponsorship.	Core
PR7	Total number of incidents of non-compliance with regulations and voluntary codes concerning marketing communications, including advertising, promotion and sponsorship, by type of outcome.	
Customer privacy		
PR8	Total number of substantiated complaints regarding breaches of customer privacy and losses of customer data.	
Compliance		
PR9	Monetary value of significant fines for non-compliance with laws and regulations concerning the provision and use of products and services.	Core

Source: Sustainability reporting guidelines, version 3.0, http://www.globalreporting.org/home.

Appendix IV: Processing routes for extraction of common metals from their ores

With the exception of those that occur in the oceans or in saline water, reactive metals always occur in oxidised form – as oxides, hydroxides and carbonates. The less reactive metals occur most commonly in oxidised form or as sulfides. Thus, it is convenient to summarise the extraction routes for metals using three generic flowsheets, one for reactive metals and two for less reactive metals (Figures IV.1, IV.2 and IV.3). Only the major stages in the production of a metal are shown in these figures. Often, several processing steps have been grouped together and the actual flowsheets are much more complex. Tables IV.1, IV.2 and IV.3 summarise the typical routes for the production of most common metals. These figures and tables closely follow those of Woollacott and Eric (1994); the flowsheets have been modified slightly and the process routes altered to reflect most recent practice.

REFERENCE

Woollacott LC and Eric RH (1994) *Mineral and Metal Extraction: An Overview.* South African Institute of Mining and Metallurgy: Johannesburg.

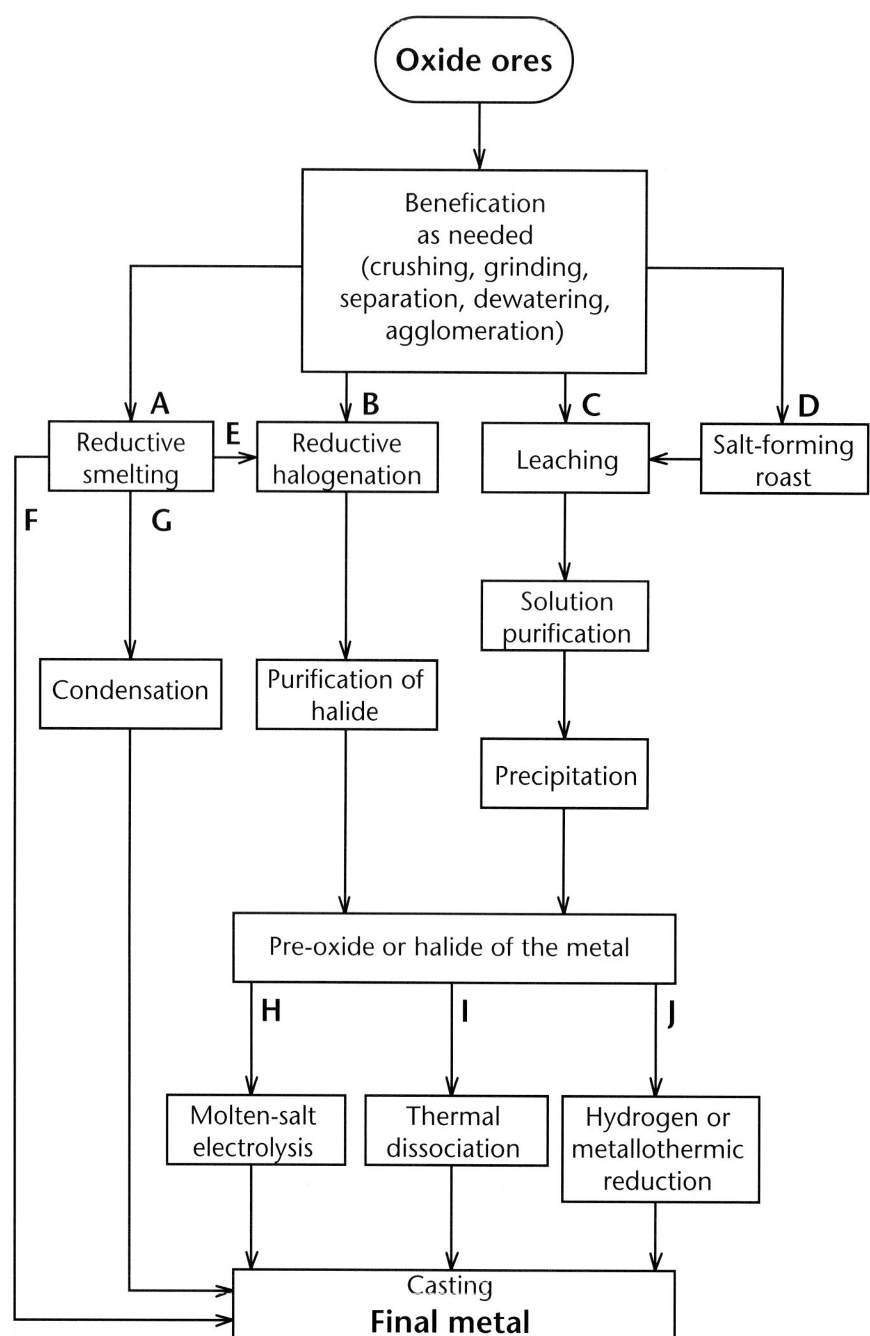

Figure IV.1: Extraction routes for reactive metals.

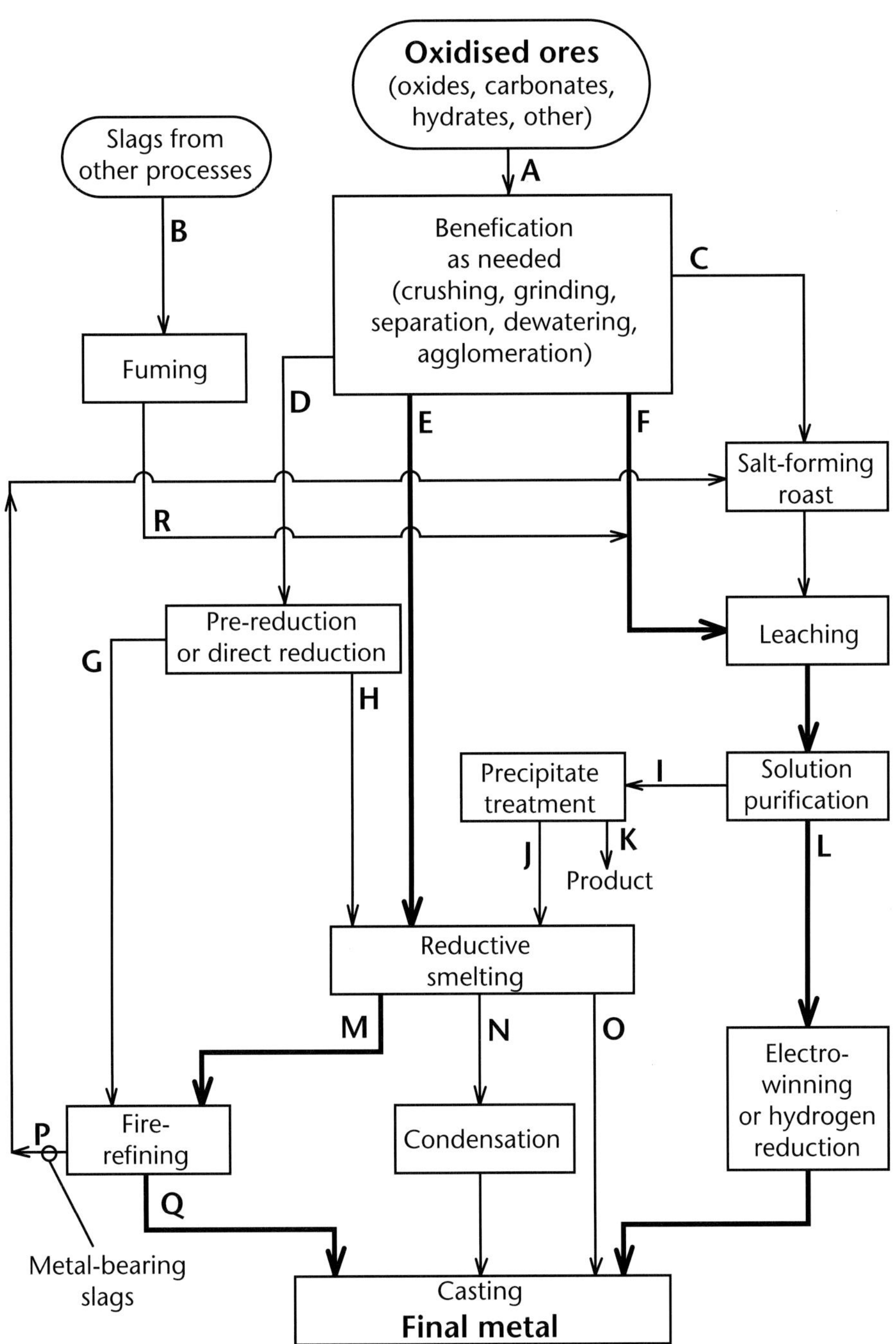

Figure IV.2: Extraction routes for less reactive metals from oxide ores. Heavy lines indicate the most frequently used routes.

Table IV.1: Extraction of reactive metals

Metal (common ore minerals)	Common extraction route (Figure IV.1)
Aluminium (bauxite)	Route C, H with no minimal beneficiation
Beryllium (beryl)	Route D, J with minimal beneficiation
Magnesium (magnesite, dolomite)	Extracted from seawater as $Mg(OH)_2$ precipitate. This is then calcined and processed via route B, H
	Extracted from carbonate via route A, G
	Extracted from carbonate via route B, H
Silicon (silica)	Extracted as ferrosilicon or metallurgical grade silicon via route A, F
	Extracted as high-purity metal via route B, J
Titanium (rutile, ilmenite)	Extracted from ilmenite as a titaniferrous slag via route A, then as metal via route E, J
	Extracted from rutile via route B, J
Zirconium (zirconia)	Extraction route B, J

Table IV.2: Extraction of less reactive metals from oxide ores

Metal (common ore minerals)	Common extraction route (Figure IV.2)
Chromium (chromite)	Ferrochromium by route A, E, O
	Ferrochromium by pre-reduction route A, D, H, O
	Chromium metal by route A, C, L
Copper (malachite, azurite, cuprite, chalcocite)	Extraction route A, F, L
Iron and steel (hematite, magnetite)	Blastfurnace route A, E, M, Q
	Direct-reduction route A, D, G, Q
Manganese (pyrolusite, manganite, psilomelane)	Ferroalloy by route A, E, O
	Manganese metal by route A, F, L
Nickel (laterite)	Nickel metal by route A, F, L
Tin (cassiterite)	Extraction route A, E, M, Q
Tungsten (wolframite, scheelite)	By classical route A, C, L
	By pressure leaching A, F, L
Uranium (uraninite, camotite)	Produced as ammonium diuranate (yellowcake) by route A, F, I, K
Vanadium (ulvospinel in titaniferous magnetite)	Production of vanadium-bearing slag by route A, D, H, M followed by either P, I, K to produce vanadium pentoxide (As V_2O_5), or P, I, J, O to produce ferrovanadium
Zinc	Extracted from slags from other processes by route B, F, L

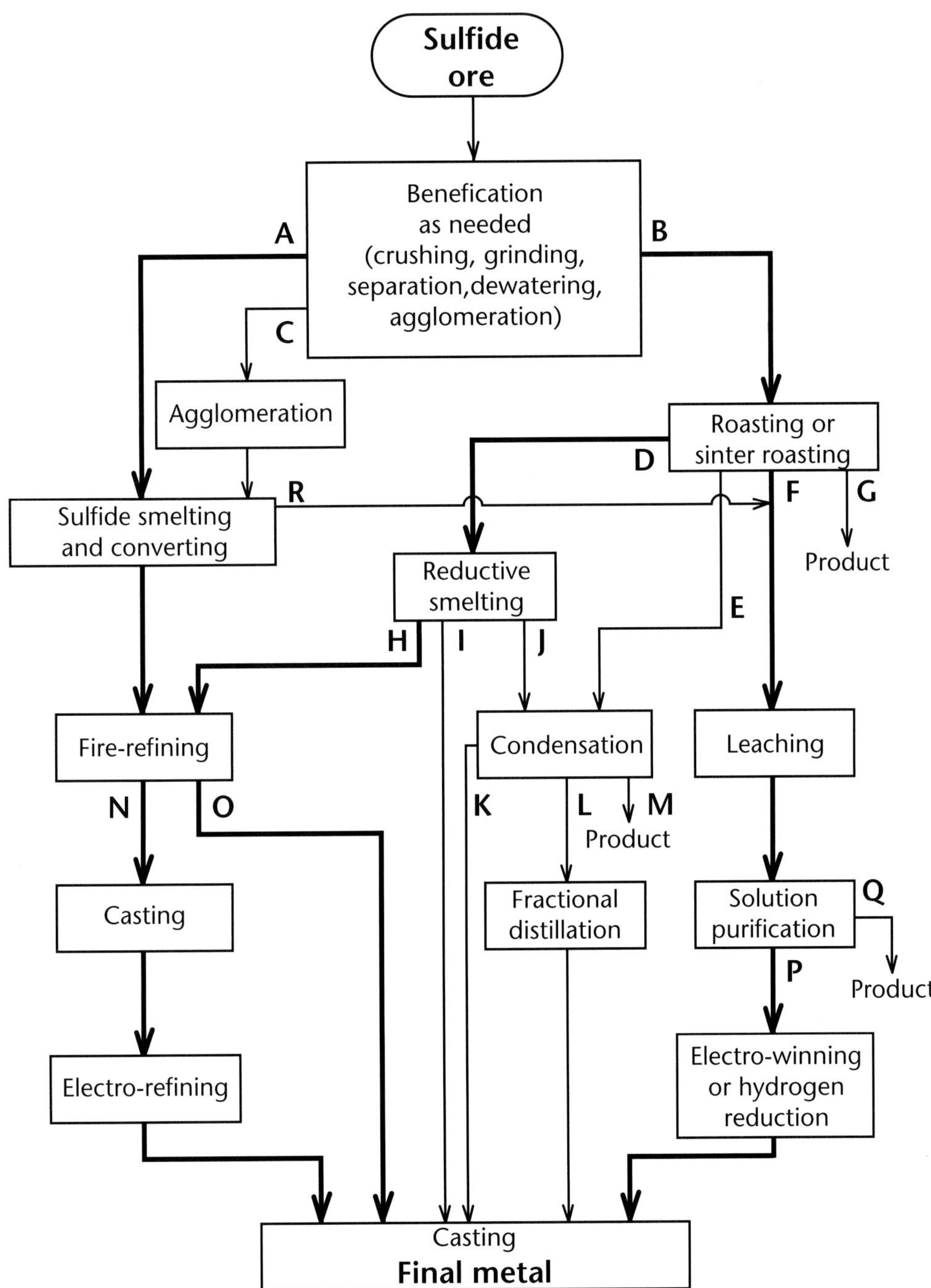

Figure IV.3: Extraction routes for less reactive metals from sulfide ores. Heavy lines indicate the most frequently used routes.

Table IV.3: Extraction of metals from sulfide ores

Metal (common ore minerals)	**Common extraction route (Figure IV.3)**
Antimony (stibnite)	Route B, D, H, O or B, F, P
Cadmium (greenokite)	Usually in association with zinc. Route B, D, J, L or B, F, P
Copper (chalcocite, chalcopyrite, bornite)	Conventional sulfide-smelting route A, N
Lead (galena)	Conventional extraction route B, D, H, O
	Direct-continuous extraction route C, O
Mercury (cinnabar)	Produced as a liquid metal by extraction route B, E, M
Nickel (pentlandite)	Route A, R, F, P
Molybdenum (molybdenite)	In association with other sulfides
	Produced as oxide by route B, G
	Produced as a ferroalloy by route B, D, I
Zinc (sphalerite)	Imperial smelting process route B, D, J, L
	Electrolytic process B, F, P

Index

acid and metalliferous drainage (AMD) 229–31, 236–7, 243, 252, 253, 256, 343
  management of 258–61
  minimisation of 258–9
  post mine closure 261
  treatment of 259–61
acid mine drainage 229–31, 236–7
acid rain 5, 39–40, 232, 234, 238
agglomerating particles 146–7
aggregate 7, 11, 100–1
Agricola, Georgius 122, 223
alkali activated materials (AAM) 341–6
alloys 7, 10
aluminium 76, 82, 83, 98, 101
  and embodied energy 192
  and motor vehicles 319–20
  production 177–8, 179, 338–9
  recycling 282–3
  red mud, utilisation of 338–9
  smelting 78, 179, 205, 232, 266, 328, 339
  solar production of 350
  spent pot lining, utilisation of 339
  traded commodity 105
  and waste production 228, 229
anode slimes 181
apatite 78
asbestos 78

backstop technology 321
ball mills 138–9
barite 78
barriers to cleaner production 359–63
base metal smelting slags 255
Basel Convention 238–9, 242–4
basalt 71, 75, 80, 89, 304
base metal deposits, distribution of 85–6
base metals 97, 105
basic materials 7, 12
basins 67, 68, 84, 85, 308, 310–11
bauxite 98, 178
beneficiating mined materials 89–94, 131, 134, 136–47, 235
  agglomerating particles 146–7
  base metal sulfide concentrates 152
  flowsheets 147–52
  and gross energy requirements 191
  mineral sand concentrating 147
  production of iron ore fines and lump 147
  separating particles 140–3
  separating solids from water 143–6
  size reduction 137–40
  and wastes 224
bentonite 78
Bhopal disaster (India) 47
Big Bang theory 63–4
bioavailability and toxicity of wastes 241–2
biogeochemical cycles 34–40
biomass 9–10, 28
  energy stored in 30
  global production of 28–30
  global stock of 30
  and waste reduction 352
biomes 28
biomimicry 55
biosphere 26–31
  thermal energy flows 32
borates 72–3
briquetting 147
Brundtland Commission 48, 49, 53
by-products 13, 94

calcium 82
capital
  financial 50, 56
  human 50, 56
  human-made 50–1
  natural 50
  reproducible (manufactured) 50
  social 50, 56
carbon cycle 35–6
carbonates 72
carbothermic reduction 166
car components 319
cars, recycling 284–7

cement 7, 11, 16, 183–5
cementation (metal recovery) 174
ceramics 7, 11, 16, 187–8
chemical fertilisers 186
chemical precipitation (refining) 171–2
chemical processing
  energy required for 203–5
  and wastes 224
chemical separation (extraction from mineral deposits) 92–3
chemical weathering (rocks) 81
chromite 78
chromium 76
classification (separating particles) 140–1
cleaner production 336–7
climate change 122
Club of Rome 47
coal-burning power stations 5, 340–1
cobalt 76
coke 319
commodities 8
  fuel 100
  manufactured mineral 100
  metallic 97–9
  non-metallic 99–100
  traded 97–100
  see also mineral commodities
comminution 137
  circuits 139–40
concentrate production 131
  separating particles 140–3
  separating solids from water 143–6
concentrates
  base metal sulfide 152
  mineral sand 147
concrete 11, 16, 183–5
conservation of matter 389
construction wastes, recycling 281
consumers 28
consumption of materials 1
contaminants 235
continental crust 25, 26, 66, 67–8, 73, 75, 296, 300, 301, 303, 308, 313
Convention on Law of the Sea (UNCLOS) 312–13
copper 6, 10
  cathode 98
  companies 118
  concentrate 98, 344–5
  deposits 83, 84, 85–9, 302–3
  distribution 85–6
  and embodied energy 192
  flow of 19–20
  gross energy requirement for production 201
  leaching 168, 169, 181, 182
  production 94, 178–81, 201, 228, 229, 255, 281
  production by-products 94
  recovery 174, 285, 290, 373
  recycling 19, 283–4, 285
  refining 172, 324
  resources and reserves 86–9
  smelting 167, 180, 234
  traded commodity 105
  uses 76, 321, 326
  and waste production 228, 229
co-product 94
corporate responses to sustainability 53
corporate social responsibility (CSR) 49
corruption and natural resources 123
covalent bonding 387
craton 67
crushing 137–8
crust, Earth 6, 10, 12, 25, 54, 63, 83, 86, 223, 298, 387
  composition 70–1
  continental 25, 26, 66, 67–8, 73, 75, 296, 300, 301, 303, 308, 313
  distribution of elements 298
  extracting value from the 89–94
  formation 66–71
  mean element composition 70
  oceanic 25, 26, 66, 68, 75, 300, 305, 308, 310
  and rock extraction 89–94, 131–6
crustal resources 297–305
crustal rocks as a source of scarce elements 304–5
cyanide 237–8
cyanide solutions, management of 257–8
cyclosilicates 73

dead-roasting 168
decomposers 28
demand for metals and minerals 315–30
  dematerialisation 322–9, 330

determinants of long-term 315–16
materials and technological substitution 318–22, 330
projections 316–18
dematerialisation 322–9, 330
demolition wastes, recycling 281
design for the environment (DfE) 291–2
development
mining as a route to 122
see also sustainable development
diamonds 308
dimension stone 11
dissolution (leaching reactions)
by acid 167
by alkali 168
by formation of complex ions 168
of a metal salt 167
by oxidisation 168
dolomite 78
domestic processed output (DPO) 17–18
drilling 132
drilling muds 78, 132
drying (separating solids from water) 146
Dutch disease and natural resources 123

Earth
and basic laws of science 31–4
biogeochemical cycles 34–40
biomes 28–30
biosphere 26–31
core 25, 26, 64–5, 67, 68, 70
distribution of elements 68–71
ecosystem services 30–1
formation of the 63–5
geological time scale 65
heat sources 64–5
hydrosphere 26–31
life on 27–8
mantle 25–6, 32, 64–5, 66, 67, 68, 70, 71, 82, 85, 310
tectonic plates 26, 27, 68, 82
see also crust, Earth; minerals; rocks
Earth Summit 48
ecological definitions of sustainability 50
ecological footprint 43–5
Ecological Rucksack 22–3
ecology, industrial 352–9
economic definitions of sustainability 50
economic impacts of mining 122–4
economic policy 121
economic value of mineral commodities 109–12
economy
and material flow analysis 16–19
material flows in the 13
ecosystem services 5, 30–1, 41, 42, 50, 51, 52, 379, 380
effect substances 7
effluent 267–8
electricity and greenhouse gas production 198
electricity generation and greenhouse gas emissions 200
electrolytic production of metals 350–1
electrostatic separation 142
electro-winning 166, 174
elements
arranged in alphabetical order 410–11
atmophile 69
chalcophile 69
crustal rocks as a source of scarce 304–5
distribution in the crust 297
distribution of 68–71, 297
geochemically abundant 83
geochemically scarce 83
lithophile 69
native 72
periodic table 69
in seawater 305–7
siderophile 69
toxic effect of 239–41
tramp 283–4
electromagnetic radiation 392–3
electrometallurgy 155
electro-refining 173
electrostatic precipitators 266
electro-winning 22, 163, 166, 172, 173, 174, 178, 182, 198, 219, 352
embankments for tailings 249–51
embodied energy 191–8
calculation of 192–4
of some common mineral commodities 195
definition 191
and global warming potential 196–8

values of 194–5
emission sustainability indicators 267–8
end-of-life products 271–3
energy
balance 16
direct/indirect 189–91
gross requirement (GER) 189–91, 194, 205–10, 278–9
required for chemical processing 203–5
required for mineral and metal commodities 189–91
required for moving materials 202
required for recycling 275–9
required for sorting and separating materials 202–3
required to produce mineral and metal commodities 193
scientific concept 389–92
sustainability indicators and reporting 205–10
see also embodies energy
energy consumption
and declining ore grade 198–200
and greenhouse gas emissions 196–8
lower limits of 200–5
in primary production 189–210
engineering materials 7, 12
environment
design for the environment (DfE) 291–2
impact of waste on 235–42
state of the 42–3
environmental awareness and recycling 273
environmental concerns 122
environmental context of sustainability 42–6
environmental degradation 1, 41, 42, 46, 48, 55, 235, 362, 375
environmental problem-solving 52
environmental resources 5, 46, 50
erosion (rocks) 82
European Chemicals Agency 243
Exxon Valdez 47
extraction
chemical 92–3
efficiency of 94
of common metals from their ores 401–6
of metals, stages 170–4
of metals from oxide ores 402, 403, 404
of metals from sulphide ores 405, 406
of mineral resources 89–94
physical 90–2
principles of metal 156–61
of reactive metals 404
of rock from the crust 89–94, 131–6
solvent 172
see also mineral material, processing mined; mining
extractive metallurgy 155

felsic magma 75
ferrous metals 97
fertilisers
artificial 7, 43, 388
chemical 186
mineral 11, 38, 100, 153, 186–7
synthetic 319
filtration (separating solids from water) 145–6
financial capital 50, 56
finex iron-making process 346–7
fire-refining 173
Five Winds stewardship model 370–1
fluorite/fluorspar 78
fly ash and slag, utilisation of 340–1
fossil fuels 1, 6, 19, 25, 30, 33–6, 100, 190, 197, 321–2, 352
froth flotation 142–3
fuel commodities 100
Fuller's earth 78

gabbro 75, 80
gas cleaning 263–6
gas cooling 262–3
gas planets 64
gaseous wastes 232–5, 262–7
gasoline 319
geochemically abundant elements 83
geochemically scarce elements 83
geological time scale 65
geopolymer concrete 341–6, 360
Gibbs energy 153–4, 155, 156–7, 159, 166, 202
glass 7, 11, 16, 185–6, 281–2
Global Mining Iniative (GMI) 124

Global Report Initiative (GRI) sustainability
  indicators 126, 395–9
  for energy consumption 205–10
  for solids, liquid and gaseous wastes 267, 267–8
  and waste management 267–8
  for water usage 219
global greenhouse gas production 198
global warming 5, 22, 196–8
globalisation 121–2
gneiss 80, 81
gold 85, 86, 228, 229, 310
Goldschmidt, V.M. 69, 71, 72
goods 7
government policy and natural resources 123
government responses to sustainability 53
granite 71, 75, 80
gravel 308
gravity separation 141
green chemistry 56, 57
green engineering 56, 57
green radicalism 52
greenhouse effect 32–3, 196–8
greenhouse gas production 198
grinding 138–9, 143–4

halides 72
Hartwick's rule 123–4
heap leaching 168–9, 196, 219, 228, 229, 346–7
heat (scientific concept) 389–92
heat recovery 262–3
heat transfer 393–4
heterogeneous reactions 157–9
HiSarna smelting reduction process 351
Hotelling's rule 109–12
  limitations 110–12
Hubbert's curve 299–300
human capital 50, 56
human-made capital 50–1
humans, impact of waste on 235–42
hydrogen reduction (metal recovery) 174
hydrometallurgical route 181
hydrometallurgy 155, 197–8
hydrosphere 26–31
  thermal energy flows 32
hydroxides 72

igneous rocks 75–80
ilmenite 78
Inconvenient Truth, An 47
indicated mineral resource 86
industrial ecology 352–9
industrial metals (classes) 100
industrial minerals 73–5, 78–9, 100
inferred mineral resource 86
information technology 122
inland sea basins 68
innovators 59
inorganic compounds 10–12
inorganic materials
  recycling 14
inosilicates 73
in-pit pre-concentration 348–50
International Bauxite Association 105
International Council on Mining and Metals
  (ICMM) 124–5, 126, 368–70, 373–4, 379
International Monetary Fund (IMF) 1, 41
International Seabed Authority (ISA) 312, 313
international system of units 383–5
ion exchange (refining) 172–3, 257
ionic bonding 387–8
IPAT equation 329
iron 64, 76, 82, 83, 98, 234–5
iron blast furnace slag 254
Iron Catastrophe 64
iron ore fines and lump, production of 147
iron-making process, Finex 346–7
island arcs 68

kaolinite 78
kerosene 319
Kwinana Industrial Area (KIA) 355–9, 361

leaching 90, 92–3, 167–70, 171, 216
  acidic and alkaline reagents 289–90, 347
  agitation 169
  bauxite 229
  copper 181, 182
  cyanide reagent 237, 258
  gold 229, 256
  heap and dump 168–9, 196, 219, 228, 229, 346–7
  in situ 89, 136, 137, 161, 168, 169

large-scale systems 169
management of toxic elements 373
operations residues 247–8, 253–4
oxide ores 402, 403
reactions 167–8, 198
sulphide ores 405
technologies 168–70
vat 169, 228, 229
and waste materials 242
lead 76, 86, 94, 98, 181, 183, 237
Lemaître, Georges 63
liberation 90–1, 93
life cycle assessment (LCA) 19–22
light metals 97
limestone 78
Limits to Growth, The 47
liquid wastes 228–31
management of 255–61
Lloyd, William Foster 46
localisers 59–60
low carbon dioxide steel-making 351––2

magma 75, 81
magnesium 76, 82, 83, 319–20
magnetic separation 141–2
Malenbaum, W. 322, 324–5
Malthus, Thomas 47
manganese 76, 78, 83
mantle (Earth) 25–6, 32, 64–5, 66, 67, 68, 70, 71, 82, 85, 310
manufactured mineral commodities 100
recycling 153
theoretical considerations 153–5
manufactured mineral products, producing 11, 153
manufacturing
and design for the environment (DfE) 291–2
marble 80, 81, 89
marginal basins 67, 68
marketeers 59
material flow analysis (MFA) 16–23
economy-wide 16–19
types of 17
material flows
in the economy 13
material groups 9–12
Material Input per Service unit (MIPS) 22
materials 6–9
aggregate 100–1
balance 16
basic 7, 12
and design for the environment (DfE) 291–2
engineering 7, 12
recovery, recycling and return rates 277
recycling of 14–15, 153, 271–2, 275, 281–91
substitution 318–22, 330
see also raw materials, secondary materials
materials balance 16
materials cycle 12–14, 131, 271, 273, 275, 336, 378, 380
closing the 367–74
quantifying the 15–24
and waste production 223, 224
matter
conservation of (scientific concept) 389
nature of (scientific concept) 387–9
measured mineral resource 86
mercury 76
metal markets 105–7
metal production 155–83
casting 350
cleaner 336–7, 359–63
electrolytic production 350–1
energy inputs required 189–91, 193
extraction from oxide ores 402, 403, 404
extraction from sulphide ores 405, 406
extraction of common metals from their ores 401–6
extraction, principles of 156–61
extraction stages 170–4
heterogeneous reactions 157–9
leaching 167–70
metal recovery 173–4
metallurgical reactions 161
phase separation 159–60
smelting 161–7
some important metals 174–83
stewardship 371–3
value-adding stages 190
and wastes 225, 226
water usage 217
metal recovery 173–4
absorption on to activated carbon 174

cementation 174
electro-winning 174
hydrogen reduction 174
metallic bonding 388–9
metallic commodities 97–9, 190
metalliferous drainage 229–31, 236–7
metalliferous sediments 310
metallothermic reduction 166
metallurgical processes, separation in extractive 159–60
metallurgical reactors 161
metals 7, 10
base 97, 105
determinants of long-term supply 295–6
embodied energy 191–8
embodied water 216–17
extraction from crustal rocks 304–5
extraction from ocean deposits 311–12
extraction from oxide ores 402, 403, 404
extraction from seawater 305–8
extraction from sulphide ores 405, 406
extraction of common metals from their ores 401–6
extraction, principles of 156–61
ferrous 97
future availability 295–313
future demand 315–30
industrial (classes) 100
life span in various applications 276
light 7
non-ferrous 97
potential sources 296–7
precious 97, 105
rare 97, 99
recycling 15, 153, 282–4
see also metal production
metamorphic rocks 75, 80–1
mica 78
middlings 90
mid-ocean ridges 68
mine sites
processing 99, 136, 372, 373, 379
and water 215–16, 228–9, 230, 231, 258, 259
mineral classes 72–5
mineral commodities 97–112
consumption 100–1
economic value of 109–12
embodied energy of some common 195
energy inputs required 189–91, 193
fuel 100
and Hotelling's rule 109–12
metallic 97–9
mineral and metal markets 105–7
non-metallic 99–100
price 101, 105
production and consumption 100–1
reserves and resources of 101–4
statistics 100–1
traded 97–100
traded, how they are 105–9
trading complexities 107–9
and wastes 225, 226
water usage 217
see also manufactured mineral commodities
mineral deposits 82–6
common forms of 84–5
discovery of 300–4
distribution of base and precious metal deposits 85–6
evaluation and development 113–14
exploration 113
extracting value from 89–94
formation of 83–4
grade 86–9
marine 309
ocean floor 300
placer 83
primary 83
profitability 295
resource and reserves 86–8
secondary 83
mineral fertilisers 11, 38, 100, 153, 186–7
mineral markets 105–7
mineral material, processing mined 89–94
agglomerating particles 146–7
base metal sulfide concentrates 152
beneficiating 136–52
by-products 94
chemical separation 92–3
effect of breakage on the surface area 93–4
efficiency of extraction 94
physical separation 90–2

rock extraction 89–94, 131–6
separating particles 140–3
separating solids from water 143–6
size reduction 91, 93, 137–40, 170, 272
stewardship 371–3
see also beneficiating mined materials; mining
mineral products, producing manufactured 11, 153
cement and concrete 183–5
and cleaner production 336–7, 359–63
commodity ceramics 187–8
glass 185–6
gross energy requirements 190
mineral fertilisers 186–7
value-adding stage 190
mineral resources/reserves 86–8
mineraloids 71
mineralogical barrier 297–9
minerals 71–5
companies 117–20
determinants of long-term supply 295–6
exploration companies 121
future availability 295–313
future demand 315–30
industrial 73–5, 78–9, 100
liberation of 90–1
ore 73, 76–7, 295
peak 399–300
potential sources 296–7
rock-forming 73–5
minerals industry 97–128
associations 120
companies 117–20
culture 120–1
economic and social impacts 122–4
hazardous nature 115–16
nature of the 115–22
non-SI units of mass 101
project cycle 112–14
reporting 126–8
size and structure 116–17
status 128
and sustainable development 124–8
trends 121–2
and water 215–16
see also mineral commodities
mining
caving methods of 136
corruption 123
definition 131
Dutch disease 123
economic and social impacts 122–4
government policy 123
Hartwick's rule 123–4
placer 134
in situ 131, 136
longwall 136
open pit 132
Resources Curse 123–4
revenue volatility 123
as a route to development 122
rock extraction 89–94, 131–6
room and pillar 134
solution 134
stope and pillar 134
strip 132–4
supported methods 134––6
surface 132
underground 134–6
unsupported methods 134
wastes 223, 224, 226
mining project cycle 112–14, 131
design, construction and commissioning 114
evaluation and development 113–14
exploration 113
life cycle 131
production 114, 131
project decline and closure 114
molybdenum 76
Moon, the 64

natural capital 50
natural capitalism 55
natural resources 5–6
non-renewable 6
renewable 5
nature of matter 387–9
nesosilicates 73
New York Mercantile Exchange 105
nickel 76, 85, 94, 98, 181, 183, 184
nitrogen cycle 37–8

nitrogen fertilisers 186
non-ferrous metals 97
non-metallic commodities 99–100

occupational health and safety 122
ocean basins 67, 68, 84–5, 308, 310–11
ocean floor mineral deposits 300, 311–12
oceanic crust 25, 26, 66, 68, 75, 300, 305, 308, 310
oil 299–300
open pit mining 132
ore grades 86–9, 297
  and energy consumption 198–200
ore minerals 73, 76–7
ore production 131, 295–6
ore reserve (ore body) 86–7
ore sorting 348
orogens 67–8
Ostrom, Elinor 46
overburden 132, 249
oxide ores, extraction of metals from 402, 403, 404
oxides 72
oxygen 71, 82
oxygen cycle 35–6

paper
  products 7, 9
  recycling 14
particles
  agglomerating 146–7
  briquetting 147
  classification 140–1
  drying (separating solids from water) 146
  electrostatic separation 142
  filtration (separating solids from water) 145–6
  froth flotation 142–3
  gravity separation 141
  magnetic separation 141–2
  pelletising 146–7
  separating 140–3
  separating solids from water 143–6
  thickening (separating solids from water) 145
Pauling, Linus 72
peak minerals 299–300
pelletising 146–7
periodic table 388, 412
perlite 78
phosphates 72
phosphorus 83
phosphorus cycle 38
phosphorus fertilisers 187
phyllosilicates 73
physical separation (extraction from mineral deposits) 90–2
physical weathering (rocks) 81
pig iron 101
placer mineral deposits 83, 308
placer mining 134
plastics 10
  recycling 14
platforms (continental crust) 67
platinum 94
pollutants 235
polymetallic massive sulfide deposits 310
Population Bomb, The 47
population growth 121
porphyry 80, 85, 344
potassium 82
potassium fertilisers 186
precious metal deposits, distribution of 85–6
precious metals 97, 105
pregnant solution/liquor 167
pretreatment (metal extraction) 170–1
primary mineral deposits 83
primary production
  management of wastes from 247–68
  waste from 223–44
processing mined mineral material 89–94
  agglomerating particles 146–7
  base metal sulfide concentrates 152
  beneficiating 136–52
  by-products 94
  chemical separation 92 3
  effect of breakage on the surface area 93–4
  efficiency of extraction 94
  physical separation 90–2
  rock extraction 89–94, 131–6
  separating particles 140–3
  separating solids from water 143–6
  size reduction 91, 93, 137–40, 170, 272
  stewardship 371–3

see also beneficiating mined materials; extraction; mining
producer price 105
producers 27, 28
production *see* mineral products, producing manufactured; metal production
pumice 78
pyrometallurgy 155197–8

quarrying 132
quartz 9, 11, 72, 73, 75, 77, 79, 80, 81, 82, 84, 187, 202

radiation, electromagnetic 392–3
rare metals 97, 99
raw materials 6, 11, 12, 13, 23, 100, 119, 153, 315, 325, 353, 355
  wastes as 335, 336, 337–46
REACH 243–4
recycling 271–92
  benefits and limitations 273–4
  cars 284–7
  construction and demolition wastes 281
  design for the environment (DfE) 291–2
  drivers of 273
  effect of repeated 279
  effect on resource life 279–81
  energy requirements 275–9
  glass 153, 281–2
  of materials 14–15, 153, 271–2, 275, 281–91
  metals 15, 153, 282–4
  terminology 274–5
  top gas 351
  waste electrical and electronic equipment (WEEE) 28791
  and wastes 224
red mud, utilisation of 338–9
refining (extraction of metals) 171–3
refractories 11
regolith 67
remanufacturing 272–3
reproducible (manufactured) capital 50
resource life cycle, effect of recycling on 279–81
Resources Curse 123
resources in the seabed 308–13
resources in seawater 305–8
revenue volatility and natural resources 123
Ricardo, David 109
rifts 68
rock(s) 71
  chemical weathering 81
  classes 75–81
  cycle 81–2
  erosion 81–2
  extraction from the crust 89–94, 131–6
  igneous 75–80
  metamorphic 75, 80–1
  physical weathering 81
  sedimentary 75, 80, 81, 82
  types, abundance in crust 80
  waste 248–9
rock-forming minerals 73–5
room and pillar mining 134
rubber 319
rutile 79

sand 308
sandstone 36, 75, 80, 81
scientific concepts, review of important 387–94
scientific notation 383–5
scrubbers 264–6
seabed resources 308–13
seawater, resources in 305–8
secondary materials 6–7, 12
  and recycling 271–94
secondary mineral deposits 83
sedimentary rocks 75, 80, 81, 82
separation in extractive metallurgical processes 159–60
separation of particles 140–3
  classification 140–1
  drying (separating solids from water) 146
  electrostatic 142
  filtration (separating solids from water) 145–6
  gravity 141
  magnetic 141–2
  solids from water 143–6
  thickening (separating solids from water) 145
services 7
settling chambers 263
shields (continental crust) 67
SI units 383–5

Silent Spring 47
silica 11, 75
silicates 10–12, 64, 73
  recycling 14
silicon 71, 77, 83
sillimanite 79
silver 85–6
sintering 146, 162, 171, 176–7
size reduction of material 91, 93, 137–40, 170, 272
  comminution circuits 139–40
  crushing 137–8
  grinding 138–9
slag
  dry granulation 350
  by-products 94, 161, 254–5
smelting 22, 90, 93, 121, 131, 146, 159, 171, 173, 179, 183
  aluminium 78, 205, 232, 266, 328, 339
  copper 167, 180
  and energy consumption 189, 196
  gaseous emissions 235, 373
  gases produced in 232–5
  high-intensity 238
  HiSarna reduction process 351
  nickle ore 183
  of ore or concentrate 161–7
  and process stewardship 368
  quantities of gas 232–4
  reactions 166–7, 171
  reductive 402, 403, 405
  slag by-products 94, 161, 254, 255
  sulfide 39, 195, 233, 234, 266, 405
  technologies 167, 180
  types of gases produced 232
  wastes 296
social capital 50, 56
social impacts of mining 122–4
sodium 82
solar production of aluminium 350
solar system 63–4
solid wastes 225–8, 229
  management of 247–55
solids, separating from water 143–6
solution mining 136
solvent extraction (refining) 172
sorosilicates 73
spent pot lining, utilisation of 339
stainless steel 175–7
steel 98, 101, 175–7, 192, 234–5, 282–3, 350
steel-making, low carbon dioxide 351––2
steel-making slag 254–5
Stockholm Declaration 47–8
stope and pillar mining 134
strip mining 132–4
substance flow analysis (SFA) 19
substitution, materials and technological 318–22, 330
sulfates 72
sulfide flotation tailings, utilisation of 343–6
sulfide ore 200
  extraction of metals 405, 406
sulfide smelting 235
sulfides 72, 76–7, 83, 85, 152, 156–7, 161, 166, 167–8, 180, 304–5, 310
sulfur 79, 310
sulfur cycle 38–40
sulfur dioxide 5, 238
  removal 266–7
Sun, the 33–4, 64
surface mining 132
survivalism 52
sustainable development 1–2, 41, 48, 52, 53, 59
  concepts of 49–52
  and ICMM 374
  and minerals industry 124–8
  principles for 56–7
  triple bottom line (TBL) 54
  World Business Council 49, 54, 375, 378–9, 380
sustainability 41–60, 367–80
  alternative definitions 49–51
  biomimicry 55
  brief history 47–9
  concepts 49–53
  corporate developments 48–9
  eco-efficiency 54
  ecological definitions 50
  ecological footprint 43
  economic definitions 50
  energy 205–10
  environmental context 42–6
  five capitals model 55–6
  frameworks 53–8

government and corporate responses 53
green chemistry/engineering 56, 57
indicators 125–8, 205–10, 218–19
international developments 47–8
interpretations of 51–2
market- and policy-based approaches to transitioning 374–5
model of 58–60
natural capitalism 55
public awareness 47
reporting 125–8, 205–10, 218–19
responses to the challenge 52–3
strong 51–2
The Natural Step 54–5
tragedy of the commons 46
triple bottom line (TBL) 54
water 218–19
weak 51

tailings 90, 236, 249–53
closure of storage facilities 252–3
conventional storage 249–51
river and sub-sea disposal 253
surface thickened, paste and dry stack methods 252
water management in storage facilities 251–2
talc 79
technological substitution 318–22
tectonic plates 26, 27, 68, 82
tectosilicates 73
terminal markets 105
terrestrial planets 64
thermodynamics, laws of 153–5, 189, 389–92
thickening (separating solids from water) 145
Thiobacillus ferrooxidans 168
tin 77
titanium 77, 82, 83, 85, 347–8
toxicity and wastes 238–41
tragedy of the commons 46
tramp elements 283–4
trenches (oceanic crust) 26, , 66, 68
triple bottom line (TBL) 54
tungsten 77

ULCORED direct reduction process 351
underground mining 134–6
underground pre-concentration 348–50
United Nations 1, 31, 41, 42, 47, 48, 126, 213, 214, 312, 316, 336
units
international system of 383–5
non-SI units of mass 101
Universe 63–4
uranium 77

value chain 13
value adding 13, 190
vanadium 77
van't Hoff isotherm 159
veins (lodes) 83
vermiculite 79
volcanic islands 68

wollastonite 79
waste electrical and electronic equipment (WEEE) 287–91
waste
and cleaner production 336–7, 359–63
definition 335
hierarchy 333–5
and industrial ecology 352–9
management strategies 333
reduction and elimination 333, 335–6
reduction through process re-engineering 346–52
rock 248–9
water 228–9
zero 333–63
Waste Framework Directive 334
wastes
acid and metalliferous drainage (AMD) 229–31, 236–7, 243, 252, 253, 256, 258–61, 343
Basel Convention 238–9, 242–4
bioavailability 241–2
disposal of 1
cyanide 237–8
fly ash and slag, utilisation of 340–1
gaseous 232–5, 262–7
geopolymer concrete 341–6, 360
GRI sustainability indicators 267, 267–8

impact on humans and the environment 235–42
impacts of mining 236–8
international regulation 242–4
lead 237
liquid 228–31, 255–61
and material flows 13–14
mining 223
and their origin 223–5
from primary production 223–44
from primary production (management) 247–68
production of 223–4, 334
as raw materials 335, 336, 337–46
REACH 243–4
red mud, utilisation of 338–9
reduction and elimination 333, 335–6
solid 225–8, 229, 247–55
spent pot lining, utilisation of 339
sulfide flotation tailings, utilisation of 343–6
sulfur dioxide 238
tailings 236
toxicity 238–41
water
consumption 214, 216–17, 219
cycle 36–7, 214
embodied content of metals 216–18
footprint 215
GRI sustainability indicators 219
information sources 220–1
management in tailings storage facilities 251–2
and mine sites 215–16, 228–9, 230, 231, 258, 259
in the minerals industry 215–16
resources, global 213–15
renewable resource 214
role in primary production 213–21
sustainability indicators and reporting 218–19
treatment and solid wastes 253–4
treatment technologies 255–7
waste 228–9
whale oil 319
wood 7, 9
recycling 14
World Bank 1, 41
World Business Council for Sustainable Development 48–9, 54, 375, 378–9, 380
World Commission on Environment and Development 48

zeolites 79
zinc 77, 83, 94, 98, 181, 183
zirconium 77, 85

The elements arranged in alphabetical order

Element	Symbol	Atomic no.	Relative atomic mass
Actinium	Ac	89	227.00
Aluminium	Al	13	26.98
Antimony	Sb	51	121.76
Argon	Ar	18	39.95
Arsenic	As	33	74.92
Astatine	At	85	210.00
Barium	Ba	56	137.33
Beryllium	Be	4	9.01
Bismuth	Bi	83	208.98
Boron	B	5	10.81
Bromine	Br	35	79.90
Cadmium	Cd	48	112.41
Calcium	Ca	20	132.91
Carbon	C	6	40.08
Cerium	Ce	58	12.01
Cesium	Cs	55	140.12
Chlorine	Cl	17	35.45
Chromium	Cr	24	52.00
Cobalt	Co	27	58.93
Copper	Cu	29	63.55
Dysprosium	Dy	66	162.50
Erbium	Er	68	167.26
Europium	Eu	63	151.96
Fluorine	F	9	19.00
Francium	Fr	87	223.00
Neon	Ne	10	20.18
Neptunium	Np	93	237.00
Nickel	Ni	28	58.69
Niobium	Nb	41	92.91
Nitrogen	N	7	14.01
Osmium	Os	76	190.23
Oxygen	O	8	16.00
Palladium	Pd	46	106.42
Phosphorus	P	15	30.97
Platinum	Pt	78	195.08
Plutonium	Pu	94	244.00
Polonium	Po	84	209.00
Potassium	K	19	39.10
Praseodymium	Pr	59	140.91
Promethium	Pm	61	145.00
Protactinium	Pa	91	231.04
Radium	Ra	88	226.00
Radon	Rn	86	222.00
Rhenium	Re	75	186.21
Rhodium	Rh	45	102.91
Rubidium	Rb	37	85.47
Ruthenium	Ru	44	101.07
Samarium	Sm	62	150.36
Scandium	Sc	21	44.96
Selenium	Se	34	78.96

Element	Symbol	Atomic no.	Relative atomic mass	Element	Symbol	Atomic no.	Relative atomic mass
Gadolinium	Gd	64	157.25	Silicon	Si	14	28.09
Gallium	Ga	31	69.72	Silver	Ag	47	107.87
Germanium	Ge	32	72.64	Sodium	Na	11	22.99
Gold	Au	79	196.97	Strontium	Sr	38	87.62
Hafnium	Hf	72	178.49	Sulfur	S	16	32.07
Helium	He	2	4.00	Tantalum	Ta	73	180.95
Holmium	Ho	67	164.93	Technetium	Tc	43	98.00
Hydrogen	H	1	1.01	Tellurium	Te	52	127.60
Indium	In	49	114.82	Terbium	Tb	65	158.93
Iodine	I	53	126.90	Thallium	Tl	81	204.38
Iridium	Ir	77	192.22	Thorium	Th	90	232.04
Iron	Fe	26	55.85	Thulium	Tm	69	168.93
Krypton	Kr	36	83.80	Tin	Sn	50	118.71
Lanthanum	La	57	138.91	Titanium	Ti	22	47.87
Lead	Pb	82	207.21	Tungsten	W	74	183.84
Lithium	Li	3	6.94	Uranium	U	92	238.03
Lutetium	Lu	71	174.97	Vanadium	V	23	50.94
Magnesium	Mg	12	24.31	Xenon	Xe	54	131.29
Manganese	Mn	25	54.94	Ytterbium	Yb	70	173.05
Mercury	Hg	80	200.59	Yttrium	Y	39	88.91
Molybdenum	Mo	42	95.96	Zinc	Zn	30	65.38
Neodymium	Nd	60	144.24	Zirconium	Zr	40	91.22

THE PERIODIC TABLE

Period	Group 1	2	3	4	5	6	7	8	9	10	11	12	13	14	15	16	17	18
1	1 H																	2 He
2	3 Li	4 Be											5 B	6 C	7 N	8 O	9 F	10 Ne
3	11 Na	12 Mg											13 Al	14 Si	15 P	16 S	17 Cl	18 Ar
4	19 K	20 Ca	21 Sc	22 Ti	23 V	24 Cr	25 Mn	26 Fe	27 Co	28 Ni	29 Cu	30 Zn	31 Ga	32 Ge	33 As	34 Se	35 Br	36 Kr
5	37 Rb	38 Sr	39 Y	40 Zr	41 Nb	42 Mo	43 Tc	44 Ru	45 Rh	46 Pd	47 Ag	48 Cd	49 In	50 Sn	51 Sb	52 Te	53 I	54 Xe
6	55 Cs	56 Ba	57 La	72 Hf	73 Ta	74 W	75 Re	76 Os	77 Ir	78 Pt	79 Au	80 Hg	81 Tl	82 Pb	83 Bi	84 Po	85 At	86 Rn
7	87 Fr	88 Ra	89 Ac															

Period														
6	58 Ce	59 Pr	60 Nd	61 Pm	62 Sm	63 Eu	64 Gd	65 Tb	66 Dy	67 Ho	68 Er	69 Tm	70 Yb	71 Lu
7	90 Th	91 Pa	92 U	93 Np	94 Pu									